Offering a comprehensive narrative of the early history of stereochemistry, Dr Ramberg explores the reasons for and the consequences of the fundamental change in the meaning of chemical formulas with the emergence of stereochemistry during the last quarter of the nineteenth century. As yet relatively unexplored by historians, the development of stereochemistry – the study of the three-dimensional properties of molecules – provides a superb case study for exploring the meaning and purpose of chemical formulas, as it entailed a significant change in the meaning of chemical formulas from the purely chemical conception of 'structure' to the physico-chemical conception of molecules provided by the tetrahedral carbon atom.

This study is the first to treat the emergence of the unique visual language of organic chemistry between 1830 and 1874 to place in context the near simultaneous proposal of the tetrahedral carbon atom by J.H. van 't Hoff and J.A. Le Bel in 1874. Dr Ramberg then examines the research programs in stereochemistry by Johannes Wislicenus, Arthur Hantzsch, Victor Meyer, Carl Bischoff, Emil Fischer and Alfred Werner, showing how the emergence of stereochemistry was a logical continuation of established research traditions in chemistry. In so doing, he also illustrates the novel and controversial characteristics of stereochemical ideas, especially the unprecedented use of mechanistic and dynamic principles in chemical explanation.

About the Author

Peter Ramberg has studied chemistry at the University of Minnesota and Indiana University and earned a PhD in history of science from Indiana University in 1993. He has taught organic chemistry at Johns Hopkins University, North Dakota State University, and Ohio University, and recently served as a Research Scholar in a group devoted to the history and philosophy of chemistry at the Max Planck Institute for the History of Science in Berlin. He currently teaches history and philosophy of science and chemistry at Truman State University in Missouri.

Science, Technology and Culture, 1700–1945

Series Editors

David M. Knight
University of Durham

and

Trevor H. Levere
University of Toronto

This new series focuses on the social, cultural, industrial and economic contexts of science and technology from the 'scientific revolution' up to the Second World War. Economic historians now cover the relations of science to technology and industrial application, while social and cultural historians have similarly recognized the realms of science and technology and, indeed, that these have helped to define culture and society. Through the agricultural and industrial revolutions of the eighteenth century, the coffee-house culture of the Enlightenment, the spread of museums, botanic gardens and expositions in the nineteenth century, to the Franco-Prussian war of 1870, seen as a victory for German science, this process has gathered momentum, while in the twentieth century the dependence of society, in both war and peace alike, on science and technology is evident. This series will provide an outlet for studies that address issues of the interaction of science, technology and culture in the period from 1700 to 1945, at the same time as including new research within the field of the history of science itself that embraces these perspectives.

Also in this series

Hewett Cottrell Watson: Victorian Plant Ecologist and Evolutionist
Frank N. Egerton

Phrenology and the Origins of Victorian Scientific Naturalism
John van Wyhe

Science, Technology and Culture, 1700–1945

CHEMICAL STRUCTURE, SPATIAL ARRANGEMENT

To my parents, who taught me to read everything and to be curious, and to Liza, my unflagging supporter

Chemical Structure, Spatial Arrangement

The Early History of Stereochemistry, 1874–1914

PETER J. RAMBERG
Truman State University, USA

ASHGATE

© Peter J. Ramberg 2003

All rights reserved. No part of this publication may be reproduced, stored in a retrieval system, or transmitted in any form or by any means, electronic, mechanical, photocopying, recording or otherwise without the prior permission of the publisher.

A portion of Chapter 5 has been reprinted by permission of the University of California Press, from *Historical Studies in the Physical and Biological Sciences*, vol. 26:1, © 1995 by The Regents of the University of California.

The author has asserted his moral right under the Copyright, Designs and Patents Act, 1988, to be identified as the author of this work.

Published by
Ashgate Publishing Limited
Gower House
Croft Road
Aldershot
Hampshire GU11 3HR
England

Ashgate Publishing Company
Suite 420
101 Cherry Street
Burlington, VT 05401-4405
USA

Ashgate website: http://www.ashgate.com

British Library Cataloguing in Publication Data
Ramberg, Peter J.
  Chemical structure, spatial arrangement : the early history of stereochemistry, 1874-1914. - (Science, technology and culture, 1700-1945)
  1. Stereochemistry - History
  I. Title
  547.1'223'09034

Library of Congress Control Number: 2001099679

ISBN 0 7546 0397 0

This book is printed on acid-free paper
Typeset by N^2 productions
Printed and bound in Great Britain by MPG Books Ltd., Bodmin, Cornwall.

Contents

List of Figures		xi
Series Editor's Preface		xxi
Acknowledgments		xxiii
1	Introduction: 'Van 't Hoff's Gold Mines'	1
2	**The Historical Development of Organic Chemistry to 1874**	11
	The Institutional Structure of Nineteenth-century German Chemistry	12
	The Development of Theoretical Organic Chemistry	15
	Methods and Methodology in Organic Chemistry	28
	The Limits of Structure Theory, 1864–1873	39
	Conclusion	50
3	**The Tetrahedral Carbon Atom, 1874–1877**	53
	Introduction	53
	Van 't Hoff and Le Bel, 1874	54
	The Evolution of Van 't Hoff's Theory, 1874–1877	66
	Conclusion	84
4	**Initial Reception of the Tetrahedron, 1874–1887**	87
	Public and Private Reaction to the Tetrahedron	87
	Adolf von Baeyer and the Strain Theory, 1885	101
	Configuration organischer Moleküle, 1886	105
	Dix années dans l'histoire d'une théorie, 1887	109
5	**Johannes Wislicenus and Molecular Dynamics**	111
	'Räumliche Anordnung', 1887	113
	Experimental Support for the Theory, 1887–1889	121
	Initial Reception	133
	Atomism and Methodology in Wislicenus' Chemistry	148

viii   *Chemical Structure, Spatial Arrangement*

6	**Victor Meyer: The New Science of Stereochemistry**	157
	Meyer and the Tetrahedral Carbon Atom, 1875–1887	159
	The Benzildioximes	162
	Spokesman for the New Science of Stereochemistry, 1889–1890	179
	The Nature of Valence and Bonding	182
	Conclusion	191
7	**Arthur Hantzsch: The Stereochemistry of Nitrogen**	193
	The Hantzsch/Werner Theory	196
	The Configuration of the Benzildioximes	214
	Other Spatial Models of the Nitrogen Atom, 1888–1893	222
	Adolf Claus versus Meyer and Hantzsch: Is Stereochemistry Actually *Chemistry*?	231
	Conclusion	238
8	**Emil Fischer and Carbohydrate Chemistry, 1884–1891**	243
	The Carbohydrates, 1884–1891	245
	Sugars and Enzymes, 1894	268
	The Significance of Sugar Chemistry for Fischer	272
9	**Alfred Werner and Coordination Chemistry, 1893–1914**	277
	The Coordination Theory, 1893–1907	280
	The Controversy with Jørgensen, 1893–1899	305
	Optically Active Coordination Compounds, 1897–1914	313
	Conclusion	318
10	**Conclusion**	321
	The Nature of Stereochemistry	322
	The Nature of Chemistry	335
	Broader Issues and Further Questions	343

Appendices

1	Tjaden Modderman to Neighbors of Van 't Hoff's Parents, 1874	353
2	Van 't Hoff's Preface to *La chimie dans l'espace*, 1875	355
3	Felix Hermann and Johannes Wislicenus to Van 't Hoff, 1875	357
4	Wislicenus' Foreword to the First Edition of *Die Lagerung der Atome im Raume*, 1877	361

5	Victor Meyer's Letters to Baeyer Concerning the Strain Theory, 1885	365
6	'Translation' of Van 't Hoff's Sign Notation for Compounds with Multiple Asymmetric Carbon Atoms into Modern Fischer Projections	369
Bibliography		371
Index		395

List of Figures

2.1 Possible formulas for benzene that represent the chemical equivalence of the six hydrogens: (a) Kekulé (b) Meyer (c) Ladenburg and (d) Loschmidt.   25

2.2 The chemical structures for propyl and isopropyl alcohol using Crum Brown's formulas.   27

2.3 (a) Wislicenus' 1859 formulas for lactic acid that emphasized different aspects of its reactivity. (b) Wislicenus' 1859 synthesis of lactic acid from ethylene chlorohydrin.   44

2.4 Structures for α and β lactic acid using Crum Brown's formulas.   45

2.5 Wislicenus' new formula for hydracrylic acid (a), based on the formula for ethylene oxide (b).   47

3.1 (a) The possible isomers for square planar geometry of a disubstituted carbon atom (IV, V and VI). (b) The enantiomorphic tetrahedron created by tetrasubstitution (VII and VIII). (c) Van 't Hoff's model of the carbon–carbon double bond (IX and X) and the triple bond (XI). From Van 't Hoff, *Voorstel*.   58

3.2 (a) Le Bel's possible structures for maleic and fumaric acid. (b) The hemihedral quadratic pyramid, one possible geometry for ethylene suggested by Le Bel.   61

3.3 Le Bel's formulas for benzene (a), and the three possible disubstitution products (b). Note that this formula also provides an alternate three-dimensional arrangement of six equivalent hydrogen atoms in benzene that is not a hexagon.   62

3.4 The illustration of the free rotation of carbon–carbon single bonds according to Van 't Hoff's second hypothesis. From Van 't Hoff, *La chimie dans l'espace*.   70

3.5 Van 't Hoff's suggested spatial isomerism resulting from the cumulation of double bonds. From Van 't Hoff (see note 48).   70

3.6 Van 't Hoff's drawing of the carbon atoms as seen by an entering ray of polarized light. From Van 't Hoff, *La chimie dans l'espace*.   71

3.7	Possible substitution products of Ladenburg's (a) and Kekulé's (b) benzene formulas. Of Ladenburg's structures, (ii) can exist in enantiomeric forms.	71
3.8	Gustav Bremer's conversion of dextrorotatory ammonium malate to the levorotatory ammonium malate.	76
3.9	Van 't Hoff's portrayal of the addition of bromine to the two faces of fumaric acid (a) and maleic acid (b). In the case of maleic acid, the two products are identical (one needs only to flip over the structure on the right to match the left). Addition to fumaric acid yields two spatially different isomers (the right-hand structure cannot be made superimposable on the left by flipping or spinning). From Van 't Hoff, *Die Lagerung* (1877).	81
3.10	Templates given in the appendix to *Die Lagerung* for constructing tetrahedral carbon atoms of varying substitution: (a) face-centered (bottom) and vertex-centered (top) models to illustrate enantiomorphism; (b) face-centered model of a single carbon–carbon bond, one of three to show rotational isomers due to Van 't Hoff's second hypothesis; (c) vertex-centered model of the double bond to show the isomerism resulting from Van 't Hoff's third hypothesis; (d) vertex-centered model of spatial allene isomers; (e) and (f) models for enantiomorphic asymmetric carbon atoms, showing the distorted tetrahedron resulting from the presence of four different groups. From Van 't Hoff, *Die Lagerung* (1877).	82
4.1	(a) Anschütz's cyclic structure for fumaric and maleic acid. (b) Fittig's structures for various unsaturated acids.	94
4.2	(a) Baeyer's synthesis of diacetylene compounds from dibromosuccinic acid. (b) Diacetylene and (c) tetraacetylenedicarboxylic acid.	101
4.3	Baeyer's drawing of ring compounds showing the relative strain (depicted as deviation from the standard bond angle for tetrahedral geometry) in rings containing two to six carbon atoms. From von Baeyer, 'Über Polyacetylenverbindungen. Zweite Mittheilung'.	103
4.4	Aemilius Wunderlich's model of the carbon atom as illustrated in Carl Bischoff and Paul Walden, *Handbuch der Stereochemie* (1894), p. 54.	107
5.1	Wislicenus' space formulas for crotonic and isocrotonic acid. From Wislicenus, 'Anordnung', p. 42.	115
5.2	Examples from 'Spatial Arrangement' showing Wislicenus' modified concept of free rotation with a 'favored' configuration. (a) shows the planesymmetric addition of chlorine to ethylene to give dichloroethane. Formulas (ii) and (iii) represent the same molecule after rotation of the upper carbon atom. (b) shows the distortion from the regular tetrahedra that results as a result of attractions between atoms on the	

	favored configuration, in this case the attraction of the positive hydrogen and the negative chlorine. From Wislicenus, 'Anordnung', pp. 15 and 16.	118
5.3	'Planesymmetric' addition, according to Wislicenus. Towards the right is addition, towards the left is elimination. Adapted from Wislicenus, 'Anordnung'.	119
5.4	The three pairs of isomeric, structurally identical unsaturated acids that Wislicenus studied in 1888 and 1889.	121
5.5	The reaction of maleic and fumaric acid with permanganate. Addition to either side of maleic acid (a) results in the same internally symmetric product that is optically inactive. Addition to either side of fumaric acid (b) leads to two different products that are non-superimposable mirror images of each other. Although the initial product from fumaric acid is optically inactive (like the product from maleic acid), it can be resolved into two optically active compounds.	123
5.6	Wislicenus' proposed mechanism for the conversion of maleic into fumaric acid. Adapted from Wislicenus, 'Anordnung'.	124
5.7	Apparent exceptions to Wislicenus' principles. In (a), maleic acid is converted to fumaric acid with bromine. In (b), bromine appears to add to acetylene dicarboxylic acid to give fumaric acid, the result of an apparent axialsymmetric addition.	125
5.8	Wislicenus' explanation for Petri's results. Bromine adds normally to maleic acid, but hydrobromic acid is generated in the formation of bromofumaric acid that then reacts with unreacted maleic acid to form fumaric acid. Adapted from Wislicenus, 'Untersuchungen zur Bestimmung der räumlichen Atomlagerung. Erste Abhandlung'.	126
5.9	Wislicenus' explanation for Bandrowski's results. A side reaction generates oxalic acid, hydrobromic acid and carbon dioxide. The hydrobromic acid generated can then react with unreacted acetylene dicarboxylic acid to form bromomaleic acid that subsequently adds bromine and loses HBr from its favored configuration to yield dibromofumaric acid. Adapted from Wislicenus, 'Untersuchungen zur Bestimmung der räumlichen Atomlagerung. Erste Abhandlung'.	127
5.10	Reactions used by Wislicenus to establish configurations of the isomeric β-chlorocrotonic acids. Adapted from Wislicenus, 'Untersuchungen zur Bestimmung der räumlichen Atomlagerung. Dritte Abhandlung'.	129
5.11	Reactions used by Wislicenus to establish the configuration of α-chlorocrotonic acid. Adapted from Wislicenus, 'Untersuchungen zur Bestimmung der räumlichen Atomlagerung. Dritte Abhandlung'.	131

xiv   *Chemical Structure, Spatial Arrangement*

5.12 Reactions used by Wislicenus to establish the configuration of α-isochlorocrotonic acid. Adapted from Wislicenus, 'Untersuchungen zur Bestimmung der räumlichen Atomlagerung. Dritte Abhandlung'.   132

6.1 (a) Meyer's synthesis of nitro compounds. (b) Thiophene as an analog of benzene.   159

6.2 Meyer's test of Van 't Hoff's theory, proposed in a letter to Adolf von Baeyer.   160

6.3 Meyer's alternative geometrical conception of cyclobutane, cyclopentane, cyclohexane and cyclooctane. The bold lines represent the ring of the molecule, while the thin lines fill in the remainder of the polyhedron. Meyer's original drawings are reproduced in the translation of his letter to Baeyer in Appendix 5.   162

6.4 Derivatives of benzylcyanide in which the remaining hydrogens proved difficult to substitute. The structure on the far right is from the anonymous report of Meyer's lecture (see note 12).   163

6.5 The reaction of hydroxylamine with geminally disubstituted carbons to yield oximes.   164

6.6 Meyer's proposed experiment, based on the reaction in Figure 6.5, to decide between the Kolbe/Fittig and Van 't Hoff theories of unsaturation.   165

6.7 The reaction of benzil with hydroxylamine.   165

6.8 Reactions of the isomeric benzildioximes.   168

6.9 The reaction of α and β benzildioximes with acetic anhydride to give different compounds.   169

6.10 Possible structural isomers containing the group $C_6H_5-C-C-C_6H_5$. One of the two isomers was assumed to have structure (a). The other isomer would have one of the other structures, except for (b), as Meyer had shown earlier that such nitro groups transformed spontaneously into an oxime group.   170

6.11 Possible reaction products of compounds (c), (d) and (e) in Figure 6.10 with potassium ferrocyanide.   170

6.12 Meyer and Auwer's configurational analysis of the benzildioximes. The top row is drawn without perspective. The middle row is drawn with perspective. The bottom two formulas (the maleinoid and fumaroid forms) were also used by Meyer to emphasize the analogy between the unsaturated acids and the benzildioxime isomers. From Meyer and Auwers, 'Untersuchungen', p. 791.   174

6.13 Meyer's preparation of phenanthrenequininonedioxime.   176

6.14 (a) Benzophenoxime and benzilmonoxime. (b) The behavior of benzophenonoxime under heat.   176

List of Figures xv

6.15 Meyer's and Auwers' new conception of the three isomeric benzildioximes, arranged in order from least stable to most stable. In the middle isomer, the two groups are perpendicular to one another.   177

6.16 Meyer's prediction of a fourth benzildioxime isomer in 1890. The configurations are drawn end-on, looking through the C−C axis (solid lines are radicals contained by the front carbon, dotted lines are held by the carbon atom behind). In the two upper drawings, the groups in front should cover completely the radicals in back. They are staggered slightly for clarity. Adapted from Meyer, 'Ergebnisse und Ziele der stereochemischen Forschung'.   181

6.17 Meyer and Riecke's illustration of their model of the carbon atom and bonding. (a) This is the carbon atom itself. One valence is behind the atom pointing away from the viewer. (b) This illustrates a single bond capable of free rotation, while (c) illustrates a single bond incapable of rotation. (d) This illustrates the formation of a carbon–carbon double bond. From Meyer and Riecke, 'Einige Bemerkungen', pp. 952–5.   186

7.1 (a) Ernst Beckmann's structures for the isomeric benzaldoximes. (b) Heinrich Goldschmidt's demonstration of structural identity for the benzaldoximes.   197

7.2 Structures for (a) oximidosuccinic acid, (b) hydroxamic acid and (c) trinitroazoxytoluenes.   198

7.3 The Hantzsch/Werner model for nitrogen–carbon bonding. Note the explicit analogy drawn between the carbon–carbon and carbon–nitrogen double bonds. From Hantzsch and Werner, 'Versuche zur Stereochemie der Stickstoffs'.   199

7.4 Hantzsch and Werner's new formulas for various oxime isomers.   200

7.5 Meyer's explanation of the isomeric benzaldoximes by a 'mobile hydrogen atom' that was located on either the nitrogen or the carbon atom.   202

7.6 The structures of (a) phenanthrenequinone, (b) diacetyl and (c) phenyltolylketone.   203

7.7 The formation of benzamide from benzaldoxime via a tautomeric intermediate.   204

7.8 The formation of an oxime either directly (a) or indirectly (b) through an intermediate addition of hydroxylamine to the ketone.   205

7.9 The structures of monoketones that formed oxime stereoisomers according to Hantzsch's theory.   206

7.10 (a) and (b) Meyer's drawings of the ball and stick models for hydroxylamine. (a) shows the attraction of the hydroxyl hydrogen for the nitrogen atom, and is meant to be pointing up out of the plane of the page. (c) and (d) The two oxime isomers resulting from Meyer's model

	of hydroxylamine. Drawn after Meyer and Auwers, 'Über die Isomeren Oxime unsymmetrische Ketone'.	208
7.11	(a), (b) and (c) Unknown nitrogen analogs of well-known compounds containing carbon–carbon double bonds. (d) and (e) Compound (d) did not react with hydroxylamine to give an oxime, and compound (e) formed only one oxime.	209
7.12	Hydrazones isolated reported by Meyer from the reaction of ketones with phenylhydrazine.	210
7.13	The formation of hydrazones from various ketone chlorides by the general reaction given at the top.	211
7.14	Hantzsch's comparison of the oximes and cinammic acids as an example of extending Meyer's theory of the oximes to other analogous structures.	212
7.15	Hantzsch's methods for determining the configuration of the oximes. (a) Elimination of water to form a nitrile. (b) The Beckmann rearrangement to form an amide. (c) The two possible isomers resulting from the Beckmann rearrangement and formed by the migration of either R_1 or R_2.	215
7.16	An example of using the Beckmann rearrangement to determine configuration. In each case, the aromatic group *cis* to the hydroxyl group migrates to the nitrogen atom. Different stereoisomers therefore result in different product amides.	216
7.17	The formation of esters via an intermediate addition product, as speculated by Meyer.	217
7.18	Van 't Hoff's model for the pentavalent nitrogen atom. After Van 't Hoff, 'Over die bindingsrichtingen'.	222
7.19	Willgerodt's spatial model for the nitrogen atom (1888). (a) The nitrogen atom's three principal coplanar valences (left) and stereoformulas for the pentavalent nitrogen atom, ammonia and ammonium chloride; (b) Willgerodt's stereoformula for hydrazine and azo compounds (top) and the possible spatial isomers resulting from disubstitution (bottom); (c) Spatial isomers for the phenylhydrazines.	224
7.20	Examples of Willgerodt's 1890 space formulas for the pentavalent nitrogen atom. (a) Hydrocyanic acid. (b) An imine group with a carbon–nitrogen double bond. (c) Various possible formulas for the oximes.	225
7.21	Robert Behrend's model for the nitrogen atom. (a) The model for the nitrogen atom itself, showing the three principal valences and the two auxiliary valences denoted with plus and minus signs. (b) A condensed version of the formula applied to the benzildioximes, showing how the position of the remaining charges would influence the	

List of Figures xvii

	favored configuration of the isomeric benzildioximes. From Behrend, 'Zur Stereochemie stickstoffhältiger Körper'.	226
7.22	S.W.U. Pickering's model of the nitrogen atom, showing only the three principal valences (a), and a view from one of the two auxiliary valences (b). From Pickering, 'Note on the Stereoisomerism of Nitrogen Compounds'.	229
7.23	Wilhelm Vaubel's stereoformulas for nitrogen-containing molecules. (a) and (b) Two views of the arrowhead-shaped nitrogen atom. (c) The cyano radical, showing how the carbon atom fits into the cleft of the nitrogen atom, satisfying all four valences, leaving one valence on the pentavalent nitrogen atom. (d) and (e) Two representative stereoformulas with Vaubel's model: (d) Pyrrole; (e) Ammonium chloride. From Vaubel, *Das Stickstoffatom*.	230
7.24	Claus' proposed structures for α-, β- and γ-benzildioximes.	232
7.25	Meyer's analysis of the reaction of α-benzylhydroxylamine with benzil using his own configurational formulas and Claus' structural formulas. (a) The conversion of α- to γ-monoximes using Claus' structural formulas, a process that Meyer called 'extremely improbable'. Meyer's structure for the product of benzil with α-benzylhydroxylamine (b) appeared more probable than Claus' structure for the product (c).	233
7.26	Claus' alternate formulas for the oximes.	236
8.1	Structures of the monosaccharides proposed by Baeyer (1870), Schiff (1870), Fittig (1871) and Tollens (1883). From Baeyer, Schiff, Fittig and Tollens (see note 10).	247
8.2	Kiliani's two-step procedure for extending the carbon chain of monosaccharides. In this case a six-carbon sugar is lengthened to a seven-carbon sugar acid.	248
8.3	The formation of phenylhydrazones and glucosazones with phenylhydrazine.	249
8.4	The removal of the osazone with hydrochloric acid and reduction of the osone to give levulose.	250
8.5	The structures of glycerin, erythritol and dulcitol.	251
8.6	The synthesis of α- and β-acrose from acrolein.	252
8.7	Possible structures for α-acrose, derived by determining the possible aldol products. From Fischer and Tafel, 'Synthetische Versuche in der Zuckergruppe III', p. 250.	253
8.8	The total synthesis of optically active levulose from the synthetic α-acrose.	256
8.9	The synthesis of gulose from saccharic acid.	259

xviii *Chemical Structure, Spatial Arrangement*

8.10 The possible configurations of the isomeric hexoses using Van 't Hoff's notation. The plus and minus signs are determined relative to the point between carbon atoms 3 and 4. Aldehyde groups are on the top. For a translation of this notation into Fischer's projection formulas, see Appendix 6. From Fischer, note 37, p. 417. Fischer adapted the chart from the first edition of Van 't Hoff, *Die Lagerung der Atome im Raume* (1877). 260

8.11 Fischer's first use in 1891 of the new projections for the tartaric acid isomers (a) and the saccharic acids (b). From Fischer, 'Über die Konfiguration des Traubenzuckers II', p. 428. 263

8.12 Fischer's 'family trees' for the synthetic and natural sugars that emphasise genetic relationships among the monosaccharides. (a) This shows the relationships between sugars as the number of carbon atoms increases from the bottom of the diagram towards the top. (b) This shows the route by which various naturally occurring sugars (and their alcohol and acid derivatives) can be derived from synthetic α-acrose. From Fischer, 'Synthesen in der Zuckergruppe I' and 'Synthesen in der Zuckergruppe II'. 266

8.13 Fischer's chart showing the relative fermentability of various sugars with various species of yeast. Three daggers indicates complete fermentation after eight days, and a dash indicates the absence of fermentation. From Fischer and Theirfelder, 'Verhalten der Zucker gegen reine Hefen'. 270

9.1 Examples of Blomstrand's formulas for the cobalt ammine complexes containing cobalt of varying valence. The luteo salt is on the far right. Blomstrand used 'a' as an abbreviation for the ammonia (NH_3) radical. From Blomstrand, *Chemie der Jetztzeit*, p. 293. 282

9.2 The structures for the luteo (a) and purpureo (b) compounds, according to Jørgensen. 284

9.3 (a) Jørgensen's structures for the cobalt series in which the number of ammonia molecules decreases. (b) The structure (according to Jørgensen's theory) that results from Werner's thought experiment of removing four ammonia molecules from the luteo compound. Werner thought this structure to be 'highly peculiar and unlikely'. 286

9.4 (a) Werner's depiction of the resulting *cis–trans* isomerism in a disubstituted octahedron. (b) The *cis–trans* isomerism resulting from the square planar arrangement of atoms in the MA_4 type. From Werner, 'Beitrag zur Konstitution anorganischer Verbindungen', p. 311, and 'Beitrag zur Konstitution anorganischer Verbindungen. II', p. 184. 289

9.5 Werner and Miolati's plots of conductivity versus ammonia content in the ammine complexes of platinum (a) and (b) and cobalt (c). Note

List of Figures xix

	the missing values for the last two complexes in the cobalt series. From Werner and Miolati, 'Beiträge zur Konstitution anorganischer Verbindungen. II'.	297
9.6	Werner's configurational assignment to the platosammine and platosemidiammine salts. From Werner, 'Beitrag zur Konstitution anorganischer Verbindungen', pp. 315–17.	300
9.7	Werner's chart showing the similarity in the genetic relationships during the synthesis of the flavo and croceo salts. The croceo salt can be derived from the praseo salt by the sequence of preparations shown in the first column. The flavo salt can be derived by a similar sequence of preparations from the carbonatotetrammine (violeo) salt at the top of the right column. Each column shows the same sequence of substitutions, and so the flavo and croceo salts must be structurally identical. From Werner, 'Beitrag zur Konstitution anorganischer Verbindungen. II', p. 184.	301
9.8	The two possible spatial isomers of the planar cobalt carbonato complexes. Werner ruled out the *trans* compound (b) because of the physical difficulty in the ability of the carbonato group to bridge the cobalt atom. From Werner, 'Beitrag zur Konstitution anorganischer Verbindungen. II', p. 185.	302
9.9	Werner's assignment of configurations to the flavo and croceo salts using octahedral formulas. C could be made only from A, while D could be made from either A or B. From Werner, 'Beitrag zur Konstitution anorganischer Verbindungen. II', p. 186.	303
9.10	Werner's 1907 synthesis of the new violeo salt portrayed in stereoformulas. The cleavage of the dicobalt compound involves simple replacement of the bridging hydroxyl groups with chloride ions resulting in the *cis* arrangement of chlorine atoms in the violeo compound. Adapted from Werner, 'Über 1.2-Dichloro-tetrammin-kobaltisalze (Ammoniakvioleosalze)', p. 4820.	304
9.11	According to Jørgensen's formulas (a) the formation of the chloroaquotetrammine chloride involved a simple insertion of the water molecule between the chlorine and cobalt atoms. Werner's spatial formulas (b), on the other hand, implied that a chlorine atom must be initially substituted to give an unknown intermediate compound ((b), middle) that then rearranged to the known chloroaquotetrammine chloride praseo chloride, with a simple insertion of an ammonia molecule. From Jørgensen, 'Zur Konstitution der Kobalt-, Chrom- und Rhodiumbasen. V', p. 146.	311
9.12	Werner's first depiction of the configuration of optically active coordination compounds, giving full structural details (a), and in	

	abbreviated form (b). (c) The structure of an analogous optically active organic compound that does not contain an asymmetric carbon atom. From Werner, 'Beitrag zur Konstitution anorganischer Verbindungen. XVII', pp. 147–8.	314
9.13	(a) The carbonato complex that Victor King initially attempted to resolve into enantiomers. (b) The two enantiomers of the cobalt complex that King succeeded in resolving using (+)-silver bromocamphorsulfonate (top right, the position of the bromine atom was unspecified by Werner). These formulas were not included in Werner's publication. Adapted from Werner, 'Zur Kenntnis des asymmetrischen Kobaltatoms. I'.	315
9.14	The formula (a) and configuration (b) of the first optically active inorganic complex that contained no carbon atoms. The stereoformula was not included in Werner's paper. Adapted from Werner, 'Zur Kenntnis des asymmetrischen Kobaltatoms. XII'.	317
A5.1	(a) Meyer's double tetrahedron from the postcard of 5 October 1885. (b) Meyer's depiction of cyclobutane (top), cyclopentane (middle) and cyclohexane (bottom) as regular polyhedra. (c) Meyer's drawing of the tetrahedral geometry of cyclobutane. (d) Meyer's drawing of cyclooctane as a cube. Reproduced from the letter from Meyer to Baeyer, 18 October 1885. Courtesy Deutsches Museum, Munich, Archiv.	366
A6.1	Eight of the possible sixteen isomers of glucose, portrayed by Fischer projections. Van 't Hoff's notation is indicated at each asymmetric carbon atom.	370

Series Editor's Preface

This is a study of the early history of structural theories in chemistry – theories that enable us to reason about the spatial position of atoms within a molecule. Peter Ramberg takes us from the context of organic chemistry up to the early 1870s, through the separate invention of the tetrahedral carbon atom by the Dutch chemist Van 't Hoff and the French scientist Le Bel, through the reception and acceptance of that theory (particularly Van 't Hoff's version) in Germany, to its application in organic and inorganic chemistry, culminating in Alfred Werner's coordination chemistry.

For anyone familiar with contemporary chemistry, it takes a real effort of imagination to conceive a world without stereochemistry. Much of chemistry today depends upon structural theory, which enables us to know where atoms are in space, how they are connected to one another, and how much freedom of movement they possess within a molecule. As a result of structural theory, we can design molecules at our desks, in a process akin to architectural and fashion design, which also need to think about spatial arrangement. Structural theory has become a tool for the invention of new molecules, and for designing the synthetic paths needed to produce them. Everyone knows that DNA is a double helix; the very definition of many molecules embodies structural models. Chemists have known for over a century that some substances may consist of either left- or right-handed molecules, i.e. may all possess the same kind of asymmetry, whereas until very recently the best efforts of chemists in the laboratory produced mixtures of left- and right-handed molecules, mirror images of one another. Since many chemical reactions in living organisms depend upon a particular asymmetry, we were shut out from producing some biologically active substances. Now, thanks to the development of chiral catalysis, we can produce many of them, with enormously important implications for medicine. Organic chemists and molecular biologists think in three dimensions, and can picture the real structure of molecules.

Until the early 1870s, however, chemists lacked the conceptual tools to picture and model three-dimensional molecules. Even apparently structural formulas indicated only the arrangement of atoms, links in a chain rather than positions in space. Speculating about what we understand by structure would have seemed to chemists like mere wool-gathering, a pointless exercise. How could one possibly know the real position of atoms in space? Chemists in England in the 1860s had debated the existence or non-existence of atoms, just at the time when physicists were developing the kinetic theory of gases, based upon the physical reality of molecules. Physicists and chemists clearly had different preoccupations. Nearer the turn of the century, the great German physical chemist Wilhelm Ostwald remained highly doubtful about the existence of chemical atoms. And yet stereochemistry, in which a belief in the reality of atoms provided the underpinning of theories about where the atoms really were, had gained widespread acceptance of the previously unnoticed 'goldmine' that Van 't Hoff had staked out.

This is a story of chemical theory and practice, and Ramberg does not shy away from technical details. But he is careful to explain the technicalities as they arise, and so readers of this book will not need expertise in chemistry to follow the argument. Using structural theory as it was transformed into stereochemistry, Ramberg shows how chemical formulas came to embody structural implications, and how molecular models moved from being heuristic devices regarded as fictions, to true representations of structure. He explores the issue of disciplinary boundaries, even though he is careful not to claim stereochemistry as a new subdiscipline. He shows how shifts in the boundaries between chemistry and physics were entailed in Van 't Hoff's development of the tetrahedral carbon atom. He looks at the factors involved in the acceptance of new ideas in science, and shows why Van 't Hoff's ideas succeeded, whereas those of Le Bel were ignored, even though it was Le Bel who worked in a laboratory in the faculty of medicine in Paris, and Van 't Hoff who worked in the obscurity of a veterinary school in the Netherlands. Historians of chemistry have long been familiar with Kolbe's inspired and vitriolic opposition to Van 't Hoff's theory. Ramberg offers an important corrective in stressing the widespread support that Van 't Hoff's theory garnered in yet another of chemistry's 'quiet revolutions'. Van 't Hoff had his bulldog in Wislicenus, just as Darwin had his in Huxley; perhaps Kolbe is the nearest approach to Bishop Wilberforce in this story. But even though the acceptance of the new theory was rapid, it was not shared equally across generations. Generational dynamics have a lot to do with the process of theory change in science, and clearly so in this case. And since, as Ramberg makes clear, stereochemists came to see their work as a natural outgrowth of earlier chemistry, there was no need for beleaguered research schools to champion the new theory. That theory was simply adopted by existing research schools, even when that meant transforming their program.

Here is rich material for the history, philosophy, and social study of chemistry.

Trevor H. Levere

Acknowledgments

Like countless authors before me, I have incurred many debts on the path to completing this book. I must first thank the faculty of the History and Philosophy of Science Department at Indiana University, especially my advisor Noretta Koertge, for their encouragement and help on a project far removed from their own personal expertise. I have also benefited from many discussions with my colleague and friend Lawrence Principe at Johns Hopkins University. In 1999, I was fortunate enough to have Ursula Klein invite me to join her new group in history and philosophy of chemistry at the Max Planck Institute for the History of Science in Berlin. I am grateful to her for making available to me the generous resources of the institute for completing the research and writing of this book. I also thank the other members of the group, John Dettloff, Sarah Vollmer and Eric Franceour, for reading draft chapters, and for their helpful commentary during group meetings.

I thank Ursula Klein, Alan Rocke and William Brock for reading various draft manuscripts, and George Kauffman for comments on Chapter 9. The book has benefited from their many suggestions, although any remaining errors are my own. I especially thank Alan Rocke and Markus Popplow for helpful discussions about the finer points of translating German into English. Eric Franceour provided much-needed help in translating French, and Geert Somsen in translating Dutch manuscripts. I thank Christoph Meinel for making available his unpublished article on the origins of molecular modeling in the 1860s.

Other versions of some portions of the book have previously appeared in *Annals of Science* (Chapter 3 – see also http://www.tandf.co.uk for further information on this journal), *Historical Studies in the Physical and Biological Sciences* (Chapter 5), *HYLE: An International Journal for the Philosophy of Chemistry* (Chapters 6 and 10), and *The Bulletin for the History of Chemistry* (Chapter 5). I am grateful to these journals for allowing me to reproduce the principal arguments of these articles.

I am thankful to The Bancroft Library, Berkeley, California, USA, for granting me permission to quote from the Emil Fischer Papers (BANC MSS 71/95z) in Chapters 5, 6, 7, 8 and 10. I also thank the Johns Hopkins University for granting me permission to publish English translations of the three letters from the Van 't Hoff papers in Appendices 1 and 3.

Research for this book was supported by grants from Indiana University, the Fulbright Foundation, the Deutsches Museum, the National Science Foundation, the American Philosophical Society, and the Max-Planck-Gesellschaft.

There have been countless helpful librarians at Indiana University, the Eisenhower Library at Johns Hopkins University, the Max Planck Institute, the Staatsarchiv Zürich, Zentralbibliothek Zürich, the Wissenschaftshistorische Sammlungen of the ETH Bibliothek, the Staatsbibliothek zu Berlin, Preussischer Kulturbesitz, the Library and Archive of the Deutsches Museum, Munich, and the Bancroft Library, Berkeley. I must also thank William Jensen of the Oesper Collection in the History of Chemistry at the University of Cincinnati, both for helpful comments and for providing copies from the collection.

Finally, I need to thank my family: Liza for her endless patience with this project, and Sonja, Paul, and Karl, who remind me every day that the present becomes the past more quickly than we think.

Chapter 1
Introduction: 'Van 't Hoff's Gold Mines'

> Our formulas (structural formulas) are pictures in a plane, but molecules are nevertheless bodies, for example [a structural formula] gives us the sequence of atoms, but by no means the three-dimensional arrangement of the molecule's atoms in space.
>
> Johannes Wislicenus, 1872

> Stereochemistry (στερεοσ, solid) deals with those chemical and physical phenomena which are believed to be due to the relative positions in space taken up by atoms within a molecule. This arrangement of atoms is termed the *configuration* of the compound; and just as we can represent the constitution of an organic compound by means of its structural formula, so we can represent its configuration by means of its space formula.
>
> Alfred Stewart, 1919

When he declared in 1872 that structural formulas did not represent the spatial arrangement of atoms in molecules, Wislicenus reflected the unanimous opinion of nineteenth-century chemists about the meaning of chemical formulas. Nevertheless, three years later he had become firmly convinced that the same structural formulas could give clues about the spatial characteristics of molecules, and that those spatial characteristics could influence the physical and chemical properties of compounds. Wislicenus' personal conversion exemplifies a general transformation in the meaning of structural formulas during the last quarter of the nineteenth century. The origins and reasons for this transformation from the concept of the 'chemical structure' of a molecule and the 'spatial arrangement' of atoms in a molecule, the last major stage in the development of traditional nineteenth-century chemistry, is the subject of this book. As defined concisely by Stewart, stereochemistry began in 1874, one of the most famous years in the history of chemistry, when Jacobus Henricus van 't Hoff and Joseph Achille Le Bel independently suggested that the four valences in a carbon atom were directed towards the corners of a tetrahedron. Van 't Hoff's version of the theory would eventually prove more influential, when in the late 1880s chemists became aware of the 'gold mines that had gone entirely unnoticed' in Van 't Hoff's theory.[1]

---

[1] Victor Meyer and Karl Auwers, 'Weitere Untersuchungen über die Isomerie der Benzildioxime', *Berichte*, 1888, *21*, 3510–29, p. 3513.

In the intervening 125 years since 1874, these 'gold mines' have proven both fundamental to the practice of chemistry and nearly inexhaustible. In the very early twenty-first century, it is difficult to comprehend the true novelty of the tetrahedral carbon atom, for the principles of stereochemistry have been seamlessly integrated into the theoretical infrastructure of chemistry, and the idea of a modern chemistry without spatial models of molecules is almost inconceivable. Organic chemists plan extensive syntheses of complex molecules or a new synthetic technique based on the specific three-dimensional orientation of atoms in molecules. The concept of an 'active site' in biochemistry relies on the three-dimensionality of an enzyme, and the structure of DNA is a three-dimensional double helix. Fullerenes, the recently discovered elemental form of carbon, have a spherical 'soccer ball' shape. *Tetrahedron* and *Tetrahedron Letters* are two major journals in organic chemistry. Part of learning chemistry is learning to conceive of molecules in three dimensions, and how the shape of molecules affects their chemical and physical behavior.

Yet for all its fame and subsequent centrality to modern organic chemistry, we know surprisingly little about the larger context of the emergence and spread throughout Europe of 'chemistry in space' (one of the original names for stereochemistry). There are, however, a number of compelling reasons for historians to look closely at the reception of stereochemistry at the end of the nineteenth century. Certainly, elucidating the spatial properties of molecules – by definition, invisible objects – from purely macroscopic manipulations of chemicals constitutes one of the outstanding accomplishments of nineteenth-century chemistry, and for this reason alone, the origin and development of one of the central features of modern chemical theory should be worth studying. The importance of the development of stereochemistry does not, however, lie solely in its subsequent central role to modern organic, inorganic and bio-chemists. Although one of the goals of history is to determine how the present is a function of the past, historians of science no longer treat past science merely as a precursor to the present, but place the emergence of scientific ideas within the intellectual and cultural context of their own time. When we look closely at the early history of stereochemistry in its own context of the late nineteenth century, we see its importance in an entirely new and much richer light.

To begin an exploration of that context, we can ask some rather simple questions. Why and how did Van 't Hoff and Le Bel assume that they could have knowledge of the three-dimensional properties of molecules? When should these properties be used in an explanation, and how can one best represent and model these properties in chemical formulas? Taking a step further back, we can ask a more fundamental question about the unique language and symbolism of chemistry. What do chemical formulas mean, and how do chemists give them meaning? More specifically, in the context of stereochemistry, we can ask how and why formulas took on a three-dimensional meaning. What did they mean before Van 't Hoff and Le Bel introduced the tetrahedral carbon atom?

These questions become more intriguing when we consider the historical circumstances surrounding the development of structural theory of organic chemistry during the 1860s. As the structure theory developed, chemical structures, and the formulas that came to represent them, took on an exclusively chemical meaning, a shorthand notation for the many chemical properties of a substance. Chemists had not intended the concept of structure to take on a physical meaning – a portrayal of the spatial arrangement of atoms in the microworld. Van 't Hoff made the unprecedented assumption that the structure *could* give a picture of the spatial arrangement of atoms by assuming that the valences of the carbon atom were directed towards the corners of a tetrahedron. The addition of this new and originally unintentional physical meaning to formulas with a previously purely chemical meaning is the central philosophical theme throughout the history of 'chemistry in space'.

Much of this book, then, concerns an historical answer to a larger philosophical problem surrounding the meaning of chemical formulas. The addition of a spatial meaning to chemical formulas, and the use of spatial properties in chemical explanation, exemplifies one of the perennially important questions in the history and philosophy of science. How does a community come to agreement on a new form of explanation? In *How Experiments End*, Peter Galison asked this question for high-energy physics, and described how physicists sort out background noise from true effects, 'carving away' the artifacts of the instrument.[2] We can see the same sort of process in stereochemistry. How did chemists sort through the data of the chemical laboratory to determine that spatial arrangement could be the true cause of the chemical effects they witnessed? Why did Van 't Hoff and other chemists conclude that knowledge of a molecule's physical appearance was necessary to explain certain chemical phenomena? How did other chemists see and exploit the fertility of Van 't Hoff's correlation between chemical structure and physical form? Can the methods used for ascertaining a spatial model of the carbon atom be transferred to other elements? The appearance of spatial properties as a necessary tool for chemical explanation is particularly pertinent, because chemical formulas prior to Van 't Hoff had no physical meaning.

The small amount of existing secondary literature on the early history of stereochemistry has centered around the co-discovery of the tetrahedral carbon atom, and for the most part has assumed that the importance of the discovery lay in its contribution to modern chemistry, not in its own nineteenth-century context.[3] The

---

[2] Peter Galison, *How Experiments End*, Chicago, IL, University of Chicago Press, 1987.

[3] Anatol Sementsov, 'The Eightieth Anniversary of the Asymmetrical Carbon Atom', *American Scientist*, 1955, *43*, 97–100. H.A.M. Snelders, 'Practical and Theoretical Objections to J.H. van't Hoff's Stereochemical Ideas', pp. 55–65 in O.B. Ramsay, ed., *Van't Hoff-Le Bel Centennial*, Washington, DC, American Chemical Society, 1975. Snelders, 'The Reception of J.H. Van't Hoff's Theory of the Asymmetric Carbon Atom', *J. Chem. Ed.*, 1974, *51*, 2. E. Fischmann, 'A Reconstruction of the First Experiments in Stereochemistry: Letters from Van't Hoff to Bremer in a New Chronological Sequence',

chemist O. Bertrand Ramsay has contributed more than anyone to our current understanding of early stereochemistry. An essential starting point for the subject is the volume edited by Ramsay containing essays by historians and chemists delivered during the proceedings of an American Chemical Society Symposium held on the centennial of the tetrahedral carbon atom. His 1981 book *Stereochemistry* is the most comprehensive treatment of the subject, from the discovery of optical activity in the early nineteenth century to the work of Derek Barton and Odd Hassel in the 1950s.[4] However useful they may be as starting points, each work has its shortcomings. The edited centennial volume offers a necessarily fragmented, incomplete history of stereochemistry, whereas *Stereochemistry* was written primarily as a textbook, with no thematic development or references.

One of the purposes of this book is to draw together the existing fragmentary accounts of stereochemistry into a more comprehensive narrative of the events surrounding the adoption of stereochemical principles as a natural, but not necessarily inevitable, outgrowth of traditional nineteenth-century chemistry. Chapter 2 will lay out the intellectual, practical and institutional background of organic chemistry in the nineteenth century, particularly the development of the concept of chemical structure during the 1860s. Much of that chapter is a synthesis of the existing historical literature on the origins of the structure theory.

Chapter 3 will lay out the immediate context surrounding the proposal of the tetrahedral hypothesis by Van 't Hoff and Le Bel. Although both chemists rightfully receive dual credit for suggesting the tetrahedral carbon atom, the scope and content of their theories proved quite different. Van 't Hoff proposed the tetrahedron to solve the unique chemical problem of isomerism, while Le Bel was interested in the relationship between optical activity and molecular form, a question framed by generations of French crystallographers. We shall explore the principal differences between the two theories in Chapter 3, and it will become clear that Van 't Hoff's ideas were applicable on a much broader scale. The subsequent development of stereochemistry derives almost entirely from him and not Le Bel, and the remainder of the book will focus on the intellectual thread begun by Van 't Hoff in 1874.

Janus, 1985, *72*, 131. Jost Weyer, 'A Hundred Years of Stereochemistry: The Principal Development Phases in Retrospect', *Ang. Chem. Int. Ed. Eng.*, 1974, *13*, 591. Weyer, 'Die Aufnahme der van't Hoff'schen Hypothese vom asymmetrischen Kohlenstoffatomen (1874) in Deutschland', pp. 311–320 in Gunter Mann and Rolf Winau, eds, *Medizin, Naturwissenschaft, Technik und das Zweite Kaiserreich*, Göttingen, Vandenhoeck and Ruprecht, 1977.

[4] O.B. Ramsay, *Van 't Hoff-Le Bel Centennial*, note 3. Ramsay, *Stereochemistry*, London, Heyden, 1981. The only other comprehensive history of stereochemistry (covering topics well into the twentieth century) is in Russian: Georgii Vladimirovich Bykov, *Istoriia stereokhimii organicheskikh soedinenii*, Moscow, Nauka, 1966.

Because the original version of Van 't Hoff's theory was printed in Dutch, most of the chemical world learned of Van 't Hoff's ideas from either the 1875 French version, *La chimie dans l'espace*, or the 1877 German translation, *Die Lagerung der Atome im Raume*. Chapter 3 will also look more closely at the evolution of Van 't Hoff's ideas from his short Dutch pamphlet to the 1877 German edition. Like many influential works in the history of science, they fit well into an established tradition, yet contained novel elements that proved immensely fertile for new research. Van 't Hoff's fundamental goal was the explanation of isomerism, a long-standing concern for organic chemists, by an appeal to the arrangement of atoms that for the first time meant a spatial arrangement. While the centerpiece of all three versions was the tetrahedral carbon atom, there was a cluster of theories that flowed from that central premise that clarified many long-standing curiosities in organic chemistry. In 1888, Victor Meyer divided this cluster into three principal hypotheses that will be explained in further detail in Chapter 3: the tetrahedral hypothesis itself, the 'free' rotation about carbon–carbon single bonds, and the restricted rotation about carbon–carbon double bonds. I have found Meyer's division useful for tracking the development of geometrical ideas in chemistry, and I will refer to Van 't Hoff's first, second, and third hypotheses throughout this book, as many of the research paths followed one of the three hypotheses.

If my first task is to present a new, more comprehensive account of the events leading to the tetrahedral carbon atom and the evolution of Van 't Hoff's theory in its various published versions, my larger goal is to lay out the development of the ideas of stereochemistry after 1874. Chapter 4 will look at the reception of the tetrahedral carbon atom in the crucial early period between 1874 and 1887. Previous analyses of the published record have assumed, largely based on the famous and violent criticism of Herman Kolbe, that the chemical community (with the important exception of Wislicenus) was indifferent or hostile to the various versions of Van 't Hoff's theory. A comprehensive analysis of the published and unpublished record shows, in contrast, a much more receptive attitude among chemists that could be best described as a cautious optimism towards the theory.

Chapter 5 will discuss the chemistry of Johannes Wislicenus, long credited as the inventor of the concept of 'geometrical isomerism', whose research on lactic acid directly inspired Van 't Hoff's theory, and who enthusiastically sponsored the translation of *La chimie dans l'espace* into German. The crucial importance of Wislicenus in disseminating the ideas of 'chemistry in space' *after* the appearance of *Die Lagerung*, however, has never been explored adequately.[5] As director of three important chemical laboratories (Zürich, Würzburg, and Leipzig), he was also one

---

[5] In 1959, Aaron J. Ihde wrote a short article on Wislicenus' 1887 essay, 'The Unraveling of Geometric Isomerism and Tautomerism', *J. Chem. Ed.*, 1959, *30*, 330. Ihde did not, however, set it into complete context, or discuss its influence.

of the major chemists of his generation, and his chemistry has so far escaped the scrutiny of historians. Wislicenus' vigorous promotion of Van 't Hoff's theory can be compared in many ways to Thomas Henry Huxley's vigorous defense of Darwin's theory of evolution, and it seems appropriate to call Wislicenus 'Van 't Hoff's Bulldog', for he more than anyone made chemists aware of the utility of Van 't Hoff's theory, especially the second and third hypotheses. Chapter 4 will look at the construction, argument and immediate reception of Wislicenus' major 1887 essay on the chemistry of the unsaturated acids.[6] Wislicenus' aim was to reorganize the known data about unsaturated acids in spatial terms and to assign three-dimensional arrangements of atoms in spatial isomers. To achieve this goal, he used intricate 'mechanisms' of chemical reactions to determine spatial configurations of the unsaturated acids. One of the novelties in Wislicenus' chemistry was his explicit and enthusiastic use of a molecular dynamics. Wislicenus had long desired a physical conception of molecules consonant with his conception of geometrical isomerism, and in 1875, when he first heard of Van 't Hoff's ideas, he was understandably excited. But it would not be until ten years later, after his 1885 move from Würzburg to Leipzig, that Wislicenus would expand Van 't Hoff's second and third hypotheses.[7] This work established Wislicenus as the principal spokesman and chief practitioner of stereochemistry.

Chapters 6 and 7 are two parts of the same story. In 1887, largely inspired by Wislicenus' work on the unsaturated acids, Victor Meyer reopened his investigation in the chemistry of the isomeric benzildioximes, and determined with his assistant Karl Auwers that these isomers were structurally identical. Meyer proposed a novel stereochemical explanation to differentiate these isomers, and although his explanation did not last long, Meyer became one of the principal spokesmen for stereochemistry in lectures and popular articles. While Meyer's theory and his enthusiasm for stereochemistry are the subject of Chapter 6, Chapter 7 will look at the alternate spatial theory of the benzildioximes, based on the stereochemistry of the nitrogen atom introduced in 1890 by Alfred Werner (in his doctoral dissertation) and his mentor Arthur Hantzsch at the Zürich Polytechnical Institute. Although Hantzsch gave Werner full credit for the idea of the tetrahedral nitrogen atom, the credit for defending the theory and firmly establishing the evidence in its favor lies almost entirely with Hantzsch.[8] Like Meyer, he also became one of the strongest advocates

---

[6] Johannes Wislicenus, 'Über die räumliche Anordnung der Atome in organischen Molekülen und ihre Bestimmung in geometrisch-isomeren ungesättigten Verbindungen', *Abh. math.-phys. Cl. kön. sächs. Ges. Wiss.*, 1887, *14*, 1–77. The essay was translated in 1901 as 'The Space Arrangement of the Atoms in Organic Molecules and the Resulting Geometrical Isomerism in Unsaturated Compounds', pp. 61–132 in G.M. Richardson, ed., *The Foundations of Stereochemistry: Memoirs by van't Hoff, Le Bel, and Wislicenus,* New York, American Book, 1901.

[7] Wislicenus, "Über die räumliche Anordnung'.

[8] Existing studies on the stereochemistry of the oximes are by George B. Kauffman: 'Alfred Werner's

for stereochemistry, writing *Grundriß der Stereochemie* in 1892, the first monograph on stereochemistry not written by Van 't Hoff. Hantzsch has generally been recognized primarily as Werner's mentor, and not as an outstanding chemist himself. Hantzsch was, in fact, one of the most innovative organic chemists of his time, employing physical measurements to supplement chemical behavior to establish the stereochemistry of the oximes, the hydrazones and diazo compounds.

In Chapter 8 we shall return to the application of Van 't Hoff's first hypothesis by Emil Fischer in his famous study of the sugars. Fischer's proposal for the configuration of glucose in 1891 has long been considered a masterpiece of chemical and logical thought, a seamless synthesis of Van 't Hoff's theory and chemical data. Fischer's chemistry of the sugars remains famous to this day, and has been summarized numerous times by chemists, but to date there has been no serious historical analysis of Fischer's papers surrounding the chemistry of the sugars, or its context within the history of stereochemistry. From a practical standpoint, Fischer's work resembles that of all stereochemists in that he attempted to match the number of theoretically possible isomers with known compounds. The origins of Fischer's famous 1894 'lock and key' hypothesis of enzyme action can also be seen in his earlier work on carbohydrates.

Chapter 9 will return to Alfred Werner, who after completing his doctoral thesis, wrote two of the most innovative and influential articles of late nineteenth-century chemistry, *Beiträge zur Theorie der Affinität und Valenz* (1891), his *Habilitationsschrift* (post-doctoral thesis required for qualification as a lecturer), and 'Beitrag zur Konstitution anorganischer Verbindungen' (1893), in which he introduced the coordination theory that incorporated elements of stereochemistry into inorganic chemistry. During the next twenty years, Werner and his students isolated hundreds of compounds predicted by his coordination theory and single-handedly reorganized inorganic chemistry, largely along explicitly spatial principles, as fundamentally as the structure theory had reorganized organic chemistry. The chemist George Kauffman has more than anyone reconstructed Werner's life and thought in over a dozen books and articles, including a full biography published over thirty years ago.[9] Yet Kauffman's analysis of Werner's career has suffered from two principal drawbacks. First, it was presented in a fragmented fashion, spread throughout many

---

Inaugural Dissertation', *J. Chem. Ed.*, 1966, *43*, 155–65. Kauffman, 'Stereochemistry of Trivalent Nitrogen Compounds: Alfred Werner and the Controversy over the Structure of Oximes', *Ambix*, 1972, *19*, 129–44.

[9] Representative of Kauffman's studies are 'Sophus Mads Jørgensen and the Werner–Jørgensen Controversy', *Chymia*, 1960, *6*, 180–204; 'Alfred Werner's Inaugural Dissertation', *J. Chem. Ed.*, 1966, *43*, 155–65; *Alfred Werner: Founder of Coordination Chemistry*, Berlin, Springer-Verlag, 1966; *Classics in Coordination Chemistry. Part 1: The Selected Papers of Alfred Werner*, New York, Dover, 1968, and 'The Discovery of Optically Active Coordination Compounds: A Milestone in Stereochemistry', *Isis*, 1973, *65*, 38–62.

different articles and translations, and lacks an overall chronological narrative of Werner's ideas. Second, Kauffman has presented a portrait of Werner and his career in near isolation from the broader context of late nineteenth-century chemistry. To be sure, Werner's introduction of the coordination concept was innovative and revolutionary for inorganic chemistry, but it was not completely without precedent. The coordination theory emerged in 1893 at a time when stereochemistry was flowering in organic chemistry. In structure and argument, and as a theoretical reorganization of known chemical facts, Werner's 1893 article strongly resembled *Die Lagerung*, and the supporting evidence for the coordination theory and for the existence of spatial isomers of inorganic compounds was of a similar kind. We should not be surprised, therefore, to find strong similarities between the arguments for stereochemistry in organic and inorganic chemistry.

I have chosen to end my narrative in 1914, the year Werner first reported the isolation of an optically active compound free of carbon, and the year after he received the Nobel Prize in chemistry. That year marks the effective end of Werner's most productive period (afterwards, his health would slowly decline until his premature death in 1919), and provides closure to the initial establishment of inorganic stereochemistry. By 1914, stereochemistry in organic chemistry had long since reached a mature stage, becoming an integral part of its theoretical infrastructure.

By now it should be clear that the focus of this study is on the reception of stereochemistry among German chemists, a choice that is both deliberate and accidental. The German chemists discussed here were concerned with similar problems and offered similar solutions, and in that sense, would by themselves comprise a naturally coherent unit for a historical study. Yet I did not initially need to make a deliberate choice to confine this study to German chemists, because before the late 1890s stereochemistry was not practiced outside of Germany. The reasons for this are not yet clear, and I do not address them here. It is important, however, to note that when discussing the overall reception of stereochemistry among chemists, I implicitly mean the reception of stereochemistry among *German* chemists, and not the international community, although many of my conclusions for Germany may be applicable elsewhere.

Nevertheless, confining this study to stereochemistry in Germany still required some choices about the specific chemists and episodes to include. The most notable exclusions are Baeyer's stereochemical models of benzene and cyclohexane derivatives, and Carl Bischoff's 'dynamic' hypothesis.[10] Also not included are

---

[10] Baeyer's benzene theory is included in Tonja A. Koeppel, *Benzene Structure Controversies 1865–1920*, PhD, University of Pennsylvania, 1973, and Koeppel, 'Significance and Limitation of Stereochemical Benzene Models', pp. 97–113 in Ramsey, *Van't Hoff-Le Bel Centennial*. Bischoff's dynamic hypothesis is discussed in G.V. Bykov, 'The Conceptual Premises of Conformational Analysis in the Work of C.A. Bischoff', pp. 114–22 in Ramsay, *Van't Hoff-Le Bel Centennial*.

discussions about the process of racemizaton (loss of optical activity) or the nature of racemic compounds, and the discovery and explanation of the Walden Inversion. These are all important episodes, but peripheral to the central story contained here.

It is perhaps surprising that so little serious extended inquiry has been made into such a fundamental change in the interpretation of chemical formula. One of the reasons is undoubtedly the technical nature of the subject, in which, for example, the rather esoteric differences between α-benzil*di*oxime and β-benzil*mon*oxime or α-isochlorocrotonic acid and β-isochlorocrotonic acid play a crucial role in arguments for and against stereochemistry. In the face of such details, the relative safety and simplicity of phlogiston versus oxygen appears quite attractive. It requires a great deal of technical knowledge – chemical formulas, nomenclature, and laboratory practice – to decipher the arguments, and some of those technical details have been included here. They are there to illustrate how chemical formulas are used to make a chemical argument in stereochemistry, and the subtlety of the arguments involved. I have done my best to introduce these details on a level suitable for non-chemists to understand. I must therefore request patience from chemists and historians of chemistry at times when the discussion becomes elementary. Historians of physics have long included much technical detail in their histories of electrodynamics, relativity and quantum theory. As a physical science with its own peculiar language, symbolism and mode of argument, the history of modern chemistry should be no more or less technical.

Finally, a few notes on nomenclature, chemical formulas and format are necessary. I have used the word 'stereochemistry' and the phrase 'chemistry in space' interchangeably throughout the Introduction. The word 'stereochemistry' was not introduced until 1889, and it replaced many other terms used to refer to the study of the spatial properties of molecules. I have used 'stereochemistry' only for convenience in this Introduction. In order to keep to contemporary terms, I will use it sparingly until I reintroduce Meyer in Chapter 5. Some of the diagrams are copies from the original source, and others, although drawn on a computer, are similar to those that chemists used for their own illustrations. The term 'chemical structure' is used exclusively in its nineteenth-century meaning of the 'sequential connection of atoms', or chemical arrangement of atoms independent of their spatial positions. The term 'configuration' will be used as a synonym for the 'spatial arrangement' of atoms in molecules (we will look at the introduction of the term 'configuration' by Aemilius Wunderlich in Chapter 4). For the isomeric compounds resulting from carbon–carbon double bonds, I have chosen to use the terms *cis* and *trans* rather than the cumbersome terms 'planesymmetric' or 'axialsymmetric' originally used by Wislicenus to indicate the two geometrical isomers in the unsaturated acids. Baeyer introduced the Latin terms in 1889 in connection with his study of benzene. I have, however, retained the terms 'planesymmetric addition' and 'axialsymmetric addition'. The meaning of all of these terms will become clear in Chapters 3 and 5. Unless otherwise indicated, all translations are my own and all emphases within quotes are by the original author.

Chapter 2

The Historical Development of Organic Chemistry to 1874

> ... we do not know what connection exists between the relative chemical effect of the atoms inside a compound molecule and their relative mechanical positions ...
>
> Aleksandr Butlerov, 1861

The emergence of 'chemistry in space' during the last quarter of the nineteenth century occurred within the relatively mature theoretical and professional culture of organic chemistry, which during the first two-thirds of the century saw two fundamental transformations – the rise of a distinct chemical profession and the accompanying pedagogical framework, and the emergence of a coherent body of chemical theory that made sense of the quickly growing number of organic compounds. During the 1860s, both changes had become relatively complete, and organic chemistry had been transformed from a relatively small area of chemistry into the premier subdiscipline within chemistry, the showcase for chemistry's success as a science and a profession. The theories of organic chemistry served as the scientific basis of a thriving and rapidly growing industry that employed increasing numbers of professionals trained specifically as chemists, and nearly all university chairs of chemistry were occupied by organic chemists who directed large teaching laboratories.

Our purpose in this chapter is to outline the development of organic chemistry to 1874, before Van 't Hoff and Le Bel introduced spatial concepts of the carbon atom. It is only by comprehending the context of the unique chemical culture as it emerged in the 1860s that we can grasp the novelty of the tetrahedral carbon atom in 1874. As we shall see below, chemical theory had been carefully and deliberately constructed according to specific ontological and epistemological constraints, from which Van 't Hoff's theory would depart. After a short summary of the institutional development of chemistry in Germany during the nineteenth century, we shall focus on the major theoretical developments in organic chemistry that relate to the meaning of chemical formulas.

The Institutional Structure of Nineteenth-century German Chemistry

The institutional history of chemistry in Germany during the nineteenth century can be divided into three periods, roughly corresponding to overall shifts in the disciplinary characteristics of chemistry.[1] The first period matches generally the emergence of organic chemistry during the first two-thirds of the century, and is characterized by the formation of various small chemical laboratories designed for training students in chemical analysis. The earliest of these was begun by Friedrich Stromeyer in Göttingen, but the most famous and most influential was developed by Justus Liebig at the University of Giessen beginning in 1826. In 1831, Liebig published a new method for chemical analysis using the famous *Kaliapparat*, a relatively simple apparatus that students could learn to use relatively quickly and enabled the growth of Liebig's laboratory during the 1830s. The *Kaliapparat* was also quickly and uniformly adopted by chemists, as it was more efficient and improved the accuracy of elemental analysis.[2] Liebig's laboratory became an example for many chemical institutes that appeared throughout the states of the German confederation. All were characterized by an instructional laboratory, and at any one time, each of these laboratories trained dozens of students in chemical analysis. The cost of these institutes was initially modest, and although each of the professors was given a small laboratory budget, it was often not sufficient to cover the total cost of running the laboratory.[3]

Initially, the size of the institutes remained small, but as the demand for professionally trained chemists increased, the capacity of these first-generation institutes quickly became inadequate. The 1860s saw the construction of a second wave of much larger, expensive institutes at universities such as Bonn, Berlin, and Leipzig to accommodate the influx of students. The state assumed more costs in directing the laboratory, and the number of assistants and extraordinary professors also increased. As a sign that organic chemistry had surpassed inorganic chemistry in sophistication of theory and practice, the directors of most institutes were organic chemists; by the 1860s, the most prestigious institutes were run by Liebig at Munich,

---

[1] This division was first suggested by Jeffrey A. Johnson, 'Academic Chemistry in Imperial Germany', *Isis*, 1985, 76, 500–24.

[2] Detailed studies of the origin and reception of Liebig's *Kaliapparat* appear in Frederic L. Holmes, 'The Complementarity of Teaching and Research in Liebig's Laboratory', *Osiris*, 1989, 5, 121–64, and Alan J. Rocke, 'Organic Analysis in Comparative Perspective', pp. 273–310 in Frederic L. Holmes and Trevor Levere, eds, *Instruments and Experimentation in the History of Chemistry*, Cambridge, MA, MIT Press, 2000.

[3] The only exceptions to this were Liebig at Giessen and Wöhler in Göttingen, in the 1840s the best-established and most prestigous chemical institutes in Germany. For the example of Bunsen and Kolbe at Marburg, see Alan J. Rocke, *The Quiet Revolution: Hermann Kolbe and the Science of Organic Chemistry*, Berkeley, CA, University of California Press, 1993, pp. 111–18.

A.W. Hofmann in Berlin, August Kekulé in Bonn, and Hermann Kolbe at Leipzig. Although the provincial universities of Bavaria (Würzburg and Erlangen) and Prussia (among others, Halle, Greifswald, Breslau and Königsberg) were smaller and less prestigious, they too possessed chemical institutes, often directed by organic chemists. The University of Würzburg in particular filled the chair of its chemical institute with a series of outstanding chemists – Wislicenus, Emil Fischer and Hantzsch – all of whom subsequently moved on to direct the institutes in Leipzig (Wislicenus and Hantzsch) and Berlin (Fischer).

These large second-generation institutes became inadequate during the 1890s as chemistry became increasingly fragmented into subdisciplines of inorganic, physical, biological and radiochemistry, with increasingly divergent methods and aims. Until the end of the century, the apparent unity of chemistry as a science had justified the appointment of a single professor, but the emerging subdisciplines put a strain on this hierarchical system. Already in Leipzig and Berlin, 'second' chemical institutes had been established, which were largely devoted to physical chemistry. In Berlin, Emil Fischer would attempt with limited success to create a single institute that would cover all subdisciplines under one director.

The University of Zürich and the Zürich Polytechnical Institute, focal points in the early history of stereochemistry, are often considered as 'German' universities in language and organization. The polytechnical institute was founded during the 1850s and modeled after the German polytechnical institutes in Stuttgart and Karlsruhe.[4] Johannes Wislicenus directed the chemical institute at the university from 1863 to 1871, when he became director of the chemical institute at the Polytechnikum. In 1872, he left Zürich for Würzburg and was succeeded by Victor Meyer, who in turn was succeeded by Arthur Hantzsch in 1885. Both Meyer and Hantzsch would also leave Zürich for German Universities, and the Polytechnikum tended, at least in chemistry, to serve as a staging ground for rising stars in German science. Alfred Werner, who became Professor of Chemistry at the university in 1893, was an exception in that he remained in Zürich until his death in 1919.

The nineteenth century also saw the emergence of chemical societies and journals for making the results of chemical research public. Chemists sometimes reported preliminary results orally, to the chemistry sections at the annual fall meetings of the Society of German Naturalists and Physicians (*Gesellschaft Deutscher Naturforscher und Ärzte*, referred to in the nineteenth century, and hereafter, as the

---

[4] Christa Jungnickel and Russel McCormmach, *Intellectual Mastery of Nature: Theoretical physics from Ohm to Einstein*, 2 vols, Chicago, IL, University of Chicago Press, 1986, vol. 2, pp. 186–7. In broad terms, the universities in Zürich were similar enough in origin and culture to German universities to consider them as 'German' universities. It is still unclear, however, if Johnson's division of institutional development (note 1) into three broad phases applies directly to the university and the polytechnical school in Zürich.

Naturforscherversammlungen), or in a meeting of a small chemical society that existed in the city where the institute was located. Summaries of these presentations were then sometimes noted in larger chemical newspapers such as the weekly *Chemiker-Zeitung*, or formal chemical journals. Most results were published in one of the several German journals devoted to chemical research. Liebig became editor of the *Annalen der Chemie und Pharmacie* in 1832; it was renamed *Justus Liebig's Annalen der Chemie* in 1873 following Liebig's death. In the 1830s, Liebig had used the *Annalen* in part as a personal outlet for the critical analysis of the chemical literature, but subsequent editors – Hermann Kopp and, after 1871, Jakob Volhard and Emil Erlenmeyer – toned down the polemical aspects of the journal. It was the publication of choice during the 1840s and 1850s, and continued thereafter to serve as a vehicle for longer, detailed papers, frequently reaching or exceeding a hundred pages, with lengthy theoretical discussions and descriptions of experimental work. Hofmann and the German Chemical Society began the *Berichte der Deutschen Chemischen Gesellschaft* in 1867. The first volume was small, but by the 1880s each volume had grown to two to three thousand pages. It became the primary outlet for the quick publication of short reports, usually of experimental results, although at times it also published short theoretical articles.

Another major chemical journal, the *Journal für praktische Chemie*, was founded in 1834 by Otto Erdmann and Franz Schweigger-Seidel. Erdmann and Schweigger-Seidel were located in Leipzig and Halle, respectively, and had assumed the production of the *Journal für Chemie und Physik*, founded in 1811 by J.S.C. Schweigger, who was at the Physikotechnische Institut in Erlangen. As the title suggests, the original focus of the *Journal* was on chemical technology, but it also became a vehicle for publishing original research, especially research that could be considered conservative. In 1870, Hermann Kolbe assumed the editorship, and emulating Liebig's use of the *Annalen* as a platform for the critical analysis of the state of chemical theory, he used his power as editor to ridicule and attack new developments in chemical theory – especially the structure theory – that he found distasteful or 'unscientific'. When Ernst von Meyer, Kolbe's son-in-law, became editor in 1884 after Kolbe's death, the *Journal* retained its conservative reputation, and many criticisms of stereochemical theory appeared within its pages. As a sign of the increasing specialization within chemistry, the *Zeitschrift für physikalische Chemie* was founded in 1887 by Wilhelm Ostwald and Van 't Hoff, and the *Zeitschrift für anorganische Chemie* first appeared in 1892.

The Development of Theoretical Organic Chemistry

In addition to the standardization of the locale (the chemical institute) and teaching methods (intensive laboratory training) of chemistry that took place by mid-century, chemical theory had reached a consensus by the 1860s. After long and sometimes bitter feuds over the direction and nature of chemical theory, the principles of structure and valence, combined with a recently established uniformity in atomic weights, gave chemists a set of tools that led to spectacular success in characterizing and classifying the constitution of molecules.

The emergence of structure theory can be broken down into two broad phases: the initial transformation during the 1830s of organic chemistry from a natural history-oriented plant and animal chemistry to the chemistry of carbon compounds, and second, the emergence of the concept of structure as an explanation of, among other things, the appearance of isomers. We will treat each of these phases in turn, focussing primarily on the second transformation, and then look at the overall character of chemical theory as it appeared in the 1860s. This chapter will then conclude by looking at the principal weaknesses or drawbacks of the structure theory.

From Natural History to Experimental Science

The term 'organic chemistry' was first used by Jakob Berzelius to refer to the combined topics of plant and animal chemistry, and as it was envisaged in the early nineteenth century, it was concerned with a wide range of issues surrounding not only the identity and composition of substances found in plants and animals, but also their physiology and anatomy.[5] Organic chemistry as such was practiced as a form of natural history, in which the identity and taxonomy of organic substances was only part of a larger scientific enterprise. This original conception of organic chemistry was transformed radically during the 1830s into an entirely different science that focussed on the chemistry of compounds containing carbon, and the concern with plant and animal anatomy and physiology was discarded in favor of studying the individual compounds isolated from natural sources. The focus also shifted from identification and isolation to the reactivity of these compounds as chemicals, and as the chemistry of these compounds unfolded, chemists found themselves dealing increasingly with *artificial* compounds, ones that did not occur naturally as a part of plant and animal chemistry. The culture of organic chemistry became increasingly experimental, rather than primarily observational and taxonomic. The seeds of this fundamental transformation were found in the application of Berzelian formulas for

---

[5] My analysis in this section is a summary of Ursula Klein's recent work, *Experimente, Modelle, Paper-Tools: Kulturen der organischen Chemie im 19. Jahrhundert*, Habilitationsschrift, University of Konstanz, 1999.

modeling organic compounds and reactions, the study and modeling of the reactions of natural and artificial organic compounds, and the general methodological guiding principle that organic compounds, like inorganic compounds, have a binary, or dualistic, constitution.

Although Berzelius introduced his famous chemical notation in 1813 and applied it to organic compounds in 1820, it would not be widely adopted by chemists for organic compounds until 1827, when it became advantageous in several respects. First, it offered a simple, algebra-like representation of organic compounds that allowed chemists to balance reactions according to the principle of the conservation of mass. Because the reactions of organic compounds usually yielded a multitude of products that often could not all be identified, the use of Berzelian formulas allowed chemists to 'discover', by manipulation of formulas on paper and the assumption of the conservation of mass, the constitution of not only isolated products of a reaction, but also products that had not yet been isolated or were present in extremely small quantities.

No less significant were the ontological consequences of using the formulas, which chemists could and did adopt independently of their personal beliefs in atomism. Using Berzelian formulas required a commitment to their representing atoms as the smallest units of chemical combination, also called chemical atoms, but not to the more controversial physical atomism of unsplittable atoms of the microworld. Formulas could, but did not necessarily, imply the structure of matter at the microscopic level. The use of Berzelian formulas therefore began a sharp epistemic distinction between knowledge at the macroscopic level – chemical equivalents and combining proportions – and the microscopic level – the nature of the ultimate structure of matter – that would continue until the appearance of 'chemistry in space' in 1874.

The use of Berzelian formulas also dovetailed nicely with chemists' efforts to establish organic compounds as having a dualistic constitution, under the same principles that had been successful in inorganic chemistry since the eighteenth century. Until the 1830s, the majority of chemists saw organic and inorganic compounds as possessing compositions that were fundamentally different. Berzelius initially found it difficult to reconcile his deep-seated belief that nature, specifically chemical behavior, was governed by a unified set of natural laws and forces, with the reality that organic and inorganic compounds obviously seemed to obey different laws of composition. Organic compounds were difficult to place within his theory of electrochemical dualism, as their composition could not be divided neatly into electropositive and electronegative components. In 1832, when Liebig and Wöhler published their now famous paper on the derivatives of the oil of bitter almonds in which they claimed that each derivative contained the benzoyl 'radical', or a grouping of atoms with the composition $C_{14}H_{10}O_2$, Berzelius found the singular example that showed that organic and inorganic compounds obeyed the same laws of composition.⁶ All chemical compounds must be dualistic and contain two components held together by the forces of electricity. Organic molecules, claimed Berzelius, contained complex radicals, while the radicals in inorganic molecules were simple.⁷

The concept of chemical radicals as it developed during the 1830s took on a number of epistemologically interesting characteristics. Because radicals were thought to play the same role in organic molecules as elements did in inorganic compounds, they were thought to be manipulable elements in the laboratory. Preferably, radicals were substances that in principle could be isolated, just as elements could be isolated from inorganic compounds. In practise, this was largely impossible, and radicals were 'discovered' simply by manipulation of formulas on paper. They were found as regularities within the Berzelian formulas, and not isolated in the laboratory directly. Finally, because the concept of radicals derived from the principle of binary constitution, it was closely related to the study of chemical reactions, that is the macroscopic manipulation of substances, and not to the study of the microworld. The concept of radical therefore was not tied to a physical or philosophical atomism, but a means of representing composition in chemical proportions which was reminiscent of Lavoisier's earlier theory of composition that was agnostic towards the atomic theory.

The Meaning of Arrangement: Radicals versus Types

As this new framework for organic chemistry evolved, chemists formulated new questions concerning the composition of organic compounds. Two of the most significant new emerging concepts, at least for this study, were those of 'isomerism' and 'arrangement'. As early as 1814, Guy-Lussac, who in 1811 had noticed the compositional identity of sugar, starch, and gum, suggested that the 'arrangement of the molecules in a compound has the greatest influence' on its properties. During the next fifteen years, the number of distinctly different compounds with identical compositions increased, and in 1830, after reviewing these compounds, Berzelius introduced the concept of isomers (same parts), for these compounds, which possessed identical chemical compositions (proportions of elements), but different chemical or physical properties.⁸ Different isomers would have a different arrangement of atoms that would explain the difference in chemical properties. This

---

⁶ Justus Liebig and Friedrich Wöhler, 'Untersuchungen über das Radikal der Benzoesäure', *Annalen der Pharmacie*, 1832, *3*, 249–87, English translation in Otto T. Benfey, ed., *Classics in the Theory of Chemical Combination*, New York, Dover Publications, 1963, pp. 15–39.

⁷ By 1833, chemists had also described the ethyl, cyanogen and cacodyl radicals, but it was the benzoyl radical that caused the most excitement for Berzelius.

⁸ Alan J. Rocke, *Chemical Atomism in the Nineteenth Century: From Dalton to Canizarro*, Columbus, OH, Ohio State University Press, 1984, pp. 168–74.

framework proved to be remarkably enduring, as the search for atomic arrangement in the effort to explain isomerism drove much of chemical theory in the first three-quarters of the nineteenth century, and would also provide the impetus for the appearance of spatial models of atoms and molecules in the last quarter of the century. Although chemists were in broad agreement about the role of 'arrangement' as an explanatory device, they did not all agree about the meaning of the term, which proved problematic in definition and explication. Prior to the emergence of the structural theory of organic chemistry in the 1860s, chemists spent enormous amounts of time and energy in ontological and epistemological debates on the meaning of arrangement.

The different approaches and assumptions chemists adopted when constructing chemical theories broke down largely, although not entirely, along national lines. Championed by Berzelius, and subsequently by Liebig, Wöhler and Hermann Kolbe, the theory of radicals dominated chemical thought in Germany. As described above, the radical theory offered several principal guiding assumptions for chemical theory. First, and foremost, in its strong form, it assumed a full analogy between inorganic and organic compounds. All substances, organic and inorganic, were dualistic in nature and contained components held together by the forces of electricity. The belief that molecules consisted of discrete components led to the view that chemists should separate molecules into their true components, or in other words, determine what pieces constituted the molecule, and such constitutions or rational formulas would serve as the endpoint of chemical theory. To serve this end, addition or elimination reactions served as their primary point of reference, and whenever possible, the components in molecules were thought to be isolable compounds themselves. For example, ethyl alcohol eliminated water to yield ether; based on this reaction, Liebig assumed alcohol consisted of two pieces, ethyl oxide ($C_4H_{10}O$) and water. Based on the same reaction, Dumas thought alcohol was ethylene combined with two molecules of water.[9] Hermann Kolbe became the strongest advocate of the radical theory in its revised form. His electrochemical treatment of acetic and valeric acids in the late 1840s resulted in the isolation of methyl and valyeryl radicals that strengthened his belief in the existence of immutable radicals within molecules, and he firmly believed that he could uncover a molecule's absolute constitution.

At approximately the same time that the radical theory was taking hold in Germany, in Paris, Jean-Baptiste Dumas announced his theory of substitution that would eventually undermine the dualistic assumptions contained within the radical theory. Dumas discovered that the electropositive hydrogen contained in acetic acid

---

[9] Alan J. Rocke, 'Convention Versus Ontology in Nineteenth-century Organic Chemistry', pp. 1–20 in James G. Traynham, ed., *Essays on the History of Organic Chemistry*, Baton Rouge, LA, Louisiana State University Press, 1987, p. 7.

could be completely replaced, or substituted, by electronegative chlorine, with little change in its chemical property: chloroacetic acid remained acidic. Dumas concluded that something fundamental in the acetic acid molecule must convey the property of acidity, and proposed that both acetic acid and chloroacetic acid belonged to the same chemical type.[10] While Dumas initially remained a dualist, the type or unitary theory would dominate chemical thought in France, and to some extent in England. In 1849, August Hofmann introduced the ammonia type, in 1851 Alexander Williamson introduced the water type, and in 1853, Charles Gerhardt began an extensive systematization and reorganization of chemical phenomena under the theory of types.

Like the proponents of radical theory, adherents to the type theory had several basic claims about the arrangement of atoms within molecules and the meaning of chemical formulas. First, each molecule possessed a particular chemical function or property, and organic compounds could be classified on the basis of these properties. Second, substitution reactions, or double decompositions, as they were also called, served as the primary point of reference. Type theorists considered the molecule as a whole (hence the alternate title of unitary chemistry), not as a group of components where one simply added or subtracted pieces.

While the proponents of the radical theory attempted to form the true or absolute constitution of a compound, the proponents of type theory practiced a more conventionalist approach to the meaning of chemical formula. Classification by type involved identifying the atom responsible for the typical properties of a compound, and chemical formulas were then constructed around this central atom in the molecule. Pigeonholing compounds in this manner relied on a subjective judgement about which property was central to the compound, which, of course, could vary depending on the reaction under consideration. Under this view, chemical formulas contained a certain degree of relativism, about which Gerhardt had no qualms.[11] According to him, a chemical formula represented the chemical properties of a compound, or in other words, the changes that it exhibited in the presence of other compounds. Since during any chemical reaction the atoms rearranged, sometimes profoundly, into new compounds, these changes could give chemists no inference about the constitution of any given molecule before it reacted with another molecule. Gerhardt therefore considered any claims about the constitution of a substance to be

---

[10] Klein, *Paper-Tools*, has emphasized the importance of the phenomenon of substitution as the key event in the conversion of the natural history-oriented organic chemistry to the chemistry of carbon compounds during the 1830s. According to Klein, the phenomenon of substitution, like that of radical theory, was also discovered and rationalized by the manipulation of formulas on paper, and not by the literal discovery of substituting atoms.

[11] N.W. Fisher, 'Organic Classification Before Kekulé, Parts I and II', *Ambix*, 1973, *20*, 106–31 and 209–33, p. 221.

epistemologically suspect. He advocated an extreme empirical approach, and denied an ability to know the constitution of molecules in any sense, chemical or physical.[12]

The 'Quiet Revolution' of the 1850s, and the Emergence of the Structural Theory of Organic Chemistry

Although I have presented chemistry as being a neat dichotomy between radicals and types, the situation was actually much more complicated, due to additional disagreements about molecular and atomic weights, and over chemical notation. During the 1850s, organic chemistry underwent a second wave of profound changes, recently described by Alan Rocke as a 'quiet revolution'. During the 1850s, nearly all organic chemists, even those in Germany (with the important exception of Kolbe), gravitated towards a more unitary conception of molecules, as dualism became less and less able to account for the increasing examples of substitution. The idea of radicals would remain, but in a substantially revised form. Williamson's ability to synthesize symmetrical and mixed ethers in 1851 gave further support for a unitary conception of molecules, and for the idea that an oxygen atom could 'link' two radicals.[13] Widely imitated in other syntheses throughout the 1850s, the Williamson ether synthesis also supported the adoption of the Laurent-Gerhardt reforms of molecular weights, leading ultimately to the consolidation of atomic and molecular weights suggested by Cannizzaro at the Karlsruhe conference of 1860.

During the 1850s it also became apparent that atoms of one element could combine with only a specific number of atoms of other elements. This was first noticed by Edward Frankland in 1849, when he discovered zinc methyl and zinc ethyl and concluded they were substitution products of zinc hydride. In attempts to oxidize these compounds further (adding atoms of oxygen), he found always that one of the organic radicals had to leave before the oxygen could add – that is, zinc could combine with only so many atoms. In light of this result, Frankland concluded:

> Without offering any hypothesis regarding the cause of this symmetrical grouping of atoms, it is sufficiently evident ... that such a tendency or law prevails, and that, no matter what the character of the uniting atoms may be, the combining power of the attracting element, if I may be allowed the term, is always satisfied by the same number of these atoms.

Frankland's 'combining power' would later come to be known also as 'atomicity', and in 1865, in an attempt to remove the 'barbarous' nature of the term 'atomicity',

---

[12] J.H. Brooke, 'Laurent, Gerhardt, and the Philosophy of Chemistry', *Hist. Stud. Phys. Sci.*, 1975, 6, 405–29, pp. 424–5.

[13] Rocke, *The Quiet Revolution*, and 'The Quiet Revolution of the 1850s: Social and Empirical Sources of Scientific Theory', pp. 87–118 in Seymour H. Mauskopf, ed., *Chemical Sciences in the Modern World*, Philadelphia, PA, University of Pennsylvania Press, 1993.

Hofmann would introduce the term 'quantivalence'. Two years later, Kekulé himself would shorten this term to 'valence', and in German, the phenomenon of 'combining power', was referred to as either *Valenz*, *Werthigkeit*, or *Atomicität*.[14]

These three factors – the movement towards unitary theories, agreement on molecular and atomic weights, and the theory of valence – would all come together in the late 1850s when August Kekulé formulated what would come to be known as the structural theory of organic chemistry. Kekulé's education as a chemist took place during the turbulent period of the 1850s, and his varied education put him in a unique position to understand and evaluate the advantages and disadvantages of both radical and type theories. He studied at several major research centers in Europe: in Giessen with Liebig (1849), in Paris with Charles Wurtz and Gerhardt (1851–2), and in London with Williamson and William Odling (1853–5). In 1856, he became *Privatdocent* in Heidelberg, and in 1857, was appointed Professor of Chemistry in Ghent, where he would develop the core principles of the structure theory.

Although Kekulé clearly had adopted Gerhardt and Williamson's chemical philosophy that favored the type theory, he charted a course between the conventionalism of type theory and the absolutism of the radical theory, building on but moving away from both. Like many other episodes in the history of science, Kekulé effected a synthesis of different traditions which previously had appeared irreconcilable. Chipping away at the radical and type theories, Kekulé retained the best of each theory and ignored the rest. First, he made explicit and more general the concept of atomicity: 'the number of atoms or radicals united with one atom [of an element or a radical], is dependent on the basicity or magnitude of affinity of the components'.[15] He noted that atoms or radicals fell into three groups, monatomic, diatomic, or triatomic, and suggested for the first time that carbon was tetravalent. Significantly, Kekulé also suggested that a carbon atom could satisfy, or 'saturate', the affinity of another carbon atom.

Second, he drew out the implications for ascertaining rational formulas contained within this revised theory of atomicity. Rational formulas would derive from 'the elements themselves which compose these compounds. I no longer regard it as the chief problem of the time, to prove the presence of atomic groups which, on the strength of certain properties, may be regarded as radicals'.[16] He had, in effect, begun this process earlier, when he introduced the idea of mixed types, but in 1858, he took it to the fullest extent and gave equal status to each atom in a chemical formula. Both the

---

[14] Colin A. Russell, *The History of Valency*, New York, Humanities Press, 1971, pp. 83–9.

[15] August Kekulé, 'Über die sogenannten gepaarten Verbindungen und die Theorie der mehratomigen Radicale', *Annalen*, 1857, *104*, 129–50. Translation in Alan J. Rocke, *Chemical Atomism*, p. 265.

[16] Kekulé, 'Über die Constitution und die Metamorphosen der chemischen Verbindungen und über die chemische Natur des Kohlenstoffs', *Annalen*, 1858, *106*, 129–59. Translation in Benfey, *Classics*, pp. 115–31.

radical and type theorists had implicitly given privileged status to certain parts of the molecule, be it an immutable radical contained within the molecule, or a 'connecting' atom such as oxygen in the water type or nitrogen in the ammonia type. Adherents to neither view had taken a logical additional step beyond a molecule's constituent pieces or its 'fundamental' property to consider the position of every atom in a molecule. According to Kekulé each atom in the molecule was therefore ontologically equivalent, and none should dominate within the molecule.[17] Kolbe explicitly recognized and deplored what he called the 'democratic' nature of molecules in structural formulas, contrasting it with his own hierarchical conception of molecules that derived from his unwavering commitment to dualistic chemistry.[18]

Kekulé's third contribution was the melding of concepts from both radical and type theories. Under his analysis, type theory lost some of its conventional character, while the concept of radical took on a more relative meaning. In fact, Kekulé considered radicals and types as relative terms that depended entirely on the reaction under consideration. A radical was simply the 'unattacked' part of any molecule during any particular reaction, just as a molecule belonged to a certain type when it exhibited behavior that belonged to that type. Simply describing constitutions in terms of the radicals that composed a molecule was not sufficient: 'we must ascertain the relation of the radicals to one another and, from the nature of the elements, deduce both the nature of the radicals and that of their compounds'.[19] By 'nature of the elements', Kekulé meant atomicity. The assumption of constant combining power and the self-linking of carbon enabled him to construct chemical formulas that truly depicted the position of each individual atom within a molecule, giving a formula that overall portrayed the arrangement of atoms.

Despite this relativism in his concept of radical, Kekulé did not merely subsume the ideas of the radical theory to the conventionalism of type theory. He insisted that chemical formulas must convey all of a compound's chemical properties, not simply the reaction of the moment. A formula must portray the unattacked residues for all known reactions and portray all the types to which the substance belonged. Although:

> rational formulas are reaction formulas and can be nothing else, in the present state of the science ... *they are intended to provide a picture of the chemical nature of a substance*. Every formula, therefore, that expresses certain metamorphoses of a compound, is rational; among the different rational formulas, however, that one is the most rational which expresses simultaneously the largest number of metamorphoses.[20]

---

[17] N.W. Fisher, 'Kekulé and Organic Classification', *Ambix*, 1974, *21*, 29–52.
[18] Rocke, *Quiet Revolution*, p. 325.
[19] Kekulé, 'Über die Constitution'. Translation in Benfey, *Classics*, p. 115.
[20] Ibid., p. 124. Emphasis added.

Kekulé had followed Gerhardt's interpretation without its agnostic component. Reaction formulas could represent only the transformations of molecules, but were also 'intended to provide a picture of the chemical nature of a substance', which meant a representation of all of its known reactions. According to Kekulé, chemical formulas were not simply conventions, but depicted the entire nature of a substance.

The Meaning of Chemical Structure

Because Kekulé's concept of rational formulas emerged as a blending of the radical and type theories, it should be evident that the epistemic and ontological status of these formulas remained unaltered. Neither radicals nor types were used for representing molecules and atoms on a microscopic level. They were, rather, attempts at modeling the reactions and composition of compounds, and many radicals were discovered simply by manipulation of formulas on paper. In a lengthy footnote to his famous 1858 paper, Kekulé clarified this epistemic status of rational formulas:

> Now it is plain that the manner in which the atoms emerge from a decomposing and changing substance cannot possibly prove how they are located in the stable and unchanging substance. Although it must certainly be considered a task of scientific research to elucidate the constitution of matter, or, if you will, the location of the atoms, yet we must admit that it is not the study of chemical reactions but rather a comparative study of the physical properties of stable substances that can supply means to that end ... it may become possible to set up 'constitutional formulas' for chemical compounds, which then of course must be unchangeable. But even when this is achieved, different rational formulas (reaction formulas) are still permissible, because a molecule with atoms in given locations can still under different conditions be split in different ways and at different places.[21]

In this instance, Kekulé was reasonably clear about the meaning of chemical formulas, but he elsewhere largely remained ambiguous and inconsistent about his conception of rational formulas. He was curiously reluctant to offer any sort of graphical representation of atomicity or rational formulas in his published papers or textbook. His famous 'sausage formulas' are restricted to footnotes in only a few pages of his textbook (and appear in only a few articles), and throughout his entire series of articles on the organic acids, presumably to ascertain their rational formulas, he used only empirical formulas. He did not give rational formulas for the unsaturated acids until 1867. Therefore, while urging chemists to search for the most rational formula, Kekulé initially scrupulously avoided giving any rational formulas! This was intentional on his part, and he explained it by noting that each chemist had his own idiosyncratic way of representing rational formulas, and that 'it is impossible to

---

[21] Ibid., p. 123.

express simultaneously the relationships and properties of the concerned substances in a simple and clear manner'.[22]

Because of Kekulé's reticence in elaborating on the specifics of his theory, it was left to others to clarify his ideas. The dominant interpreter and popularizer of Kekulé's ideas was the Russian chemist Aleksandr Butlerov, who had become aquainted with many Western chemists during a tour of laboratories in Western Europe. Butlerov's strong advocacy for Kekulé's theory has led some historians, especially those in the former Soviet tradition, to argue that Butlerov was the true founder of the structural theory of organic chemistry.[23] The absolute priority of Kekulé or Butlerov is not at issue here, but it is useful to differentiate between Kekulé's ideas and Butlerov's clarification of them. It is clear that Kekulé was not consistent in his use of atomicity, did not vigorously promote his own ideas, and refrained from offering any rational formulas. As a result, his reliance on Gerhardt's conventionalism appeared to undercut his desire for true constitutional formulas. This is the primary reason given by historians who argue that Kekulé's theory was purely classificatory, or that Butlerov should be given full credit for the concept of structure. As I argued earlier, Kekulé stressed that he was interested in chemical constitutions (rational formulas), and not strictly classification.[24]

Butlerov certainly grasped the essence of Kekulé's theory of atomicity, and was able to express it more clearly and succintly than Kekulé ever did. In an 1861 lecture at the Speyer *Naturforscherversammlung*, he suggested that the consistent application of Kekulé's concept of atomicity would lead to a unique 'chemical structure' (*chemische Struktur*) for each compound:

> Starting from the assumption that each chemical atom possesses only a definite and limited amount of chemical force (affinity) with which it takes part in forming a compound, I might call this chemical arrangement, or the type and manner of the mutual binding of the atoms in a compound substance, by the name of 'chemical structure'.[25]

Only one rational formula was possible for each compound, and when the general laws governing the dependence of chemical properties on chemical structure had been derived, this formula would express all of these properties.

---

[22] August Kekulé, 'Über einige organische Säuren', *Zeit. Chem.*, 1861, *4*, 613–23, in Richard Anschütz, *August Kekulé*, 2 vols, Berlin, Verlag Chemie, 1929, vol. II, 224–33. Kekulé, 'Untersuchungen über organische Säuren', *Annalen Suppl.*, 1862, *2*, 85–116, pp. 274–98 in Anschütz, *August Kekulé*, vol. II, p. 232.

[23] Alan J. Rocke, 'Kekulé, Butlerov, and the Historiography of the Theory of Chemical Structure', *Brit. J. Hist. Sci.*, 1981, *14*, 27–57.

[24] Fisher, 'Kekulé and Classification', and D.F. Larder, 'A Dialectical Consideration of Butlerov's Theory of Chemical Structure', *Ambix*, 1971, *18*, 26–48.

[25] Alexander Butlerov, 'Einiges über die chemische Structur der Körper', *Zeit. Chem.*, 1861, *4*, 549. Translation in Russell, *Valency*, pp. 147 and 149.

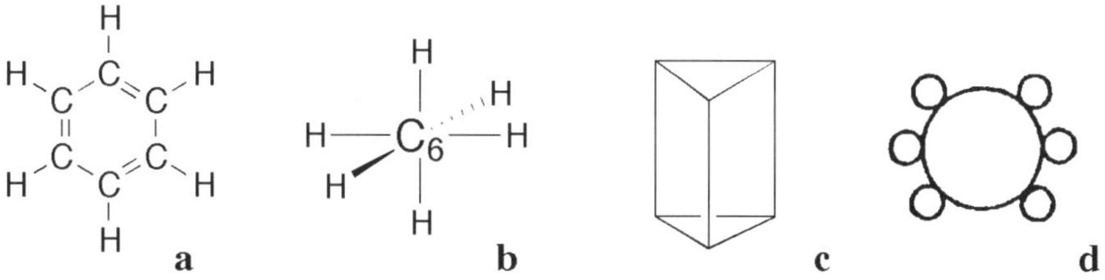

2.1 Possible formulas for benzene that represent the chemical equivalence of the six hydrogens: (a) Kekulé (b) Meyer (c) Ladenburg and (d) Loschmidt.

The emphasis in the phrase 'chemical structure' was on the word 'chemical', and Butlerov offered one of the clearest conceptions of what this meant. The rules of atomicity, according to Butlerov, certainly led to the assignment of a constitution to each compound, but to a *chemical* and not a physical constitution. Like Kekulé, Butlerov adopted Gerhardt's fundamental principle – the study of chemical reactions would never lead to 'the positions of the atoms in the interior of a molecule'. '[I]t seems quite obvious,' Butlerov went on:

> that chemistry, which only deals with bodies in a state of transformation, is powerless to judge this mechanical structure, as long as physical investigations are not brought to bear on this question ... To be sure, we do not know what connection exists between the relative chemical effect of the atoms inside a compound molecule and their relative mechanical positions; *we do not even know whether, in such a molecule, two atoms which directly affect each other chemically are in fact situated next to one another* ...[26]

Butlerov therefore made a sharp epistemological distinction between what could and could not be ascertained from the study of chemical behavior. Ascertaining an atom's 'chemical position' and 'mechanical position' within the molecule were two separate issues. The study of chemical properties alone could locate an atom's *chemical* position within its structure (or constitution), but not its *mechanical* (physical) location in the molecule as a physical object. This distinction therefore created a clear dichotomy between the concepts of chemical structure and spatial arrangement.

The difference between chemical and physical positions of atoms in molecules can easily be illustrated by the example of benzene. The famous benzene ring introduced by Kekulé in 1865 (Figure 2.1(a)) meant not that the ring itself was hexagonal, but that the six hydrogen atoms were chemically equivalent: in a substitution reaction,

---

[26] Butlerov, 'Einiges über die chemische Structur'. Translation in Rocke, 'Historiography'. Emphasis added.

any of the six could be lost without preference. The ring symbol did not, however, entail that spatial arrangement of the hydrogen atoms in the benzene molecule was physically hexagonal. This is substantiated by at least three different alternate conceptions of the benzene molecule that maintained the equivalency of hydrogens. In 1861, Josef Loschmidt supposed that benzene possessed a hexavalent six-carbon unit surrounded by hydrogen atoms (Figure 2.1(d)), and eleven years later, Lothar Meyer supposed that the cyclohexatriene structure:

> shall not express the opinion that the atoms be arranged spatially in a circle, but only show, with which other atoms each of them are assumed to be. For example [in benzene] the atoms can be placed spatially to one another as in the corners of a regular octahedron [Figure 2.1(b)].[27]

Albert Ladenburg's famous prism formula also initially proved a plausible alternative to Kekulé's cyclohexatriene structure (Figure 2.1(c)). All three models satisfied the chemical requirement for the chemical equivalence of the six hydrogens, but did not have a hexagonal distribution of carbon or hydrogen atoms.[28] In short, the structure theory in its original meaning implied no correlation between a molecule's chemical structure and the physical, spatial arrangement of its atoms. The two may in fact be identical (for instance, benzene may be hexagonal), but the use of chemical properties or transformations alone did not justify this assumption.

Once the concept of structure took hold, however, the line between its purely chemical intentions and physical ideas of molecules could easily become blurred, despite the explicit effort of chemists to keep them distinct. One such blurring occurred with the introduction of the term 'bond' by Edward Frankland. Frankland, one of the leading supporters of German chemistry in England, exemplified the chemical epistemological tradition when he wrote to his friend Hermann Kolbe that 'graphic formulae are intended to represent neither the shape of the molecules nor the relative positions of the constituent atoms',[29] and are 'merely symbolic expressions of atomicity', yet in his 1866 textbook, *Lecture Notes for Chemical Students*, he introduced the loaded term 'bond' to describe these chemical connections.[30] He intended it to be a neutral expression of the fact that two atoms were held together by some unknown force:

---

[27] Lothar Meyer, *Die Modernen Theorien der Chemie und ihre Bedeutung für die chemische Statik*, 2nd edn, Breslau, Maruschke and Berendt, 1872, p. 182.

[28] J. Loschmidt, *Konstitutionsformeln der organischen Chemie in graphischer Darstellung*, Leipzig, Engelmann, 1861, reprinted by Aldrich Chemical Company, 1989.

[29] Russell, *Valency*, p. 165.

[30] Edward Frankland to Hermann Kolbe, 3 December 1871, Deutsches Museum München, Archiv, HS 3566.

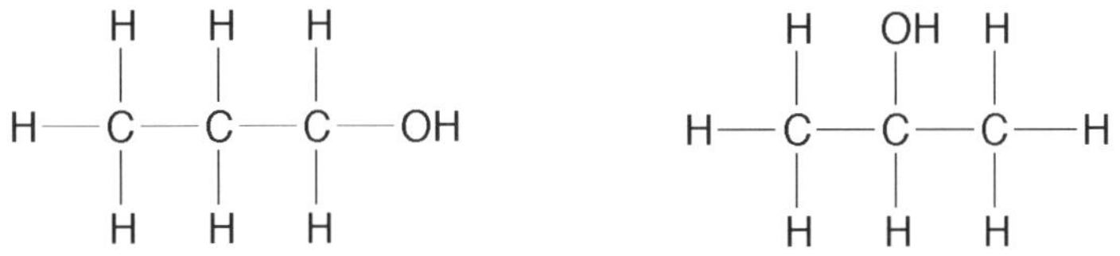

```
      H   H   H                    H   OH  H
      |   |   |                    |   |   |
  H — C — C — C — OH           H — C — C — C — H
      |   |   |                    |   |   |
      H   H   H                    H   H   H
```

2.2   The chemical structures for propyl and isopropyl alcohol using Crum Brown's formulas.

> To avoid any such dubious expressions [of chemical constitutions] I have long used the very neutral word 'bond' to express the connection between two hypothetical atoms in a compound. No one can deny that the elements of a compound are bound together and that the thing which binds them together must be either matter or force. It appears to me therefore that this thing can be named a bond with perfect propriety, whether it be like the mortar which binds bricks together, or the force which binds sun and planets. For the same reason, instead of the 'structure of a chemical compound' I prefer to speak of 'the disposal of the bonds of a chemical compound'.[31]

Despite these disclaimers and the precautions Frankland took to emphasize the purely chemical intentions behind the term 'structure', in his *Lecture Notes* he also referred to a bond as a 'point of attachment', a phrase that introduces an unmistakable physical concept into his intention. The word 'bond' itself implies the linking of two things together, and is difficult to separate from its physical connotations, unless one uses it in a purely metaphorical sense in the way that people become 'bonded' to one another. Frankland does not seem to be using it in that sense, but as a notion for describing a linkage between atoms.

The attempts to depict chemical structures by graphical notation would also eventually undermine the distinction between chemical and physical arrangement, despite the best attempts of chemists to maintain it. Kekulé's sausages and Loschmidt's circles, among others, all gave pictures of the molecules that could be interpreted in a physical sense. Of all the methods for graphically depicting structures, by far the most influential was the remarkably simple notation invented by Alexander Crum Brown, first published in 1864 and still used today. Using straight lines radiating from a central atom to depict the valence, the construction of a molecule that satisfied all affinities was simple. Using his novel notation, the two known alcohols with the molecular formula C_3H_8O, for example, could be represented by the formulas in Figure 2.2, in which one alcohol has an oxygen at the end of the carbon chain, and the other has an oxygen in the middle. Crum Brown was unequivocal in the meaning of his graphic formulas. They were:

---

[31]   Ibid.

used to express constitutional formulae, and by which, it is scarcely necessary to remark, I do not mean to indicate the physical, but merely the chemical position of the atoms ... and while it is no doubt liable, when not explained, to be mistaken for a representation of the physical position of the atoms, this misunderstanding can easily be prevented.[32]

The resulting line between atoms provided a clear depiction of the bond between atoms, giving Frankland's term still more physical significance. Formulas under Brown's notation created one of the clearest pictures of the structure of a molecule, depicting unequivocally the position of each atom. But these formulas could easily be interpreted (or misinterpreted) in a mechanical, physical sense, and made it 'harder to maintain the intellectual outlook that there was possibly (or probably) nothing in the "real" molecule that did hold one pair of atoms together, even if one conceded any physical meaning of the term "molecule" '.[33] The story of exactly how and why chemists in various countries quickly adopted Brown's notation through the 1860s has yet to be written, but it did become commonplace by the late 1860s, popularized at first in Britain, where Frankland recognized its pedagogical value and adopted it for his lectures.[34] Its adoption in Germany seems to have been slower, perhaps because of genuine doubts about the reality behind the representation. Kekulé himself rarely used it. It is not unimportant that Van 't Hoff and Le Bel used Brown's notation when they suggested the tetrahedral carbon atom.

Methods and Methodology in Organic Chemistry

Laboratory Practise

Having laid out the general theoretical character of chemistry during the first three-quarters of the nineteenth century, it is useful to summarize the experimental methods employed by chemists, and some characteristics of their underlying methodology.[35] As we have already seen, chemists developed the concepts of arrangement, valence and structure in order to differentiate between the hundreds of organic compounds with wildly different properties, but which were composed of only a few elements. The experimental basis for this differentiation lay in the synthesis, isolation and characterization of organic compounds in the laboratory. Until the late nineteenth

---

[32] Alexander Crum Brown, 'On the Theory of Isomeric Compounds', *Trans. Roy. Soc. Edin.*, 1864, *23*, 707–19, p. 708.
[33] Russell, *Valency*, pp. 101–2.
[34] Colin R. Russell, *Edward Frankland*, Cambridge, Cambridge University Press, 1996, pp. 281–6.
[35] I have adopted here Brooke's useful distinction between 'method' and 'methodology' in chemistry. J.H. Brooke, 'Methods and Methodology in the Development of Organic Chemistry', *Ambix*, 1987, *34*, 146–55.

century, chemical synthesis in the sense of rational, designed production of compounds was rare, and chemists focussed primarily on isolating and characterizing organic compounds that were produced in chemical reactions. As anyone who has spent even a short time in the laboratory will attest, chemical reactions, especially of organic compounds, often result in a complex mixture of compounds along with tars and oils that defy analysis. The techniques for isolating and identifying these compounds in the mixture require great skill that is acquired only through extensive practise with an accomplished master, and since the days of alchemy, chemists have always learned their craft in an apprentice-like way.

Even though these techniques required great skill, they are not difficult to understand in principle. Mixtures of reaction products could be extracted with organic solvents and acidic or basic aqueous solutions, either separately or in sequence, to separate compounds on the basis of their different solubilities. The mixture could be subjected to simple or fractional distillation, separating compounds by boiling point. Solids could be separated and purified by either simple or fractional crystallization. Distillation and crystallization served as techniques for purification, a process which in itself can be considered as a type of separation. A great deal of the skill required by chemists is the ability to make educated guesses about the number and state of given products in a chemical reaction – that is, the purity of the initial product(s), or if the product(s) were solids or liquids, acids or bases, water- or fat-soluble – factors that in turn would suggest a mode of separation and purification. The source of this educated guesswork was and is largely tacit knowledge acquired through years of experience in the laboratory, allied with knowledge of those reactions reported in the literature.

A chemist's day-to-day work in the laboratory involved several manual tasks that characterized traditional chemical research. He would purify compounds for reaction by distillation or crystallization, characterize them by melting point or boiling point, and judge their purity by the precision of the melting point or boiling point. He would react compounds (or allow compounds to react), taking care to maintain a constant temperature using an ice bath, water bath or flame; often reactions were conducted at several temperatures to determine the conditions that would produce the highest yield. These methods, of course, were constrained to those compounds that were relatively easy to crystallize or that did not decompose rapidly under the temperatures required for distillation. The majority of compounds described in nineteenth-century chemical literature possessed precisely those properties that allowed a relatively easy crystallization or distillation. The technique of crystallization required the greatest skill and patience of all chemical techniques, because creating the correct conditions to produce crystals required skill, intuition and luck.[36]

---

[36] For more about the methods of organic chemistry, see Rocke, *Quiet Revolution*, Chapter 10.

Having isolated a compound and decided that it was pure, the next step was to identify its elemental composition, calculate its empirical or rational formula, and after the 1860s, construct a structural formula. Those products that yielded to standard purification techniques were subjected to elemental analysis in which they were burned using a variant of Liebig's *Kaliapparat* to determine their relative proportions of carbon, hydrogen and oxygen. Ascertaining nitrogen content required a separate analysis. The combustion analysis would yield only the *relative* proportions of these elements, however, and there remained the task of converting these numbers to the empirical formula by application of the atomic weights, and then conversion to the molecular formula, if the molecular weight was known. Finally, the molecular formula could then be transformed, using the rules of valence, into a structural formula. These relative proportions and molecular formulas were then sometimes compared to theoretical values calculated from a presumed composition, and these comparisons would often indicate the existence of a new compound. This might seem like an extraordinarily long and tedious process, but to a large extent, chemists did not need to carry out every analysis from scratch, as the structures of the starting materials was usually already known. Using the assumption of least chemical change, much of the structure of the starting material would be retained in the product, so the structure of the product would be relatively easy to determine. Significantly, much of the process for establishing a structure was not at the laboratory bench, but at the desk, manipulating numbers and formulas on paper.[37]

One notable aspect of the methods in organic chemistry is the sparing use of physical properties by chemists to construct their chemical theories. The primary properties employed by chemists were melting and boiling points, indices of refraction (for liquids) and the weights of starting materials and products.[38] Chemists would also often describe the appearance of crystals, sometimes providing a drawing of the crystal itself. Other physical properties were used: for example, the presence of isomorphism in the crystals of two different compounds indicated that the compounds could share analogous molecular formulas, and the Dulong-Petit law of specific heats allowed the calculation of molecular weights, but chemists on the whole tended to be reluctant to give physical and chemical properties equal status.

As a physical property highly relevant to the appearance of 'chemistry in space', optical activity was also only slowly incorporated by chemists into the realm of relevant properties of compounds.[39] Jean-Baptiste Biot first noticed in 1815 that

---

[37] This practice of working on paper originated with the widespread adoption of Berzelian formulas during the 1830s. See Klein, *Paper-Tools*.

[38] The first use of each of these properties has not been documented, but it is known that Wilhelm Heintz, during the 1850s, was one of the first chemists to use melting points as a criteria for identification.

[39] This summary of optical activity is derived from Robert Ward, *The Development of the Polarimeter in Relation to Problems in Pure and Applied Chemistry: An Aspect of Nineteenth Century Scientific Instrumentation*, PhD, University of London, 1980.

solutions of certain organic compounds would rotate the plane of polarized light, and by 1818 had formulated the correlation between the magnitude of rotation, length of the sample, and the wavelength of light for both mineral species and organic compounds in solid and vapor states and in solution. In a second period of intense investigation into optical activity during the 1830s, Biot continued to refine the construction and use of the polarimeter and techniques for making polarimetric measurements. He set up standards for measuring and calculating angles of rotation, and defined the concept of molecular rotatory power (known as the [α] value), and introduced the use of monochromatic light. Biot himself was convinced from the beginning that optical activity had both great theoretical and practical value. He assumed that optical rotation gave access to molecular properties, but he also used it for defining and identifying various plant sugars, for following the chemical process of the reaction of potato starch with sulfuric acid, for understanding the processes of growth in plants by measuring the optical activity of their components at various stages in their development, and for detecting sugar in urine at the onset of diabetes.

Other chemists were uncertain about the role optical activity could play in organic chemistry. Gerhardt, for example, was reluctant to regard any physical properties useful for constructing reaction formulas. Chevreul was more receptive, but was not convinced that Biot had made a compelling case for optical activity as an important or defining property of organic compounds, as there were often doubts about the purity of the samples derived from plants and animals. Certainly, part of this doubt was related to the complexity of the apparatus and the technically demanding measurements that were highly dependent on concentration, wavelength, sample length and purity. It was a significantly more complex operation to make accurate rotation measurements than to measure a melting or boiling point, and it was not clear that measuring optical rotation would supply any additional information. Nor were polarimeters readily available until the 1840s, when they came into common use in the sugar industry. Only in the 1860s did optical activity appear to play any role in the identification and classification of chemicals. Even then, its appearance in the literature appears to have been sporadic, and the exact reason chemists chose to make such measurements remains unclear.[40]

The reluctance of chemists to use physical properties is tied to how they used them. The specific magnitude of the property – that a compound melted at 75° C, for example – was not as important as the fact that 75° C was *different* from the melting

---

[40] For example, Justus von Liebig and Hermann Kopp reported the results of Pasteur's 1848 paper on the optical activity of the tartrates in the physics section of the *Jahresberichte über die Fortschritte der Chemie*, under 'Optics'. See N.W. Fisher, 'Wislicenus and Lactic Acid: The Chemical Background to van't Hoff's Hypothesis', pp. 33–54 in O.B. Ramsay, ed., *Van't Hoff-Le Bel Centennial*, Washington, DC, American Chemical Society, 1974.

point of another compound. Physical properties were used as 'markers' to determine whether a compound was in fact different, or that a chemical transformation had taken place. Chemists were not interested, for example, in *explaining* either the presence or magnitude of a compound's physical properties. There was therefore no compelling need to multiply the number of physical properties used for identification when only a few would suffice.

Another important aspect of chemists' methods was their extraordinary stability: while chemical theory saw enormous changes throughout the nineteenth century, laboratory practise in chemistry remained remarkably static. After the introduction of Liebig's *Kaliapparat*, there was almost no new instrumentation that would contribute to the changes in chemical theory. One might argue that the polarimeter was a new instrument that enabled the practise of stereochemistry, but as we shall see, its introduction was not crucial to the success of 'chemistry in space'. The relative stability of chemical practise in the midst of significant theoretical change is one of the unique features of nineteenth-century chemistry, and the spectacular success of chemical theory during this period suggests that significant changes in theory can occur without simultaneous changes in laboratory practise.[41]

Crystallography and Chemistry

As mentioned above, chemists paid attention to the overall shape of crystals for the identification of organic compounds, but curiously the science of crystallography played only a small role in the elucidation of atomic arrangement in molecules, despite offering a promising technique. During the first half of the nineteenth century, the flourishing French crystallographic tradition influenced the shape of French unitary chemistry, offering a different insight into ascertaining the arrangement of atoms in molecules. In the 1830s, A.E. Baudrimont, drawing on the extensive French crystallographic tradition begun in the eighteenth century by René Just Haüy and Jean-Baptiste Romé de l'Isle, suggested that the arrangement of atoms in a molecule could be revealed by crystal form. In the 1840s, August Laurent took Baudrimont's suggestion and created the most significant fusion of crystallography with unitary chemistry, drawing detailed analogies between the realm of crystal structure and molecular structure. Because a series of substitution products were of the same chemical type, according to Laurent they must possess the same fundamental internal radical, which in turn must be discernable in the crystalline structure. These internal

---

[41] Recently, historians looking at scientific practice have implied that changes in theory must be accompanied by a change in practice, usually by the introduction of a new instrument. This is best exemplified by Steven Shapin and Simon Schaffer, *Leviathan and the Air Pump: Hobbes, Boyle, and the Experimental Life*, Princeton, NJ, Princeton University Press, 1985.

radicals were analogous to isomorphic crystals, in which substitution did not alter the fundamental unit that gave the molecule its principal chemical property.[42]

In the 1840s, Louis Pasteur would draw on this extensive crystallographic-chemical tradition for his chemistry and physics theses of 1847, and his famous study of the tartrates published in 1848.[43] For his chemistry thesis, Pasteur studied the crystallographic properties of the arsenic acids, discovering a new arsenious acid whose novelty was confirmed by its crystalline form. For his physics thesis, he looked at the rotatory polarization of liquids and the relationship between optical activity and crystalline form. His principle conclusion, suggested earlier by Laurent, was that compounds with the same crystalline form would have the same optical activity. In early 1848, Pasteur began to study the various crystalline forms exhibited by the salts of tartaric acid, first isolated in 1769 by the Swedish chemist Karl Wilhelm Scheele from the solid residue remaining from fermenting grapes. In 1820, another acid, with similar composition but differing properties, was isolated in small quantities from the large-scale production of tartaric acid. Named paratartaric acid, it was soon recognized as an isomer of tartaric acid. Both tartrate isomers also could be converted easily into over a dozen salts, each of which exhibited a unique crystalline form. The most important of these salts for our purposes were the sodium-ammonium salts of tartaric and paratartaric acid. In 1844, the German chemist Eilhard Mitscherlich had noticed that these two salts exhibited identical external crystalline forms. In fact, the two ammonium sodium salts were virtually identical except for their behavior towards polarized light. A solution of tartrate rotated polarized light, while a solution of the paratartrate had no effect. This proved puzzling, because the influence on polarized light was not reflected by a difference in crystal forms.

Initially, Pasteur wanted to characterize the various crystalline forms of the tartrate salts and elucidate how the differing amounts of waters of crystallization were expressed in differing crystal forms. For reasons that we probably will never know, he perceived a crucial difference, overlooked by Mitscherlich, in the crystals of the sodium-ammonium tartrate and paratartrate. The crystals of tartrate and paratartrate, Pasteur noticed, were hemihedral and asymmetric, but the paratartrate consisted of two types of asymmetric crystals that were non-superimposable mirror images of one another. Using Pasteur's words, the tartrate salt consisted of 'right-handed'

---

[42] Seymour H. Mauskopf, *Crystals and Compounds: Molecular Structure and Composition in Nineteenth Century French Science*, Philadelphia, PA, American Philosophical Society, 1976.

[43] Excellent detailed studies of Pasteur's work on the tartrates are Gerald Geison, *The Private Science of Louis Pasteur*, Princeton, NJ, Princeton University Press, 1995, and Dorian B. Kottler, 'Louis Pasteur and Molecular Dissymmetry, 1844–1857', *Stud. Hist. Biol.*, 1978, *2*, 57–98. A recreation of Pasteur's experiment has shown both the contingent nature of Pasteur's results and the expertise required to recognize 'right-handed' and 'left-handed' crystals. George B. Kauffman and Robin D. Myers, 'The Resolution of Racemic Acid: A Classic Stereochemical Experiment for the Undergraduate Laboratory', *J. Chem. Ed.*, 1975, *52*, 777–81.

crystals, and the paratartrate salt consisted of an equal mixture of 'right-handed' and 'left-handed' crystals. Despite his experience with optical activity, Pasteur had never intended to use it to characterize the crystals, but this discovery, and perhaps Mitscherlich's report that paratartrate was optically inactive, prompted him to further characterize the difference between the two salts on the basis of their effect on polarized light. Pasteur painstakingly separated the two sets of crystals found in the paratartrate, making a pile of right-handed crystals and a pile of left-handed crystals. He tested each for their effect on polarized light, and found that:

1. a solution of the right-handed crystals rotated light to the right with a magnitude equal to that of the tartrate;
2. a solution of the left-handed crystals rotated light to the left, with a magnitude equal to that of the tartrate; and
3. a solution containing equal amounts of the two crystals, as Mitscherlich had already shown, had no effect on polarized light.

He concluded that the left-handed and right-handed crystals were composed of left-handed and right-handed asymmetric molecules of tartrate. The optical inactivity of Mitscherlich's paratartrate, beforehand considered an unexplained anomaly, Pasteur could now attribute to an equal mixture of left-handed and right-handed molecules, each of which cancelled the effect of the other. In the final form of his theory, Pasteur concluded that there must exist four kinds of symmetry or asymmetry at the molecular level:

1. an optically active dextrorotatory isomer;
2. an optically active levorotatory isomer;
3. an optically inactive mixture of dextro- and levorotatory isomers that was capable of resolution into the two optically active isomers; and
4. an optically inactive isomer that was 'untwisted' and incapable of resolution into two optically active compounds.

Pasteur also drew his famous conclusions that it would be impossible to create asymmetric compounds by chemical synthesis without asymmetric starting materials, and that asymmetric molecules were essential components of living systems.

This now-famous experiment ensured Pasteur's place in the Parisian scientific community, and by the mid 1850s, his results had been incorporated into textbooks in France, England and Germany.[44] Pasteur was the first to show unequivocally a distinct relationship – long maintained by Biot in his initial studies of optical activity –

---

[44] Ward, *Polarimeter*, pp. 215–21.

between crystalline form, optical activity, and asymmetry at the molecular level. Pasteur did not simplify or make improvements to the techniques in measuring optical activity, but demonstrated that optical activity could be a useful property for chemists to consider, thereby justifying its greater application. In 1856, for example, Gerhardt changed his opinion about the value of optical activity for chemistry, and included extensive examples of optical activity data in his *Traité de chimie organique*.[45] Pasteur's results therefore stimulated interest in the use of optical activity by chemists, and its importance increased gradually until, as described above, it became more prevalent in the research literature of the 1860s.

Conclusions or Speculations? Chemistry and the Hypothetico-deductive Method

In addition to the laboratory methods of organic chemistry, we can also identify some fundamental methodological guiding principles for the construction of chemical theories. First among these is the process of transdiction, in which the macroscopic properties of substances can tell us something about the nature of the molecule – about the 'arrangement' of atoms, however it was defined. It was (and still is) assumed that the arrangement of atoms in the molecule directly caused the observed properties at the macroscopic level.

Another readily apparent trend in nineteenth-century chemical methodology was a shift in chemists' attitude towards the role of hypothesis. Early in the century, organic chemists closely followed the traditions and techniques of natural history, following a methodology that could loosely be termed as 'inductive' fact-gathering and classification of substances by their animal or vegetable origin. In other words, during this period, chemists focussed on deriving conclusions from the 'facts' of nature, or at least crafted their rhetoric to emphasize this approach. The fundamental transformation of chemistry in the 1830s away from this natural history approach towards an experimental organic chemistry resulted in an increased emphasis on the role of theories as predictive tools, although the large number of new compounds generated required chemists also to continue focussing at least some of their attention on problems of classification. In this sense, chemistry retained much of its 'fact-gathering', natural history nature, yet moved towards the explicit recognition of theories as tools for generating new facts. As the structure theory emerged in the 1850s and 1860s, it contained explicit hypothetico-deductive characteristics, along with the classificatory devices in the tradition of natural history. This dual character resulted in a bifurcation of what chemists considered the appropriate role of a theory. Kekulé, for example, used theory to make active predictions, while Rudolf Fittig gathered data to construct formulas.

---

[45] Ibid., p. 220.

This shift from inductive to deductive methodology was not unique to chemistry, as other sciences during the nineteenth century also moved towards explicitly hypothetical concepts. Maxwellian electrodynamics, optical theories and Darwin's theory of natural selection were considered at the time models of the 'hypothetico-deductive' method, and were actively criticized as such. At the turn of the nineteenth century, Dalton's physical atomism and Avogadro's hypothesis were rejected as hypothesis and speculation. The overwhelming use early in the century of chemical equivalents rather than atoms, and the use of the non-ontologically demanding Berzelian formulas, reflects chemists' overall antipathy towards hypothesis. However, by the 1850s at least, the role of hypothesis became more explicit in scientists' pronouncements about their methodology. For example, we can note Cannizzaro's adoption of Avogadro's hypothesis in advocating his set of atomic weights, Williamson's famous ether synthesis, and Kekulé's benzene theory.[46] Yet this dichotomy between induction and deduction must be considered carefully, as many pronouncements about methods are often chosen to reflect currently fashionable ideas, and do not accurately reflect the writer's own methodology, which may be a unique blend of different approaches. What appears as a significant change from inductive to deductive methodology can therefore also be regarded as a change in what was rhetorically considered appropriate – what became allowable in scientific argument. We shall see more explicit use of hypothesis and confirmation throughout the early history of 'chemistry in space'. Indeed, most of the criticisms were disagreements with the speculative methods of Van 't Hoff.

Physicalist versus Materialistic Explanation in Chemistry

The general form of explanation in nineteenth-century chemistry displays a tension between physicalist and materialist, or reductionist and nonreductionist, forms of explanation.[47] This tension is displayed not necessarily between different camps of chemists committed to different modes of explanation, but within the structure of the theory itself, as many chemists, and many theories display this tension within themselves. The physicalist tradition derived from Newtonian mechanics and strongly influenced natural philosophy, including chemistry, during the first half of the eighteenth century. According to the physicalists, chemical phenomena should be traced to the presence of, and attractions between, hard, massy and impenetrable atoms. Although influential, the reductionist view never completely supressed the older materialist tradition in chemistry, exemplified by Stahl's phlogiston theory.

---

[46] Alan J. Rocke, 'Methodology and Its Rhetoric in Nineteenth Century Chemistry: Induction versus Hypothesis', pp. 137–55 in Elizabeth Garber, ed., *Beyond History of Science: Essays in Honor of Robert E. Schofield*, Bethlehem, PA, Lehigh University Press, 1990.

[47] Alan J. Rocke, 'Convention Versus Ontology'.

Materialist forms of explanation would eventually dominate the late eighteenth century with Lavoiser's oxygen theory and Dalton's atomic theory that explained certain chemical properties such as combustion or combining ratios by the presence of certain material components or elements.[48]

The compounds of organic chemistry, because they contained a limited number of elementary components, forced chemists to employ explanations with a greater physicalist character. The concept of a rational formula, especially in the radical theory, was physicalist, in that it attempted to explain phenomena by the arrangement of atoms, however defined, rather than by their material nature. Formulas created under the type theory were equally physicalist, as they relied on the phenomena of substitution in which the type remained constant despite changes in elementary composition. With the emergence of the structure theory, we can see the presence of both physicalist and materialist explanation. It is the molecule as a whole, together with the material nature of the atoms that compose it, that creates the compound's unique set of properties. The unitary and quasi-mechanical picture of molecules under the structure theory tended to supress the materialistic component, however, and as we shall see, this tendency would increase as spatial properties of molecules became more important in chemical explanation.

Modeling Practises in Chemistry

One of the signal characteristics of chemical arguments is the prominent use of symbolic formulas that give the concepts of valence and structure a graphic clarity, or *Anschaulichkeit*. Beginning in the 1830s, chemists used nearly exclusively Berzelian formulas in some form, and by the late 1860s, notation stabilized around that introduced by Crum Brown, itself a modification of Berzelian formulas that allowed the graphic depiction of the distribution of valences in a molecule. Beginning in the 1850s and 1860s, however, chemists also increasingly employed hand-held molecular models – three-dimensional versions of Crum Brown formulas – that were also intended to give a graphic clarity to the concepts of valence and structure.[49] Two

---

[48] Dalton's own view of his theory was also physicalist, but his primary influence on other chemists was the material differences (atomic weights) between elements.

[49] This discussion of modeling is based on Christoph Meinel, 'Modelling a Visual Language for Chemistry, 1860–1875', unpublished manuscript, and Eric Francoeur, 'The Forgotten Tool: the Design and Use of Molecular Models', *Social Studies of Science*, 1997, *27*, 7–40, and Francoeur, 'Beyond Dematerialization and Inscription: Does the Materiality of Molecular Models Matter?', *HYLE*, 2000, *6*, 63–84. Other studies on the use of physical modeling during the 1860s and 1870s are by O.B. Ramsay, 'Molecular Models in the Early Development of Stereochemistry. I: The van't Hoff Model. II: The Kekulé Models and the Baeyer Strain Theory', pp. 74–96 in *Van't Hoff-Le Bel Centennial*, O.B. Ramsay, ed., Washington, DC, American Chemical Society, 1974; and Ramsay, 'The Early History and Development of Conformational Analysis', pp. 54–77 in Traynam, *Essays*.

of the most famous of these models were the 'glyptic' models of A.W. Hofmann, and the tetrahedral models developed by Kekulé, who introduced them primarily for use in chemical lectures. Chemists regarded these models as heuristic aids – extensions of the paper formulas – to see the implications of valence and structure, not as research tools or graphic depictions of the molecules themselves.

There are two important aspects of the meaning and use of paper and hand-held models that we need to consider here. First, molecular models, like all physical means of representation, are cultural and not natural objects – that is, they are invented for certain purposes with specific techniques of representation that do not depict the molecule as it 'really' is, but only certain aspects that are important at the moment. For Hofmann, his models were developed for a public lecture at the Royal Institution, in which he attempted to convince his audience that chemistry was a synthetic science that constructed molecules at will. The many possibilities in the assembled models told the chemist where to direct his efforts. Kekulé's lecture demonstration models, created as a modification of Hofmann's, more easily allowed the direct representation of triple bonds between carbon atoms. Whereas Hofmann's models had placed the atoms surrounding the carbon atom in a plane, Kekulé had brass rods pointing towards the corners of a tetrahedron. Single, double and triple bonds could then be modeled easily by matching up the ends of one, two or three of the brass rods. Kekulé's models therefore reflected his own belief in the constancy of valence by allowing him to illustrate more accurately the phenomena of multiple bonding.[50]

The second point to make concerns the ontological significance of these models. Before he created the glyptic models, Hofmann had simply used stacked cubes of standard volumes to show combinatorial relationships in the various chemical types. The glyptic models were invented with precisely the same intentions – as representations of chemical combination and construction templates, not representations of an object.[51] Kekulé developed his models to illustrate graphically multiple bonds without bending or breaking brass rods in Hofmann's planar formulas. Like their paper counterparts, chemists regarded these hand-held models with a great deal of skepticism. Frankland warned that 'students, even when warned against such an interpretation, will be liable to regard them as representations of the actual physical position of atoms of compounds'. Kolbe also called structural formulas and the models based on them 'dangerous, as they leave too much scope for the imagination ... We must therefore take care to avoid drawing a picture [of the spatial arrangement of atoms] for ourselves, just as the Bible warns us from making a visual representation

---

[50] The relationship between Kekulé's tetrahedral models and Van 't Hoff's theory of the asymmetric carbon atom will be discussed in Chapter 3.

[51] Meinel ('Modelling Visual Language') has used the analogy of architectural models, which do not represent, but suggest a building's construction.

of the Godhead'.[52] Other chemists, including Benjamin Brodie, Carl Schorlemmer and Adolphe Wurtz, issued explicit warnings about the overinterpretation of such models. They played a similar role as the mechanical models used by Maxwell in constructing his laws of electrodynamics. Despite these epistemological disclaimers, however, these modeling practices became commonplace and firmly established during the 1860s, such that in Kekulé's lab, at least, they went beyond their original pedagogical intentions and became tools for research that served theoretical purposes.

The Limits of Structure Theory, 1864–1873

The Nature of Valence and Affinity

The structure theory was undoubtedly successful in providing a set of axioms for classifying and organizing the large and growing number of organic compounds, yet in its initial form it also gave rise to theoretical curiosities. First, the success of structure theory in solving the chemical problem of isomerism relied on the principle of valence, which raised the inevitable question about what it actually was. Strictly speaking, valence, as a concept developed to account for chemical transformations, was a number that possessed no physical significance. Nevertheless, the postulation of a tetravalent carbon atom or divalent oxygen atom begged the physical question about how an atom's chemical affinity could be 'split' into different parts. The concept of affinity itself had always remained an elusive, slippery idea. Chemists had used it much as physicists had used the concept of force as an entity whose existence is known (because of its effects), but whose exact origins were nearly totally unclear.[53] The traditional means of explaining chemical affinity had been to assume an analogy to the known physical attractive forces. Electrical and gravitational forces could not adequately explain valence, however, because these forces never displayed an ability to split into different parts, and therefore the concept of a unit of chemical affinity was absurd in physical terms.[54] We will see later how spatial models of the carbon atom attempted, at least partially, to reconcile the concept of valence with the traditional concepts of attractive forces.

---

[52] Kolbe to Frankland, 27 May 1866. Translation in Rocke, *Quiet Revolution*, p. 314.
[53] Mary Jo Nye, 'Explanation and Convention in Nineteenth-century Chemistry', pp. 171–86 in R.P.W. Visser, ed., *New Trends in the History of Science*, Amsterdam, Editions Rodepi, 1989.
[54] Alan J. Rocke, 'Convention versus Ontology', p. 14.

Absolute Isomerism and Multiple Bonding

Another curiosity raised by the application of valence rules was the introduction of multiple bonds to explain the phenomenon of unsaturation, and the difficulty in representing the known differences between some isomeric unsaturated compounds. Kekulé himself developed a theory of unsaturation in the early to mid-1860s, first in his study of the organic acids, and later in his theory of benzene and the aromatic compounds. Already in 1859, Kekulé noted the limitations of his principles in providing a unique rational formula for all molecules:

> But in other cases, different properties in bodies are observed that in the current state of our knowledge it appears that we are required to assign the same rational formula. For now, this latter case can be designated as isomers in the absolute sense [*im engeren Sinne*].[55]

The concept of isomers in the 'absolute sense' became the preferred term during the 1860s for isomeric compounds that defied differentiation by the axioms of structure theory. Josef Loschmidt, for example, most likely borrowed the phrase from Kekulé in his *Konstitutionsformeln der organischen Chemie in graphischer Darstellung* (1861).

The first examples of isomers 'in the absolute sense' mentioned by Kekulé were maleic and fumaric acid (formula $C_4H_4O_4$), members of the large family of organic acids. They were derived from the dehydration of malic acid ($C_4H_6O_5$), another organic acid commonly found in fruits such as apples. In a series of articles on the organic acids that appeared between 1861 and 1864, Kekulé looked at the reactivity of malic, maleic, fumaric and related organic acids with the aim of obtaining their rational formulas. A rational formula for malic acid was easy to obtain, but removing the elements of water from it without replacing the atoms resulted in a formula that lacked atoms to satisfy all of carbon's valences. To complicate matters further, the two acids were clearly isomeric. Maleic and fumaric acid both acted as if the affinity of carbon atoms were 'unsaturated', meaning that Kekulé found that these acids rapidly absorbed hydrogen to form succinic acid, and combined easily with hydrobromic acid or elemental bromine to form bromosuccinic ($C_4H_5O_4Br$) or dibromosuccinic acids ($C_4H_4O_4Br_2$) respectively. The addition of hydrobromic acid or bromine to either maleic or fumaric acid appeared to saturate the affinity of the carbon atoms, meaning that its valency requirement was fulfilled. In 1862, Kekulé proposed a possible explanation for this affinity towards hydrogen or bromine that came to be known as the 'gap' theory:

---

[55] August Kekulé, *Lehrbuch der organische Chemie*, 2 vols, Erlangen, 1861–1866, vol. I, p. 187.

At the position in the molecule where both hydrogen atoms are missing, two affinity units of the carbon are not saturated; there is, so to speak, a gap [*Lücke*] at this position. This explains the exceptional facility with which, as it were, deficient substances combine with hydrogen or bromine. The free affinity units of the carbon strive to be saturated and so fill the gaps.

In a footnote, Kekulé proposed an alternate explanation, what has come to be known as a 'double' bond:

One can of course just as well assume the carbon atoms are shoved together (*zusammengeschoben*) so that two carbon atoms bind themselves with two affinity units each. This is only another form of the same idea.[56]

In 1867, using the concept of the double bond and free affinities, Kekulé was able to offer rational formulas for maleic and fumaric acid that would explain their isomerism. The removal of water from adjacent carbon atoms would lead to gaps on each carbon. These gaps would then mutually saturate each other, producing a double bond. If both elements of water were removed from the same carbon atom, the 'gaps' would be unable to saturate themselves and remain on one carbon. Because of its greater ability to absorb hydrogen, maleic acid contained the two 'free' affinities, whereas fumaric acid contained a double bond.[57]

Kekulé's answer to the problem of unsaturation was by no means definitive, and the rational formula for the unsaturated acids would prove problematic until the acceptance of Van 't Hoff's theory that the two isomers both contained a double bond and were spatially different. In some respects, Kekulé's own conviction that carbon must always be tetravalent made the postulation of double bonds or free affinities rather *ad hoc*. One could easily imagine that in maleic and fumaric acid, carbon was trivalent, although at the cost of a plausible explanation for the rapid addition of hydrogen. Strictly speaking, Kekulé's idea of unsaturated affinities violated the tetravalency of carbon, and it is not clear that Kekulé had realized this or was prepared to admit it, for the saturation process could be driven by the need to fill the two vacant valences. Proposing multiple bonds, however, would save the constant valence of carbon, and the idea of a double bond would become more concrete and difficult to ignore with the introduction of Crum Brown's notation. Crum Brown in fact assumed that carbon was tetravalent and disallowed any 'gaps'. He gave fumaric and maleic acid the same structure, designating them as true 'absolute' isomers.

---

[56] August Kekulé, 'Untersuchungen über organishen Säure', *Annalen Suppl.*, 1862, *2*, 85–116, p. 31, 114–15.

[57] August Kekulé, report on 'Sur les dérivés par addition de l'acide itaconique et de ses isomères' by M. Swarts, *Bull. Acad. Roy. Belg.*, 1867, *24*, 8–14.

Johannes Wislicenus and Lactic Acid, 1860–1873

Although the unsaturated acids provided the easiest case for establishing the existence of 'absolute' isomers, the difficult chemistry of the isomeric lactic acids would provide the direct inspiration for Van 't Hoff. The primary figure in the study of lactic acid during the 1860s was Johannes Wislicenus, who was born in 1835 in Klein-Eichstadt bei Querfurt in Saxony, to Gustav and Emilie Wislicenus.[58] Gustav Wislicenus was a pastor trained in conventional protestant theology who, after becoming familiar with the theology of David Friedrich Strauß, gradually came to doubt the authority of the official Church, and regarded one's own 'residing inner spirit of truth', not the Bible, as the highest authority for religious belief.[59] In 1853, he published *Die Bibel in Lichte der Bildung unserer Zeit*, in which he attempted to liberate people from 'superstitious adoration' (*aberglaublichen Verehrung*) of the Bible.[60] Sentenced to two years in prison and the destruction of his book for blasphemy and for ridiculing the Bible, Gustav took his family to Boston, Massachusetts, where the 18- year-old Johannes worked for two years as an assistant to Eben Horsford, a student of Justus Liebig, at Harvard University. The attempt at emigration was short-lived, however, and the family left the United States for Zürich in 1856.[61]

Soon after their return, Johannes left for Saxony to study at the University of Halle (the city where his father had become a pastor in 1841) under Wilhelm Heintz, with whom he had already started to work as an assistant in the spring of 1853 shortly before his family left for America. By Wislicenus' own account, Heintz had a profound influence on him. During his two years as private assistant to Heintz, they engaged in many discussions of theoretical chemistry, in which Wislicenus was forced to sharpen his knowledge by a thorough study of the chemical literature. According to Wislicenus, Heintz did not adopt the premises of the type theory until Wislicenus convinced him of its validity in 1858.[62] In 1859, Wislicenus had completed his studies at Halle, but as a stipulation for granting him a PhD, the university asked him to renounce the religious views of his father. He refused, and returned to Zürich, where on 7 January 1860, he received a PhD from the University

---

[58] The best and most sympathetic account of Wislicenus' life is by his assistant, colleague and friend at Leipzig, Ernst Beckmann: 'Johannes Wislicenus', *Berichte*, 1904, *37*, 4861–946. Wislicenus' most famous English student, William Henry Perkin Jr, also wrote a substantial account of his life: 'Wislicenus Memorial Lecture', *J. Chem. Soc.*, 1905, *87*, 501–34.

[59] G. Frank, 'Gustav Wislicenus', *Allgemeine Deutsche Biographie*, vol. 43, 542–5, Dunecker and Humboldt, Leipzig, 1898. The words are Frank's, not Wislicenus'.

[60] Ibid., p. 544.

[61] For more about the Wislicenus family in America, see Paul R. Jones, 'The Young Johannes Wislicenus in America', *Bull. Hist. Chem.*, 1997, *20*, 28–32.

[62] Johannes Wislicenus, 'Wilhelm Heintz', *Berichte*, 1882, *16*, 3121–40, p. 3133.

of Zürich for his essay on 'The Theory of Mixed Types'. In February, the Zürich faculty accepted the same manuscript as his *Habilitationsschrift*, and in March he was appointed to the faculty of the University of Zürich.

In his *Habilitationsschrift*, Wislicenus attempted to find a middle ground between the radical and type theories.[63] Although he firmly adhered to many of the assumptions of the type theory, he wanted a more solid ontology in which chemical formulas expressed 'the chemical unity, I would almost say, the individuality of a molecule'.[64] As his next major independent project, he chose to investigate the behavior of lactic acid under the constraints of his theory of mixed types. The study would last for fifteen years, during which he isolated several isomers of lactic acid and struggled in vain to express accurately their 'individuality'.

The story of Wislicenus' lactic acid research has been told in detail by Nicholas Fisher, but it is important to recount it here in the context of the emerging structure theory outlined above.[65] Lactic acid, with its dual behavior as acid and alcohol, provided an ideal case for Wislicenus to determine and express what he meant by a molecule's individuality. Its behavior placed it in a particularly difficult position. In some reactions it behaved like an alcohol (for example, it formed ethers), and in others it behaved like a carboxylic acid (for example, it formed acid salts with bases). The attempt to represent both of these properties in terms of absolute radicals or in terms of types resulted in problems for both theories, since both properties could not be expressed simultaneously by either the radical or type theories.

On the basis of his theory of mixed types, Wislicenus presented a new formula for lactic acid that suggested an analogy between lactic acid and propionic acid. He therefore attempted to synthesize it in a similar manner from ethylene chlorohydrin via a cyanide intermediate (Figure 2.3). Based on the product's crystal form and the amount of water of crystallization in its zinc salts, he concluded that he had not actually made lactic acid, but an isomer of it. Wislicenus named this new compound ethylene lactic acid and the old one ethylidene lactic acid, and suggested that the differences between the two were located within the carbon nucleus C_2H_4/H. He also suggested that the two major radicals in lactic acid, the carbonyl, CO, and the alcohol, C_2H_4, could be switched to emphasize either the alcoholic or acidic nature of lactic acid.

These relative expressions, and his admitted uncertainty about the difference between ethylene and ethylidene lactic acid, did not mean Wislicenus had given

---

[63] Johannes Wislicenus, 'Theorie der Gemischten Typen', *Zeit. Ges. Naturw.*, 1859, *14*, 96–175, p. 97. Wislicenus' *Habilitationsschrift* appeared in August 1859, slightly more than a year after the appearance of Kekulé's landmark 1858 paper. Like Kekulé, Wislicenus wanted to reconcile radical and type theories, but he did not cite these papers, suggesting they were not influential in the development of Wislicenus' own theory. Wislicenus cited Gerhardt more often than any other chemist.
[64] Ibid., p. 101.
[65] Fisher, 'Wislicenus'.

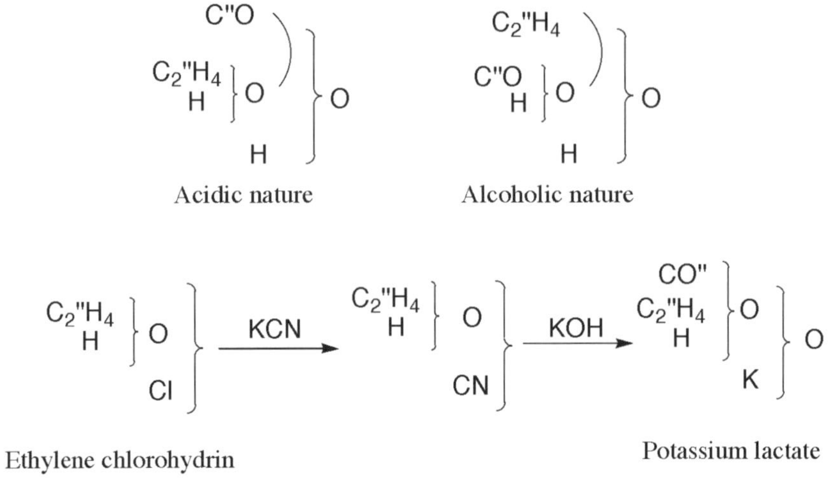

$$\left.\begin{array}{c}C"O\\ C_2"H_4\\ H\end{array}\right\}O\Bigg\}O \qquad \left.\begin{array}{c}C_2"H_4\\ C"O\\ H\end{array}\right\}O\Bigg\}O$$

Acidic nature Alcoholic nature

a

$$\left.\begin{array}{c}C_2"H_4\\ H\end{array}\right\}O\Bigg\}Cl \xrightarrow{KCN} \left.\begin{array}{c}C_2"H_4\\ H\end{array}\right\}O\Bigg\}CN \xrightarrow{KOH} \left.\begin{array}{c}CO"\\ C_2"H_4\\ H\end{array}\right\}O\Bigg\}O\Bigg\}O$$
$$\qquad\qquad\qquad\qquad\qquad\qquad\qquad\qquad\qquad\qquad K$$

Ethylene chlorohydrin Potassium lactate

b

2.3 (a) Wislicenus' 1859 formulas for lactic acid that emphasized different aspects of its reactivity. (b) Wislicenus' 1859 synthesis of lactic acid from ethylene chlorohydrin.

up trying to find the individuality of lactic acid. Rather, he admitted an inherent weakness in chemical notation's ability to express all of its properties. Any representation of molecules, he said, must necessarily be a one-sided, distorted picture, even if our goal is to represent the true nature of a substance:

> ... but one and the same body, as Kolbe certainly correctly notes, can have only *one* constitution, its elements can only be combined in a unique style and manner into a molecular unity. The above lactic acid formulas do not indicate different, but one and the same *kind* of union of all the components necessary for the molecular being of the lactic acid molecule; the radicals in them are all the same, the *kind* of its mutual saturation remains exactly the same, only the *sequence* is exchanged. But one such change in the sequence will always remain justified, even if we actually should become acquainted with the *spatial arrangement* of atoms in a compound, because our formulas can at most be *pictures of bodies depicted in a plane*. But if we want to portray all aspects of the visible properties of a body, *many pictures belong to it, taken from different perspectives*. As long as chemical formulas are only such pictures taken from different sides of one and the same chemical body, that is, only show the same proximate components united in a different *sequence*, and not by a different *kind* of mutual combination, then in my opinion the most rigorous chemical conscience can have no objections.[66]

---

[66] Johannes Wislicenus, 'Studien zur Geschichte der Milchsäure und ihrer Homologen', *Annalen*, 1863, *125*, 41–70, pp. 45–6. Wislicenus' emphasis.

$$\underset{\substack{\text{Ethylidene lactic acid} \\ (\alpha\text{-hydroxypropionic acid})}}{\text{H}-\overset{\overset{\displaystyle\text{H}}{|}}{\underset{\underset{\displaystyle\text{H}}{|}}{\text{C}}}-\overset{\overset{\displaystyle\text{OH}}{|}}{\underset{\underset{\displaystyle\text{H}}{|}}{\text{C}}}-\overset{\displaystyle\text{O}}{\underset{\displaystyle\|}{\text{C}}}-\text{OH}} \qquad \underset{\substack{\text{Ethylene lactic acid} \\ (\beta\text{-hydroxypropionic acid})}}{\text{HO}-\overset{\overset{\displaystyle\text{H}}{|}}{\underset{\underset{\displaystyle\text{H}}{|}}{\text{C}}}-\overset{\overset{\displaystyle\text{H}}{|}}{\underset{\underset{\displaystyle\text{H}}{|}}{\text{C}}}-\overset{\displaystyle\text{O}}{\underset{\displaystyle\|}{\text{C}}}-\text{OH}}$$

2.4   Structures for α and β lactic acid using Crum Brown's formulas.

This lengthy analysis of the meaning of chemical formulas reflected clearly his desire to express the individuality of a molecule. In his article on Wislicenus and lactic acid, Fisher argued that this statement implied that Wislicenus was already thinking in three dimensions in 1863. At that time, however, Wislicenus did not have any concrete ideas about the nature of the physical form of molecules, and he continued to use the old type theory notation which did not represent a molecule's physical characteristics. When Wislicenus elaborated on the meaning of chemical notation, he intended a representation of chemical properties. He may have been convinced that a molecule had a physical form, but he did not yet make any claims about it.

Although Wislicenus could emphasize either the acidic or alcoholic behavior in lactic acids by juggling radicals in the formula, he was still unable to represent visually the difference between the ethylene and ethylidene isomers. Nevertheless, chemists quickly adopted his general discovery of the second lactic acid, and in the mid-1860s, Crum Brown's new formulas could express the difference precisely as a difference in the location of the hydroxyl group (Figure 2.4). Wislicenus went on to investigate the relationship between the two lactic acids and hydracrylic acid, first prepared in 1862 by Beilstein, and concluded that hydracrylic acid was identical to ethylene lactic acid, and importantly, reported that paralactic acid (*Paramilchsäure*) from meat extract was identical to ethylidene lactic acid except that it was optically active, whereas fermentation lactic acid (*Gährungsmilchsäure*) was not. It is not clear why Wislicenus initially chose to measure the optical activity of either lactic acid, but he used this difference in properties to conclude that the two compounds were different, and therefore isomeric.

Because structural formulas allowed only two possible arrangements of atoms containing both an acid and an alcohol (the α- and β- hydroxypropionic acids in Figure 2.4), Wislicenus was unable to depict graphically the difference between paralactic and fermentation lactic acid. He reported this strange result to the September 1869 Innsbrück *Naturforscherversammlung*, where he first suggested publicly that the cause of the isomerism could lie in a difference in the spatial arrangement of atoms. The text of the lecture itself no longer exists, but Kekulé, in a report of the meeting to the German Chemical Society, described the essence of Wislicenus' statement as follows: 'Such finer cases of isomerism would probably be interpreted by spatial conceptions concerning the grouping of atoms, that is, by model formulas.'[67] The phrase 'model formulas' (*Modellformeln*) was rather cryptic, as Kekulé said nothing further about what Wislicenus meant by 'models'.

Two months later, Wislicenus made a more precise statement about what he meant at a meeting of the Zürich chemical *Harmonika*.[68] Like the earlier lecture, no transcript or text exists, and the only record of its contents is a report to the German Chemical Society by O. Meister, a student in Wislicenus's Zürich laboratory. The following quote, often cited in histories of chemistry, is usually attributed to Wislicenus himself.[69] Whether or not it is actually a direct transcription of portions of Wislicenus' lecture is not certain, but the general content of Wislicenus' talk remains quite clear, and it contained the first vague insights into how these isomers could produce different optical properties:

> Facts like these will force us to explain the difference of isomeric molecules of equivalent structure by different positions of their atoms in space, and to look for possible ideas about these positions. Possibly an exact determination of the density of the modifications of lactic acid will bring to light a difference in the spatial materialization of molecules [*molekülare Raumerfüllung*], perhaps such that the optically active meat lactic acid, which the lecturer considers a modification of ethylidene lactic acid ... does not contain the atoms arranged together in the smallest possible space.[70]

Wislicenus' idea about 'smallest possible space' remained vague, but it is clear that he was convinced that some sort of physical cause was necessary to explain this observed difference in optical rotation. By 1869, therefore, Wislicenus had recognized a limitation in the structure theory's power of representation, but he still could not suggest anything further than this uncertain guess about the density of atoms within the molecule.

During the next few years, Wislicenus' theoretical problem became more urgent when he discovered that Beilstein's hydracrylic acid was not the same as ethylene lactic acid, but apparently was a fourth lactic acid isomer. Wislicenus communicated these results first to the Zürich Chemical Society and then submitted them in 1872

---

[67] August Kekulé, Report of the 1869 *Naturforscherversammlung*, *Berichte*, 1869, *2*, 550–1.

[68] Wislicenus founded the Zürich 'Chemische Harmonika' to 'form Zürich as a central point of chemical life'. Ernst Beckmann, 'Wislicenus', p. 23.

[69] For example, see J.R. Partington, *A History of Chemistry*, 4 vols, London, MacMillan, 1961–70, vol. 4, p. 761. Ramsay incorrectly mentions Kekulé, not Meister, as the reporter. O.B. Ramsay, *Stereochemistry*, London, Heyden, 1981, p. 79.

[70] O. Meister, 'Report of Wislicenus' address to the Zürich chemical club, Nov. 2, 1869', *Berichte*, 1869, *2*, 619–21, p. 620.

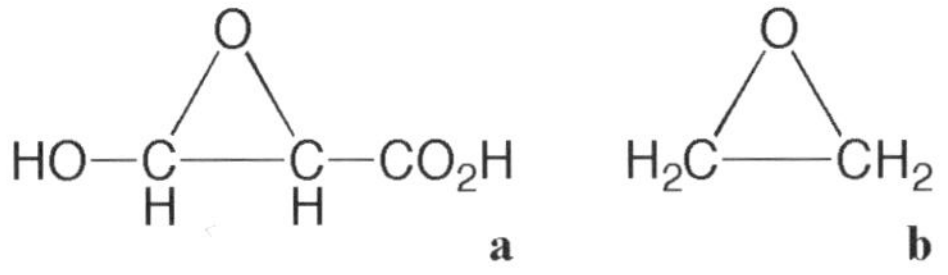

2.5 Wislicenus' new formula for hydracrylic acid (a), based on the formula for ethylene oxide (b).

and 1873 to the *Annalen* in two long, detailed articles totaling over a hundred pages. Much of his discussion in these articles centered around clarifying the confusing chemistry of the lactic acids and their derivatives, laying out their specific chemical properties, the techniques and apparatus for making measurements of optical rotation, and proposing various possible structural formulas for the four isomeric lactic acids. At one point, Wislicenus suggested that hydracrylic acid and meat lactic acid (*Paramilchsäure*) were structurally similar and contained a three-membered ring unit based on the formula for ethylene oxide (Figure 2.5), but ruled this out because hydracrylic acid was not optically active. This led him to speculate again on the cause of the rotation:

> Then we should not hesitate to admit that the cause that enables the rotation of the plane of oscillation of light travelling through paralactic acid can also be strong enough to determine the relatively small peculiarities of their salts in relation to their solubility and water of crystallization, and subsequently there is *as yet no fact at hand that forces us to give paralactic acid a different structural formula than fermentation lactic acid*. Because of this it seems therefore reasonable to assume at least tentatively that paralactic acid is *structurally identical* [*struktur-identisch*] to fermentation lactic acid, meaning that the sequence of mutual connections among the composed atoms in both can be expressed by the formula $HOOCCHOHCH_3$.

Wislicenus therefore first suggested that the cause of optical activity lay in a particular structural unit, the three-membered ring, but the absence of optical activity in hydracrylic acid ruled that out, leaving him with four isomers and only two possible formulas. He concluded:

> Once we recognize the possibility that equally composed molecules, structurally identical but somewhat divergent in their properties, it cannot of course otherwise be explained except by assuming that the reason for this difference rests only in a *different kind of spatial arrangement* of the atoms bound in a fixed sequence with one another.

He elaborated further on the difference between physical and chemical phenomena:

> That the chemical properties of a molecule are most decisively [*am Entschiedensten*] determined by the nature of the atoms that compose it *and* by the sequence of their mutual

combination, *the chemical structure* of the molecule, is now a generally shared conviction. No less justified, it appears to me, is the assumption that differences in the *geometrical* arrangement of chemically structurally identical molecules, that can primarily give rise to deviations in the molecule's size and form, must above all also become noticeable in physical properties, among which could quite possibly be included certain differences in those properties lying on the border areas of physical and chemical relationships, such as solubility, crystal form, water of crystallization, and so forth.[71]

Wislicenus made an important distinction here between two sets of cause and effect. First, the chemical properties of a substance were expressed by the 'structure' of the compound or the sequence of linked atoms given in its constitution. Deviations in the magnitude and form of molecules, independent of the connected sequence of their atoms, manifested themselves in different physical properties such as optical rotation. Therefore, the nature of the atoms and the sequence of their connections caused the individual chemical behavior (its unique set of transformations), and the physical characteristics of molecules caused their individual physical characteristics.

To clarify this new concept more precisely, Wislicenus reviewed and named the different types of isomerism. Propyl and isopropyl alcohol, which differed in the location of the hydroxyl group, were a case of 'positional isomerism' (*Orts-Isomerieen*). Because they differed in the carbon skeleton, butane and isobutane would be a case of 'core isomerism' (*Kern-Isomerieen*). The relationship between the two pairs of lactic acids, he suggested, then would be best indicated by the term 'geometric isomerism' (*geometrische Isomerie*). Wislicenus chose the term 'geometric' because it conveyed what he believed to be the inner cause of the difference between isomers, and because he wished to distinguish his theory from that of Ludwig Carius. In 1863, Carius had used the term 'physical isomerism' to describe the difference between isomers that appeared structurally identical, and had explained the difference between the two isomers as 'different aggregations of the composed molecules', and meant to distinguish this cause from polymerism and metamerism.[72] Wislicenus found 'geometric' more suitable to his purposes, '… if the way of expressing a type of isomerism should express not only the fact of the difference in quantitatively equivalent composed molecules, but rather the inner cause of this fact'.[73] Although he was sure that the difference could be explained by a 'geometrical' cause, Wislicenus could not say anything further about it, except that he was 'occupied with experimental studies concerning the particular "how" of this explanation'.[74]

---

[71] Johannes Wislicenus, 'Über die isomerische Milchsäure. II. Abhandlung. Ueber die optisch-activ Milchsäure der Fleischflüssigkeit, die Paramilchsäure', *Annalen*, 1873, *167*, 302–46, pp. 343–4. Wislicenus' emphasis.

[72] Ludwig Carius, 'Über Addition von Unterchlorigsäurehydrat und von Wasserstoffsuperoxyd', *Annalen*, 1863, *126*, 195–217, p. 216.

[73] Wislicenus, 'Über die isomerische Milchsäure. II', pp. 343–6.

[74] Ibid.

As Fisher has argued, in 1873 Wislicenus had reached a theoretical dead end, in part because of experimental difficulties in preparing pure samples of lactic acids. This certainly seems true, but if we also consider his work on lactic acid in light of the ontological and epistemological constraints of structural chemistry, another major source of his frustration was most likely the epistemological division between the concepts of chemical and physical arrangement of atoms in molecules. In 1859, as a representative of the type theory, he had remarked that chemical formulas were not representations of the microscopic world. Chemical formulas, Wislicenus noted, 'should, moreover, only be expressions of a sum of chemical relationships. They are double decomposition formulas: not constitutional formulas, but reaction formulas'.[75] Further on in the same paper, he made the common distinction between physical and chemical atoms:

> I understand by [atoms] not physical atoms, that is, not a concrete corporeal body, but quite specifically only the smallest weight–volume relationships in which elements or radicals enter into combination with one another. Even less do I consider the concept of a molecule in the sense of the atomistic hypothesis, but rather a relationship in Gerhardt's sense.[76]

This reluctance to claim knowledge of a molecule's physical form continued until 1875, when he first encountered Van 't Hoff's theory. In his famous 1869 statement, he acknowledged that formulas did not make statements about the three-dimensional arrangement of atoms – it was one of their principal weaknesses. His 1873 distinction between the two worlds of chemical and physical properties and their sets of distinctly different causes also indicates that he assigned a particular meaning for chemical formulas.

The most explicit statement confirming this traditional view of the meaning of structure, however, is found in a copy of lecture notes taken during one of Wislicenus' courses at the Zürich Polytechnical Institute in the winter semester of 1871–1872, during the period in which his work on lactic acid was reaching a peak. The notes were taken by Robert Gnehm, later an assistant to Victor Meyer at the polytechnic, for a class entitled *Bau der Kohlenstoffverbindungen*, and were deposited in the archives of the polytechnic (now the ETH). On the first page of the notebook, dated 8 November 1871, Gnehm wrote:

> Our formulas (structural formulas) are pictures in a plane, but molecules are nevertheless bodies, for example [Gnehm gives the structure for ethanol with Crum Bown notation] gives us the sequence of atoms, but by no means the three-dimensional arrangement of the molecule's atoms in space. We still lack clues [*Anhaltspunkte*] about the mathematical form

---

[75] Wislicenus, 'Gemischten Typen', p. 96.
[76] Ibid., p. 174.

of the molecule. But certain essential properties of the molecule depend on these spatial properties, at least, for example, the effect of light, [and] the crystal form.[77]

Although these were probably not Wislicenus' direct words, his meaning was absolutely clear to Gnehm. Like his contemporaries, Wislicenus expressed an explicit epistemic caution when considering the meaning of chemical formulas.

Wislicenus was certainly the closest among his contemporaries to describing a new level of meaning for chemical structures, and we can see clearly in his work on lactic acid the epistemological divide between chemical structure and physical form. He did reach an intellectual breakthrough, although incomplete, when he made clear the weaknesses inherent in chemical formulas, but he was not successful in redressing them. Before 1875, it appears as if Wislicenus regarded the meaning of chemical structures under Kekulé's and Butlerov's constraints.

Conclusion

Historians have commonly described those chemical formulas written before 1874 as being 'two-dimensional', and held that Van 't Hoff 'extended' them into three dimensions. Servos, for example, described chemistry before Van 't Hoff as a 'flat, two-dimensional affair'.[78] As I have tried to show in this chapter, this characterization is misleading, because while structural formulas were two-dimensional drawings on paper, they were not meant to portray molecules with *any* dimensionality. From the various statements on the meaning of chemical formulas given by Kekulé, Butlerov, Brown and Wislicenus, and numerous similar statements in the literature made by other chemists at the time, it seems clear that chemical structures were not originally meant to be pictures of the microworld.

What, then, did these structural formulas mean? In order to answer this question, it is useful to employ some basic concepts from semiotics, or the study of signs. The nineteenth-century American philosopher Charles Pierce identified three types of representational signs: symbols, icons and indices. We only need consider the first two here. Symbolic signs represent objects or concepts by convention, and do not necessarily resemble that object or concept. For example, eagles are often a symbol of freedom or nobility, and mathematics is used to represent concepts in physics. Iconic signs mimic or physically resemble the object. Road maps or geological stratigraphic

---

[77] Robert Gnehm, lecture notes, 'Bau der Kohlensoffverbindung', Hs 633:6, Wissenschaftshistorische Sammlung, ETH-Bibliothek, Zürich.

[78] John W. Servos, *Physical Chemistry from Ostwald to Pauling: The Making of a Science in America*, Princeton, NJ, Princeton University Press, 1990, p. 25.

maps are an excellent example of iconic representation, in that they consciously mimic certain physical characteristics of the landscape.

Returning to structural formulas, we can see that they were originally intended to be *symbolic* forms of representation, not iconic images.[79] Formulas and hand-held models were not meant to be 'windows' into the physical reality of the molecule, nor to mimic its physical characteristics, but represented the abstract concepts of chemical combination implied by valence. Chemists took great pains to emphasize their purely symbolic meaning. When Kolbe denounced the use of structural formulas and models as similar to creating graven images of God, he recognized precisely this distinction between the symbolic and iconic meaning of formulas.

Understanding structural formulas as symbols rather than icons helps somewhat in the tricky business of understanding what chemists 'saw' when looking at chemical formulas. Such formulas were not a representation of an entity in the microworld, but of chemical reactivity and constitution. A structural formula provided a convenient shorthand way of representing and explaining the chemical behavior of a substance by displaying the arrangement of atoms in a molecule, where 'arrangement' meant simply the sequence of connections between atoms. Formulating a structure was a matter of linking atoms together on paper until all had satisfied their valence. Each pure organic compound could be assigned a unique structure, and conversely, each sequence of connections possible on paper should correspond to a real substance, known or unknown. The ability to assign structures under the rules of valence fulfilled the long-standing goal of differentiating isomers, or any group of compounds, by the arrangement of atoms in the molecule. This one-to-one correspondence of formulas to compounds offered great predictive capability well beyond that of earlier chemical theories.

Because chemical concepts such as atomicity and valence were dependent on convention, chemists often displayed reticence about claiming that structural formulas represented the molecule as it 'really' was. Yet most chemists were realists, convinced that their chemical picture of the sequence of atoms in molecules was 'real', in the sense that it represented, however faintly, some aspect of nature. Lothar Meyer, for example, thought the 'templates' (*Schablonen*) of structural formulas 'will correspond so vaguely to reality as a charcoal drawing of a sunny landscape on a coarse wall'.[80] The purpose of chemical theory was to explain the behavior of organic

---

[79] Mary Jo Nye and Stephen Weininger have conducted semiotic analysis of chemical symbols, including the difference between symbols, icons and indices, but have not noticed the difference between structural and stereoformulas as the difference between symbolic and iconic signs. Mary Jo Nye, *From Chemical Philosophy to Theoretical Chemistry: Dynamics of Matter and Dynamics of Disciplines, 1800–1950*, Berkeley, CA, University of California Press, 1993; Stephen J. Weininger, 'Contemplating the Finger: Visuality and the Semiotics of Chemistry', *HYLE*, 1998, *4*, 3–27.

[80] Lothar Meyer, 'Über die Constitution des Benzols', *Annalen*, 1888, *247*, 251–4.

substances on the basis of the arrangement of atoms within the molecule, and the structure theory provided a means for modeling that arrangement, with the proviso that 'atom' meant a chemically indivisible unit, not a discrete part of matter in space (the hard, massy, impenetrable atoms of Newton and Dalton). Chemists took great pains to ensure that the chemical properties of substances – tangible, sensible transformations detectable in external properties or elemental composition – played the central role in constructing chemical formulas. The mechanistic tradition in chemistry persisted, exemplified by the belief in physical atoms, but this belief had little concrete relation to the development of chemical theory.[81] For those chemists such as Kekulé who believed that chemical formulas described the actual constitution of molecules, any realism they professed must be qualified, because the reality described by the structure theory was not a *physical* reality, but a *chemical* reality.

In summary, I have described organic chemistry before 1874 as taking place in two major stages: the transformation of plant and animal chemistry into an experimental science based on the increasing number of artificial compounds, and the emergence of the structure theory during the 1860s. During both of these stages, chemists concentrated on constructing a symbolic language for representing the composition and reactivity of organic compounds that had no intended correspondence to the microscopic molecule. When Van 't Hoff introduced the idea that the carbon atom had the physical shape of a tetrahedron, he initiated a third fundamental transformation in chemical theory, during which this carefully constructed chemical reality was joined to physical reality, and in which chemists began to interpret those same chemical formulas in a true physical sense. In doing so, they transformed structural formulas from *symbolic* into *iconic* images.

---

[81] There is also a strong pragmatic component in the adoption of atoms and other principles of organic chemistry, including stereochemistry. I shall describe this in more detail in the Conclusion.

Chapter 3

The Tetrahedral Carbon Atom, 1874–1877

> Modern chemical theory has two weak points. It discusses neither the relative positions that the atoms take up in a molecule, nor their types of motions.
>
> Jacobus Henricus van 't Hoff, 1875

Introduction

If chemists did create a conventional *symbolic* language for chemical structures, as argued in Chapter 2, how was that same language transformed into an *iconic* language? How did Van 't Hoff and Le Bel argue that the physical characteristics of the molecule could be known, and what purpose did their new theories serve? In this chapter we must turn to the origins and content of the two famous works in which Van 't Hoff and Le Bel took the first steps across the epistemological divide so carefully constructed by chemists during the first three-quarters of the nineteenth century. Long considered to be the founding documents of stereochemistry, these papers are quite different, and by themselves proved not to have any significant influence. Le Bel did not revise or extend his original ideas, and would never create a significant research program into his ideas, and his paper might well have drifted off into obscurity to be discovered by a later generation of historians if in the later versions of his own theory Van 't Hoff had not repeatedly given credit to Le Bel for the same idea. Van't Hoff's own 1874 Dutch pamphlet would suffer the same fate of obscurity, and chemists would learn of Van 't Hoff's ideas only through his extensively revised editions in French (*La chimie dans l'espace*, 1875) and in German (*Die Lagerung der Atome im Raume*, 1877). The early literature on 'chemistry in space' frequently mentions the Le Bel-Van 't Hoff theory, reflecting the appearance of Le Bel's French paper in a well-known journal before the appearance of *La chimie dans l'espace*.[1] Although Van 't Hoff would also publish a second edition of *La*

---

[1] For example, Hans Landolt (1877), Kekulé (1877), Adolf von Baeyer (1885) and Emil Fischer (1889).

chimie dans l'espace in 1887 (*Dix années dans l'histoire d'une théorie*), and two more editions of *Die Lagerung* in 1894 and 1908, his contribution to the theoretical development of 'chemistry in space' ended in 1877, when he turned his attention to broader issues in organic and physical chemistry.

All of these various editions of Van 't Hoff's theory are well known, but curiously, they have never been critically examined for their own internal arguments, for their relationship to general issues in nineteenth-century chemistry, or as texts written within the intellectual pathway that Van 't Hoff began to follow in 1874. This last point is extremely important, for the previous, fragmented accounts of the origins of stereochemistry have looked at the various versions of Van 't Hoff's theory from far in the future, and therefore tended to emphasize the elements of modern stereochemistry within them. Nor have these editions been considered in sequence, treating the revisions and additions Van 't Hoff made to an initially very short pamphlet. A more accurate historical analysis requires that we look at these documents within the context of the personal intellectual and professional journey of the very young Van 't Hoff as he attempted to establish himself within the chemical community. The purpose of this chapter is therefore threefold. First, we will examine the circumstances surrounding the appearance of both Van 't Hoff's and Le Bel's 1874 papers, and examine their arguments, focussing on the differences and similarities between their approaches for ascertaining the spatial characteristics of molecules. As the documentation for Van 't Hoff's student years is significantly more plentiful than that for Le Bel, and in light of the subsequent greater influence of his theory, our story after 1874 will then follow the evolution of Van 't Hoff's ideas to 1877 in the context of Van 't Hoff's own personal research trajectory.

Van 't Hoff and Le Bel, 1874

Voorstel tot Uitbreiding (September 1874)

Van 't Hoff's now famous Dutch pamphlet was printed in Utrecht, financed by his father, and dated 5 September 1874, with the lengthy title 'Proposal for Extending the Currently Employed Structural Formulae in Chemistry into Space, Together With a Related Remark on the Relationship Between Optical Activating Power and

---

all mentioned Le Bel's article as appearing first. Hans Landolt, 'Untersuchungen über optische Drehungsvermögen. 1. Abhandlung', *Annalen*, 1877, *189*, 241–337, August Kekulé, 'Die Wissenschaftlichen Ziele und Leistungen der Chemie', in Richard Anschütz, *August Kekulé*, 2 vols, Berlin, Verlag Chemie, 1929, vol. 2, pp. 903–17; Adolf von Baeyer, 'Über Polyacetylenverbindungen. Zweite Mitteilung', *Berichte*, 1885, *18*, 2269–75; Emil Fischer and Josef Hirschberger, 'Über Mannose II', *Berichte*, 1889, *22*, 365, 294–305 in Emil Fischer, *Gesammelte Werke*, vol. 6, *Untersuchungen über Kohlenhydrate und Fermente*, Berlin, Springer-Verlag, 1908.

Chemical Constitution of Organic Compounds' (henceforth referred to as the *Voorstel*, the first word in the Dutch title).[2] When Van 't Hoff completed this 11-page pamphlet, it was six days after his twenty-second birthday, and a few months before he would complete his eclectic and peripatetic formal education. Van 't Hoff's path to this publication reflects his broad interests in both physics and chemistry, and from the time he entered the technical school at Delft, he followed as much an autodidactic path as that of formal instruction. At Delft, he studied chemistry, and after a stint in a sugar factory, his interests turned to pure, rather than applied science. He seemed from an early age to want to create his own education, and on his own, he read and studied closely the biographies of great men for guidance, studied the philosophy of Auguste Comte and Willliam Whewell, and read the poetry of Lord Byron and Robert Burns. He found himself strongly attracted to Comte's outline of the sciences in his *Cours de philosophie positive*, in particular the primacy Comte gave to mathematical physics. According to his biographer Ernst Cohen, Van 't Hoff's 'mathematical needs' directed him to a formal study of physics, mathematics and analytical geometry at the University of Leiden, where he completed his candidate's exam in June of 1872.[3]

As Leiden had an undistinguished chemistry faculty, Van 't Hoff left after completing his exams to begin a formal study of organic chemistry elsewhere, and during the next two years, he alternated his studies between the Netherlands, Germany and France. Because he was smitten with the idea of the 'heroic individual' (from the poetry of Byron and Burns), he first chose to work for August Kekulé, the founder of structure theory, at his Bonn laboratory. He was disappointed with his experience in Bonn, but he did manage to complete work for his first publication on a new synthesis of propionic acid.[4] He returned to the Netherlands, where he matriculated at the University of Utrecht to study for his doctoral exams, which he passed in December 1873. According to Van 't Hoff's own later account, it was during this period in Utrecht that he encountered Wislicenus' articles on the constitution of lactic acid. While taking a break in his reading, Van 't Hoff conceived of the idea of the tetrahedral carbon atom as a way to solve Wislicenus' problem, although he let the

---

[2] Jacobus Henricus van 't Hoff, *Voorstel tot Uitbreiding der tegenwoordig in de scheikunde gebruikte Structuur-Formules in de ruimte; benevens een daarmeê samenhangende opmerking omtrent het verband tusschen optisch actief Vermogen en Chemische Constitutie van Organische Verbindingen*, Utrecht, Greven, 1874. E. Fischmann, 'A Reconstruction of the First Experiments in Stereochemistry: Letters From Van't Hoff to Bremer in a New Chronological Sequence', *Janus*, 1985, 72, 131–56.

[3] This sketch of the early part of Van 't Hoff's career is taken from Peter J. Ramberg and Geert J. Somsen, 'The Young J. H. van 't Hoff: The Background to the Publication of his 1874 Pamphlet on the Tetrahedral Carbon Atom, Together with a New English Translation', *Ann. Sci.*, 2001, 58, 51–74. Reproduced with kind permission of the *Annals of Science*. Further information on this journal can be found on http://www.tandf.co.uk.

[4] For more on the detailed relationship between Van 't Hoff and Kekulé, see Jost Weyer, 'Van't Hoff, Kekulé, und die Stereochemie: Zwei unveröffentliche Briefe von J.H. van't Hoff an A. Kekulé', *Janus*, 1977, 64, 217–30.

idea gestate for some time. He passed his exams on 22 December 1873, and moved to Paris to study in Adolphe Wurtz's laboratory at the *Ecole de Médecine* during the first half of 1874. There he met Le Bel, a fellow student in Wurtz's lab. His stay in Paris was more pleasant and productive, as Wurtz was not only a strong proponent of the structural theory in France, but also encouraged new ideas, personal initiative and cooperation among his students.[5] Some time in the summer of 1874, after returning to Utrecht, Van 't Hoff collected his ideas for the pamphlet, a means of publication chosen most likely to gain priority.[6] The awkwardness of the title reflects the pamphlet's style as a whole, which reads as if Van 't Hoff was in need of a good (or at least a better) editor. In introducing his German translation, Cohen remarked that he was struck by the 'wretchedness' (*Dürftigkeit*) of Van 't Hoff's style.[7]

A number of direct and indirect factors converged in the person of Van 't Hoff and allowed the publication of the *Voorstel*. First, Van 't Hoff had been steeped not only in the tradition of German structural chemistry in Bonn and Paris, but also in mathematics, mathematical physics and geometry at the newly created Rotterdam *Hoogere Burgerschole* (secondary school) and the University of Leiden. He took Comte's statement seriously that geometry 'offers the advantage of exercising this faculty of the human mind, what we call imagination itself, in the highest degree and in the most sure and precise manner'.[8] Van 't Hoff's conception of nature in the *Voorstel* is reminiscent of Galileo's conception of geometry in nature: nature and molecules *are* geometric, and geometrical concepts are not mere abstractions of mathematicians.

When he studied in Bonn, the young Van 't Hoff no doubt also became acquainted with Kekulé's lecture demonstration models, mentioned in Chapter 2, which allowed more easily the direct representation of triple bonds between carbon atoms. The importance of Kekulé's models in Van 't Hoff's thinking has been noticed before, but the great similarity between them and Van 't Hoff's drawings in the *Voorstel* cannot be overemphasized. The two principal differences between them – which resulted in the true innovations of the *Voorstel* – are that Van 't Hoff, first, used the tetrahedral arrangement to explain the appearance of the physical property of optical activity, and second, considered the tetrahedron as a graphic, literal representation of the arrangement of valences around the carbon atom, and drew the tetrahedron surrounding the valences. Finally, according to Van 't Hoff's own account, the idea

---

[5] For detailed studies of Wurtz, see Ana Carneiro, 'Adolphe Wurtz and the Atomism Controversy', *Ambix*, 1993, *40*, 75–95, and Alan J. Rocke, *Nationalizing Science: Adolphe Wurtz and the Battle for French Chemistry*, Cambridge, MA, MIT Press, 2000.

[6] Ramberg and Somsen, 'The Young J. H. van 't Hoff'.

[7] Ernst Cohen, *Jacobus Henricus van't Hoff: Sein Leben und Werken*, Leipzig, Akademische Verlagsgesellschaft, 1912, and 83–4.

[8] Stellingen 30, J.H. van 't Hoff, *Bijdrage tot de Kennis van Cyanzijnzuur en Malonzuur*, Utrecht, P.W. van de Weijer, 1874. Translation published in Ramberg and Somsen, 'The Young J. H. van 't Hoff'.

of the tetrahedral carbon atom was directly inspired by Wislicenus' extensive 1873 papers on lactic acid.

After exploring the background to the appearance of the *Voorstel*, we are ready to look at its content. First of all, the *Voorstel* was a completely theoretical work, meaning that it provided no new empirical evidence for Van 't Hoff's theory. Rather, Van 't Hoff offered a reinterpretation of the known facts of organic chemistry. The purpose of the pamphlet was to present 'a few thoughts' that would 'lead to a discussion'. These thoughts were conceived because 'It is more and more apparent that the current constitutional formulas are incapable of explaining certain cases of isomerism; perhaps this is due to the lack of a more definite pronouncement about the actual arrangement of atoms'.[9] His method for making this pronouncement on the 'actual arrangement' of atoms was surprisingly simple. Implicitly, he assumed that the bonds between atoms must be rigid and not capable of changing places, a concept that could very well have come from the rigidity of the bonds in Kekulé's physical models. He then proposed particular three-dimensional arrangements for the valences, counted the number of theoretically possible isomers with those arrangements, and compared that number to the number of known isomers. For example, the four groups attached to a central carbon atom could occupy the same plane. With this arrangement, a disubstituted carbon atom, dichloromethane for example, could exist in two spatially different forms (Figure 3.1(a)). Because only one dichloromethane had ever been isolated, this arrangement must be incorrect. At this point Van 't Hoff introduced the tetrahedron:

> A second assumption brings theory and fact into agreement, that is, by imagining the affinities of the carbon atom directed towards the corners of a tetrahedron whose central point is the atom itself ... *in cases where the four affinities of the carbon atom are saturated with four mutually different univalent groups, two and not more than two different tetrahedra can be formed, which are each other's mirror images, but which cannot ever be imagined as covering each other, that is, we are faced with two isomeric structural formulas in space*. (Figure 3.1(b))[10]

Van 't Hoff named this tetrasubstituted carbon atom 'asymmetric' and ascribed the appearance of optical activity in a compound to the presence of at least one asymmetric carbon atom in its structure. He justified this hypothesis by listing optically active compounds that possessed such carbon atoms (the first on his list was lactic acid), and noted that in several reactions of optically active compounds, the removal of an asymmetric carbon atom resulted in the loss of optical activity.

The second and third parts of the *Voorstel* were a consequence of two assumptions: (1) Kekulé's theory of multiple bonding, and (2) the assumption that all carbon atoms,

---

[9] J.H. van 't Hoff, translation in Ramberg and Somsen, 'The Young J. H. van 't Hoff', p. 67.
[10] Ibid., pp. 67–8. Van 't Hoff's emphasis.

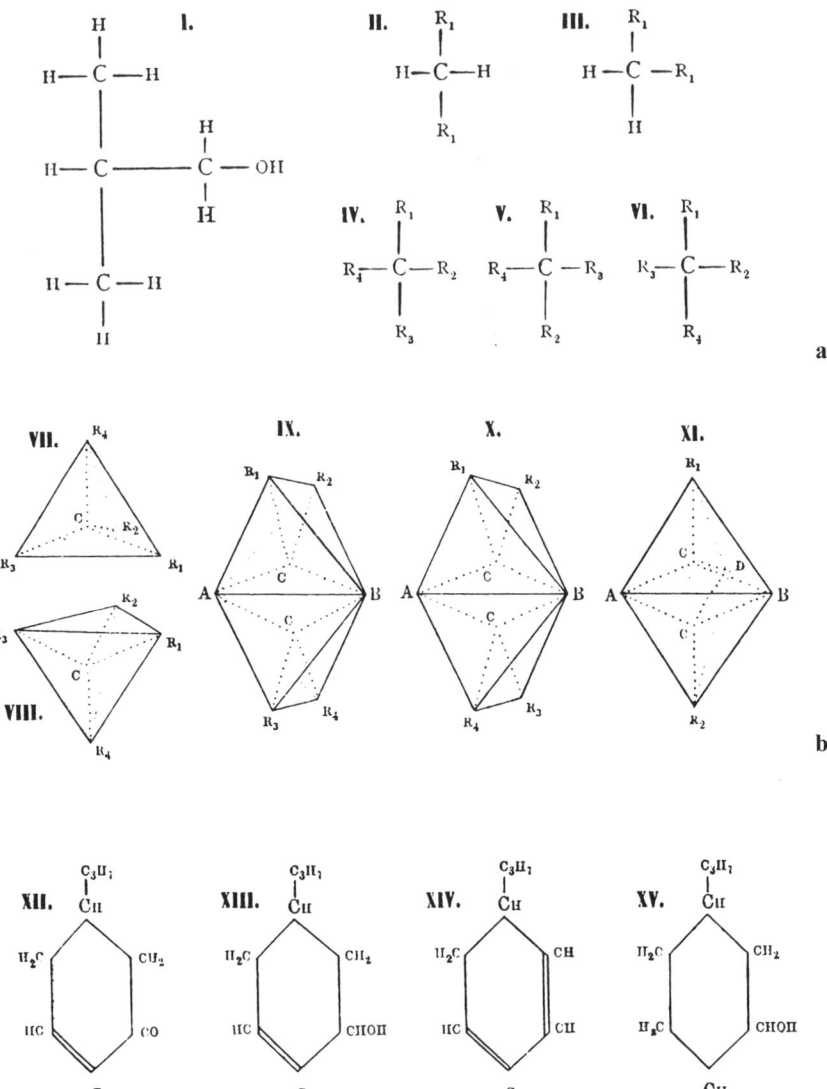

3.1 (a) The possible isomers for square planar geometry of a disubstituted carbon atom (IV, V and VI). (b) The enantiomorphic tetrahedron created by tetrasubstitution (VII and VIII). (c) Van 't Hoff's model of the carbon–carbon double bond (IX and X) and the triple bond (XI).

even those that were not asymmetric, had their affinities directed toward the corners of a tetrahedron. The second part considered 'The influence of the new idea on compounds with doubly bound carbon atoms'.[11] The double bond could be modeled by supposing that the two tetrahedra shared an edge (Figure 3.1(c)). Should R_1 and R_2, and R_3 and R_4 be different at the same time, two isomers with the same structure would be possible. As examples, Van 't Hoff mentioned maleic and fumaric acid and the other unsaturated acids. According to his theory, these acids, which had been difficult to explain in structural terms, became spatial isomers. A carbon–carbon triple bond could be modeled by assuming that two tetrahedra shared a face. Unlike the other two parts, this model did not lead to a reinterpretation of known chemistry, and Van 't Hoff wrote only one paragraph about it. In the Conclusion, Van 't Hoff noted that the arrangement of the atoms in the active molecule was entirely analogous to the arrangement of molecules in the optically active crystals (as evidenced by enantiomeric forms) discovered by Pasteur, and recently by the crystallographic chemist Karl Rammelsberg. This was the only portion of the *Voorstel* in which Van 't Hoff mentioned either planes of symmetry or Pasteur.

In comparison to the 1875 and 1877 revisions and expansions of his theory that contained numerous elements of molecular dynamics, the *Voorstel* appeared as a modest request, thoroughly grounded in the goals of nineteenth-century chemistry, to extend the available means of explaining isomerism. Like all chemists since Berzelius, Van 't Hoff wanted to explain the existence of isomers by appealing to the difference in the arrangement of atoms in the molecule. But for the first time, Van 't Hoff supposed that 'arrangement' could mean *spatial* arrangement, instead of merely the sequence of connections between atoms, and that spatial arrangement was directly related to the structural formulas, which became iconic images. Like a topographic map, in which the lines of elevation allow the visualization of certain physical characteristics of a landscape, Van 't Hoff supposed that the physical characteristics of the molecule could be inferred from structural formulas by assuming the tetrahedral arrangement of valences in each carbon atom. He could therefore 'translate' a molecule's chemical structure into its physical form, and the epistemological basis for this claim was similar to that of the original structure theory – the ability to account for cases of isomerism.

Le Bel's Article (November 1874)

While we can recount Van 't Hoff's early life in some detail, we lack a comparable account of background material from Le Bel's student years. The only significant, but still sketchy biography was written in 1947 as an introduction to Le Bel's collected

---

[11] Ibid., p. 71. This passage was emphasized in the original.

papers.[12] Le Bel was born in 1847, five years before Van 't Hoff, to a wealthy family in Pechelbronn, an Alsatian city in France. He studied at the Ecole Polytechnique between 1865 and 1867, and successively at the Collège de France and the Ecole de Médicine, where he was assistant to Wurtz and met Van 't Hoff. Unlike Van 't Hoff, Le Bel never held an academic post, and wrote and worked in a variety of areas including petroleum chemistry, crystallography, fermentation, prehistory and cosmogony.

We cannot precisely determine when Le Bel conceived of the relationship between optical activity and chemical structure, but we can only guess that it was shortly before the publication of his paper in November 1874. It was titled 'On the Relationships that Exist Between the Atomic Formulas of Organic Bodies and the Rotatory Power of their Solution' (henceforth referred to as 'Rotatory Power').[13] As the title of the paper suggested, Le Bel wanted to illustrate a correlation between atomic formulas (not structures explicitly) and optical rotation, and second, he wished to suggest a plausible means of ascertaining the specific form of molecular asymmetry that could cause the optical activity. As a student of Wurtz, a lone advocate for the structural theory in France, Le Bel adopted the axioms of structural chemistry, and used Crum Brown's representation throughout the paper. His approach to molecular morphology would combine this structural perspective with the extensive tradition of French crystallography described in Chapter 2, in which molecular form was intimately tied to external crystal form. Unlike Van 't Hoff, Le Bel did not suggest a ubiquitous form for the carbon atom and construct the molecular form; rather, he considered the molecule as a whole.

Le Bel began by describing a thought experiment on a molecule containing a formula of type MA_4, and offered two general principles. According to the first general principle, if three of the groups on MA_4 were replaced by three different groups, resulting in four different groups on the carbon atom, the 'body obtained will be asymmetric', and the groups will 'form a structure which is enantiomorphous with its reflected image'.[14] Such compounds derived from the type MA_4 would then be optically active. This is one of the more cryptic statements in 'Rotatory Power', for the production of an asymmetric object by this method is not at all obvious from the text – it is made as a declarative statement without diagrams, justification or clarification. It seems reasonable that Le Bel was *assuming*, and not concluding, that the resulting molecule was asymmetric, and would point out the final form later. The

---

[12] M. Marcel Delépine, ed., *Vie et Oeuvres de Joseph-Achille Le Bel*, Paris, Dupont, 1947.

[13] Joseph Achille Le Bel, 'Sur des relations qui existent entre les formules atomiques des corps organiques et le pouvoir rotatoire de leurs dissolutions', *Bull. soc. chim.*, 1874, *22*, 337–47. The English translation by G.M. Richardson is on pp. 161–71 in Otto T. Benfey, ed., *Classics in the Theory of Chemical Combination*, New York, Dover Publications, 1963. All subsequent translations are from this volume.

[14] Le Bel in Benfey, *Classics*, p. 162.

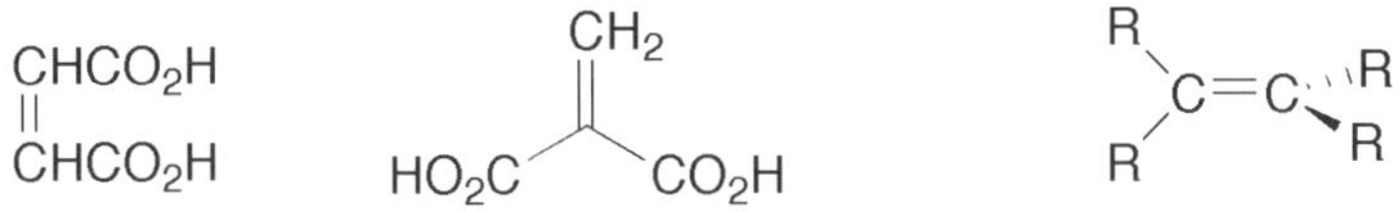

3.2 (a) Le Bel's possible structures for maleic and fumaric acid. (b) The hemihedral quadratic pyramid, one possible geometry for ethylene suggested by Le Bel.

first principle, he noted, would have two exceptions: first, if all four groups contained a plane of symmetry, or second, if one of the groups was composed of the same group as that with which it was combined, the compound would be inactive. The second exception was an oblique reference to the meso form of tartaric acid described by Pasteur, and also is not clear until Le Bel mentions the isomers of tartaric acid.

According to the second general principle, substitution of only one or two groups on MA_4 would lead to either a symmetric or asymmetric product, depending on whether symmetry was present originally. If symmetry was present before the substitution, disubstitution would create a molecule that was inactive. Once again, Le Bel stated this second principle as an assumption without any clarification, but at this point, three pages into the paper, Le Bel concluded that if these two principles are valid for the type MA_4, then 'we are obliged to admit that the four atoms A occupy the angles of a regular tetrahedron, whose planes of symmetry are identical with those of the whole molecule MA_4'.[15] Having introduced the molecular form that could produce asymmetry in the type MA_4, Le Bel applied his principles specifically to organic chemistry, using methane (CH_4) as the fundamental type MA_4, and listed many examples of methane substitution derivatives that fell under the first general principle. Notably, Le Bel also chose lactic acid as his first example of an optically active compound.

After establishing a geometry for saturated compounds, Le Bel attempted to apply his principles to unsaturated compounds as derivatives of ethylene with the general formula C_2H_4. If the four hydrogens were planar, trisubstitution derivatives would be inactive (by the first exception to the first principle), but Le Bel admitted that there were no known examples of trisubstituted ethylene derivatives to support this conclusion. Disubstitution of ethylene gave more uncertain results, as it could result in structurally different derivatives (either a 1,1 or a 1,2 substitution, Figure 3.2(a)). Therefore, the second general principle was not applicable, unless the hydrogen atoms were arranged in a 'hemihedral quadratic pyramid', that is, the two hydrogens

---

[15] Ibid., p. 163.

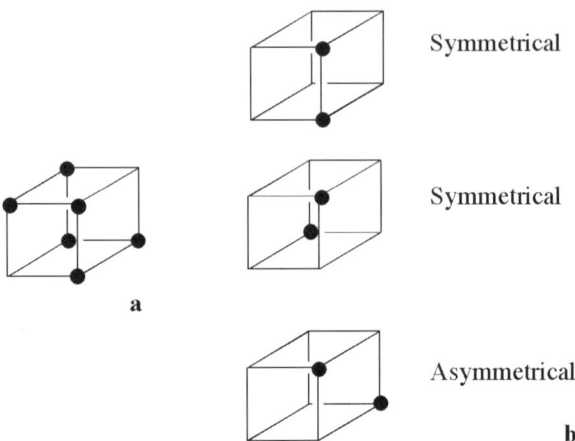

3.3 Le Bel's formulas for benzene (a), and the three possible disubstitution products (b). Note that this formula also provides an alternate three-dimensional arrangement of six equivalent hydrogen atoms in benzene that is not a hexagon.

on each carbon atom were arranged perpendicular to each other (Figure 3.2(b)).[16] In this case, disubstitution would lead to either a symmetrical (inactive) or an unsymmetrical (active) molecule.

Le Bel also attempted with less success to apply his principles to the aromatic compounds derived from benzene (C_6H_6), whose isomerism had been accounted for 'elsewhere' by the geometrical models in which the six hydrogen atoms are situated either at 'six equivalent vertices of a rhombohedron' or 'those of a vertical pyramid with an equilateral triangular base'.[17] Unfortunately, Le Bel gave no diagrams and made no reference to where these two alternatives were discussed, but as Paoloni has recently pointed out, the latter is likely a mistake, and should read 'prism' rather than 'pyramid', making it the Ladenburg formula.[18] The first possibility is more difficult to explain, but Paoloni's plausible suggestion is a cube or a rhomboid, in which the six peripheral corners are taken as the location of the hydrogens (Figure 3.3(a)).

---

[16] Because Le Bel did not include diagrams, the meaning of 'hemihedral quadratic pyramid' remains unclear. The arrangement given in Figure 3.2(b) is the most plausible arrangement, and was suggested by Benfey, *Classics*.

[17] Le Bel, in Benfey, *Classics*, p. 169.

[18] Leonello Paoloni, 'Stereochemical Models of Benzene, 1869–1875', *Bull. Hist. Chem.*, 1992, *12*, 10–24. In the standard English translation (in Benfey, *Classics*, p. 169), Richardson also recognized this mistake. He wrote 'pyramid', but placed 'prism' immediately afterwards in parentheses.

Substitution of two of the hydrogens on either the rhombohedron or the prism predicted the same result: three possible products, of which one was asymmetrical and capable of enantiomorphism, and two were symmetrical and optically inactive (Figure 3.3(b)). Unfortunately, Le Bel's argument here was not clear. First, it is not evident how the principles derived from the type MA_4 could be applied to benzene derivatives with a molecular formula C_6H_6. Second, Le Bel used a somewhat unorthodox meaning of substitution in benzene derivatives that also makes his argument confusing.[19] It also seems certain that the success of his principles was hindered by the lack of existing optically active aromatic compounds.

Simultaneous Discovery?

The striking similarities between Van 't Hoff's and Le Bel's papers, together with their near simultaneous and independent appearance, have made them a prime example of a 'simultaneous' discovery in science. The existence of simultaneous discoveries, such as energy, valence, or oxygen, has been used in support of scientific realism, that such discoveries are the uncovering of how nature truly is, and therefore an 'inevitable' outcome of historical events. In 1975, Gay used the specific example of Van 't Hoff and Le Bel for such a purpose.[20] Arguing for or against realism would take us far from our purposes here, but we can explore what exactly the phrase 'simultaneous discovery' means in the context of Van 't Hoff and Le Bel.

First of all, we must ask ourselves what was actually discovered at the same time: what characteristics – theoretical or methodological – do the *Voorstel* and 'Rotatory Power' possess in common? First, and most important, Van 't Hoff's 'asymmetric carbon atom' and Le Bel's 'first and second general principles' were different ways of making the same correlation between the physical property of optical activity and chemical structure. Both theories allowed them to predict whether any existing or non-existing saturated organic compound would have an effect on polarized light. Both explained this correlation by the existence of a tetrahedral, asymmetric arrangement of atoms or radicals surrounding the carbon atom. The number of organic compounds whose optical rotation was known was still small, but had become large enough that both Van 't Hoff and Le Bel, nearly simultaneously, could make a reasonable correlation between optical activity and chemical structure. We should also note that both Van 't Hoff and Le Bel suggested the tetrahedral

---

[19] Paoloni ('Stereochemical Models', p. 16) argues that Le Bel did not understand the difference between substitution and addition reactions. Although Le Bel did not use 'substitution' in the way that Kekulé had intended it for the aromatic compounds, there is an internal logic to how he used the term.

[20] Hannah Gay, 'The Asymmetric Carbon Atom: (a) A Case Study in Independent Discovery; (b) An Inductivist Model for Scientific Method', *Stud. Hist. Phil. Sci.*, 1978, *9*, 207.

arrangement before introducing examples to support it, and therefore – at least for saturated compounds – both made a primarily deductive argument.

Yet, even after acknowledging this significant similarity, on the whole the two articles presented quite different theories.[21] Van 't Hoff introduced and justified the tetrahedron as a theory that more accurately accounted for the number of possible isomers. The *Voorstel* offered essentially an expansion of the explanatory power of structural chemistry, and a method for the direct visualization of the three-dimensional properties of all molecules containing carbon atoms. Van 't Hoff's assumption that *all* carbon atoms were tetrahedral, whether asymmetric, unsaturated or neither, and his concern for explaining cases of isomerism by 'extending the structural formulas into space' led also to the prediction of spatial isomers in unsaturated molecules. This predictive characteristic gave the *Voorstel* a thoroughly deductive character that is absent in 'Rotatory Power'. Finally, it is important to recognize how Van 't Hoff used the phenomenon of optical activity. Although he made a correlation between the 'asymmetric carbon atom' and optical activity, optical activity itself was not important, except as a marker for distinguishing isomers, a characteristic physical property like a melting or boiling point. Pasteur's study of optical activity, and therefore molecular asymmetry in the broad sense, was therefore only of peripheral interest in the *Voorstel,* and Van 't Hoff mentioned Pasteur only in passing in the last paragraph. This is substantiated by his initial rejection of the planar formulas for the carbon atom; Van 't Hoff did *not* reject them because they were symmetric and would have no effect on polarized light, but because they predicted too many isomers.

All of these characteristics of the *Voorstel* contrast sharply with those of 'Rotatory Power'. Le Bel introduced the tetrahedron by arguing from the possible symmetries or asymmetries in the molecular type, not from a concern with accounting for or predicting cases of isomerism. His principal aim was to develop a correlation between structure and optical activity, and Pasteur's conclusion that optical activity was the primary indicator of asymmetry at the molecular level was the crucial starting point for, and central to, Le Bel's paper. 'Rotatory Power' was therefore an extension of the traditional French concern with crystallography. Le Bel initially introduced the tetrahedral arrangement for saturated compounds in a deductive manner, but then considered the other classes of compounds separately for possible symmetries or asymmetries in the molecular type. The *overall* character of 'Rotatory Power' is therefore more 'inductive' than the *Voorstel*, although labels such as inductive and deductive do not apply well to this case. 'Rotatory Power' has been described as more 'abstract' than the *Voorstel*, meaning that there is less of a sense of immediate applicability or predictability. The difficulty in understanding 'Rotatory Power'

---

[21] These differences are also outlined in H.A.M. Snelders, 'J.A. Le Bel's Stereochemical Ideas compared with Those of J.H. van't Hoff', pp. 66–73 in O.B. Ramsay, ed., *Van't Hoff-Le Bel Centennial*, Washington, DC, American Chemical Society, 1974, and Gay, 'The Asymmetric Carbon Atom'.

also derives from the complete lack of diagrams for visualizing the asymmetries Le Bel wanted to express. The abstract character results, I think, because Le Bel was struggling towards a general *method* for establishing molecular asymmetry, and not arguing for a ubiquitous atomic form. That is, Le Bel did not intend to regard all structural formulas automatically as iconic representations. Formulas with the compositions MA_4 may be iconic, but M_2A_4 or aromatic compounds may not be. His method required treating the molecule as a single unit, without a universal shape for the carbon atom.

These differences in approach resulted in differences in the predictive power of each theory that had a significant effect on their future development. In the long run, the deductive theory offered by Van 't Hoff proved more successful, as it made significant predictions that could be tested. Van 't Hoff and his Dutch colleague Bremer would test the asymmetric carbon atom in 1875 (described below), and the model for multiple bonding would create large, successful research programs in the 1880s and 1890 (the subject of the following chapters). Le Bel's inductive approach required the accumulation of more structural data, so a conclusion about the form of molecules could be induced.

After elaborating on the similarities and differences between Van 't Hoff and Le Bel's papers, we can return to the question of simultaneous discovery. Was the tetrahedron a natural outcome of existing historical conditions? I think the answer to this question is a qualified 'No'. Both Van 't Hoff and Le Bel offered a convenient correlation between chemical structure and optical activity, but they offered divergent techniques for arriving at the three-dimensional properties of molecules. That is, they agreed on one fundamental principle – the correlation between optical activity and structure in saturated compounds – but the inherent theoretical and methodological differences in the remainder of both papers do not suggest an 'inevitable' conclusion that the valences in all carbon atoms were tetrahedral. Van 't Hoff offered a beautifully conceived hypothesis that correlated many known but anomalous facts of organic chemistry by treating chemical structures as iconic images. But given the still relatively low number of known compounds in 1874 that displayed optical activity (particularly among the aromatic and unsaturated compounds), Le Bel's approach seems equally promising. Looking ahead from 1874, molecules of organic compounds belonging to different classes could conceivably *not* have had a tetrahedral geometry.

The Evolution of Van 't Hoff's Theory, 1874–1877

The Dissertation, 1874

Although the exact number of printed copies of the *Voorstel* is now unknown, it was certainly small. The fact that it was written in Dutch must also have limited the distribution of the existing copies within the Netherlands, and it seems likely that Van 't Hoff sent copies of the pamphlet to chemists at Dutch universities and technical schools. Buys-Ballot at Utrecht and Tjaden Modderman at Groningen received copies some time before October 1874.[22] During the remainder of 1874, Van 't Hoff completed the work for his dissertation in Eduard Mulder's laboratory, and successfully defended it on 22 December 1874. The dissertation, *Contributions to the Knowledge of Cyanic and Malonic Acid* (*Bijdrage tot de kennis van cyanazijnzuuren en malonzuur*), was on the whole unremarkable, an ordinary nineteenth-century Dutch chemistry dissertation, entirely typical of those directed by Mulder, with a dry and straightforward experimental account of the synthesis and properties of one or more compounds. Unlike the *Voorstel*, the entire text was extensively referenced, as any good dissertation should be, to demonstrate Van 't Hoff's familiarity with the current literature. The text, although extensive, was limited to a discussion of the background literature and the empirical results of Van 't Hoff's own research, and the text did not mention the ideas in the *Voorstel*.

The last four pages of Van 't Hoff's dissertation, however, consisted of a series of 32 *stellingen* (theses) to be defended. A requirement for Dutch dissertations, the *stellingen* section was simply a set of numbered statements resembling in form Newton's queries. Van 't Hoff's *stellingen* covered a variety of topics from organic chemistry to gravitation to education, and of the 32, 6 related directly or indirectly to the ideas in the *Voorstel*.[23] It seems most likely that, given the poor reception of the *Voorstel* and the priority that Mulder gave to empirical work, Van 't Hoff simply wanted to complete his dissertation requirements as quickly as possible, and avoided any controversy by choosing an unremarkable topic for his dissertation, relegating any speculations to the *stellingen* section.

After completing his dissertation, Van 't Hoff returned to Rotterdam to live with his parents, and looked unsuccessfully for a teaching position at a *Hoogere Bergerschole* (secondary school) in the towns of Dordrecht, Leeuwarden and Breda. An impression

---

[22] Tjaden Modderman to neighbors (unknown) of Van 't Hoff's parents, 17 October 1874, J.H. Van 't Hoff Papers, Ms. 74, Special Collections, Milton S. Eisenhower Library, Johns Hopkins University.

[23] Thirteen of the 32 *Stellingen* are given in English translation in Ramberg and Somsen, 'The Young J. H. van 't Hoff'. Four of these were also published previously: Theses 14 and 15 in H.A.M. Snelders, 'The Reception of J.H. Van't Hoff's Theory of the Asymmetric Carbon Atom', *J. Chem. Ed.*, 1974, *51*, 2–7; Theses 7 and 32 in H.S. van Klooster, 'van't Hoff (1852–1911) in Retrospect', in *Proceedings of the International Symposium on the Reactivity of Solids, 1952*, Goetberg, 1954, 1095–100, p. 1096.

of Van 't Hoff's presence during a job interview in December 1875 was reprinted in Cohen:

> Yesterday, Dr. J. H. van 't Hoff of Rotterdam was here. He had sent me earlier his dissertation and other publications, from which it is evident that he is chiefly concerned with organic chemistry. He is much less interested in natural history. As far as I can tell immediately, he gives the impression of an inventor. He broods, he is immersed in his discovery that in the carbon atoms that rotate polarized light, the carbon atom is probably a tetrahedron whose corners indicate the directions of the affinities. He looks sloppy. Colleagues who saw him in the 'club' protested his election and said that is no man for Breda. I fear that he will be very absent-minded and have difficulties with the pupils.[24]

This passage tells us much about the youthful Van 't Hoff, who throughout 1875 had a very uncertain career in teaching and in chemistry. His youth and apparently taciturn, brooding personality and unkempt appearance certainly contributed to an overall negative impression on the faculty at Breda. After searching for work for a year in Rotterdam, he returned to Utrecht with a stipend from his father, and advertised as a private tutor in chemistry and physics. Finally, in March 1876, Van 't Hoff's future became more certain when the Director of the Imperial Veterinary School in Utrecht, Thomas MacGillavry, suggested appointing Van 't Hoff to a new position teaching chemistry and physics.

La chimie dans l'espace, 1875

While 1875 was a frustrating year for Van 't Hoff's job prospects, it was exciting in the development of his theory. Because of Buys-Ballot's interest, a full French translation of the *Voorstel* had appeared in the *Archives Neérlandaises* in the Fall of 1874 with the shorter title 'Sur les formules de structure dans l'espace', along with a new set of diagrams.[25] This translation was published in an abridged version without diagrams or formulas in March 1875 in the more accessible *Bulletin de la société chimique de France*, with minor revisions, most notably an acknowledgement of Le Bel's November article that recognized the more general approach Le Bel took to understanding the relationship between molecular form and optical activity.[26] This version received a favorable response from Wurtz, and a critique from Berthelot, but still these versions did not attract the attention Van 't Hoff desired. Some time during the first part of 1875, he expanded and rewrote the ideas of the *Voorstel* into a book that appeared in May with the still shorter title *La chimie dans l'espace*, published at

---

[24] Cohen, *Jacobus Henricus van't Hoff*, p. 117. Translated by Cohen from Dutch into German. My translation from the German.

[25] J.H. van 't Hoff, 'Sur les formules de structure dans l'espace', *Archives Neérlandaises des sciences Exactes et Naturelles*, 1874, *9*, 445–54.

[26] J.H. van 't Hoff, 'Sur les formules de structure dans l'espace', *Bull. soc. chim.*, 1875, *23*, 295–301.

his own expense in Rotterdam by an old schoolfriend.[27] In order to further publicize the theory, he mailed copies of the book and sets of cardboard molecular models, presumably handmade by Van 't Hoff, to prominent chemists. In a footnote on page 3, he mentioned that model sets had been sent to Adolf von Baeyer, Marcelin Berthelot, Aleksandr Butlerov, Frankland, Louis Henry, August von Hofmann, Kekulé, Adolphe Wurtz and Wislicenus, and if the reader did not have access to these models, additional sets were available on request. The surviving correspondence in the Van 't Hoff archives suggests that the models were sent in June 1875.

Comparing *La chimie dans l'espace* with the *Voorstel*, we immediately see many differences. It was clearly a more mature work that expanded on the implications of the tetrahedron, and resulted in a work that was nearly four times longer at 43 pages with three plates of drawings. This time, Van 't Hoff placed his name on the title page, along with an epigraph containing Wislicenus' 1869 statement about the need for a spatial conception of molecules given at the Innsbruck *Naturforscherversammlung*. He expanded the first few sentences of the *Voorstel* that had introduced the initial justification for his theory into a two-page preface (translated in full in Appendix 2), written in a very forceful, almost defensive style that makes plain his frustration at waiting 'in vain' for a discussion of the ideas in the *Voorstel*, even after publication of the French translations. The preface also makes clear the importance Van 't Hoff gave to the 'indisputable' nature of 'proofs' expressed in terms of numbers. For organic chemistry, this meant the explanation and prediction of the possible number of isomeric compounds. In a passage emphasized in italics, Van 't Hoff noted both that the 'number of existing isomers surpasses those predicted by theory', and that 'current theory is powerless to predict certain isomers'. He also emphasized the importance of hypotheses in directing scientific inquiry:

> each new hypothesis, if I may say so here, ought to pass through two very distinct phases; at first it is a question of seeing if it presents in its interpretation of the known facts an advantage over that which exists. Then, if it has received this support it is still necessary that experiment demonstrate the truth of its predictions.[28]

As he did in the French version of the *Voorstel* that appeared in the *Bulletin de la société chimique*, he recognized that in November 1874, Le Bel had 'pronounced himself in favor of one part of my views'.

In the first two paragraphs of Chapter 1, Van 't Hoff emphasized what he saw as the two principal shortcomings of chemical theory: 'Modern chemical theory has two weak points. It discusses neither the relative positions that the atoms take up in a molecule, nor their types of motions.' Van 't Hoff then explicitly discussed his

---

[27] Cohen, *Jacobus Henricus van 't Hoff*, p. 88.
[28] J.H. van 't Hoff, *La chimie dans l'espace*, Rotterdam, Bazendijk, 1875, pp. 3–4.

assumptions about the molecular dynamics of molecules that had been left implicit in the *Voorstel*:

> As a result, doubt still prevails over all questions of chemical statics and dynamics; approaching this rationally indicates that we must initially work in the first direction. One could object, however, that each movement changes the form of a system that is fixed. But the atoms must have a periodic movement in a molecule (if the quality of the molecule is a function of the movement of the atoms, we cannot have equality in one without periodicity in the other). We could therefore represent the relative positions of these atoms in one of the phases of their movement.[29]

Given this assumption about the set of periodic atomic motions about an equilibrium position, Van 't Hoff proceeded to his principal argument for ascertaining the spatial arrangement of atoms. The argument given in *La chimie dans l'espace* repeats closely the 'isomer counting' system given in the *Voorstel*. Van 't Hoff introduced the asymmetric carbon atom and its consequences for isomerism in the alkenes. In addition, he offered three new hypotheses that flowed from considering the tetrahedron. First, he recognized that the exact number of possible isomers could be calculated from the number of asymmetric carbon atoms. Because every asymmetric carbon atom doubled the number of possible isomers, the presence of n asymmetric carbon atoms predicted 2^n isomers. The possibility of internal compensation, however, could reduce the number of isomers by an amount that could also be determined mathematically. Second, it seemed clear that carbon–carbon single bonds must be capable of free rotation about the axis containing the carbon atom: 'It is then evident that [those in Figure 3.4] do not represent isomers, but the same combination in two phases of movement around the axis uniting the carbon atoms.'[30]

If the rotation were not possible, Van 't Hoff argued, the model would predict an infinite number of spatial isomers. Because these isomers had never been isolated, they must be capable of interconverting between themselves. Third, the extension, or cumulation, of double bonds in a carbon chain led to an additional series of isomers (Figure 3.5). The *cis-trans* isomerism exhibited by maleic and fumaric acid would also occur in any compound containing an odd number of cumulated alkenes, and these compounds would be optically inactive. Any compound containing an even number of cumulated dienes could exist in two enantiomeric forms, and by implication would be optically active. Van 't Hoff referred to this as the 'second case' of optical activity.

Chapters 2 and 3 contained extensive discussion of examples to support the theory presented in the first chapter. The basic argument given in Chapter 2 was identical to that in the *Voorstel*, correlating the property of optical activity with the presence of an

---

[29] Ibid., p. 5.
[30] Ibid., p. 8.

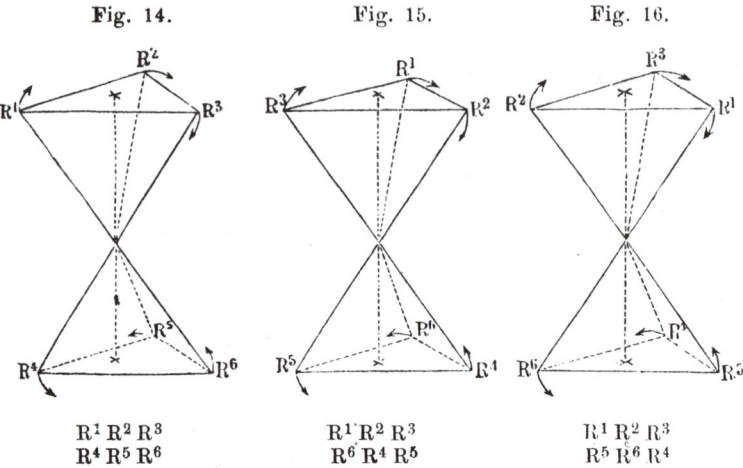

3.4 The illustration of the free rotation of carbon–carbon single bonds according to Van 't Hoff's second hypothesis.

asymmetric carbon atom. However, Van 't Hoff added an intriguing explanation of polarization that expanded on the analogy between asymmetric atomic arrangements in the molecule and asymmetric molecular arrangements in the crystal discussed briefly in the last paragraph of the *Voorstel*. Van 't Hoff argued that as the light passed through the tetrahedral arrangement of atoms, it went through a helical arrangement of the atoms surrounding the central carbon atom, producing a modification of the light wave that resulted in optical rotation (Figure 3.6). Chapter 3 examined the

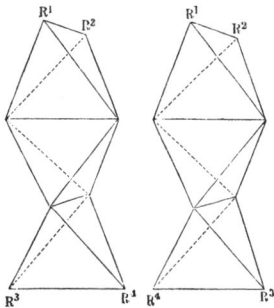

3.5 Van 't Hoff's suggested spatial isomerism resulting from the cumulation of double bonds.

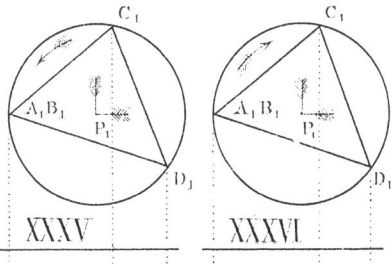

3.6 Van 't Hoff's drawing of the carbon atoms as seen by an entering ray of polarized light.

known cases of isomeric unsaturated acids as examples to illustrate Van 't Hoff's theory of unsaturation.

In the fourth and last chapter, Van 't Hoff discussed the implications of the tetrahedral carbon atom for aromatic compounds. His aim conformed to that of the rest of the book: he intended to show both how the tetrahedral distribution of valences could predict certain cases of isomerism in benzene derivatives, and how his predictions could differentiate between the two principal rival theories for the structure of benzene: Kekulé's cyclohexatriene, and Ladenburg's prism formulas. Van 't Hoff offered two modes of reasoning based on the substitution and addition products of benzene. Disubstitution of the Ladenburg formula led to three possible products (Figure 3.7). The cyclohexatriene formula predicted five disubstitution products, but by adopting Kekulé's 'oscillation hypothesis', in which the double bonds alternated between carbon atoms, the number was also reduced to three. One of Ladenburg's formulas contained two asymmetric carbon atoms, and could therefore exist as two enantiomorphic forms, called the 'third case' of optical activity by Van 't

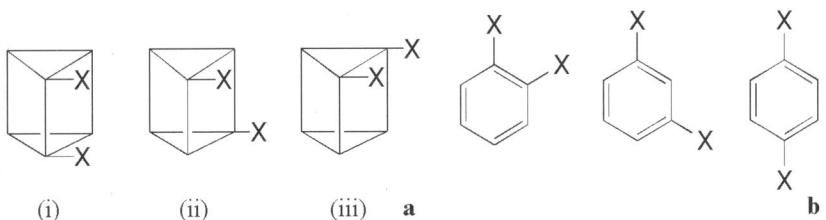

3.7 Possible substitution products of Ladenburg's (a) and Kekulé's (b) benzene formulas. Of Ladenburg's structures, (ii) can exist in enantiomeric forms.

Hoff. Finally, Van 't Hoff noted that the two models predicted different addition products.

Although different in details and much more explicit, Van 't Hoff's treatment of the substitution and addition products of aromatic compounds seems too close to Le Bel's discussion of aromatic compounds to attribute to coincidence. On the basis of these similarities, Paoloni has recently argued that the significant revisions that Van 't Hoff made between the *Voorstel* and *La chimie dans l'espace* only occurred after he read Le Bel's paper, which convinced him to consider the broader conceptions of asymmetry at the *molecular* level in addition to that at the atomic level.[31] Paoloni makes two main arguments. First, Van 't Hoff's consideration of the cumulated dienes bears a close resemblance to Le Bel's first principle for the type MA_4, in which M is an even or odd number of carbon atoms with four groups. Second, the inclusion of a chapter on aromatic compounds, its placement at the end of the book, and the exclusion of aromatic compounds from discussion of the asymmetric carbon atom in Chapter 1 mean that Van 't Hoff was led to revise his ideas only at the end of writing the book, as the earlier chapters could no longer be revised. As additional support, Paoloni cites a passage in Chapter 4 of *La chimie dans l'espace*:

> ... during this work we have come to predict the possibility of optical activity in *two more cases*; and these cases refer to *unsaturated* (carbon) *compounds* ... in order for my observation to keep all its value, the sentence [sic] 'any carbon compound' must be replaced by 'any *saturated* carbon compound'.[32]

Paoloni interprets this passage as a *correction* to the rules given in previous chapters, an unintended consequence of considering Le Bel's ideas written after the previous chapters had been completed and couldn't be revised.

Paoloni's general claim certainly seems correct, as Van 't Hoff's prediction of optical activity in the allene derivatives, and the explicit recognition of the meso compounds (optical inactivity by internal compensation) both indicate that Van 't Hoff now considered molecular asymmetry as a cause for isomerism. The similarities between Van 't Hoff's and Le Bel's treatment of aromatics is also difficult to ignore, and it also seems likely that Van 't Hoff decided to include a chapter on aromatic compounds after seeing Le Bel's paper. Aromatics were a major class of compounds, albeit less well understood, and if his theory were to be comprehensive and successful, they should have been included.

Determining unambiguously the influences on a published work is notoriously difficult, however, and is made even more difficult in the case of *La chimie dans*

---

[31] Paoloni, 'Stereochemical Models'. Paoloni's argument was prompted by and expands on a comment given in an earlier article by R.B. Grossman, 'Van't Hoff, Le Bel, and the Development of Stereochemistry: A Reassessment', *J. Chem. Ed.*, 1989, 66, 30–3.

[32] Van 't Hoff, *La chimie dans l'espace*, p. 42, Paoloni's translation and emphasis.

l'espace, a work in which Van 't Hoff gave few references. Although we cannot definitively exclude the possibility that the chapter on aromatics was unintended even as Van 't Hoff was writing the first three chapters, or that the passage Paoloni cites is a correction, we do not know from the existing documentation the order in which Van 't Hoff wrote the chapters, or when he began making revisions to the *Voorstel*. The earliest mention in print of Le Bel's paper by Van 't Hoff is in the March 1875 version that appeared in the *Bulletin de la société chimique de France*. This paper was presented at the 5 March meeting of the society by Van 't Hoff's colleague and friend in Wurtz's laboratory, Arthur Henninger, and published in the *Bulletin* shortly thereafter. It seems likely that Van 't Hoff learned about Le Bel's paper from Henninger (who also collaborated with Le Bel for a publication on a distillation apparatus) some time between November 1874 and February 1875. Van 't Hoff may also have learned about Le Bel's theory, and may have incorporated the general ideas about molecular asymmetry and aromatic chemistry in *La chimie dans l'espace* from Le Bel himself, as the two had been in correspondence with each other. It also seems likely that Van 't Hoff did not begin revising the *Voorstel* in earnest until January of 1875, after he had completed his dissertation and moved back to Rotterdam. In other words, given the chronology, it seems equally plausible that Van 't Hoff did not start writing *La chimie dans l'espace* until well after becoming acquainted Le Bel's article.

Moreover, the book's organization suggests a greater coherence than Paoloni has suggested. The first chapter is an introduction to how the tetrahedral carbon atom can give rise to cases of isomerism – either the enantiomorphism of the asymmetric carbon atom, or the isomerism created by alkenes and cumulated double bonds. Van 't Hoff does not mention optical activity until Chapters 2 and 3, in which he 'matched the theory with the facts' by providing the examples of known compounds that fit under the new theory, correlating optical activity with the presence of an asymmetric carbon atom, and predicting activity in a compound with an even number of cumulated alkenes. Chapter 3 predicts that a compound with an even number of cumulated dienes will be optically active (although no examples existed), and that compounds with an odd number of cumulated double bonds will be inactive, but be capable of *cis-trans* isomerism. Van 't Hoff then gave examples of compounds with double bonds that would fall into this category. Chapter 4 on aromatic compounds does seem more out of place, as aromatics were not considered in the first chapter, but as Van 't Hoff noted at the beginning of Chapter 4, the constitution of aromatic compounds was still in dispute, and it was unclear whether carbon atoms in the aromatic system could be asymmetric – those in the prism formula can be asymmetric, whereas those in the cyclohexatriene system cannot. Although Van 't Hoff ultimately came down in favor of Kekulé's cyclohexatriene formula, he nevertheless decided to consider aromatic compounds separately, as the constitution had not been completely settled, and he could not provide examples as he did in Chapters 2 and 3.

In the absence of Van 't Hoff's dated draft manuscripts or the correspondence between Van 't Hoff and Le Bel, any attempt to establish the chronological sequence of the writing of *La chimie dans l'espace* remains difficult to substantiate, even if it seems highly likely that significant portions of *La chimie dans l'espace* bore a large debt to Le Bel. It also seems likely, as Ramsay has noted, that Van 't Hoff's prediction of spatial isomers in the cumulated dienes is not inherently obvious without using three-dimensional models, and it seems more plausible that Van 't Hoff would stumble onto these asymmetries while manipulating his cardboard models, rather than from an abstract idea of molecular asymmetry gained from Le Bel.[33] It may be that once he discovered the isomerism in the cumulated dienes, he saw that it correlated nicely with Le Bel's first principle.

Early Experimental Work

At about the same time that *La chimie dans l'espace* was completed, Gustav Bremer began a series of experiments to test the relationship between optical activity and constitution outlined in the *Voorstel*. As the first piece of new empirical evidence in favor of van 't Hoff's theory, Bremer's results would appear in his October 1875 doctoral dissertation, *A Dextrorotatory Malic Acid*. Bremer and Van 't Hoff had been fellow students in Mulder's laboratory in Utrecht during the fall of 1874, but had both since returned to their parents' homes, Van 't Hoff in Rotterdam and Bremer in Assan, a small town in the northern Netherlands. Bremer's experiments were chronicled in their correspondence, from which seven of Van 't Hoff's letters have survived, dating from December 1874 to late 1875. In 1927, W.P. Jorrissen printed transcriptions of the letters, and in 1985, Fischmann arranged the letters in a new chronological sequence, translated portions into English, and used the letters to reconstruct Bremer's experiments.[34] It is useful here to recount the experiments, for they shed light on how Bremer and Van 't Hoff initially conceived of the scope of the theory in 1875.

At Van 't Hoff's urging, Bremer first attempted to synthesize the artificial dextrorotatory malic acid from dextrorotatory tartaric acid using hydrogen iodide, and the resulting malic acid was treated with ammonia to facilitate purification by crystallization as the ammonium salt. He then compared the optical rotation of this ammonium salt with the ammonium salt of the naturally occurring levorotatory malic acid. As they were enantiomorphous molecules, they should have had rotations of equal magnitude but opposite signs. Bremer found the two salts to differ in sign

---

[33] O.B. Ramsay, 'Molecular Models in the Early Development of Stereochemistry. I: The van't Hoff Model. II: The Kekulé Models and the Baeyer Strain Theory', pp. 74–96 in Ramsay, *Van't Hoff-Le Bel Centennial*.

[34] W.P. Jorissen, 'Eenige brieven van van't Hoff (1874–1875)', *Chemische Weekblad*, 1924, *21*, 495–501, and Fischmann, 'Reconstruction'.

as expected, but he also found a slight difference in magnitude. The unnatural dextrorotatory salt had a rotation of +7.935°, while the natural levorotatory salt had a rotation of −5.939°. In a letter to Bremer, Van 't Hoff explained the observed rotation by noting that in malic acid *two* ammonium acids salts were possible, because it contained two acid groups:

> Your determinations were carried out with the acid ammonium salt, and it will be clear that each malic acid has *two* acid ammonium salts:
>
> $CO_2H \cdot CH(OH)CH_2 \cdot CO_2NH_4$ (a)
> and $CO_2NH_4 \cdot CH(OH) \cdot CH_2 \cdot CO_2H$ (b)
>
> Now it is conceivable that one acid gives ammonium salt (a), and the other (b); the prediction of equal and opposite acivities will then no longer apply, as the surrounding groups are no longer the same …[35]

The difference in magnitude therefore resulted from the presence of two different asymmetric carbon atoms that differed in the structural position of the ammonium group. The difference in magnitude was of little consequence, however, for the difference in sign between the natural and unnatural malic acid salts was enough for a provisional demonstration of Van 't Hoff's principle of the tetrahedral carbon atom. Bremer published a short article in the *Berichte* in June 1875, omitting the magnitude of the sign and ignoring the potential conflict with the theory.[36]

In July, Bremer succeeded in producing optically inactive malic acid from racemic acid. In an intriguing experiment, he converted this inactive malic acid into an ammonium salt, predicting that this compound would have an activity equal to the sum of the activities of the enantiomeric malic acid salts, or about +2°. The new ammonium salt from racemic malic acid had a rotation of +2.036°, apparently confirming Van 't Hoff's explanation for the discrepancy between the natural and unnatural ammonium salts of malic acid. Finally, Bremer conducted an experiment with even more striking results. He converted the ammonium salt of levorotatory malic acid into a calcium salt, and he then neutralized the remaining acid group with ammonia to give an ammonium-calcium salt (Figure 3.8). This double salt was then treated with oxalic acid to remove the calcium and release the ammonium salt. The net result was a movement of the ammonia from one acid group to the other. The resulting ammonium salt had an optical rotation of −7.816°, remarkably close to the expected value of −7.912°. By this ingenious sequence of reactions, Bremer was therefore apparently able to convert one isomeric ammonium salt of malic acid with a rotation of −5.939° into the other isomeric ammonium salt. Through Henninger,

---

[35] Van 't Hoff to Bremer, 1875. Translation by E. Fischmann, 'Reconstruction', p. 137.
[36] G.J.W. Bremer, 'Vorläufige Mittheilung über eine neue Aepfelsäure, welche die Polarisationsebene rechts dreht', *Berichte*, 1875, *8*, 861–3.

```
HOOC—CH₂—CHOH—COONH₄  ⟶      (HOOC—CH₂—CHOH—COO)₂Ca
                                              │ NH₃
                        Oxalic acid            ▼
H₄NOOC—CH₂—CHOH—COOH  ◀———     (H₄NOOC—CH₂—CHOH—COO)₂Ca
```

3.8 Gustav Bremer's conversion of dextrorotatory ammonium malate to the levorotatory ammonium malate.

Van 't Hoff reported Bremer's results in Paris to the *Société de chimique de France* on 19 November 1875, where they 'caused a considerable heehaw' among those present, including Marcelin Berthelot, Emile Jungfleisch and Eduard Grimaux.[37]

Bremer's synthesis of dextrorotatory malic acid provided some evidence for the asymmetric carbon atom, but the other results were somewhat ambiguous. The first major problem, raised by those attending the *Société* meeting, was sample purity. In the letter that related the events of the meeting, Van 't Hoff noted that 'Jungfleisch thinks that your [starting] malate 5.9° [in the last experiment] contains small amounts of inactive malate, which are eliminated during the transformation to calcium malate and the other acid ammonium salt.'[38] In other words, Bremer's conversion of one ammonium salt into another was merely a purification. Van 't Hoff also told Bremer the numerous questions Berthelot had asked about the experimental technique:

> Preparation of the acid l and d ammonium salt: how is this done, and under what conditions? Did the acid itself from mountain ash berries really have an activity of 3° 29'? Was no increase in temperature involved in the treatment? Was the same l malic acid used for the preparation of the two ammonium salts? Do the two l acid ammonium salts differ in *solubility, melting point, appearance* and *crystal form*? Where they analyzed? Are their reactions identical? Has Bremer prepared the l malic acids from the two different l acid ammonium salts, to compare them? For which calibration was the rotation determined?[39]

From Bremer's reported optical rotations, and the number and kind of questions Berthelot asked, we can ascertain some idea of the experimental complexities involved in ascertaining a precise optical rotation.

Another curious aspect of this work, touched on by Fischmann in his reconstruction,

---

[37] Van 't Hoff to Bremer, 1875, in Fischmann, 'Reconstruction', p. 146.

[38] Jungfleisch's explanation is rather confusing, as the presence of an *inactive* malate (either racemic or internally inactive) would neither increase nor decrease the optical activity of the sample. Perhaps he meant to say the dextro isomer. Van 't Hoff to Bremer, 1875, in Fischmann, 'Reconstruction', p. 147.

[39] Ibid., p. 147.

was the contradiction between Van 't Hoff's and Bremer's assumptions with Pasteur's fundamental theory of optical activity. First, Van 't Hoff explained the results of Bremer's initial experiment that the two enantiomorphic malic acids may give different ammonium salts. Two enantiomorphic molecules should, however, be in all respects identical except for the optical rotation: that is, enantiomorphic molecules should always give identical chemical products. Second, in the second experiment described above, Bremer attempted to create, apparently successfully, an optically active compound from racemic (optically inactive) malic acid, which, according to Pasteur (and still today), was impossible without the action of another asymmetric compound.

This apparent lack of knowledge of Pasteur's rules for optical activity is difficult to understand completely. Elsewhere in his letters to Bremer, Van 't Hoff clearly mentioned Pasteur's theory that all optically active molecules must exist in four isomeric forms – dextrorotatory, levorotatory, an inactive resolvable (racemic) form, and an inactive non-resolvable (meso) form. He must therefore have been somewhat familiar with it. Fischmann explained the ignorance of Pasteur's theory as a result of the initial excitement of obtaining confirming results for Van 't Hoff's theory. This might certainly be the case; no doubt this initial excitement spurred both Van 't Hoff and Bremer to continue their research. We must also remember, however, that Van 't Hoff's theory was not a descendant of Pasteur's work on the tartrates, but a derivative of structure theory. Van 't Hoff looked for explanations of isomerism in terms of structure, connectivity and arrangement, and was not concerned with confirming, refuting or extending Pasteur's theory of optical activity.

In late 1875, Van 't Hoff himself undertook his first project in 'chemistry in space', and chose to examine the optical activity of the compounds derived from storax, a resinous medicinal extract from the Asian *Liquid amber orientalis* tree. Hermann Kopp had found that the compound cinnamol (C_8H_8), derived from cinammic acid by distillation, was identical to styrol (today styrene), isolated from storax. In 1866, Marcelin Berthelot had reported that styrol was optically active, a property of which Van 't Hoff noted that 'for chemists who aren't looking for a connection between optical activity and constitution, these results aren't very peculiar'.[40] According to Van 't Hoff's theory, styrol contained no asymmetric carbon atoms, and should be optically inactive; any observed activity would be caused by an impurity. Late in 1875, Van 't Hoff began an investigation to show that styrene derived from storax was not optically active, as Berthelot had reported. He submitted his first articles in Dutch to the *Mandblaad voor Natuurwetenschappen*, and in German to the *Berichte*. Extracting 10 kilograms of storax gave 40 grams of an oil that was optically active as

---

[40] J.H. van 't Hoff, 'Die Identität von Styrol und Cinnamol, ein neuer Körper aus Styrax', *Berichte*, 1876, *9*, 5–6, p. 5.

Berthelot had reported, but with a larger rotation (-5°; Berthelot had reported -3°). The difference in rotation would be due to differences in concentration of an impurity, and 'increased my suspicion that there was a possibly a mixture present'.[41] Removing the styrene by polymerization and distilling the remainder gave a product with different chemical properties and elemental composition than styrene.[42] This distillate was optically active, and as he continued the process of removing styrene by polymerization and distillation, maximizing the purity (judged by constant boiling point), the magnitude of optical rotation increased to -30°. The new optically active compound, named styrocamphene by Van 't Hoff, possessed the formula $C_{10}H_{16}O$ or $C_{10}H_{18}O$.

Berthelot responded in a short note that repetition of his preparation of styrene from storax gave a 'pure carbide' (*carbure pur*) with a constant boiling point and an optical rotation between -3.1° and -3.4°. The variation could be ascribed to the production of a small amount of inactive styrene during distillation. He concluded by saying that the 'rotatory power of styrene is certain, and all theories incompatible with this property are convicted of inaccuracy'.[43] Van 't Hoff submitted a response on 18 August to the *Berichte*, in which he noted that heating 'so-called optically active' cinnamol precipitated 'metastyrol' and the rotation of the solution did not decrease, as the supposedly active metastyrol was removed from solution. 'Herr Berthelot's numbers,' Van 't Hoff concluded confidently, 'speak in favor of a mixture in varying amounts.'[44]

Die Lagerung der Atome im Raume, 1877

The changes between the *Voorstel* and *La chimie dans l'espace* were the largest and most significant between any two versions of Van 't Hoff's books, but they were not the last. The translation into German would create the last additions and revisions to Van 't Hoff's theory before he abandoned the subject. The translation, as is well known, was instigated by Wislicenus, who in 1872 had moved to the University of Würzburg. By 1875, Wislicenus had nearly abandoned the problem of lactic acid, or had exhausted his own interest in its chemistry. His last publication on the subject appeared in the *Berichte* in September 1875. It did not mention Van 't Hoff's theory,

---

[41] Ibid., and J.H. van 't Hoff, 'Styro-kamfer, een nieuw lichaam uit styrax', *Mandblaad voor Natuurwetenschappen*, 1876, 6, 71.

[42] The elemental composition of the oil (rotation -5.59°) was reported by Van 't Hoff as 79.82% carbon and 9.90% hydrogen. Styrene, C_8H_8, contains 92.31% carbon, and 7.69% hydrogen. Van 't Hoff to Bremer, 1875, in Jorissen, 'Eenige brieven', p. 499.

[43] Marcelin Berthelot, 'Sur le pouvoir rotatoire du styroléne', *Compte rendus hebdomadaires des séances*, 1876, 82, 441–2. J.H. van 't Hoff, 'Een rechtsdraaeind lichaam in styrax', *Mandblaad voor Natuurwetenschappen*, 1876, 7, 4.

[44] J.H. van 't Hoff, 'Beiträge zur Kenntniss des Styrax', *Berichte*, 1876, 9, 1339–40, p. 1340.

and in a November 1875 letter to Van 't Hoff, he indicated that he had received the copy of *La chimie dans l'espace* the previous June, but because of the 'various fates which have befallen me since its arrival, along with the desire to study your work thoroughly', he did not read the pamphlet until the following fall.[45]

Once he read Van 't Hoff's new theory, however, he greeted it with the greatest enthusiasm of all those chemists who had received a copy of the book. In his letter to Van 't Hoff, he described the theory as:

> not only an extraordinary intellectual attempt to explain previously incomprehensible facts but [I] also believe that it provides a great host of entirely new stimuli for our science, and subsequently will be of epoch-making importance. It therefore deeply satisfied me, and I welcomed it with the greatest joy. In a short time, you will see in my own projects, so I hope, the highest approving interest I take in your work.

He also asked Van 't Hoff if his assistant Felix Hermann, 'a clever, mathematically educated young chemist', could translate the book into German. Hermann, under Wislicenus' direction, was creating 'explanatory expansions, that I think are desirable for the easier understanding of the greater part of the scientific chemical community, and, by the look of it, may gain your approval (which will certainly be obtained first)'.[46] He also noted that he was having the cardboard models made on a larger scale for his older students.

Wislicenus also enclosed a letter from Hermann, in which Hermann made some initial suggestions regarding the translation. Wislicenus had suggested that Hermann present the contents of *La chimie dans l'espace* at the 8 November meeting of the local chemical society, during which:

> your theory raised in no small way the interest of all those concerned with chemistry in the scholarly sense; and without doubt this interest became still greater on further acquaintance with the beautiful consequences of your theory, if I can be permitted to judge by my personal experience.

Hermann was equally laudatory of the 'truly fundamental importance' of Van 't Hoff's theory, and went on to describe the changes he had in mind to make the argument more 'palatable' (*mundgerecht machen*) to the 'general chemical public'. 'Developments of a mathematical nature,' Hermann continued,

> which for an audience trained in this area are easily comprehensible with only fleeting

---

[45] Johannes Wislicenus, 'Mittheilungen aus dem Universitätslaboratorium Würzburg', *Berichte*, 1875, *8*, 1206-9. One of the 'various fates' could have been attending to the affairs surrounding the death of his father in October 1875.

[46] Johannes Wislicenus to Van 't Hoff, 9 November 1875, Van 't Hoff Papers, transcription printed in Cohen, *Jacobus Henricus van't Hoff*, p. 114. The letter is translated in its entirety in Appendix 3.

clues, must be recast explicitly to achieve the full understanding of a readership deficient in mathematical preparation who only rarely considers mathematical thoughts.[47]

According to Hermann's suggestion, they rewrote the original argument in German while retaining the overall length, structure and argument of *La chimie dans l'espace*, and the final version contained a number of important additions and deletions. Most obviously, they changed the title from 'Chemistry in Space', to the clearer, more specific 'The Arrangement of Atoms in Space' (*Die Lagerung der Atome im Raume*). For the first time, extensive literature references were given and the figures were included within the text, and not as separate plates. Significantly, they dropped the section on aromatic compounds. Wislicenus added a four-page foreword, in which he described Van 't Hoff's 'mathematical development and application to the steadily increasing number of cases of geometrical isomerism' as 'astounding' and 'gripping'.[48]

Two other additions are worth a closer look. The first and most significant was Van 't Hoff's implicit claim that molecules not only had spatial characteristics, but that actual spatial arrangement of the atoms in a molecule could be determined: that is, whether maleic acid was *cis* or *trans*. Van 't Hoff did not develop this claim completely, but he did give some indications about how it could be done. For example, the facility with which maleic acid formed an anhydride argued that it was the *cis* isomer. In his further discussion of the conditions for the appearance of optical activity, Van 't Hoff supposed that the addition of bromine to maleic and fumaric acid would not proceed equally; maleic acid would produce an inactive dibromide, similar to *meso*-tartaric acid, while fumaric acid would give an inactive dibromide similar to racemic tartaric acid: it could be split into two isomers that had the opposite effect on polarized light (Figure 3.9).[49]

Because these bromides had not yet been resolved into their enantiomers, Van 't Hoff recognized that this was not yet a definitive argument for his theory of optical activity. But it does illustrate two important, inter-related thoughts. First, Van 't Hoff assumed that the consideration of chemical reactivity could generate a picture of a molecule's physical form. Second, he implicitly proposed a kind of reaction process, or mechanism, in the addition of bromine to multiple bonds in which the addition of both atoms of bromine occurred from the same side of the molecule. Whether Van 't Hoff conceived this idea himself is not clear. It could have been one of the 'expansions' that Wislicenus had mentioned in his letter, in which case the idea could

---

[47] Felix Hermann to Van 't Hoff, 9 November 1875, Van 't Hoff Papers.
[48] Johannes Wislicenus, 'Foreword', in Jacobus Henricus van't Hoff, *Die Lagerung der Atome in Raume*, translated by Felix Hermann, Braunschweig, Vieweg, 1877, p. ix. A full translation of the Foreword is given in Appendix 4.
[49] Ibid., pp. 40–1.

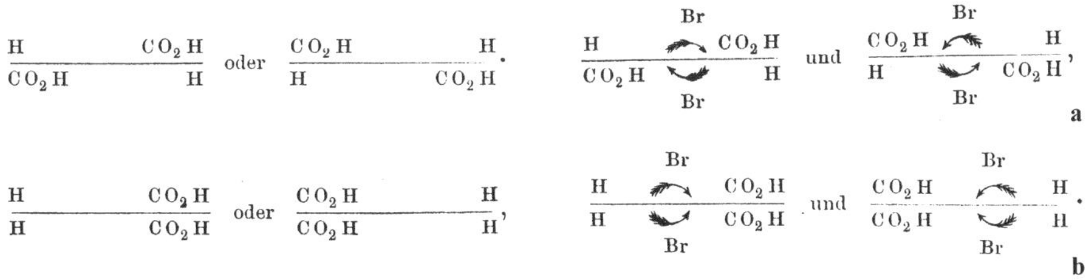

$$\begin{array}{c}\dfrac{H\qquad CO_2H}{CO_2H\qquad H}\;\text{oder}\;\dfrac{CO_2H\qquad H}{H\qquad CO_2H}\end{array}$$

(fumaric acid bromine addition diagrams with Br above and below)

a

$$\dfrac{H\qquad CO_2H}{H\qquad CO_2H}\;\text{oder}\;\dfrac{CO_2H\qquad H}{CO_2H\qquad H}$$

(maleic acid bromine addition diagrams)

b

3.9 **Van 't Hoff's portrayal of the addition of bromine to the two faces of fumaric acid (a) and maleic acid (b). In the case of maleic acid, the two products are identical (one needs only to flip over the structure on the right to match the left). Addition to fumaric acid yields two spatially different isomers (the right-hand structure cannot be made superimposable on the left by flipping or spinning).**

be due to either Hermann or Wislicenus. Certainly, in 1887 Wislicenus would later explicitly adopt and exploit this idea, to fill in the details left out by Van 't Hoff and ascertain the physical form of the unsaturated acids.

The second important addition to *Die Lagerung* was an appendix that contained templates for constructing cardboard models of various tetrahedra. The first set (Figure 3.10(a)) was to demonstrate the non-superimposability of enantiomeric tetrahedra. This could be done be coloring either the vertices or the faces in four different colors. Additional templates (Figure 3.10(b)) would model the carbon–carbon single bond, including the different rotational isomers. Gluing the edges of two of the first models would produce a model of a double bond, and the colored vertices would show the resulting isomerism (Figure 3.10(c)). Gluing three or more tetrahedra by their edges would demonstrate the effect of cumulated alkenes (Figure 3.10(d)). In the last two pages, Van 't Hoff offered models of increasingly unsymmetrical tetrahedra in which one, two, three and four different groups were attached to the tetrahedron (for example, Figure 3.10(e) and 3.10(f)). Changing the groups on the carbon atom changed the length of the bonds, and therefore the overall appearance of the tetrahedron. The last case resulted in a completely asymmetric tetrahedron that was non-superimposable on its mirror image. These last templates were the most explicit depiction of what Van 't Hoff meant by the 'asymmetric carbon atom', in that they showed the atom as an asymmetrically shaped object.

Van 't Hoff and Molecular Modeling

We have so far passed lightly over this crucial component of Van 't Hoff's campaign for the adoption of 'chemistry in space': the use of physical models to represent

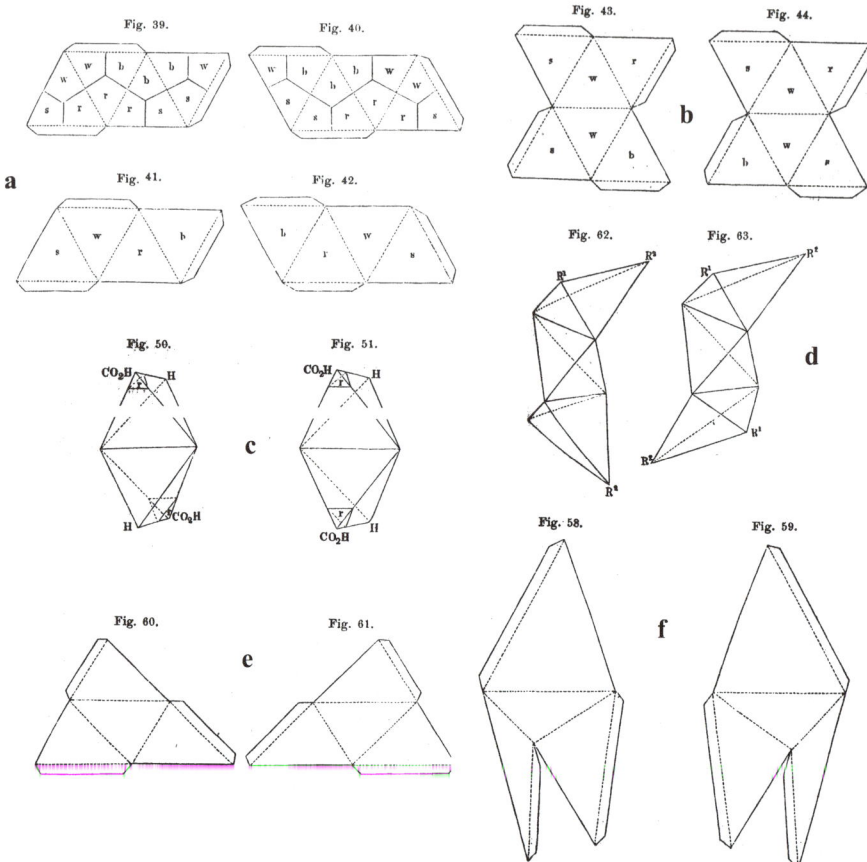

3.10 Templates given in the appendix to *Die Lagerung* for constructing tetrahedrel carbon atoms of varying substitution: (a) face-centered (bottom) and vertex-centered (top) models to illustrate enantiomorphism; (b) face-centered model of a single carbon–carbon bond, one of three to show rotational isomers due to Van 't Hoff's second hypothesis; (c) vertex-centered model of the double bond to show the isomerism resulting from Van 't Hoff's third hypothesis; (d) vertex-centered model of spatial allene isomers; (e) and (f) models for enantiomorphic asymmetric carbon atoms, showing the distorted tetrahedron resulting from the presence of four different groups.

molecules and the resultant isomerism created by the tetrahedron. When Van 't Hoff prepared and sent his paper tetrahedral models to prominent chemists in 1875, this was done within an established tradition of using such models for pedagogical purposes, only he was attempting to convince his colleagues of an unintended epistemological possibility for chemical structures, rather than using them to teach students the existing meaning of structures.

Van 't Hoff clearly seems to have been influenced by Kekulé's models, but the precise physical meaning of his drawings and tetrahedra remained ambiguous, at least from his published works. From most of the drawings given in the *Voorstel, La chimie dans l'espace* and *Die Lagerung*, the immediate parallel to Kekulé's models is obvious: 'bonding sites' are at the corners of the tetrahedron, and the groups attached to the asymmetric carbon atom are attached to the vertices of the model. As Ramsay and Snelders have pointed out, however, Van 't Hoff actually had two complementary models of the carbon atom. In the fall of 1875, Van 't Hoff wrote an open letter, published in the Dutch *Maandblaad vor Natuurwetenschapen*, as a response to an earlier letter written by Buys-Ballot as a critique of *La chimie dans l'espace*.[50] In this letter, Van 't Hoff suggested that the points of attachment in the carbon atom were actually the *faces*, and not the vertices of the tetrahedron. Thus four groups on an asymmetric carbon atom would attach to each face, a single bond resulted when the faces of two carbon atoms joined, and double and triple bonds resulted when either two or three faces on each atom lined up. Because in multiple bonds the faces could not physically join simultaneously using this model, multiple bonds would be much weaker, relatively speaking, than a single bond: that is, a triple bond would not simply be three times as strong as a single bond. This alternative description of the tetrahedron provided both a physically more satisfying description of bonding, as the atoms attracted one another by straight lines of attraction, rather than around a corner, as was implied in the Kekulé models. Because Van 't Hoff's description of the 'face-centered' model remained exclusive to the 1875 open letter written in Dutch, it has remained a largely unknown aspect of Van 't Hoff's theory. Why Van 't Hoff did not describe or advocate the face-centered model outside this obscure letter published in Dutch remains unclear.

Significantly, however, the cardboard tetrahedra constructed and sent by Van 't Hoff in 1875 used both the face-centered and the vertex model of bonding. The template given in *Die Lagerung* for the fully asymmetric carbon atom could *only* be made if bonding took place at the vertices. The choice of cardboard as the medium for the models in itself has always been unexplained – Van 't Hoff did not mention that the Kekulé ball and stick models would serve equally well for modeling his theory.

---

[50] Jacobus Henricus van 't Hoff, 'Isomerie en atoomligging: Antwoord op den openbaren brief van Dr. C.H.D. Buys Ballot', *Mandblad vor Natuurwetenschappen*, 1875, *6*, 37, German translation in Cohen, *Jacobus Henricus van 't Hoff*, pp. 104–13.

Although it probably did not escape his attention that cardboard models would be inexpensive to construct, the choice of cardboard with flat surfaces to show the bonding sites theoretically makes more sense in many cases if he was trying to model the 'face-centered' carbon atom, and if the cardboard tetrahedra were large enough, they could be easily be modified for the vertex model of bonding. Many of the templates in the appendix to *Die Lagerung* were also designed with the face-centered model in mind, although nowhere did Van 't Hoff describe the face-centered model in the text.

It seems clear that these two types of models had their origins in two different phenomena that required modeling, and in the constraints of publication and distribution. The drawings, based on Kekulé's ball and stick models, illustrated the new cases of isomerism and showed bonding most clearly, while the face-centered models represented the new model for carbon–carbon bonding. The designs of both models were also dictated by printing and construction techniques. In the diagrams for the text, it would be much easier for the illustrator, and more legible for the reader, if the groups were labeled at the corners of the tetrahedron, rather than on the faces. In the cardboard models, unless the tetrahedra were large enough, the most convenient place to label the groups is on the face of the tetrahedron. The use of different forms of representations of the tetrahedron makes it clear that what Van 't Hoff meant by the 'tetrahedral carbon atom' was not the atom itself, but the spatial distribution of valences *around* the atom.[51] The arrangement was defined only in reference to the atoms bound to carbon, and not in the atom itself. That he used different models to emphasize different aspects of his theory illustrates a pragmatic use of paper formulas and cardboard models without an explicit interest in the actual appearance of the carbon atom.

Conclusion

Van 't Hoff's Three Hypotheses

In 1888, Victor Meyer classified the ideas in *Die Lagerung* into three principal hypotheses that will organize the remaining chapters of our study.[52] The first hypothesis was the tetrahedral arrangement of valences itself (the asymmetric carbon atom), while the second was the free rotation about the carbon–carbon single bond, and the third was the restricted rotation about a carbon–carbon double bond. These

---

[51] Van 't Hoff to Wilhelm Ostwald, 20 January 1888, in Hans-Günther Körber, ed., *Aus dem wissenschaftlicher Briefwechsel Wilhelm Ostwalds*, 2 vols, Berlin, Akademie-Verlag, 1969, vol. 2, p. 213.

[52] Victor Meyer and Karl Auwers, 'Untersuchungen über die zweite van't Hoff'sche Hypothese', *Berichte*, 1888, *21*, 790–817.

three hypotheses and the implications within them were created by Van 't Hoff as a fusion of three important traditions within chemistry: the long-standing concern with isomerism, the growing awareness of the relationship between optical activity and chemical structure, and the increasing use of three-dimensional molecular models. Although *Die Lagerung* and the works that preceded it broke new epistemological ground and contained novel elements of mechanism, they also fit well into traditional nineteenth-century chemistry, because in them Van 't Hoff explained previously unexplained cases of isomerism by an appeal to the arrangement of atoms that for the first time meant a spatial arrangement.

It is also important to recognize the role that optical activity played in Van 't Hoff's theory. The increasing awareness of optical activity as an identifying property during the 1860s by chemists such as Wislicenus and Hans Landolt no doubt encouraged its use as a physical property for distinguishing isomers, and it was recognized by Biot and Pasteur that the physical property of optical activity indicated that a compound possessed a molecule that was asymmetric. Van 't Hoff certainly recognized the relationship between optical activity and molecular asymmetry in general, but the phenomenon of optical activity in itself was important to him mostly as a way to distinguish isomers, much like a melting or boiling point.

Van 't Hoff's theory has sometimes been characterized as the fusion of Kekulé's structure theory with Pasteur's theory of optical activity.[53] This is strictly speaking not true, for as we have seen, in his correspondence with Bremer Van 't Hoff either deliberately or inadvertently contradicted Pasteur's central claims about molecular asymmetry, and there is little reference to Pasteur in the *Voorstel* or it subsequent revisions. As Van 't Hoff himself described the origins of his theory, the asymmetric carbon atom was a natural outgrowth of structure theory, not a continuation of Pasteur's work. No less important in Van 't Hoff's synthesis was the extensive use of physical models as pedagogical devices. Van 't Hoff's use of paper models was then a direct continuation of the tradition begun during the 1860s, with the important innovation that his models were intentionally iconic, meant to mimic directly certain physical characteristics of the molecule.

The Meaning of the Tetrahedron, and its Role in Chemical Theory

There are a number of meanings that Van 't Hoff gave to chemical formulas. First, as we have already noted, he regarded structural formulas as iconic representations, because giving them such a meaning offered an explanation of certain cases of isomerism. Second, he depicted the tetrahedral carbon atom pragmatically by

---

[53] Wilhelm Ostwald, 'Jacobus Henricus van't Hoff', *Berichte*, 1911, *44*, 2219; Emil Fischer, 'Gedächtnisrede'.

different means – face-centered or vertex bonding – for representing different aspects of the tetrahedra, and importantly, Van 't Hoff was not interested in the shape of the carbon *atom*, but in the distribution of valences around the atom. It also seems important to note extensive references to Van 't Hoff's theory as 'mathematical', as Wislicenus specifically pointed out that Hermann was 'mathematically inclined', and Hermann wanted to created the German version for a mathematically less astute audience. This would seem to reinforce the notion that Van 't Hoff had given a new meaning to chemical formulas, which previously presumably contained no mathematical character.

Finally, even though Van 't Hoff largely abandoned work on his own ideas after 1877, the promotion and expansion that he did accomplish between 1874 and 1877 is in large part responsible for the eventual formation of the large research programs in 'chemistry in space' that would emerge in the late 1880s and 1890s.[54] In *Die Lagerung*, Van 't Hoff laid extensive groundwork for the innovative extension of the structural theory by grafting an additional level of theoretical meaning onto chemical formulas, and overall we can identify three distinct different theoretical levels. The first was the level of structure theory, and was the least controversial. All chemists, with the important exception of Kolbe (who died in 1884), were committed to this level. Although most chemists would also agree that, under the constraints of the structure theory, chemical formula contained no implication of molecules as mechanical objects, it did contain some latent physical aspects. Especially because of the adoption of Crum Brown's formulas, they could easily be interpreted (or misinterpreted) mechanically.

Above the structure theory were two additional levels of theoretical commitment. According to Van 't Hoff, certain properties of molecules, such as optical rotation, could be explained by an appeal to their spatial characteristics. Because this level involved molecules with distinct physical shape and form, it contained explicit elements of mechanism, and could be considered a true 'mechanization' of structure theory. This was the least controversial aspect of Van 't Hoff's theory. More controversial was the claim that the three-dimensional properties of molecules could be known in an absolute sense – whether maleic acid, for example, was *cis* or *trans*. Of the three theoretical levels (structure, spatial isomerism, absolute configuration), the third usually involved the greatest amount of speculation, and the greatest reliance on a mechanistic conception of molecules, including their intra- and inter-molecular motions. Instead of relying solely on molecular statics, it also relied on a new conception of the dynamics of molecules and chemical reactions. This conception, only latent in *Die Lagerung*, would become Wislicenus' major effort in 1887.

---

[54] O.B. Ramsay, 'Molecular Models', p. 75.

Chapter 4

Initial Reception of the Tetrahedron, 1874–1887

Who ever answers a lunatic?!
Wilhelm Lossen, 1877

Public and Private Reaction to the Tetrahedron

Although publication of the *Voorstel* did not generate the discussion Van 't Hoff desired, he did receive some response. In October 1874, friends of Van 't Hoff's parents in Rotterdam received a letter from Tjaden Modderman, Professor of Chemistry in Groningen, who had recently received a copy of the pamphlet. Modderman had read it with 'interest', noting that 'one thing is certain, there is something about your young friend'.[1] As we saw in Chapter 3, Buys-Ballot was also sympathetic with the young Van 't Hoff, and initiated the publication of the French translation of the *Voorstel* for the *Archives Neérlandaises*. These two responses, both favorable, are the only known reaction to the initial publication of the *Voorstel* in 1874.

Publication of *La chimie dans l'espace* in 1875 resulted in a significant increase in the awareness of Van 't Hoff's theory, and responses to it. Between 1875 and 1877, Van 't Hoff certainly received acknowledgement, if not always praise, but without stimulating the full discussion that he wanted. Buys-Ballot continued to encourage Van 't Hoff by writing the open letter to him in the fall of 1875 (described in Chapter 3). In this lengthy letter, Buys-Ballot pointed out to Van 't Hoff that the idea that molecules were spatial objects was not new, but directed the young Van 't Hoff to:

---

[1] Tjaden Modderman to neighbors of Van 't Hoff, 17 October 1874, J.H. van 't Hoff Papers, Ms. 74, Special Collections, Milton S. Eisenhower Library, Johns Hopkins University (hereafter cited as 'Van 't Hoff Papers'). Appendix 1 contains a full translation of the letter from the original Dutch.

Go further on an already somewhat trodden path, your merit lies only in that you believe you must rather attempt an account that is not certain, than to follow one that certainly isn't right ... Strongly defend Wislicenus' thesis and develop it in its details, as you have done already.[2]

Buys-Ballot was equally encouraging about Van 't Hoff's explanation for optical activity in compounds, 'that we previously possess no better hypothesis than yours, as I hope, for seeking the cause of rotation of the plane of polarization in solutions'.[3] Buys-Ballot was certainly supportive, but he was not instrumental in spreading Van 't Hoff's theory to a broader audience. That role would fall to Wislicenus, who with Hermann offered the most enthusiastic public and private response to *La chimie dans l'espace*. They wrote the enthusiastic letters to Van 't Hoff seen in Chapter 3, and presented the contents of *La chimie dans l'espace* to a favorable reception to a fall 1875 Würzburg chemical society meeting. Most importantly, they initiated the translation into German, containing the foreword by Wislicenus, and most chemists in Germany would become acquainted with Van 't Hoff's ideas from *Die Lagerung*.

Between 1875 and 1877, Van 't Hoff received several supportive letters and requests for copies of *La chimie dans l'espace* or the accompanying models. In his letter to Van 't Hoff concerning the translation, Hermann mentioned that an assistant in the Würzburg laboratory, a Dr Fabingi, was interested in translating *La chimie dans l'espace* into Hungarian. In 1876, Van 't Hoff received six letters from lesser-known chemists in Russia, Switzerland and India, requesting either copies of *La chimie dans l'espace* or the model sets, and in 1877 he received a letter requesting a copy of *Die Lagerung*.[4] Reaction from prominent chemists was also supportive, if not always public. Although Adolf Baeyer would not publicly come out in favor of Van 't Hoff's ideas until 1885 and the publication of his strain theory (discussed below), he expressed privately his enthusiasm for Van 't Hoff's theory on at least two occasions. Emil Fischer recalled that after Baeyer read *La chimie dans l'espace*, he declared to the students in the Strasbourg laboratory that 'There is truly again a new, good thought in our science that will bear ripe fruits'. Baeyer also expressed his enthusiasm in a September 1875 letter to his student Victor Meyer:

---

[2] Ernst Cohen, *Jacobus Henricus van't Hoff: Sein Leben und Werken*, Leipzig, Akademische Verlagsgesellschaft, 1912, p. 96. Translated from the original Dutch into German by Cohen. My translation from the German.

[3] Ibid., pp. 96 and 101. Translated from the original Dutch into German by Cohen. My translation from the German.

[4] F.V. Spitzer to Van 't Hoff, 19 March 1876, Adolph Krätz to Van 't Hoff, 15 May 1876; J. Barbieri to Van 't Hoff, 19 March 1876; L. Conen to Van 't Hoff, 14 July 1876; Chattoprachlyay to Van 't Hoff, 8 March 1877; N. Leÿ to Van 't Hoff, 14 August 1876. Another letter, dated 19 August 1885, from A. Bartoli in Florence, requested the model sets from *La chimie dans l'espace*. Van 't Hoff Papers.

I am enchanted by Van 't Hoff's study, whether it is true, who knows? But I think the idea is quite nice, it's very clear to me, and in the aliphatic series everything is right. Rebukes from the aromatics have really always been somewhat uncertain.[5]

A month earlier, Meyer had written to Van 't Hoff from Zürich for a copy of *La chimie dans l'espace*, 'of which I have heard so much in addresses as well as in notable reviews'. Theodor Zincke, Kolbe's successor at Marburg, wrote in 1876 asking for a copy of *La chimie dans l'espace*. In 1877, Adolphe Wurtz told Van 't Hoff that he read *La chimie dans l'espace* 'with attention and interest', and Otto Wallach asked 'whether perhaps the necessary [models] are prepared under your direction and available commercially there', because he did not completely trust his bookbinder with the construction of the models.[6] Walter Spring in Liege, himself a supporter of Van 't Hoff's theory, reported that Jean-Servais Stas in Brussels spoke:

> *With the greatest praise* of your 'Chemistry in space'. He finds that you have there a pleasant idea that will contribute greatly to true understanding in general chemistry. I was all the happier for you with M. Stas' words, given the fact that his appraisals are not usually generous. When he praises a work, you can be sure it is a work of merit.[7]

When Hans Landolt, at the Aachen Polytechnical Institute, first read *La chimie dans l'espace* in 1875, he had been occupied since 1868 with an extensive cataloging of known optically active organic compounds and evaluation of polarimeters to establish rigorous, reproducible standards for making optical activity a defining characteristic for organic compounds.[8] Because of this familiarity with optical activity, Landolt was probably in one of the most advantageous positions to evaluate Van 't Hoff's explanation of optical activity, and therefore took a great interest in the contents of *La chimie dans l'espace*. He wrote to Van 't Hoff in July 1876 to request a set of models:

> I have just been occupied in the study of your brochure 'La chimie dans l'espace' with great interest. But in reading it, I acutely sense the need for the cardboard models, and would wish to request that you procure for me the entire collection as soon as possible ... From what I

---

[5] Emil Fischer, 'Gedächtnisrede auf Jacobus Henricus van't Hoff', pp. 891–902 in Emil Fischer, *Gesammelte Werke*, Bd. 4, *Untersuchungen aus verschiedenen Gebieten: Vorträge und Abhandlungen Allgemeinen Inhalts*, Max Bergmann, ed., Berlin, Springer Verlag, 1924, p. 893; Adolf von Baeyer to Victor Meyer, 27 September 1875, Deutsches Museum München, Archiv, HS 7027.

[6] Victor Meyer to Van 't Hoff, 31 August 1875; Theodor Zincke to Van 't Hoff, 4 November 1876; Wurtz to Van 't Hoff, 26 June 1877; Otto Wallach to Van 't Hoff, 30 May 1877. Van 't Hoff Papers.

[7] Walter Spring to Van 't Hoff, 16 August 1876. Van 't Hoff Papers. Spring's emphasis.

[8] The only significant study of Landolt and optical activity is given in two short sections in Robert Ward, *The Development of the Polarimeter in Relation to Problems in Pure and Applied Chemistry: An Aspect of Nineteenth Century Scientific Instrumentation*, PhD, University of London, 1980, pp. 262–7 and 288–9. The entry for Landolt in the *Dictionary of Scientific Biography* does not mention Landolt's study of optical activity.

have seen so far, your theories are of the greatest importance, and it is quite incumbent upon me to understand it completely.[9]

He also sent Van 't Hoff reprints of two 'small papers' on the 'rotatory power of dissolved substances' that had appeared in the 1876 volume of the *Berichte*.[10] These papers were a prelude to his first comprehensive review of optical activity published in the *Annalen* in 1877, in which he devoted eight pages to the plausibility of Van 't Hoff's and Le Bel's correlation between optical activity and chemical constitution. 'In the inherent interest of Van 't Hoff's hypothesis', Landolt listed structures of known aliphatic compounds arranged by the number of carbon atoms, that were active and contained an asymmetric carbon atom the structures of those compounds that were inactive and contained no asymmetric carbon atom, and those structures for compounds that possessed an asymmetric carbon atom but did not exhibit rotatory power. Aromatic compounds were all optically inactive, Landolt noted, unless a side chain possessed an asymmetric carbon atom, or unless one or more double bonds of the benzene ring were saturated to create asymmetric carbon atoms. Landolt concluded his short analysis by noting that:

> until now there is no demonstrable case that contradicts Van 't Hoff's hypothesis and contradicts the following statements:
>
> 1) *Bodies lacking an asymmetric carbon atom show no rotatory power.*
> 2) *Active substances always contain one or several asymmetric carbon atoms.*[11]

The conclusions were also repeated in Landolt's comprehensive 1879 book on optical activity. Landolt's largely supportive analysis of Van 't Hoff's first hypothesis was one of the first to appear in print after the appearance of *Die Lagerung*. His response to the tetrahedral carbon atom was, next to those of Wislicenus and Baeyer, one of the most enthusiastic.

While most of the positive comments on Van 't Hoff's theory seem to have been private, the few known early criticisms of his theory were public. The earliest negative criticism came from Marcelin Berthelot at the March 1875 meeting of the

---

[9] Hans Landolt to Van 't Hoff, 10 July 1876, Van 't Hoff Papers, also reprinted in Cohen, *Jacobus Henricus van 't Hoff*, p. 578.

[10] These papers were most likely 'Zur Kenntniss des specifischen Drehungsvermögen gelöster Substanzen', *Berichte*, 1876, *9*, 901–14, and 'Über das specifische Drehungsvermögen des Camphers," *Berichte*, 1876, *9*, 914–17. Both had been submitted in June of 1876, and neither mention Van 't Hoff's hypothesis, explicitly or implicitly.

[11] Hans Landolt, "Untersuchungen über optische Drehungsvermögen. 1. Abhandlung', *Annalen*, 1877, *189*, 241–337, pp. 260–7, and *Das optische Drehungsvermögen organischen Substanzen und die praktische Anwendungen desselben*, Vieweg, Braunschweig, 1879.

Société chimique de France, where Arthur Henninger first presented Van 't Hoff's theory. While Berthelot recognized the 'general interest' in Van 't Hoff's three-dimensional formulas, he noted that 'a complete representation of chemical compounds could not take place without an idea of the rotatory and vibratory motions which animate each atom in particular and each groups of atoms in the molecule'. The essentially static model of the carbon atom given in the *Voorstel* and 'Rotatory Power' could not explain the existence of Pasteur's four molecular types – dextrorotatory, levorotatory, racemic and inactive – because they can only differentiate between the first two types. Berthelot went on to describe how the four molecular types of Pasteur could be differentiated by the intra-molecular atomic motions:

> The existence of rotatory power clearly explains itself by the different orientation of the molecules; or more exactly, by the orientation of their vibratory motions. Given the same molecular system, it is indeed conceivable that certain of these atoms can vibrate along the same plane: these would be inactive bodies; or also in another plane, inclined symmetrically to the right or to the left relative to the plane of the fundamental atoms: these would be left and right bodies. One could yet conceive of two symmetrical systems juxtaposed in such a way so as to give an intermediate state of movement analogous to the one of the inactive bodies; this is the case for neutral bodies. The four types of bodies endowed with rotatory power are explained in such a way. One could equally conceive that there could exist a multitude of isomeric bodies of identical atomic structure, but differ by the unequal orientation and nonsymmetric vibratory movements of their atoms: This seems to be the case between the isomeric camphenes, active and inactive to polarized light.[12]

While Van 't Hoff would later attempt to prove false Berthelot's 1866 claim that styrene was optically active (described in Chapter 3), he never responded directly to this theoretical objection, although Henninger no doubt made him aware of it. It is perhaps significant that *La chimie dans l'espace*, which appeared only a few months after Berthelot's critique, contained a significantly increased discussion of the dynamic properties of molecules, and an explanation of optical activity. Berthelot's own explanation of optical activity rested, of course, on his assumption, derived from Pasteur, that all optically active compounds must have four isomers, a requirement that both Van 't Hoff and Le Bel had abandoned. Le Bel had explained the existence of the inactive tartrate by noting that tartaric acid had *two* asymmetric carbon atoms, and therefore could exist in four different forms. Both Van 't Hoff and Le Bel predicted the existence of only two isomers when a compound had only one asymmetric carbon atom. Following Van 't Hoff's publication on the isolation of the optically active component in styrene derived from styrax, Le Bel wrote to Van 't Hoff concerning Berthelot:

---

[12] Report of Berthelot's comments at the meeting of the March 1875 meeting of the *Société chimique de France*, quoted in French in Cohen, *Jacobus Henricus van 't Hoff*, pp. 91–3.

I received your letter and I congratulate you on your results with styrol: this amuses me especially because you punish M. B. very well for meddling in our affairs. Furthermore, what he has said is not exactly transcendent, for to state that the cause of the rotatory power must be sought only in the orientation of the atoms – and this after having read your work and mine, seems to me most inflexible.[13]

In 1880 and 1881, Adolf Claus at the University of Freiburg and Wilhelm Lossen at the University of Königsberg critiqued the idea of the tetrahedral carbon atom. Their complaints were not specifically aimed at Van 't Hoff, but towards the general conception of valence offered in textbooks written by Kekulé, Emil Erlenmeyer, Lothar Meyer, Hermann Kolbe, and August Hofmann. All of these authors had defined valence, implicitly or explicitly, as a divided portion of an atom, either as a subatomic particle or as a discrete portion of its chemical affinity. Not surprisingly, the idea of the tetrahedral carbon atom, in which valences were directed into four different places, possessed much in common with this view.

Claus regarded the attempts to establish experimentally the equivalence or inequivalence of the four affinity units on the carbon atom as 'irrelevant' (*gegenstandslos*).[14] If the valence of a hydrogen atom, or the attractive power necessary to hold one hydrogen atom was chosen as a unit, it was justified to conclude that the carbon atom was tetravalent and that it could possess four equivalent affinities. But it was *not* justified, said Claus, to assume that the carbon atom possessed four valences divided *a priori*. It was entirely 'arbitrary, indeed it appears unnatural to me' to assume the existence of four separate affinity units on the carbon atom independent of its combination:

> Moreover, I cannot conceive of the chemical attractive power that belongs to a so-called polyvalent atom any other way than as an *a priori* coherent, unified whole, that only when the atom meets other atoms in chemical combination, according to the latter's value, can it split itself into a different (but limited) number of parts, and according to its nature into equal or different size parts.[15]

Lossen went one step further, and denied the existence of chemical energy at all, either divided or undivided. Chemical affinity only existed when atoms were in the presence of other atoms. 'We often say,' Lossen wrote:

> the carbon atom attracts the oxygen atom, or the oxygen atom attracts the carbon atom; both expressions rest on an imprecise conception. In the binding of atoms both attract each other mutually; an effect of the carbon atom on the oxygen is inconceivable without a simultaneous active effect, equal in size and opposite in direction, of the oxygen atom on the

---

[13] Le Bel to Van 't Hoff, undated, in Cohen, *Jacobus Henricus van 't Hoff*, p. 93.
[14] Adolf Claus, 'Zur Frage nach den Äffinitätsgrossen des Kohlenstoffs', *Berichte*, 1881, *14*, 432–5.
[15] Ibid., p. 433.

carbon atom. The force with which both atoms attract one another depends on the nature of each; a force given to the individual atom as such does not ever exist.[16]

Lossen defined valence simply as the number of atoms that were located in the 'binding zone' of another atom.[17] It was a simple number indicated by the chemical structure, and he made no commitment to a single valence number for any given atom (for example, carbon had a valence of either two, three or four). He found the concept of a multiple bond an absurd idea, since he interpreted this to mean that an atom could find itself in the 'binding zone' of another atom *twice*; an atom was either there and bound, or it was not. Presumptions of multiple bonds were an effort, in Lossen's view, to save the theory of constant valence, and rested furthermore on the assumption that atoms were divisible, since two different parts of an atom must attract two different corresponding parts on another atom. To be able to divide atoms in such a way, however, one must know what the atoms themselves were like, and at the current state of science, this was not accessible to observation. Therefore, Lossen conceived of atoms as simple points.

Van 't Hoff's conception of the tetrahedron proved incompatible with this provisional assumption that atoms were material points. Lossen did not object to Van 't Hoff's theory of optical activity, since it actually presented the position of atoms in space. The theory of multiple bonding, however, presented not only a position of atoms in space, but also affinities. For example, Van 't Hoff's model of acetylene compounds showed four atoms in a straight line:

> If the tetrahedra are regular, as Van 't Hoff draws them, the four atoms lie in a straight line; their position in space is determined (*präcisirt*) as soon as their distance from one another is known. Thus the depiction of the molecule C_2H_2 by two tetrahedra *now* is by no means a presentation of the *position of atoms in space*; in this representation what do the touching corners of the tetrahedra mean?[18]

The corners of the tetrahedron could not possibly indicate the direction of the forces, because at the corners of Van 't Hoff's model for double or triple bonds, there were no atoms. Lossen found it difficult to accept Van 't Hoff's theory because Van 't Hoff went further than the position of atoms in space, and proposed a position of affinities in space.

The presumed restriction of rotation about the carbon–carbon double bond, crucial to Van 't Hoff's explanation of the isomerism between the unsaturated acids, was actually not possible if one followed Lossen in considering the position of point-mass

---

[16] Wilhelm Lossen, 'Über die sogenannte Verschiedenheit der Valenzen eines mehrwehrtiges Atoms', *Berichte*, 1881, *14*, 760–5, p. 762.
[17] Wilhelm Lossen, 'Über die Vertheilung der Atome im Raum', *Annalen*, 1880, *204*, 265–364.
[18] Ibid., p. 337.

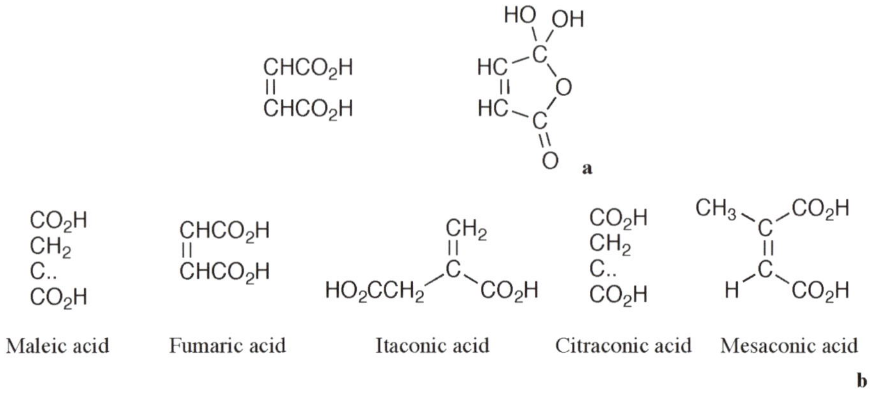

4.1 (a) Anschütz's cyclic structure for fumaric and maleic acid. (b) Fittig's structures for various unsaturated acids.

atoms in space. Nor did the model for the double bond make physical sense to Lossen, since the lines of bonding did not lie along a straight line between the carbon atoms. Van't Hoff's model had assumed that the atom had a shape, and therefore had parts:

> In my opinion [Van 't Hoff's] conception leads necessarily to the assumption that multivalent atoms cannot be considered as material points at all, that rather there are parts of them to distinguish, from which emanates their influence on other atoms.[19]

Lossen then inquired how Van 't Hoff could know anything of these parts. This particular criticism, we should note, could have been avoided had Van 't Hoff been more explicit about his face-centered model of bonding, in which all attractions lay along straight lines

Rudolf Fittig and Richard Anschütz also saw little reason for adopting Van 't Hoff's model for the double bond. Anschütz, assistant to Kekulé at the University of Bonn, recognized that both maleic and fumaric acid contained a double bond, but still assigned the difference between them to structural causes. Fumaric acid contained two carboxyl groups and maleic acid, a cyclic lactone structure (Figure 4.1(a)).[20] Fittig, as part of his extensive experimental study of the unsaturated acids, concluded that maleic and fumaric, and citraconic and mesaconic acids were structural, and not geometric isomers. Fittig invoked Kekulé's structure for maleic acid and a double bond in fumaric acid (Figure 4.1(b)), and declared that 'as long as we are able to

---

[19] W. Lossen, 'Ueber die Lage der Atome im Raume', *Berichte*, 1887, *20*, 3306–10.
[20] Richard Anschütz, 'Zur Geschichte der Isomerie der Maleinsäure und Fumarsäure', *Annalen*, 1887, *239*, 161–85.

explain the facts in such a simple unforced manner, we do not yet need, in our opinion, the hypothesis proposed by Van 't Hoff for the explanation of these isomers'.[21]

But all of these critiques were mild in comparison to Hermann Kolbe's famous, or infamous, reaction to *Die Lagerung*. Kolbe used his role as editor of the *Journal für praktische Chemie* to publish a vicious attack with the title 'Signs of the Times' that declared Van 't Hoff to have returned to the dangerous speculations of *Naturphilosophie*:

> Whoever thinks this worry [about the return to *Naturphilosophie*] seems exaggerated should read, if he is capable of it, the recent phantasmagorically frivolous puffery of Messrs. van 't Hoff and Hermann on 'The Arrangement of Atoms in Space' ... A Dr J.H. van 't Hoff, of the Veterinary School of Utrecht, finds, it seems, no taste for exact chemical research. He has considered it more convenient to mount Pegasus (apparently loaned by the veterinary school), and to proclaim in his 'La chimie dans l'espace', how, during his bold flight to the top of the chemical Parnassus, the atoms appeared to him to be arranged in cosmic space. The prosaic chemical world had no taste for these hallucinations, so Dr F. Hermann, assistant at the Heidelberg Agricultural Institute, undertook a German edition to give the work a wider audience ... It is typical of these uncritical and anti-critical times that two virtually unknown chemists, one of them at a veterinary school and the other at an agricultural institute, pursue and attempt to answer the deepest problems of chemistry which probably will never be resolved (especially the question of the *spatial* arrangement of atoms) and moreover with an assurance and an impudence which literally astounds the true scientist.[22]

He also declared that Wislicenus had 'left the rank of exact scientists', because he had contributed to such 'puffery'. For years, Kolbe had attacked proponents of the structure theory, particularly Kekulé, always vociferously and sometimes personally, for reverting to *Naturphilosophie*. It should not be surprising, then, that Van 't Hoff's further claim that spatial arrangements, always threatened by the structure theory, were now *necessary* for chemical theory incited Kolbe's wrath.

Kolbe's attack became the most famous reaction to *Die Lagerung*, and not surprisingly, has usually been explained away as the musings of a crank, of someone who simply could not accept new ideas. H.A.M. Snelders, for example, called it 'no serious criticism'.[23] It is true that by 1874, Kolbe had, by virtue of his obstreperous personality and penchant for personal attacks on fellow chemists, become a virtual outsider in the field that he had helped create in the 1850s and 1860s. His violent

---

[21] Rudolf Fittig, 'Über die Beziehungen zwischen Fumar- und Maleinsäure und zwischen Citraconsäure und Mesaconsäure', *Berichte*, 1877, *10*, 516–18.

[22] Hermann Kolbe, 'Zeichen der Zeit', *J. prak. Chem.*, 1877, *123*, 473–7, translation in Alan J. Rocke, 'Kolbe vs. the "Transcendental Chemists": The Emergence of Classical Organic Chemistry', *Ambix*, 1987, *34*, 156–68.

[23] H.A.M. Snelders, 'Practical and Theoretical Objections to J.H. van't Hoff's Stereochemical Ideas', pp. 55–65 in O.B. Ramsay, ed., *Van't Hoff-Le Bel Centennial*, Washington, DC, American Chemical Society, 1974.

polemics did not usually generate any response, and the usual response to Kolbe's tirades was silence.[24] In a postcard to Van 't Hoff, J.D. van der Waals indicated as much:

> In the matter of Kolbe: some (Baeyer, Hofmann, etc.) have planned to make an open protest to convince foreigners that Kolbe's opinion is not that of German chemists. On second thought, however, this was considered unnecessary. When I spoke with [Wilhelm] Lossen about the best way to answer Kolbe, Lossen said 'Who ever replies to a lunatic?!!'.[25]

Wurtz also encouraged Van 't Hoff saying that 'whatever M. Berthelot and also M. Kolbe, who has attacked you, have to say about it, I see there a new way that is good to enter, no doubt with prudence, but with a perseverance justified by the results that have already been glimpsed'.[26]

If we strip away Kolbe's unfair criticisms based on Van 't Hoff's and Hermann's current academic positions and the vitriolic character and focus on Kolbe's declaration that Van 't Hoff had 'no taste for exact chemical research', and that Wislicenus was no longer an 'exact' scientist, we do find something substantial in Kolbe's remarks. Kolbe correctly perceived, and profoundly disagreed with, the explicitly hypothetical character of Van 't Hoff's methodology, in which experiment did not pave a cautious road to theoretical speculation. The tetrahedral hypothesis was not derived from the slow accumulation of facts and observations, but was a bold set of assertions about the physical nature and actual appearance of all carbon atoms and organic molecules that could not be confirmed by direct experience. In part, Kolbe's criticism was pure rhetoric, for he had used hypotheses himself to direct his own research, and Van 't Hoff's methodology was partly an excuse for Kolbe to attack a theory that he simply didn't like by making it look 'unscientific'.

Lossen had also noticed this method of hypothesis in Van 't Hoff's theory, and offered a similar criticism in the critique mentioned above. It was unacceptable, according to Lossen, simply to speculate regarding the form of the carbon atom and its parts, and then construct the spatial properties of molecules. Unlike Kolbe, Lossen did not argue that the spatial distribution of atoms was unknowable, but he did object to Van 't Hoff's claim to knowledge of the spatial distribution of affinity units on the atom: the shape of the atom itself. Lossen was quite specific about the sequence of events chemists should follow to gain knowledge of a molecule's spatial properties.

---

[24] Alan J. Rocke, *The Quiet Revolution: Hermann Kolbe and the Science of Organic Chemistry*, Berkeley, CA, University of California Press, 1993, pp. 325–39.

[25] 'In zake Kolbe: Men (Bäyer, Hofmann enz) hebben plan gehad een protest openbaar te maken, om het buitenland te overtuigen, dat Kolbe's meening niet die der Duitsche Chemici is. By nader inzien vond men zooiets echter onnoodig. Hoen ik met Lossen sprak over de beste wijze om Kolbe te antwoorden, zy L. "Wer antwortet dann einen Verrückten?!!" ', J.D. van der Waals to Van 't Hoff, undated postcard, Van 't Hoff Papers.

[26] Adolphe Wurtz to Van 't Hoff, 26 June 1877, Van 't Hoff Papers.

One could only consider the position of atoms in space *after* the determination of the specific atomic form, and after the location of the seats of chemical affinity were located.[27] Van 't Hoff had addressed the problem exactly the other way around. He had first assumed a spatial distribution of affinities, bypassing completely even a preliminary definition of affinity unit (or valence bond), and then constructed the form of molecules that in turn led to certain observable predictions, namely the appearance of different spatial isomers.

Van 't Hoff seemed particularly affected by Kolbe's criticism. Of the criticisms described above (those by Berthelot, Lossen, Claus, Anschütz, Fittig and Kolbe), Kolbe's critique is the only one to which Van 't Hoff specifically replied on two occasions. The first was at the beginning of a short 1877 article on the relationship between optical activity and constitution. Mentioning Kolbe explicitly, but without citing the actual location of Kolbe's article, Van 't Hoff defended his methodology:

> A theory that is not yet contradicted by a single fact can only be tested further experimentally for additional judgement. Now if anyone, even a man like Kolbe who has served chemistry well, says that a chemist should not concern [*plagen*] himself with theories, because he is still unknown and is employed at a veterinary school, if he does not think it worthy to greet the representative of a new (possibly wrong) view as Homeric heroes before the battle, so I consider this kind of behavior, luckily, not as a sign of the times, but as contributing to the understanding of an individual.[28]

The second occasion on which Van 't Hoff addressed Kolbe was his 1877 inaugural lecture to the University of Amsterdam entitled 'Imagination in Science', a subject suggested by Van 't Hoff's father. Kolbe was not mentioned directly or indirectly, nor did Van 't Hoff even speak on chemistry, but the choice of imagination (*Phantasie*) as his subject was meant to oppose directly Kolbe's condemnation of the importance given to hypothesis in Van 't Hoff's work. The goal of science, according to Van 't Hoff, was the 'elucidation in every detail' of the relationship between cause and effect, and the topic of his lecture was role of imagination in uncovering this specific relationship. 'Imagination,' Van 't Hoff said, was 'the capacity to visualize a particular thing so clearly, that all its properties can be recognized with the same certainty as if the object were directly observed'.[29] Drawing on his earlier extensive

---

[27] Lossen's methodological critique appeared in 1887, as a response to Wislicenus' essay on the unsaturated acids. W. Lossen, 'Über die Lage', note 19. See also Peter J. Ramberg, 'Johannes Wislicenus, Atomism, and the Philosophy of Chemistry: A Translation and Commentary', *Bull. Hist. Chem.*, 1994, *15/16*, 45–53.

[28] J.H. van 't Hoff, 'Über den Zusammenhang zwischen optische Activität und Constitution', *Berichte*, 1877, *10*, 1620–3, p. 1620.

[29] O.T. Benfey, 'The Role of Imagination in Science: Van't Hoff's Inaugural Address', *J. Chem. Ed.*, 1960, *37*, 467–70, p. 467. Benfey's translation. Another translation is by G.F. Springer, J.H. van 't Hoff, *Imagination in Science*, Berlin, Springer-Verlag, 1967.

98   *Chemical Structure, Spatial Arrangement*

study of the biographies of great scientists, Van 't Hoff recounted that imagination and an 'artistic sense' was always present in the most creative scientists. In what was perhaps a sly dig at Kolbe, he noted that with the generally increasing numbers of scientists, their average ability has decreased, and the 'rare gifts, among them imagination, move into an unfavourable position with respect to those generally present. This has changed the pattern by which science proceeds'.[30]

Significantly, Van 't Hoff also noted a change in scientific methodology, from cautious speculation based on the 'facts' to the use of hypothesis to generate predictions. He used a military metaphor:

> But scientific discovery has now become something other than it formerly was: It now resembles the shooting up of a fortress from several sides, the cautious ascent of the ruins, and the battle to hoist the flag as all reach the summit together. It used to resemble the simple shift of a single battery, like that with which Napoleon managed to overcome the British fleet at Toulon. Although the imagination can now be replaced by the sacrifice of a great amount of labor, it is not excluded; the role it plays has changed; it is not the role that it is capable of playing.[31]

Van 't Hoff's 'single battery' and 'shooting the fortress' metaphors would also become apparent in later stereochemistry, especially in the dispute between Wislicenus and Arthur Michael over the spatial properties of the unsaturated acids. The incorporation of the spatial arrangement of atoms as an explanatory device, with its implied physical atomism, into chemical theory was coincident with a transformation of chemistry from an inductive, experimentally based science, to a theory-driven science that depended increasingly on the empirical confirmation of predictions.[32]

From this overview of the known responses to the early versions of Van 't Hoff's theory, what conclusions can we draw about its 'reception' amongst chemists? The prevailing conception within the secondary literature on this very early period in the life of 'chemistry in space' has been that chemists, with the exception of Wislicenus, were either critical or indifferent towards it. Snelders, for example, wrote in 1974 that 'only Wislicenus was very enthusiastic', and that 'most chemists were critical of [Van 't Hoff's theory]'.[33] Weyer also wrote of a 'hesitant' acceptance of Van 't Hoff's

---

[30] Benfey, 'The Role of Imagination in Science', p. 470.
[31] Ibid.
[32] Alan J. Rocke, 'Methodology and its Rhetoric in Nineteenth Century Chemistry: Induction Versus Hypothesis', pp. 137–55 in Elizabeth Garber, ed., *Beyond History of Science: Essays in Honor of Robert E. Schofield*, Bethlehem, PA, Lehigh University Press, 1990.
[33] H.A.M. Snelders, 'Practical and Theoretical Objections to J.H. van 't Hoff's Stereochemical Ideas', pp. 55–65 in Ramsay, *Van 't Hoff-Le Bel Centennial*.

ideas among German chemists. There is an element of truth to this claim: with the exception of Landolt's 1877 article (the culmination of work begun before the appearance of *La chimie dans l'espace*), the only published works in 'chemistry in space' before the appearance of Baeyer's strain theory in 1885 were written by Van 't Hoff himself, his friend Gustav Bremer, or his brother Herminus Johannes.[34] The results of their experimental work, moreover, were published in Dutch and only communicated in a shorter form to journals in other languages, usually the *Berichte*. The chemical literature contains almost no mention – favorable or unfavorable – of either these experimental articles or Van 't Hoff's monographs.

Yet, as we have seen, the surviving private reaction to the theory, either by word of mouth or in letters, indicates that chemists were very interested in considering closely and carefully Van 't Hoff's hypothesis, although it is difficult to make an accurate assessment of how chemists as a group accepted Van 't Hoff's theory from the existing documents. There is only one known surviving letter to Van 't Hoff that is at least mildly critical of his theory, for example, and it is conceivable that Van 't Hoff did not keep other negative letters.[35] But the number of letters that do exist definitely indicate an interest in pursuing 'chemistry in space', therefore describing chemists as indifferent certainly appears incorrect. Equally incorrect is the claim that most chemists were opposed to Van 't Hoff's theory. Kolbe's famous, dramatic diatribe – repeated in even the shortest recountings of the early history of stereochemistry – has had the effect of reinforcing Wislicenus as the only enthusiastic supporter of Van 't Hoff's theory (Baeyer and Landolt ran a close second in their enthusiasm), and silencing the other, more valid criticisms offered by Lossen and Claus.[36] The lack of publications in this early period, either supportive or critical of Van 't Hoff, has also tended to reinforce the notion that chemists were indifferent.

We should ask, then, what was the reason for the lack of public enthusiasm for *La chimie dans l'espace* and *Die Lagerung*? Certainly, the prevailing epistemological caution surrounding the meaning of chemical formulas played a role. Fischer recalled that Baeyer had used three-dimensional Kekulé models to explain isomeric hydromellitic acids 'exactly in the sense of today's stereochemistry, but did not have the courage to include the drawings in the article about these compounds'.[37] Another

---

[34] Landolt's study of optical activity appeared in 1877, but it was not so much a work in 'chemistry in space' proper, but a study of the relationship between optical activity and constitution in the existing literature.

[35] Alexander Basarow to Van 't Hoff, 17/29 February 1876, Van 't Hoff Papers. This letter consists of 14 large pages on the experimental weaknesses in Van 't Hoff's theory of optical activity. Baserow was a student of Kolbe.

[36] It also had the effect of reinforcing the opinion of Kolbe as a crank, and not a chemist of the first order, a view that has only recently been redressed in the recent biography by Alan J. Rocke, *Quiet Revolution*.

[37] Emil Fischer, 'Gedächtnisrede', p. 894.

more general reason perhaps was a general sense of cautious optimism surrounding the application of the theory, requiring a waiting period in which the ideas settled in, or for Van 't Hoff, as the principal architect of the theory, to follow through on an intriguing but speculative theory with the experimental work to substantiate it. Modderman's letter expressed precisely this attitude:

> When the author develops as well practically as theoretically, he shows more than a little promise for the future. All the same, it is very difficult to make a definite judgement about the value of his 'proposal'. One could argue for or against it, but *summa summarum*, experiment must decide ... Mr Van 't Hoff can do no better than to demonstrate by experiment the value of his speculations. In this sense, his hypothesis will always be useful, whatever the outcome may be.[38]

In his only known reaction to Van 't Hoff's theory in his 1877 lecture, 'The Scientific Aims and Achievements of Chemistry', Kekulé also expressed this cautious attitude towards the tetrahedron, 'a hypothesis that does not perhaps earn the unreserved praise with which Wislicenus has revered it, but in any case, still less the bitter ridicule that Kolbe has poured upon it'.[39]

We shall return to the early reception of Van 't Hoff's theory in the Conclusion, and compare it to the reception of other theories in chemistry, but here we can note that skepticism displayed by chemists likely had its origin in three areas. First, the prevailing ontological status of structural formulas as symbols perhaps made chemists hesitant to adopt Van 't Hoff's theory. But the complete lack of commentary, positive or negative, about Van 't Hoff's theory that addressed the implied shift from symbolic to iconic formulas makes this reason less important than it might seem at first. Second, as Lossen and Claus noted, Van 't Hoff's model for double and triple bonds was not compatible with known physical forces, and implied an *a priori* division of valence. The third reason for skepticism involves the provisional nature of Van 't Hoff's theory, an aspect that has received little attention in the secondary literature. The number of optically active compounds Van 't Hoff and Le Bel listed in support of their theories whose structures were known with certainty was rather small. The only optically active compounds common to both papers were lactic, aspartic, malic and tartaric acids, and the structures for other active compounds they listed were not yet known with certainty. As Ward has suggested, this made their theories rather speculative, despite Landolt's more comprehensive 1877 and 1879 study that made a stronger connection between constitution and activity.[40] What

---

[38] Tjaden Modderman to neighbors of Van 't Hoff, 17 October 1874, Van 't Hoff Papers.
[39] August Kekulé, 'Die Wissenschaftlichen Ziele und Leistungen der Chemie', pp. 903–17 in Richard Anschütz, *August Kekulé*, 2 vols, Berlin, Verlag Chemie, 1929, vol. 2, p. 912.
[40] Ward, *Polarimeter*, p. 288.

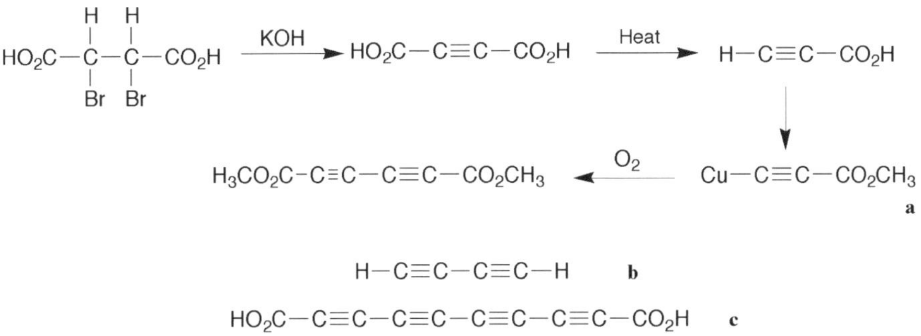

$$HO_2C-\underset{\underset{Br}{|}}{\overset{\overset{H}{|}}{C}}-\underset{\underset{Br}{|}}{\overset{\overset{H}{|}}{C}}-CO_2H \xrightarrow{KOH} HO_2C-C\equiv C-CO_2H \xrightarrow{Heat} H-C\equiv C-CO_2H$$

$$H_3CO_2C-C\equiv C-C\equiv C-CO_2CH_3 \xleftarrow{O_2} Cu-C\equiv C-CO_2CH_3$$

a

$$H-C\equiv C-C\equiv C-H \quad \mathbf{b}$$

$$HO_2C-C\equiv C-C\equiv C-C\equiv C-C\equiv C-CO_2H \quad \mathbf{c}$$

4.2 **(a) Baeyer's synthesis of diacetylene compounds from dibromosuccinic acid. (b) Diacetylene and (c) tetraacetylenedicarboxylic acid.**

Weyer has called the 'hesitant acceptance' of the tetrahedral carbon atom in Germany could be described then as a period of cautious optimism towards a theory that was promising but not yet substantiated by its principal author. If this was the case, waiting for Van 't Hoff to pick up the experimental trail would be fruitless, for after 1877 he would leave all experimental consequences and theoretical development to other chemists.

Adolf von Baeyer and the Strain Theory, 1885

In a two-part article published in the *Berichte* in 1885, Adolf von Baeyer would introduce his famous 'strain theory' as the first application of the tetrahedral carbon atom not discussed by Van 't Hoff. The subject of the articles – the chemistry of the polyacetylene compounds – seems at first glance to be a highly unlikely source for spatial models of the atom, and in the opening pages of the first article, Baeyer focussed on a particular problem in chemical synthesis – producing compounds whose structures contained chains of pure carbon atoms. The synthetic target in the first paper was diacetylene dicarboxylic acid, made by the known oxidative coupling of copper acetylene compounds discovered by Karl Glaser in the 1860s. Dibromosuccinic acid was converted into acetylene dicarboxylic acid (Figure 4.2(a)), which under heat yielded propargylic acid. This acid was then converted to a copper compound and oxidized to give the diacetylenedicarboxylic acid, which was found to be a reasonably stable compound (although light-sensitive), that upon heating exploded 'quite violently with a bang and under discharge of a voluminous amount of charcoal, to my knowledge the first example of an explosive substance made of only carbon, hydrogen, and oxygen'.[41] Despite the thermal instability of the diacetylene, this promising synthesis of the simplest polyacetylene compound indicated:

> that in all probability we will achieve the construction of still longer chains of carbon atoms, if our plans are not thwarted by the facile decomposition already revealed in the diacetylene compounds. With luck we can expect interesting results, because that kind of accumulation of pure carbon atoms in all probability may lead to new insights into the nature of free carbon, about which so little is still known despite the immeasurably large number of carbon compounds.[42]

Baeyer had good reason to worry about the stability of these polyacetylene compounds, although he did have some luck in preparing longer chains of pure carbon. In the second paper, he described the preparation and properties of tetraacetylene dicarboxylic acid (Figure 4.2(c)), isolated in only 'small quantities, crystallized in beautiful needles', but 'which are extraordinarily explosive'.[43] The isolation of diacetylene itself (Figure 4.2(b)) proved to be problematic, and Baeyer attempted to characterize it by preparing terminal diiododiacetylene, which also 'On heating in the tube ... exploded with great violence and produced a red flash of light, a very peculiar appearance'.[44]

The explosive characteristics of these polyacetylene compounds, noted repeatedly as peculiar by Baeyer, prompted him to end the second article with four pages of 'theoretical considerations', in which he attempted to explain this *chemical* property of polyacetylenes – their tendency to explode and release large amounts of heat – by postulating a particular *physical* characteristic of bonding within the molecule. Specifically, he related the explosive properties of acetylenes to the 'laws of ring closure', which he derived from seven 'general axioms (*Sätze*) on the nature of the carbon atom'. The first six axioms contained both the general laws of chemical combination from conventional structure theory and the basic principles of the tetrahedral carbon atom:

1 the tetravalence of carbon;
2 the equivalence of valence;
3 the equal distribution of valences in space, 'the corners corresponding to those of a regular tetrahedron inscribed in a sphere';

---

[41] Adolf von Baeyer, 'Über Polyacetyleneverbindungen. Erste Mittheilung', *Berichte*, 1885, *18*, 674–81; Adolf von Baeyer, 'Über Polyacetylenverbindungen. Zweite Mittheilung', *Berichte*, 1885, *18*, 2269–75, reprinted in Adolf von Baeyer, *Gesammelte Werke*, 2 vols, Brauschweig, Vieweg, 1905, vol. 2, pp. 685–91 and 692–703.
[42] Baeyer, *Gesammelte Werke*, vol. 2, p. 691.
[43] Ibid., p. 694.
[44] Ibid., p. 698.

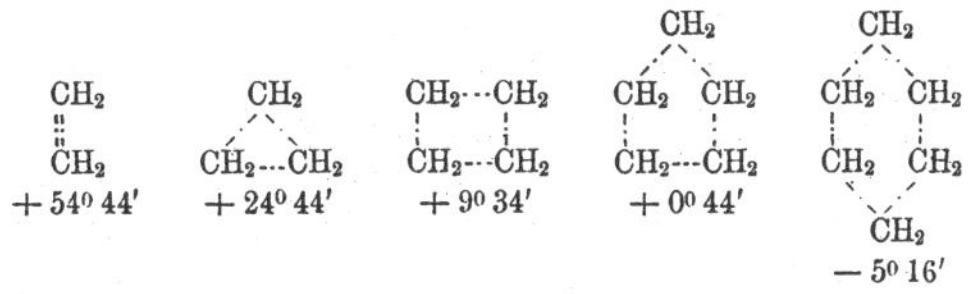

$$
\begin{array}{ccccc}
CH_2 & CH_2 & CH_2\text{---}CH_2 & \overset{CH_2}{CH_2\ \ CH_2} & \overset{CH_2}{CH_2\ \ CH_2} \\
\overset{\|}{CH_2} & CH_2\text{---}CH_2 & CH_2\text{---}CH_2 & \underset{CH_2\text{---}CH_2}{|\quad\quad|} & \underset{CH_2\ \ CH_2}{|\quad\quad|} \\
+54^{\circ}\,44' & +24^{\circ}\,44' & +9^{\circ}\,34' & +0^{\circ}\,44' & \underset{CH_2}{\diagdown\ \diagup} \\
& & & & -5^{\circ}\,16'
\end{array}
$$

4.3 Baeyer's drawing of ring compounds showing the relative strain (depicted as deviation from the standard bond angle for tetrahedral geometry) in rings containing two to six carbon atoms.

4   the 'Le Bel-Van 't Hoff law' (*Le Bel-Van 't Hoff' sche Gesetz*) that the groups on carbon atom cannot exchange places and that a carbon atom with four different groups has two isomers;
5   the multiple bonding of carbon atoms; and
6   the existence of open or closed chains of carbon atoms.[45]

To these six axioms, Baeyer added a seventh, clearly derived from the 'Le Bel-Van 't Hoff law':

> The four valences of the carbon atom act in directions that join in the middle point of the sphere with the corners of the tetrahedron and which make an angle of 109° 28' with one another.
> The direction of attraction can, however, experience a deviation [*Ablenkung*] that consequently has a strain [*Spannung*] that increases with the magnitude of the deviation.[46]

In one of the earliest explicit mentions of physical molecular modeling, Baeyer noted that this concept of strain could be easily seen with the 'Kekulé's ball models', and assuming:

> that the wires resemble an elastic spring and are mobile in all directions. Combining this with the assumption that the direction of the attraction always coincides with the direction of the wires, we attain a true picture of the hypothesis proposed in the seventh axiom.[47]

Forming a five-membered ring with these models resulted in an angle that nearly matched the tetrahedral bond angle, but to produce larger or smaller rings, 'one must bend the wires, that is, a strain occurs in the sense of the seventh axiom'. Baeyer calculated the angular distortion in rings containing two to six carbons, which he provided in a diagram (Figure 4.3). The largest amount of strain was found in ethylene,

---

[45] Ibid., p. 700. Only those portions in quotation marks are direct translations of the axioms.
[46] Ibid. This entire theorem was emphasized by Baeyer.
[47] Ibid.

considered by Baeyer as a 'special case' of methylene rings, 'that according to the seventh axiom ... both axes experience the same deviation, the latter are bowed so far until they are parallel'.[48] The double bond was therefore the 'loosest' (*lockerste*), and as was experimentally well known, would react under the mildest of conditions.

Baeyer could then explain the explosive nature of acetylenes by analogy to the strain in the double bond: the formation of a triple bond from a double bond required an additional 'bending' of a valence to create an even greater strain. This was substantiated by the known values for the heats of reaction ascertained by Julius Thomsen, in which the conversion of a single to a double bond required less energy than the conversion of a double to a triple bond. When coupled with Baeyer's calculation of the deviation in bond angles required for a triple bond, he was led to conclude: 'The strain in acetylene must therefore also be much more considerable than in ethylene.'[49] The concept of strain therefore was tied closely to the thermal properties of various rings. The study of heats of combustion would correlate with the amount of strain indicated using the Kekulé models. Baeyer therefore suggested that the physical property of heat of combustion, like optical activity, would be correlated to the physical characteristics of the molecule.

The strain theory was clearly related to Van 't Hoff's first and third hypotheses, yet Baeyer offered an alternative method for making claims about the three-dimensional properties of molecules. He did not offer an argument based on isomer counting, or on the explanation of a physical property such as optical activity, but from the chemical process of ring closure:

> Ring closure is apparently the one property that can give the most information about the spatial arrangement of atoms. If a chain of 5 and 6 members can be closed easily, and one with a smaller or larger number of members closes only with difficulty or not at all, there must be a spatial reason for this. A theory of carbon compounds that considers the spatial arrangement will therefore naturally begin from ring closure.[50]

Baeyer also recognized that the use of the 'spatial relationships of the carbon atom' had been employed only 'in a few cases, as for example in the Le Bel-Van 't Hoff law, and has raised numerous objections'. Although Baeyer concluded the article by downplaying the significance of his theory, which he did not give the 'value of a completed theory confirmed by experience',[51] his discussion was consciously intended to stimulate interest in spatial conceptions of molecules:

---

[48] Ibid.

[49] Ibid., p. 702.

[50] Ibid., p. 699. Although ingenious, Baeyer's stress on ring formation as a central property in giving information on the spatial properties of molecules was somewhat misleading, as his initial assumptions about the carbon atom included the 'Le Bel-Van 't Hoff law', which was based not on chemical reactions, but the physical property of optical activity.

[51] Ibid., p. 703.

I subsequently hope that the simple connection that can be obtained from these spatial considerations between the explosive nature of acetylene compounds and the laws of ring closure will prompt colleagues to overcome the aversion to this sort of speculation and to recognize the necessity of a further expansion of our discipline in this direction.[52]

It also is important to note that Baeyer used Kekulé's and not Van 't Hoff's molecular models, and the extensive use of these models itself is also an important characteristic of his argument. Van 't Hoff's models did not represent bonds, and could not suggest the idea of a bond as a 'spring', the essential characteristic of the bond in Baeyer's theory. Kekulé's ball and stick models, on the other hand, easily suggested elasticity and spring in bond angles. Clearly, Baeyer considered the ball and stick models as iconic representations – the 'spring' of the wires not only indicated the direction of the attraction between atoms, but modeled directly the amount of strain in a bond between atoms. Baeyer therefore manipulated his models with the assumption that he was mimicking the the physical characteristics of carbon rings, which illustrates that he had implicitly accepted the iconic meaning of structural formulas implied by Van 't Hoff's theory.

Configuration organischer Moleküle, 1886

In 1886, Aemilius Wunderlich, a recently promoted organic chemist at the University of Würzburg, published a 32-page pamphlet entitled *Configuration organischer Moleküle* that contained a novel theory about the form of the carbon atom.[53] In contrast to Van 't Hoff and Le Bel, Wunderlich focussed on the form of the carbon *atom*, rather than simply the arrangement of groups around it, in order to show 'how we can achieve far-reaching results, if we ascribe solid, although unknown form to the atoms'.[54] Wunderlich assumed that atoms existed with specific dimensions out of a

---

[52] Ibid., p. 699.
[53] Aemilius Wunderlich, *Configuration organischer Moleküle*, Würzburg, Leitholdt, 1886. Very little is known about Wunderlich himself, including his birth and death dates. He completed a PhD from Würzburg in 1886 with dissertation entitled *Über Carbamincyanid*, in which he stated his birthplace as Neuenmörbitz in Saxony. He dedicated *Configuration* to Otto Hecht, a student of Emil Erlenmeyer and *ausserordentlicher* Professor of Chemistry at Würzburg at the time. Although *Configuration* could have served as a *Habiliationsschrift*, there is no indication that it did. After the appearance of *Configuration*, Wunderlich's name disappears as an author from the chemical literature, and he is not listed in such comprehensive sources as Poggendorf's *Handwörterbuch* or in Partington's *History of Chemistry*. Two letters from Wunderlich to Heinrich Caro in the Caro Papers, however, indicate that Wunderlich obtained his PhD under Emil Fischer, and perhaps eventually obtained a position as a schoolteacher. A. Wunderlich to Heinrich Caro, 12 and 15 November 1886, Caro Nachlaß, Deutsches Museum München, Archiv, Standnummer N93.
[54] Wunderlich, *Configuration*, p. 6.

'finite quantum of matter that also be distributed in a finite space'.[55] It was therefore impossible in a bond between two atoms to have all points on those two atoms equidistant from one another, and specific parts of each atom may be close to each other while others were far away. Each atom possessed a number of binding sites (*Bindestellen*) equal to its valence number, and that atom is saturated when each of these binding sites is close proximity to another atom's binding site.[56]

To these fundamental assumptions, Wunderlich added definitions of 'central binding point' and 'form':

> 1 Whatever the form of the binding site may be, there is always a designated point that lies in relation to all its points like no other, that is, a focal point in the space of the binding site; it shall be abbreviated as a 'central binding point' [*Bindeschwerpunkt*].
> 2 While I understand form, as usual, to be shape, figure, configuration, the cavity [*Hohlform*] of a molecule shall indicate the space filled by the matter of a molecule and into which it can be thought of as placed as an undivided whole, like a jewel in its case.[57]

Wunderlich then turned to ascertaining the form of the carbon atom, the location of its binding sites, and formulating a model of carbon–carbon bonding. Using a complicated geometrical argument without diagrams, similar to but more rigorous than that used by Van 't Hoff and Le Bel, Wunderlich concluded that the carbon atom possessed a tetrahedral shape, with the four central binding sites occupying the center of each face or 'binding plane' (*Bindeebene*) of the tetrahedron (Figure 4.4). The central binding sites of the carbon atoms would be saturated with atoms of other elements or by other carbon atoms.

A single bond resulted from the simple 'parallel' matching of two binding planes. A double or triple bond would occur when two or three binding sites came as close as physically possible to each other, but in such a bond the binding planes could never simultaneously join as in a single bond:

> The solidity of the atoms makes it impossible for these four faces to position themselves pairwise parallel to one another; for the central binding sites, as we have seen, are unable to approach one another if the binding sites are so inclined instead of parallel to one another, and so in ethylene the binding sites are incompletely saturated ... Finally, should two carbon atoms come as close together as possible, as in acetylene, with three binding sites, so the inclination of these surfaces becomes still greater, the distance between the saturating central binding sites also becomes greater, the saturation smaller, as experience teaches us, the connection is exceedingly modest despite the triple bond.[58]

---

[55] Ibid.
[56] Ibid., p. 7.
[57] Ibid., pp. 7–8.
[58] Ibid., p. 13.

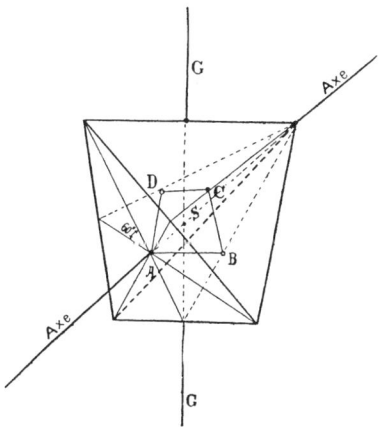

4.4 Aemilius Wunderlich's model of the carbon atom as illustrated in Carl Bischoff and Paul Walden, *Handbuch der Stereochemie* (1894).

Because the binding planes were not parallel in double and triple bonds, the central binding points could not join, and the resulting bond between carbon atoms was weaker than a single bond. This explained the reactivity of alkenes and alkynes towards addition reactions, and also why the strength of double and triple bonds was not simply two or three times the strength of a single bond.

So far, Wunderlich's proposed form for the carbon atom was not significantly different from Van 't Hoff's theory as it appeared in 1877, as Wunderlich used his model to demonstrate the existence of Van 't Hoff's first, second and third hypotheses. Wunderlich's derivation of the form of the carbon atom and his concept of the binding planes, however, allowed him to offer an intriguing alternative model for Baeyer's strain theory. Using the tetrahedral atoms, the construction of rings of increasing size decreased the angle between the binding planes, until in a five-membered ring the binding planes were nearly parallel. In rings with a smaller or larger number of carbon atoms (including ethylene as a two-membered 'ring'), the binding planes would not be parallel and the central binding points could not be completely saturated. The rings therefore increased in stability from ethylene to cylopentane, and the stability would decrease again with larger rings. Wunderlich's model for the 'strain' in small rings was not based on the analogy to a spring, but the abililty of the binding sites to attract one another as the angles between binding planes changed with the size of the ring.

Wunderlich also used his model of the atom to support Kekulé's cyclohexatriene structure of benzene. He first ruled out the Ladenburg prism formula as a geometrically 'impossible thing', but the 'only objection' to the Kekulé cyclohexatriene formula, Wunderlich noted, was the non-existence of isomeric 1,2 disubsituted benzenes predicted by the theory. Kekulé had supposed that the double bonds in the benzene ring could 'oscillate' between carbon atoms, making any given sample a mixture of both isomers. He had even speculated unsuccessfully on the mechanical nature of that oscillation. Wunderlich's model, on the other hand, suggested a plausible mechanical representation of Kekulé's oscillation hypothesis based on the vibrational movements of the carbon atoms that increased and decreased the distance between the central binding points:

> We can conceive of the contours of the binding tetrahedra that cut perpendicularly the axis of benzene as rotational axes around which the tetrahedra make oscillating rotations of 60° from a neutral position, and indeed in following one another in the same sense as ball bearings. Then a beautiful wave motion presents itself in the axis of benzene, which will not affect the side chains, but has the effect that two carbon atoms are bound to one another first with a double bond, then with a single bond.[59]

Although almost unknown now, *Configuration* was well known and often cited during the 1880s and 1890s as a significant contribution to 'chemistry in space'.[60] For all practical purposes, Wunderlich's model of binding planes was identical to Van 't Hoff's obscure face-centered model of bonding given in his open letter to Buys Ballot in 1875, but there is no evidence that *Configuration* derives from that letter. Wunderlich's treatment of the face-centered bonding model is much more sophisticated than Van 't Hoff had ever offered, and the most significant characteristic of *Configuration* was its plausible physical model of the carbon atom and bonding that avoided the misgivings Lossen and Claus had voiced about the problems inherent in the tetrahedron. This model provided an ingenious way to reconcile Van 't Hoff's requirement of non-rotation about a double bond to maintain *cis-trans* isomerism with the need for an attractive force that operated in straight lines.

Wunderlich's other contribution in *Configuration* was the introduction of the word 'configuration' itself to distinguish 'space formulas' from constitutional formulas. Although Wunderlich used the word sparingly, its prominence in the title prompted chemists to adopt 'configuration' as a standard term almost immediately, used in preference to, or sometimes interchangeably with, 'geometrical isomer' or 'space formula' (*Raumformel*). A molecule's constitution would specifically indicate the sequence of chemical connections of atoms in molecules, while its configuration would indicate the spatial arrangement of its atoms.[61]

---

[59] Ibid., p. 22.

[60] As an indication of its rarity and obscurity today, I have been able to locate only one existing copy of the pamphlet in the United States (Emil Fischer's copy in the Fischer archives in California).

[61] The term 'configuration' also contained the modern concept of 'conformation'. Rotational isomers, today called conformations, were called configurations by chemists until 1929, when Howarth coined the term 'conformation'. Jost Weyer, 'A Hundred Years of Stereochemistry: The Principal Development Phases in Retrospect', *Ang. Chem. Int. Ed. Eng.*, 1974, *13*, 591–8, p. 597.

Dix années dans l'histoire d'une théorie, 1887

In 1887, on the tenth anniversary of the publication of *Die Lagerung*, Van 't Hoff reflected on the fortunes of 'chemistry in space' in *Dix années dans l'histoire d'une théorie*. Dedicated to Le Bel 'as a token of my respectful affection', and subtitled as a second edition of *La chimie dans l'espace*, *Dix années* contained both a recounting of the principal arguments from the three previous versions of his theory and a historical section containing Le Bel's article, the initial French translation of Van 't Hoff's pamphlet from *Archives Néerlandaises*, Wislicenus' foreword to *Die Lagerung*, and, in full, Kolbe's critique from the *Journal für praktische Chemie*. The effect of the historical section was to cement the theory of the tetrahedral carbon atom as the 'Van 't Hoff-Le Bel (or Le Bel-van 't Hoff) theory', a label that had already appeared in Landolt's 1877 paper on optical activity, and to make Kolbe's attack famous.

The remainder of *Dix années* was a new presentation of the main theoretical principles of 'chemistry in space', and the inclusion of new chemical facts and theories that had appeared since the publication of *Die Lagerung*. Van 't Hoff reviewed the basis for the asymmetric carbon atom, the rotation about carbon–carbon single bonds, and the restricted rotation around carbon–carbon double bonds. In a new chapter, he discussed the known attempts at cleaving racemic compounds into their optically active components. Van 't Hoff was confident in the success of his theory, and noted that 'at the present moment this opposition has disappeared in Germany with the death of Kolbe, and is now only sustained in France in the person of Berthelot'.[62] The last section dealt with the recent extensions of the dynamic properties of molecule first described in *Die Lagerung*, including the concepts of planesymmetric addition and elimination and residual affinity. The application of these two concepts had been introduced by Wislicenus earlier in 1887 in a major theoretical essay, to which we shall turn in the next chapter.

---

[62] J.H. van 't Hoff, *Dix années dans l'histoire d'une théorie*, Rotterdam, Bazendijk, 1887, p. 24.

Chapter 5

Johannes Wislicenus and Molecular Dynamics

> One cannot avoid imagining [atoms] as spatial objects.
> Johannes Wislicenus, 1888

In 1887, Wislicenus stood at the peak of his career as the Director of the Chemical Institute at the University of Leipzig, one of the largest and most prestigious chemical institutes in Germany. In an ironic twist of fate, Wislicenus had received the call to Leipzig in 1885 to succeed Hermann Kolbe, who had died the previous year. Shortly thereafter, to add insult to injury (at least to the spirit of Kolbe), Wislicenus' research interests turned explicitly to establishing the spatial isomerism in the unsaturated acids. The results of this research were published in March of 1887 as a 77-page essay entitled 'On the Spatial Arrangement of Atoms in Organic Molecules and its Elucidation in the Geometrically Isomeric Unsaturated Acids' (hereafter referred to as 'Spatial Arrangement').[1] In 'Spatial Arrangement', Wislicenus breathed new life into 'chemistry in space' by showing chemists the rich potential of Van 't Hoff's theory of unsaturation, which had 'still not come to full fruition'.[2] Central to his argument was an unprecedented 'physico-chemical' conception of the behavior of molecules that gives 'Spatial Arrangement' a unique position in the history of chemistry. According to Wislicenus, the chemical *and* mechanical characteristics of molecules governed their behavior – he therefore envisioned an unprecedented molecular dynamics to explain chemical reactions.

---

[1] Johannes Wislicenus, 'Über die räumliche Anordnung der Atome in organischen Molekülen und ihre Bestimmung in geometrisch-isomeren ungesättigten Verbindungen', *Abh. math.-phys. Cl. kön. sächs. Ges. Wiss.*, 1887, *14*, 1–77 (hereafter referred to as 'Anordnung'). An English translation was published in 1901: 'The Space Arrangement of the Atoms in Organic Molecules and the Resulting Geometrical Isomerism in Unsaturated Compounds', pp. 61–132 in G.M. Richardson, ed., *The Foundations of Stereochemistry: Memoirs by Pasteur, Van 't Hoff, Le Bel, and Wislicenus*, New York, American Book, 1901. I have retranslated the title, and all subsequent translations are mine; subsequent page numbers refer to the German text.

[2] Wislicenus, 'Anordnung', p. 4.

It is not entirely clear why, despite his obvious enthusiasm for Van 't Hoff's theory, Wislicenus waited until 1887 before he began his own intensive study of the spatial properties of the unsaturated acids. Like his colleagues, Wislicenus may have simply deferred to Van 't Hoff, expecting him to pursue further the development of his own hypotheses. Another possibility was the theoretical atmosphere in Germany, which prevented chemists from fully considering molecules in spatial terms, such that Wislicenus was slow to grasp the implications of Van 't Hoff's second and third hypotheses. This seems unlikely, for the available evidence suggests that Wislicenus grasped immediately the significance of Van 't Hoff's new iconic meaning for structural formulas. In the winter semester of 1878, Wislicenus considered Van 't Hoff's theory important enough to add to his regular lectures in organic chemistry.[3] At the 1877 Munich *Naturforscherversammlung*, a lively discussion took place in which Wislicenus suggested to Wilhelm Lossen that he could solve a theoretical problem in hydroxylamine chemistry assuming that the nitrogen atom had spatial properties. Carl Liebermann's report of the meeting to the German Chemical Society made clear how committed Wislicenus was to the importance of spatial properties for chemical theory, and how developed it was already in the fall of 1877:

> Following [the discussion] Wislicenus indicated that this and many similar facts increasingly make compelling an expansion of chemical concepts to the spatial position of the atoms in the molecule. Despite Kolbe's accusation of spiritualism on account of his endorsement of Van 't Hoff's booklet, he considers Van 't Hoff's theory of the asymmetric carbon atom a fully justified mathematical expansion of our chemical views, and considers its extension to pentavalent nitrogen necessary. Then the facts observed by Herr Lossen could also be explained by considering the proposition that all atoms forming the molecule still exert chemical attractions on one another, including those atoms not directly bound to one another.[4]

The idea that indirectly bound atoms could still attract one another would become a principal feature of 'Spatial Arrangement', and we will return to it later. Finally, in 1883, Wislicenus reported the results of an experiment concerning Van 't Hoff's 'first hypothesis' to the Physical-Medical Society of Würzburg.[5]

The most likely reason for the ten-year period between the appearance of *Die Lagerung* and 'Spatial Arrangement' involves the changes in Wislicenus' professional career. After his 1885 move to Leipzig, his research changed from the synthesis of acetoacetic esters to the chemistry of the unsaturated acids. In an 1872 letter to Emil

---

[3] Wilhelm Sonne, *Erinnerungen an Johannes Wislicenus aus den Jahren 1876–1881*, Leipzig, Wilhelm Engelmann Verlag, 1907, p. 19.

[4] Carl Liebermann, report of the 1877 Deutscher *Naturforscherversammlung*, *Berichte*, 1877, *10*, 2224.

[5] Johannes Wislicenus, 'Abhängigkeit der optischen Drehung von Constitution', *Sitz. der phys.-med. Ges. Würz.*, 1883, 37–40.

Erlenmeyer, Wislicenus had indicated that, on the eve of his move to Würzburg from Zürich, he wanted to complete his work on lactic acid and leave Zürich with a 'clean slate', and perhaps on moving from Würzburg to Leipzig, he intended to do the same.[6] Considering that Wislicenus was also continually occupied with official duties in Würzburg, not the least of which was as rector of the university during its three-hundredth anniversary, this appears the most plausible explanation.[7] Moreover, in 1885 more was known about the chemistry of the unsaturated acids, and it seemed clear that Van 't Hoff was no longer interested in pursuing their chemistry, so the time was therefore propitious for Wislicenus to undertake it himself.[8]

'Räumliche Anordnung', 1887

Although we cannot tell when Wislicenus completed writing 'Spatial Arrangement', it appears that its overall content and argument were completed by the fall of 1886, when he intended to begin a campaign for its contents. He was scheduled to present portions of it in September to the Berlin *Naturforscherversammlung*, but illness prevented him from traveling. His first public presentations therefore occurred in November 1886 and February 1887, in two lectures before sessions of the Royal Saxon Academy of Sciences in Leipzig. In September 1887, he made two major addresses: to the Manchester Meeting of the British Association for the Advancement of Science, and before a general session of the Wiesbaden *Naturforscherversammlung*, at which he advocated his theory as the latest and necessary step in the development of isomerism.[9] In printed form, 'Spatial Arrangement' appeared in spring of 1887 in the *Proceedings of the Royal Saxon Society*, and it was apparently so successful that the society issued a second printing in 1889.[10]

The clarity and organization of Wislicenus' scientific papers facilitates well an analysis of his thought. It is easiest to separate the essay into three different areas,

---

[6] Johannes Wislicenus to Emil Erlenmeyer, 2 May 1872, Deutsches Museum München, Archiv, HS 1968, 476/1.

[7] William Henry Perkin, 'Wislicenus Memorial Lecture', *J. Chem. Soc.*, 1905, *87*, 501–34.

[8] Another reason could be the appearance of Wunderlich's 1886 pamphlet discussed in Chapter 4, which offered a simple model for addition and elimination reactions. As Wislicenus would later explicitly point out that he thought the centers of affinity were at the *corners* of the tetrahedron, and not the faces, the direct influence of Wunderlich on Wislicenus does not therefore seem likely.

[9] Percy Faraday Frankland, 'Johannes Wislicenus', *Proc. Roy. Soc.*, 1907, *A78*, iii–xii, p. vi; Johannes Wislicenus, 'Entwickelung der Lehre von der Isomerie chemischer Verbindungen', *Verh. Vers. Ges. Deut. Naturf. Ärtze*, 1887, *60*, 47–56.

[10] The only previous second printing of a treatise from the society was Wilhelm Weber's work on electrodynamics. Wilhelm Ostwald, 'Johannes Wislicenus', in *Abhandlungen und Vorträge allgemeine Inhalt*, Leipzig Akademische Verlagsgesellschaft, 1904, 444–5, p. 452.

114   Chemical Structure, Spatial Arrangement

roughly corresponding to how Wislicenus himself divided the paper: aims and intentions, fundamental principles used to achieve these aims, and the application of these principles to specific examples.

Establishing Configurations

As we saw in Chapter 4, Van 't Hoff had introduced, but not fully developed, a second level of theory, in which he claimed that the absolute configuration of a molecule could be determined, that the chemist could know whether maleic acid was *cis* or *trans*. Van 't Hoff gave a preliminary argument for making an assignment to maleic and fumaric acid, but had not extensively or explicitly discussed any comprehensive method for assigning configurations to the other pairs of unsaturated acids. In 'Spatial Arrangement', Wislicenus attempted to fill in the details of this second level, and his primary and explicit goal, reflected in the essay's title, was the fulfilment of Van 't Hoff's implicit claim about ascertaining the configuration of molecules, specifically the unsaturated acids.

Wislicenus therefore not only claimed that the differences in the unsaturated acids could be explained by the spatial arrangement of atoms, but also that it was possible to know exactly which configuration each isomer possessed. 'The number of stereometric formulas conceived in this way [geometrically]', according to Wislicenus, 'is completely sufficient to designate the constitutional differences in the "abnormal" isomers; it only remains whether there are means to establish which of the possible symbols corresponds with each individual modification.'[11] In other words, Wislicenus wanted to establish a method or methods for assigning the configurations to each of the unsaturated acids, analogous to methods used to assign specific chemical structures to unknown compounds. For example, Wislicenus noted that deciding which of the two 'space formulas' (*Raumformeln*) for crotonic acid (Figure 5.1) belonged to the two acids was 'not easy to find' (*nicht leicht zu treffen*).[12] But the means for establishing their configurations 'nevertheless follows by itself, if one follows with physical models all processes of formation for the geometrically isomeric unsaturated compounds, as well as their transformations into one another, or of course under constant consideration of their geometric relationships'.[13]

Ideally, this method would yield not only the configurations, but would also show that the 'genetic' relationships – the set of chemical transformations within a related group of compounds – was a natural consequence of these principles:

---

[11] Wislicenus, 'Anordnung', p. 11.
[12] Ibid., p. 12.
[13] Ibid.

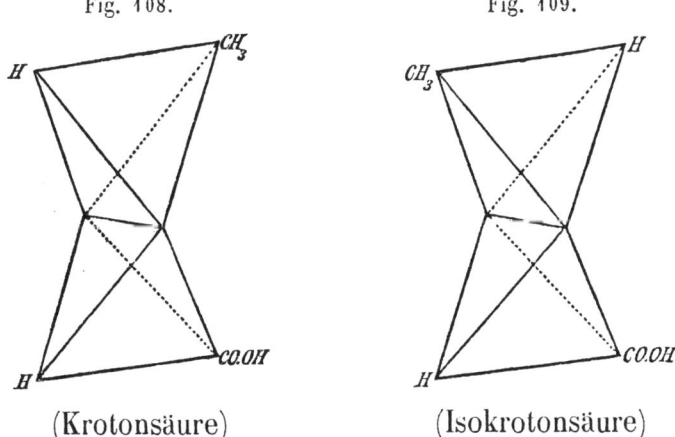

5.1 Wislicenus' space formulas for crotonic and isocrotonic acid.

> This attempt will generally be successful not only if it manages for any reason to assign plausible formulas, spatially different but structurally identical, to the relevant isomers, but also if these formulas are rational in the greatest sense, that is, when from these formulas the totality of their chemical and especially genetic [*genetische*] relationships, unexplained in any other way, follow as a simple consequence.[14]

Wislicenus therefore had two inter-related goals in mind, and two inter-related questions to answer. The primary and explicit goal, reflected in the title of the work itself, remained the determination of the configuration of molecules – what do molecules look like? But he also wanted a theoretical system from which the observed chemical transformations flowed as a 'simple consequence' – what underlying process or processes relates these compounds together?

Wislicenus' did not base his argument for configurational assignments by an appeal to new experimental evidence. Although he noted a few preliminary results from his own laboratory, the bulk of his argument consisted of a theoretical reinterpretation of existing chemical literature on the unsaturated acids. Wislicenus was confident that the 'available observations' would be completely sufficient to justify Van 't Hoff's second and third hypotheses, to assign configurations to the unsaturated acids, and to explain 'facts previously considered as "abnormal" in a surprisingly simple way'.[15] Wislicenus considered the explanation of these 'abnormal' facts as the strongest proof of his theory.

---

[14] Ibid., p. 6.
[15] Ibid.

Fundamental Principles

In order to establish the configurations of the unsaturated acids, Wislicenus employed three fundamental tools; one representational, and two theoretical. The first tool was Wislicenus' elaborate use of 'space formulas' to represent the spatial characteristics of molecules. These space formulas in 'Spatial Arrangement' and subsequent publications were a simplified version of Van 't Hoff's drawings that depicted the tetrahedron literally, with no central carbon atom or valence lines (for example Figure 5.1). The carbon atom was the tetrahedron itself, and the corners were the points of chemical affinity. From his later theoretical statements, it is clear that Wislicenus regarded these tetrahedra as a reasonable approximation of the appearance of a carbon atom, although in 1887 he made no explicit statements to that effect. The text of 'Spatial Arrangement' was liberally sprinkled with such formulas (the title page mentioned 186 total figures), with one on almost every page. He used space formulas in a much more elaborate manner than Van 't Hoff, drawing various reaction intermediates and rotational configurations in addition to the space formulas for the unsaturated acids themselves. Wislicenus' argument was therefore as much visual as verbal. It seems unlikely that Wislicenus could have reached his conclusions without using physical hand-held models of some sort, but curiously, he mentioned 'physical models' only once and in passing, and largely remained silent about the details of the models themselves.

Wislicenus also introduced two additional theoretical premises necessary for establishing configurations. The first was a modification of Van 't Hoff's second hypothesis. Wislicenus assumed that in a carbon–carbon single bond there was a free and 'law-like' (*gesetzmäßig*) rotation of the atoms about their common axis, and that in a double or triple bond, this rotation did not exist. According to Wislicenus, however, the rotation about the carbon–carbon single bond was not entirely free, but subject to the forces of chemical affinity present in the molecule. Atoms in molecules did not consume all their affinity when they formed bonds with other atoms: 'In a compound molecule the elementary atoms not directly united to one another still exert an attractive effect on one another – not only by gravity but by an actual chemical attraction.'[16] This force of attraction was caused by the ordinary elective or differential chemical affinities between atoms: 'The mutual intramolecular effects of indirectly bound elementary atoms must have the same cause as the effects that atoms belonging to different molecules have on one another – it is the effects of their specific affinities.'[17] Wislicenus regarded the residual existence of 'energies of affinity' (*Affinitätsenergien*) as 'the essential part' of the new theory, and 'just as

---

[16] Ibid., p. 14. Wislicenus emphasized this entire passage.

[17] Ibid., pp. 14–15. Wislicenus emphasized this entire passage.

necessary as the assumption that when different molecules approach each other, the mutual chemical effect between their atoms causes the chemical conversions among them'.[18]

The theoretical concept of residual affinity was not invented by or unique to Wislicenus. The idea had surfaced in at least three short theoretical articles in the *Berichte* between 1876 and 1881, and in the British chemical literature in 1885 and 1886.[19] In 1876, Heinrich Kommrath supposed that in a chemical reaction, the 'available force' in the starting materials was always greater than that in the products.[20] In 1881, Lossen postulated that the oxygen and sulfur atoms exerted an attractive force on one another in OCS (carbon dioxide with one oxygen atom replaced by sulfur), even though they did not contain a direct bond between them, and in the same year, Heinrich Klinger suggested that the given quantity of affinity in an atom was not necessarily matched by the affinity contained in the atom of another element (that is, the given quantities were not always exact multiples of one another).[21] In Britain, the concept of residual affinity appeared in the context of explaining the presence of water of crystallization. Because certain compounds contained water as a component, and this water could be added or removed without altering the fundamental properties of the compound, S.U. Pickering supposed that the molecules of water were held to the molecule by a residual affinity because the total valence for the central atom did not correspond to a whole number.[22] The concept of residual affinity was therefore somewhat commonplace, but its precise origin in Wislicenus' theory remains unclear, as he did not offer any citations to articles on the concept. It may be that the use of residual affinity was commonplace enough that he thought a reference unnecessary.

Whatever the origins of his use of residual affinity, it is clear that Wislicenus did not postulate a purely mechanical rotation about the axis of the carbon–carbon single bond, but a dynamic rotation mediated by the chemical attraction of non-bound atoms. If the radicals on each carbon system were the same (such as in ethane), the rotation would be relatively free and unhindered. If the radicals differed, however, the chemical attractions between the atoms or radicals attached to each carbon

---

[18] Ibid., p. 14.

[19] The origins of the concept of residual affinity have not yet been fully explored, although Russell has described some of its manifestations in the mid-nineteenth century. Colin A. Russell, *The History of Valency*, New York, Humanities Press, 1971, pp. 201–23.

[20] H. Kommrath, 'Beitrag zur Theorie der chemischen Verwandschftskraft', *Berichte*, 1876, *9*, 1392–5.

[21] W. Lossen, 'Über die sogennante Verschiedenheit der Valenzen eines mehrwehrtigen Atoms', *Berichte*, 1881, *14*, 760–5; H. Klinger, 'Zur Frage nach den Affinitätsgrössen des Kohlenstoffs', *Berichte*, 1881, *14*, 783–5. Both articles were written in response to Adolph Claus, 'Zur Frage nach den Äffinitätsgrossen des Kohlenstoffs', *Berichte*, 1881, *14*, 432–5, discussed in Chapter 4.

[22] S.U. Pickering, 'Atomic Valency', *Proc. Chem. Soc.*, 1885, *1*, 122–5, and 'On Water of Crystallization', *J. Chem. Soc.*, 1886, *49*, 411–32.

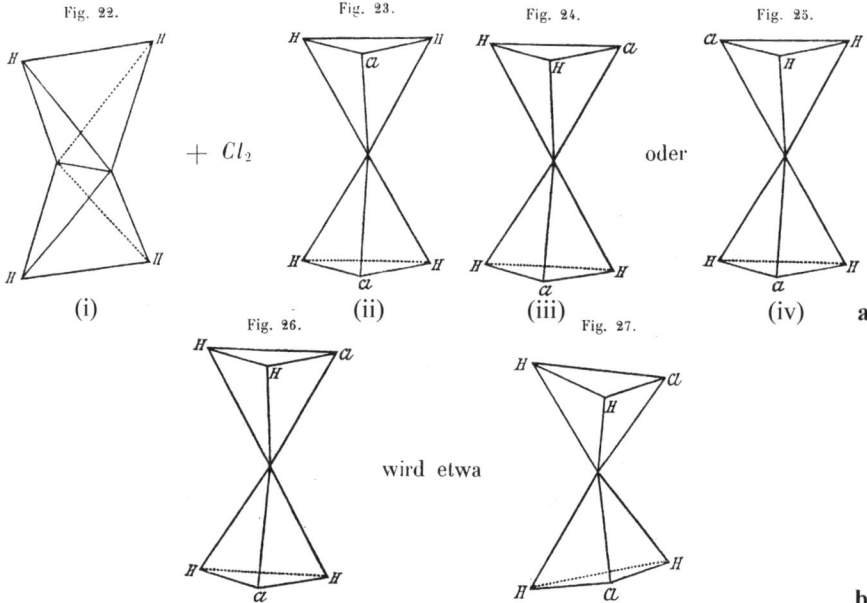

5.2 Examples from 'Spatial Arrangement' showing Wislicenus' modified concept of free rotation with a 'favored' configuration. (a) shows the planesymmetric addition of chlorine to ethylene to give dichloroethane. Formulas (ii) and (iii) represent the same molecule after rotation of the upper carbon atom. (b) shows the distortion from the regular tetrahedra that results as a result of attractions between atoms on the favored configuration, in this case the attraction of the positive hydrogen and the negative chlorine.

atom would force the rotation to occur in such a way that the atoms or groups would position themselves to satisfy greatest their chemical affinity. Some spatial configurations were more stable than others, and the rotation would eventually reach an equilibrium, or 'favored' (*begünstigte*) configuration (Figure 5.2):

> Under the influence of [chemical affinities] *the rotation of both univalent carbon atom systems bound to one another ... must take place in a way such that elementary atoms with the greater affinity for one another approach each other as closely as possible*, and their directions become parallel to a common axis.[23]

---

[23] Wislicenus, 'Anordnung', note 1, pp. 14–15. Wislicenus' emphasis.

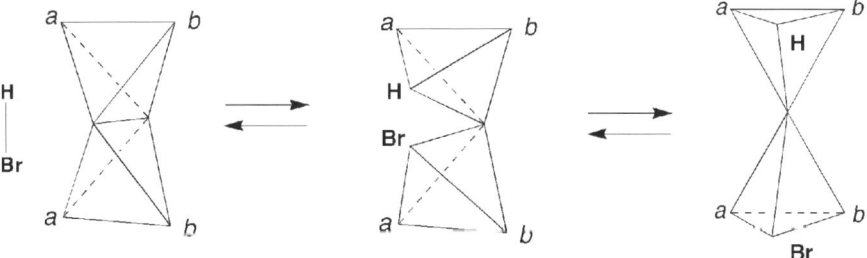

5.3 'Planesymmetric' addition, according to Wislicenus. Towards the right is addition, towards the left is elimination.

Under the influence of heat, this stability would decrease, and more of the less favorable configurations would occur among the population of molecules:

> *At sufficiently high temperature, in a molecular aggregate configurations will always exist that do not correspond to the strongest attractions. Their number will increase with the rising mean temperature of the sample.* But the positions determined by the strongest attractive forces always become the most favorable, and are still present at high temperature in greater amounts than those configurations produced only by thermal impulses [*Wärmestosse*].[24]

While Wislicenus' first assumption was a novel modification of Van 't Hoff's second hypothesis, his second assumption was an explicit adoption of the mechanism of addition and elimination suggested in *Die Lagerung*. When a diatomic molecule (such as chlorine or hydrochloric acid) added to an unsaturated compound, or eliminated from it, the process always occurred from the *same side* of the molecule, a process Wislicenus called 'planesymmetric' (*plansymmetrisch*) addition (Figure 5.3):

> First of all it is obvious *that in the conversion of a trivalent to a divalent bond between two carbon atoms*, presuming that no process occurs other than simple addition and the dissolution of only one of the three valence pairs, *the two radicals that are taken up by the carbon atoms at the outset must fall on the same side of the common axis of both systems*, for this can occur as long as the common bond between the two other valence pairs is unchanged.[25]

Wislicenus assumed that the opposite process, where addition or elimination of diatomic molecules occurs from opposite sides of the molecule (an 'axial-' or

---

[24] Ibid., p. 16. Wislicenus' emphasis.
[25] Ibid., p. 13. Wislicenus' emphasis.

'centrisymmetric' addition) could never occur, and he did not even mention this process as a possibility. His reasons for choosing (or assuming) the planesymmetric process over the axialsymmetric process were twofold. First, given the form of his 'space formulas', the idea of an axialsymmetric addition was itself inconceivable, since an addition would involve the dissolution and reformation of both bonds joining the carbon atoms, and they somehow would then have to rejoin to form the product. The planesymmetric process, on the other hand, required breaking only one of the two bonds, and involved the least amount of change in the molecule. Given the means of representation and the premise of least structural change, the assumption of planesymmetric addition does not appear unreasonable.

These two principles – favored configuration and planesymmetric addition – rested on different foundations. Deciding which configuration was favored relied on the principles of electrochemistry and the recognition that certain atoms and radicals were positive or negative. Attributing a positive or negative charge to an element, in turn, derived from that element's migration in an electrolytic solution towards the positive or negative pole of a battery. Wislicenus did not make an explicit list of negative and positive elements, or rank them by their degree of charge, and as we shall see, assigning a residual affinity was as arbitrary as it was useful.[26] Nevertheless, despite the ambiguity in absolute values, assigning a positive and negative charge did have a reasonably secure empirical basis. This was clearly not true for the assumption of planesymmetric addition. Although it proved extremely useful for Wislicenus' purposes, it had no empirical foundation. It was a simply logical consequence of considering Van 't Hoff's model of multiple bonding together with the phenomena of addition reactions, and an assumption that helped Wislicenus achieve his goals. Just as Dalton's assumption of simplest chemical combination allowed him to calculate a consistent set of atomic weights, Wislicenus' assumption of planesymmetric addition allowed him to make consistent assignments of configurations.

When we compare Wislicenus' conception of molecules given in 'Spatial Arrangement' with his earlier theoretical publications (described in Chapter 2), it is quite clear that in 1887 Wislicenus considered molecules and atoms in a profoundly different manner. Drawings on paper were clearly iconic in character, mimicking the physical characteristics of molecules. Chemical formulas were not simply reaction formulas, but represented objects subject to the laws of physics and mechanics, in addition to the known laws of chemical affinity. With the combination of rotations, attractions, favored configurations and the influence of heat, Wislicenus considered

---

[26] Like the concept of residual affinity, Wislicenus provided no references for the positive and negative character of atoms and radicals, and their relative amounts. This would seem also to be an instance of common knowledge which required no references. In his critique of Wislicenus, Arthur Michael would cite different charges and magnitudes than Wislicenus, also without providing references.

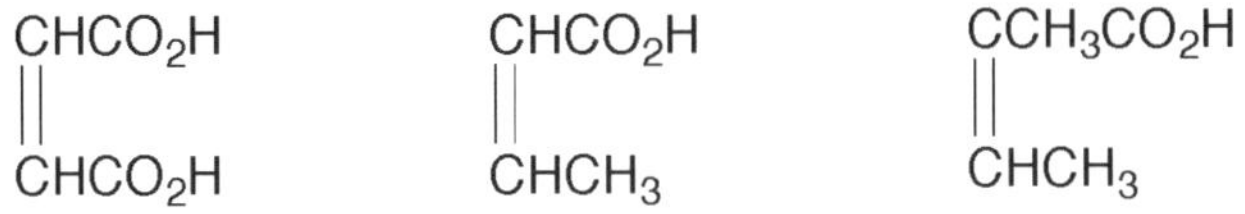

$$\begin{array}{c}\text{CHCO}_2\text{H}\\||\\\text{CHCO}_2\text{H}\end{array} \qquad \begin{array}{c}\text{CHCO}_2\text{H}\\||\\\text{CHCH}_3\end{array} \qquad \begin{array}{c}\text{CCH}_3\text{CO}_2\text{H}\\||\\\text{CHCH}_3\end{array}$$

Maleic and   Crotonic and   Angelic and
Fumaric acid   Isocrotonic acid   Tiglic acid

5.4 The three pairs of isomeric, structurally identical unsaturated acids that Wislicenus studied in 1888 and 1889.

the molecule as a complex object with a unique set of inner dynamics.[27] Also visible is a continuity with his earlier theoretical goals. Van 't Hoff's theory enabled him to visualize in a more precise manner the individuality of molecules. The unique combination of physical form, intramolecular motions, and intramolecular chemical attractions determined its chemical behavior.

Experimental Support for the Theory, 1887–1889

Although the argument in 'Spatial Arrangement' was largely theoretical, throughout the text Wislicenus referred to preliminary results from his laboratory that supported his configurational assignments. The full details of these results were published in a series of four lengthy articles published in the *Annalen* between 1887 and 1889. Three of the articles attempted to establish configurations for sets of isomeric unsaturated acids (Figure 5.4), and the fourth (by the student A. Blank) was on the addition products of stilbene derivatives. For our purposes, the two articles on maleic and fumaric acid, and crotonic and isocrotonic acid (parts one and three in the series) are sufficient for demonstrating the nature of the chemistry behind Wislicenus' configurational assignments.

Maleic and Fumaric Acids

In 1887, chemists still largely disagreed about the explanation for the difference between maleic and fumaric acid, and Van 't Hoff's spatial theory was by no means the only explanation for their isomerism. Rudolf Fittig continued to advocate the existence of a double bond in fumaric acid and free affinities in maleic acid. Richard Anschütz, in a comprehensive review of maleic and fumaric acid that appeared almost

---

[27] Wislicenus also invoked other motions, such as migrations of radicals to 'vacant' valences, and 'loosening' of bonds, to explain certain reactions.

simultaneously with 'Spatial Arrangement', advocated a cyclic lactone structure for maleic acid described in the last chapter, and continued to deny the validity of Van 't Hoff's spatial explanation.

Wislicenus drew together the known chemical relationships between maleic and fumaric acid, previously considered 'entirely incomprehensible', to demonstrate that 'in the light of the new theory, [they] are no longer puzzling'. He recounted ten known transformations of the two acids that could be explained on spatial terms. Of those ten, four will serve to summarize the overall form of his argument:

1 At temperatures less than 150° C, malic acid lost water to form fumaric acid, but between 170 and 180° C it formed larger quantities of maleic acid. This indicated the increased presence of the less favored configuration at higher temperatures.
2 Maleic acid was transformed relatively easily in the presence of strong acids into fumaric acid, but fumaric acid remained unchanged or underwent addition of the mineral acid.
3 Oxidation of maleic acid with potassium permanganate resulted in optically inactive tartaric acid, whereas fumaric acid yielded racemic tartartic acid.
4 Upon heating, maleic acid easily lost water to form an anhydride, while fumaric acid 'on heating sublimes with no essential change'.[28]

Although it came last in Wislicenus' list of reactions, the permanganate oxydation of maleic and fumaric acid would seem most crucial for demonstrating the assumption of planesymmetric addition as well as the proposed configurations of maleic and fumaric acid. In 1882, Kekulé and Anschütz found that fumaric acid reacted with potassium permanganate to give racemic tartaric acid: optically inactive, but capable of separation by crystallization into two different tartrates that had the opposite effect on the plane of polarized light – that is, the product was composed of two enantiomers. The reaction of maleic acid with permanganate also gave an optically inactive tartaric acid, but one that could *not* be split by crystallization into two different tartrates (Figure 5.5).[29]

Wislicenus started from the supposed configurations of the two isomeric tartrates obtained by oxidation, and worked backwards to the starting acids. Maleic acid was therefore the *cis* isomer, and fumaric acid the *trans* isomer. It is important to note here the unprecedented way in which Wislicenus interpreted the process of addition. Since the atoms of maleic and fumaric acid lay in one plane, neither side would be favored and the oxygen atoms could add to both sides randomly. In the case of maleic acid, both sides were geometrically identical – addition to either side resulted in the

---

[28] Wislicenus, 'Anordnung', p. 12.
[29] August Kekulé and Richard Anschütz, 'Über Tanatar's Bioxyfumarsäure', *Berichte*, 1880, *13*, 2150, and 'Über Tanatar's Trioxymaleinsäure', *Berichte*, 1882, *14*, 713–17.

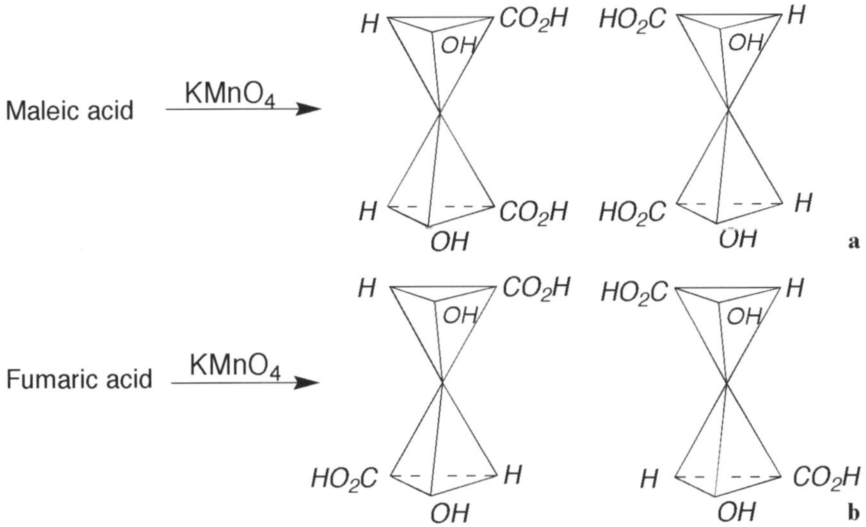

5.5 The reaction of maleic and fumaric acid with permanganate. Addition to either side of maleic acid (a) results in the same internally symmetric product that is optically inactive. Addition to either side of fumaric acid (b) leads to two different products that are non-superimposable mirror images of each other. Although the initial product from fumaric acid is optically inactive (like the product from maleic acid), it can be resolved into two optically active compounds.

same molecule. 'But geometrically,' noted Wislicenus, 'fumaric acid behaves entirely differently.'[30] Addition of oxygen to the two sides of permanganate resulted in two different molecules that were enantiomorphic.

Reaction (2) rested on the empirical observation of the stability of the two acids in the presence of mineral acids. In the presence of acids, maleic acid transformed into fumaric acid, but under the same conditions, fumaric acid remained intact. The purely mechanical property that in a *cis* acid the two acid groups were close to one another predicted that under the right conditions, one group should move to the other side so that both groups did not occupy the same space. This assumption led to the conclusion that maleic acid was the less stable of the two acids, and therefore the *cis* acid. Wislicenus went further and suggested a mechanical explanation (or mechanism) for this transformation, in which he equated the role of the mineral acid with that of a 'ferment' (Figure 5.6). This would become one of the weakest explanations in

---

[30] Wislicenus, 'Anordnung', p. 36.

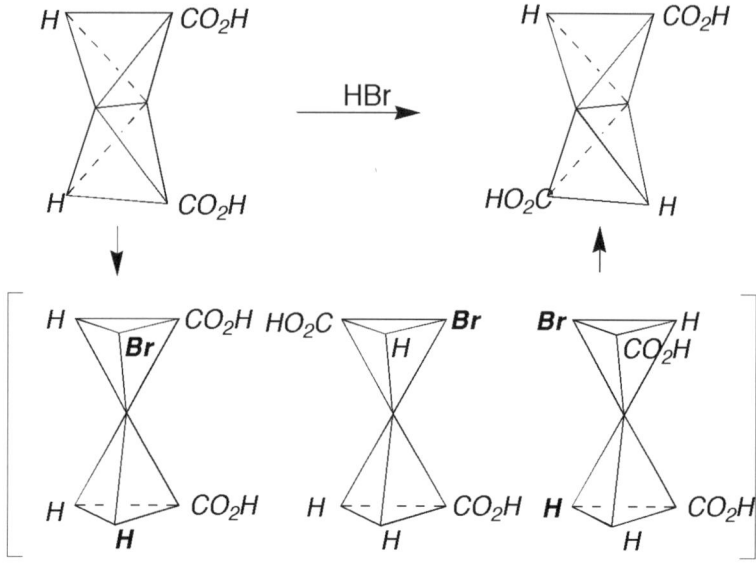

5.6 Wislicenus' proposed mechanism for the conversion of maleic into fumaric acid.

Wislicenus' theory, and we will return to it later, but it illustrates how he used his principles to explain a chemical transformation.

The formation of maleic acid anhydride supported the configurational assignments most securely, since it depended on an empirical observation, and not on the more complex theoretical assumptions of the other two arguments. Using the assumption that only acid groups spatially close to one another could lose water easily, maleic acid must be the *cis*, and fumaric acid the *trans* isomer. These three arguments – anhydride formation, permanganate oxidation, and the conversion of maleic to fumaric acid – together had a mutually reinforcing effect. The cleavage of water from maleic acid to form an anhydride was empirically the most secure argument for the *cis* configuration. Given this configuration, the results of the permanganate oxidation were in turn reinforced, and gave credence to the assumption of planesymmetric addition. The proof of planesymmetric addition then supported the presumed mechanism for the catalytic transformation of maleic acid to fumaric acid, and therefore also the presumed favored configurations in his other cited transformations.

The most convincing evidence for these configurations arose from Wislicenus' investigation of exceptions to his theory of maleic and fumaric acid, the full results of which were published as the first of the supporting articles in the *Annalen*

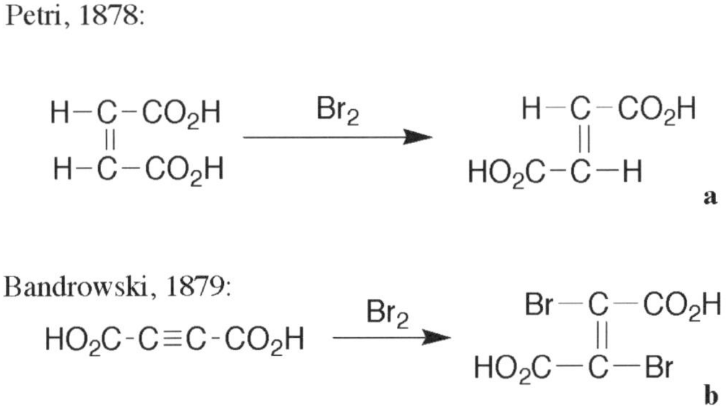

Petri, 1878:

$$\begin{array}{c}H-C-CO_2H \\ \| \\ H-C-CO_2H\end{array} \xrightarrow{Br_2} \begin{array}{c}H-C-CO_2H \\ \| \\ HO_2C-C-H\end{array} \quad \mathbf{a}$$

Bandrowski, 1879:

$$HO_2C-C\equiv C-CO_2H \xrightarrow{Br_2} \begin{array}{c}Br-C-CO_2H \\ \| \\ HO_2C-C-Br\end{array} \quad \mathbf{b}$$

5.7 Apparent exceptions to Wislicenus' principles. In (a), maleic acid is converted to fumaric acid with bromine. In (b), bromine appears to add to acetylene dicarboxylic acid to give fumaric acid, the result of an apparent axialsymmetric addition.

(summarized in Figure 5.7). In 1878, Camille Petri, a student of Rudolf Fittig, had found that in the presence of bromine, maleic acid transformed into fumaric acid, a transformation that Wislicenus found 'simply inexplicable'. In 1879, Bandrowski had found that bromine added to acetylene dicarboxylic acid (ADCA) to give dibromo*fumaric* acid, a result that 'directly contradicts the theory'.[31] Bandrowski's results forced one of two conclusions: 'either my theory is wrong, or the process is not as we have perceived it, that is, it is not a simple addition process.'[32] If planesymmetric addition was universal, as Wislicenus claimed, the reaction of bromine with ADCA could not take place in a single addition, but must involve a sequence of several steps.

Wislicenus studied each of these reactions closely, isolating and identifying as many products as possible. He repeated Petri's experiment, isolating and identifying the various products of the reaction, and found that hydrobromic acid formed during the reaction. A careful separation of reaction products showed mostly isodibromosuccinate, but also small amounts of a substance that melted at 175–177° C, close to the known melting point of bromofumaric acid. This indicated to Wislicenus that the

---

[31] Wislicenus assumed it was a derivative of fumaric acid because it did not form an anhydride. Rudolf Fittig, 'Untersuchungen über die ungesättigte Säuren. II. Abhandlung', *Annalen*, 1879, *195*, 56–179; E. Bandrowski, 'Weitere Beiträge zur Kenntniss der Acetylendicarboxylsäure', *Berichte*, 1879, *12*, 2212–16.

[32] Johannes Wislicenus, 'Untersuchungen zur Bestimmung der räumlichen Atomlagerung. Erste Abhandlung: Beiträge zur Geschichte der Fumarsäure und Maleinsäure', *Annalen*, 1888, *246*, 53–96, p. 69.

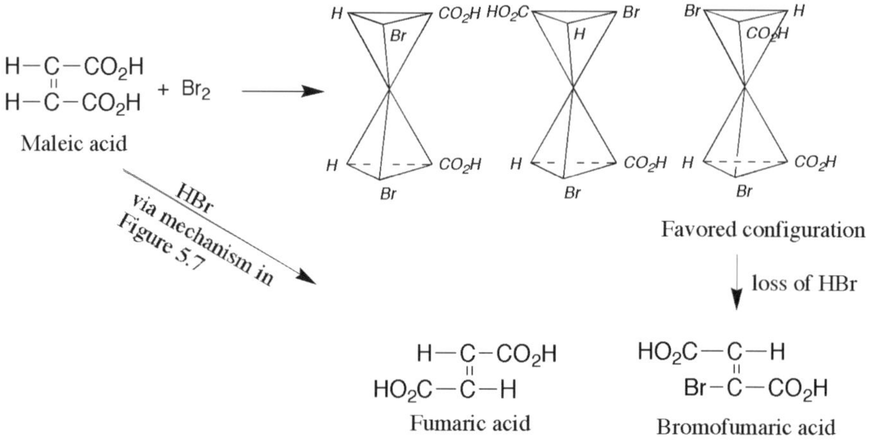

H−C−CO₂H
‖
H−C−CO₂H   + Br₂   →

Maleic acid

via HBr mechanism in Figure 5.7

[Diagram: three tetrahedral structures showing bromine addition stereochemistry, leading to "Favored configuration" and then "loss of HBr"]

Favored configuration
↓ loss of HBr

H−C−CO₂H   HO₂C−C−H
‖   ‖
HO₂C−C−H   Br−C−CO₂H

Fumaric acid   Bromofumaric acid

5.8 Wislicenus' explanation for Petri's results. Bromine adds normally to maleic acid, but hydrobromic acid is generated in the formation of bromofumaric acid that then reacts with unreacted maleic acid to form fumaric acid.

intermediate dibromo addition compound had decomposed to bromofumaric acid and hydrobromic acid, which could then add to unreacted maleic acid, causing the conversion to fumaric acid by the mechanism he had proposed earlier (Figure 5.6). Given these two facts, the mechanism of addition became clear (Figure 5.8):

> The formation of fumaric from maleic acid when the latter encounters bromine is … only the consequence of the decomposition of part of the initially formed isodibromosuccinic acid and the normal effect of the produced hydrobromic acid on the still unchanged maleic acid.[33]

Wislicenus presented a similar analysis of Bandrowski's results. By repeating Bandrowski's experiment and closely observing the progress of the reaction, he noted that the reaction developed 'considerable' (*erhebliche*) amounts of carbon dioxide. Among the products he found oxalic acid, and noted again the formation of hydrobromic acid. In a further extraction of the liquor taken from the crystallization of the main product dibromofumaric acid, he found successively higher amounts of monobromofumaric acid, and smaller amounts of bromomaleic acid and dibromomaleic acid. The major products of the reaction then were dibromofumaric acid, monobromofumaric acid and hydrobromic acid, along with considerable amounts of

---

[33] Ibid., p. 68.

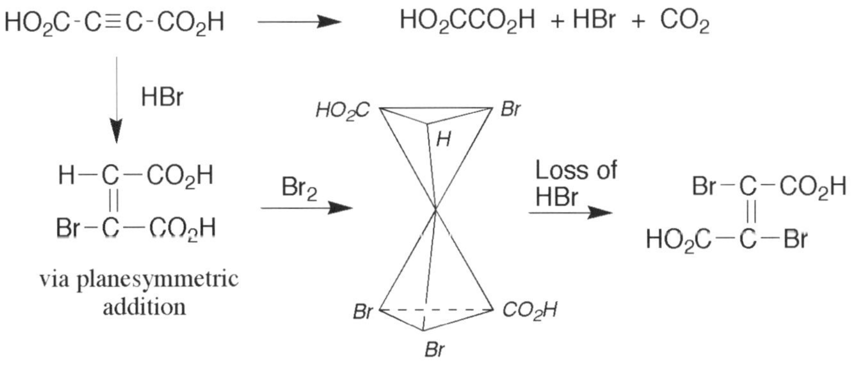

$$HO_2C\text{-}C\equiv C\text{-}CO_2H \longrightarrow HO_2CCO_2H + HBr + CO_2$$

$$\downarrow HBr$$

$$\begin{array}{c} H-C-CO_2H \\ \parallel \\ Br-C-CO_2H \end{array} \xrightarrow{Br_2} \quad \xrightarrow{\text{Loss of HBr}} \begin{array}{c} Br-C-CO_2H \\ \parallel \\ HO_2C-C-Br \end{array}$$

via planesymmetric addition

Favored configuration

5.9 **Wislicenus' explanation for Bandrowski's results. A side reaction generates oxalic acid, hydrobromic acid and carbon dioxide. The hydrobromic acid generated can then react with unreacted acetylene dicarboxylic acid to form bromomaleic acid that subsequently adds bromine and loses HBr from its favored configuration to yield dibromofumaric acid.**

oxalic acid and carbon dioxide. From the various products of the reaction, Wislicenus reconstructed a plausible mechanism for the production of dibromofumaric acid (Figure 5.9). Hydrobromic acid was generated and added to any remaining unreacted acetylene dicarboxylic acid (ADCA), and produced bromomaleic acid. Bromine added to this intermediate to give a tribromosuccinic acid, which then eliminated hydrobromic acid to form dibromofumaric acid.[34]

Wislicenus' proposed mechanism led to the conclusion that the exclusion of hydrobromic acid from the reaction mixture would result in dibromomaleic acid as the major product. However, hampered by the lack of a suitable reaction solvent that would dissolve both bromine and ADCA and not be prone to generation of hydrobromic acid (the previous experiments were done in water), Wislicenus attempted to minimize the influence of hydrobromic acid by diluting the reaction mixture. He subsequently found that the concentration of hydrobromic acid significantly decreased on dilution, and that the amount of dibromomaleic acid increased.[35] Wislicenus felt safe to conclude: 'These observations definitely confirm

---

[34] Ibid., p. 77. The two steps could also occur in the opposite order, leading to the same result.
[35] Ibid., p. 80. Arthur Michael solved the solubility problem more effectively by using the esters, and not the free acids of ADCA. He could therefore use carbon tetrachloride, not water, as solvent. It is not clear why Wislicenus did not conduct this experiment.

the supposition that it is the hydrobromic acid that hindered the direct addition of bromine to acetylenedicarboxylic acid.'[36]

The ingenious and painstaking set of experiments (their description took up 40 pages in the *Annalen*) illustrates Wislicenus' experimental skill in separating and identifying various products from an extremely complex mixture of reaction products. This skill complimented fully his grasp of the theoretical issues and his use of theory to generate his research program. He had determined to his own satisfaction that the addition process was not simple, and to his credit, he turned an experimental result that seemingly most contradicted his theory into results that *supported* it. He therefore felt justified in concluding: 'to my knowledge, in the genetic relationships of fumaric and maleic acid there no longer exists a single fact that cannot be explained by geometrical considerations, and no fact that does not serve decisively to support the new hypothesis.'[37]

The Crotonic and Isocrotonic Acids

The assignment of configurations to maleic and fumaric acid was relatively simple compared to the analysis of the crotonic acids, a family of β-methylacrylic acids. The earliest and best known member of this series was crotonic acid itself, a solid that melted at 72° C. Isocrotonic acid existed only as a liquid (considered an impure compound) that boiled at 170–174° C. In addition, there were also α and β chlorocrotonic and isocrotonic acids, not all of which had been isolated:

$CH_3CH=CHCO_2H$ crotonic acid and isocrotonic acid
$CH_3CH=CClCO_2H$ α-chlorocrotonic acid and α-isochlorocrotonic acid
$CH_3CCl=CHCO_2H$ β-chlorocrotonic and β-isochlorocrotonic acid

Van 't Hoff's third hypothesis suggested that crotonic and isocrotonic acid were spatial isomers, and also predicted two α-chlorocrotonic acids and two β-chlorocrotonic acids. The two β-chloroacids had been isolated by Wislicenus' student A.R. Friedrich in 1883, who had declared them structurally identical on the basis of their reaction with potassium hydroxide. In 1887, there was only one known α-chlorocrotonic acid, but Wislicenus would soon report the existence of several new isomers and attempt to establish their configuration.

Because all of the crotonic acids were monobasic and could not form anhydrides, Wislicenus lacked an obvious, empirically secure way to draw a preliminary conclusion about which of the two compounds was *cis*. Establishing configurations

---

[36] Ibid.
[37] Ibid., p. 37.

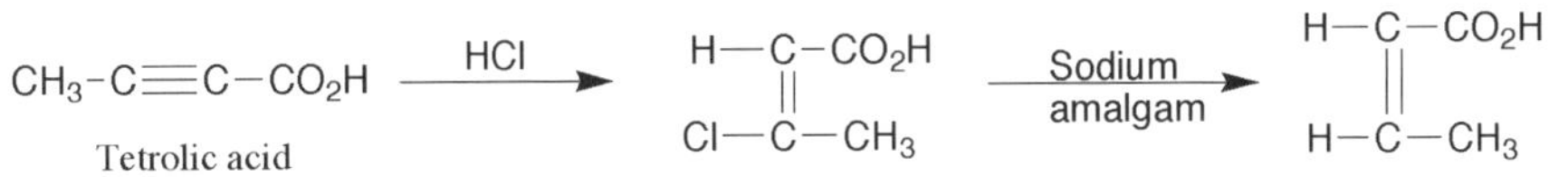

$$CH_3-C\equiv C-CO_2H \xrightarrow{HCl} \begin{array}{c} H-C-CO_2H \\ \parallel \\ Cl-C-CH_3 \end{array} \xrightarrow{\text{Sodium amalgam}} \begin{array}{c} H-C-CO_2H \\ \parallel \\ H-C-CH_3 \end{array}$$

Tetrolic acid ⟶ β-chlorocrotonic acid, m.p. 95° C ⟶ Solid crotonic acid
β-chloroisocrotonic acid must
therefore be the opposite configuration.

5.10 Reactions used by Wislicenus to establish configurations of the isomeric β-chlorocrotonic acids.

of the crotonic acids would therefore require a greater amount of theoretical sophistication and a longer, less direct chain of reasoning. Wislicenus' argument relied heavily on his principles of favored configuration and planesymmetric addition for the analysis of the transformations between the two acids and their α and β halogen derivatives. In 'Spatial Arrangement', Wislicenus assigned configurations to the crotonic and isocrotonic acids, and the two β-halogen derivatives, basing the argument on Friedrich's results. His reasoning for the β-chlorocrotonic acids went as follows (summarized in Figure 5.10):

1 Hydrochloric acid added to tetrolic acid to give β-chlorocrotonic acid, a solid that melted at 95° C. Given the assumption of planesymmetric addition, this acid must be β-chlorocrotonic acid with the *cis* configuration.
2 The other acid (β-chloroisocrotonic acid) with a melting point of 40° C must have the *trans* configuration.
3 These configurational assignments were also confirmed by the reaction velocities of the two β-chlorocrotonic acids. Since β-chlorocrotonic acid was *cis*, it should eliminate hydrochloric acid faster in the presence of alkali. This was the case.
4 When reduced with sodium amalgam, β-chlorocrotonic acid gave solid crotonic acid. Assuming the process of least structural change, the hydrogen atom must literally take the same place occupied by the chlorine atom. Therefore, solid crotonic acid had the *cis* configuration, and isocrotonic acid had the *trans* configuration.

Wislicenus followed a similar path in establishing configurations for the α-chlorocrotonic acids. In 1887, Wislicenus gave a preliminary assignment of configurations and reported the synthesis of the previously unknown α-chlorocrotonic acid predicted on the basis of Van 't Hoff's theory. He published the complete argument for his assignment of configurations in the third part of his series of papers in the *Annalen* that appeared in 1888. In this article, he described the full genetic relationships and proposed configurations for all of the members of the chlorocrotonic and bromocrotonic acids.[38] The configurations of the α-chlorocrotonic acids were assigned with the following argument (summarized in Figure 5.11):

1 Solid crotonic acid (assumed to have the *cis* configuration, for the reasons given above) reacted with chlorine to give a dichloroaddition product that melted at 63° C.
2 This addition product reacted with alkali (usually potassium hydroxide) to give α-chloroisocrotonic acid, that melted at 66° C. This was a new acid, and possessed the *trans* configuration (based on the favored configuration of the intermediate dichloro succinic acid). Sodium amalgam reduced this acid to give isocrotonic acid.
3 The addition product (from point 1 above) also reacted with sodium bicarbonate to give a great deal of carbon dioxide and a liquid that boiled at 33° C with the empirical formula C_3H_5Cl. Its boiling point was lower than an already known liquid with the same composition, α-chloropropylene, and therefore the new liquid was an isomer of the first, iso-α-chloropropylene, with a *cis* configuration (assigned on the basis of the favored configuration of the intermediate acid salt).
4 A similar argument followed for a sequence of reactions beginning with liquid isocrotonic acid (summarized in Figure 5.12). Isocrotonic acid yielded an oily addition product that Wislicenus was unable to purify.
5 The addition product reacted with alkali to give α-chlorocrotonic acid that melted at 97° C.
6 The addition product also reacted with sodium bicarbonate to give a liquid that boiled at 36° C, matching the boiling point of the already known α-chloropropylene.

The proof of the assigned configurations therefore rested ultimately on comparing the two spatially isomeric α-chloropropylenes, whose configurations were assigned using the assumption of planesymmetric addition and favored configuration of intermediates. The chloropropylene from isocrotonic acid should be *trans* (Figure 5.11), and that from crotonic acid should be *cis* (Figure 5.12). Therefore, the new α-chloropropylene should react with potassium hydroxide faster than its isomer, by virtue of the spatial proximity in the *cis* grouping of the hydrogen and chlorine atoms. The iso-α-propylene isomer, as predicted, decomposed nearly twice as fast with alkali as its isomer.

The apparently tight-knit relationship between theoretical prediction and experimental results demonstrated quite forcefully that Wislicenus had found a method that would actually elucidate the spatial properties of molecules. In other words,

---

[38] Johannes Wislicenus, 'Untersuchungen zur Bestimmung der räumlichen Atomlagerung. Dritte Abhandlung: Geometrische Constitution der Crotonsäure und ihrer Halogensubstitutionsproducte', *Annalen*, 1888, *248*, 281–355.

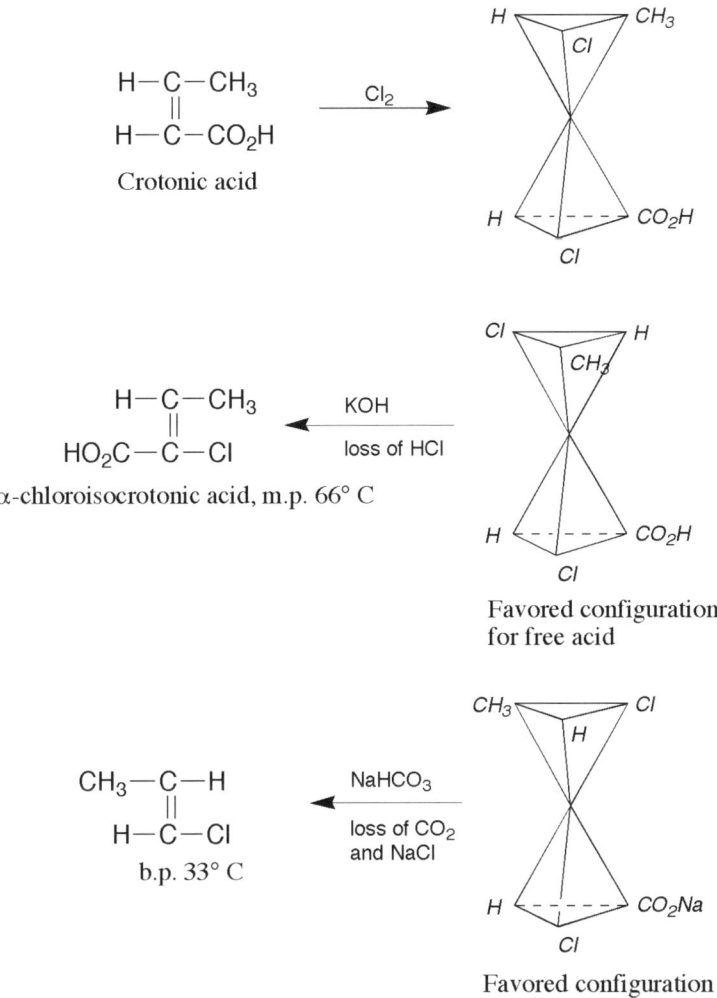

5.11 Reactions used by Wislicenus to establish the configuration of α-chlorocrotonic acid.

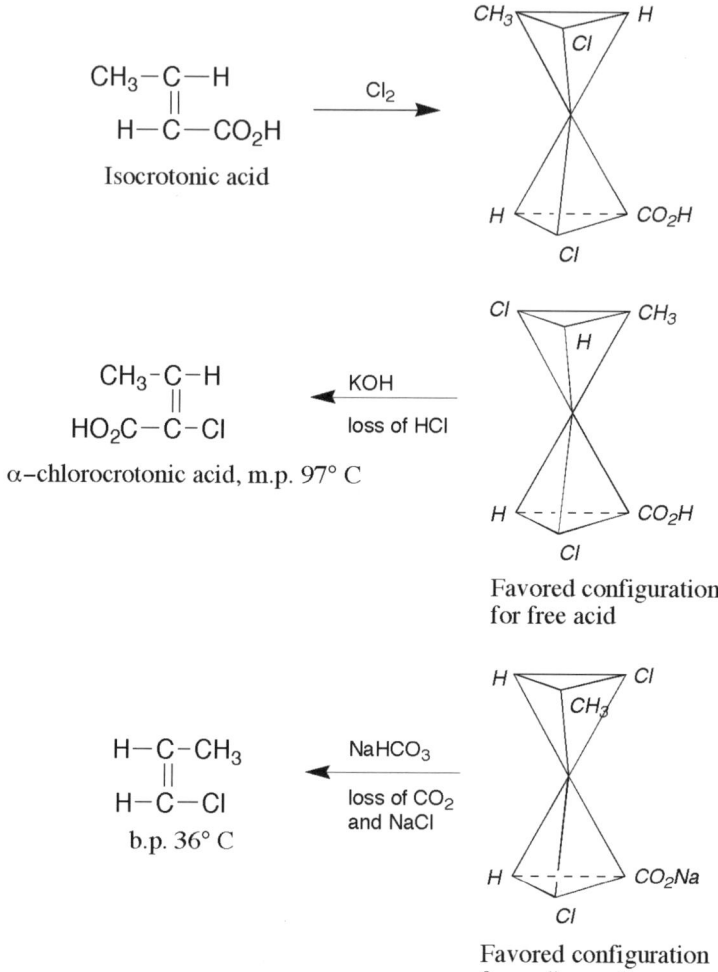

5.12 Reactions used by Wislicenus to establish the configuration of α-isochlorocrotonic acid.

these seemingly intractable, inaccessible properties of molecules had now become accessible to experimental test:

> ... in spite of the still existing uncertainty in applying [these principles] to many special cases, by confining them to the simplest circumstances of spatial arrangement [*räumliche Lagerungsverhältnisse*], they have the value of a true theory, for their simple hypothetical foundation not only provides a uniform explanation for a great number of previously absolutely incomprehensible facts, but moreover is also capable of experimental test.[39]

He also considered his method of using the genetic relationships between isomeric compounds as superior to Baeyer's use of ring closure for ascertaining the spatial arrangement of atoms:

> At the moment, I cannot entirely accept Baeyer's statement that 'Ring closure is apparently the one property that can give the most information about the spatial arrangement of atoms'. As I believe I have shown in the preceding argument, of still far greater significance for the time being, are the simplest conversion processes of unsaturated into saturated compounds in which systems of only two or three carbons are considered, and in which issues such as the occurrence of directional changes in affinity, like those from a state of strain, can as yet remain unconsidered.[40]

Whereas Baeyer's approach employed a mixture of chemical properties (ease of ring formation) and physical properties (heats of formation), Wislicenus relied exclusively on chemical properties – the transformations of the unsaturated acids – to ascertain configurations. Wislicenus therefore made unprecedented inferences from chemical behavior to statements about the physical properties of molecules that cannot be overemphasized. Chemists had consistently denied that *chemical* relationships of compounds could give provide evidence for claims about the *physical* position of atoms in molecules. Making claims about chemical and physical arrangements were two separate issues that required separate methodologies. Wislicenus ignored this claim and inferred spatial properties directly from the chemical data, using an intricate interplay between chemical reactions, chemical attractions, and the spatial configuration of molecules.

Initial Reception

Between 1887 and 1892, several responses appeared to the arguments given in 'Spatial Arrangement'. Several of these were short notes to which Wislicenus did not reply, but which reveal the principal factual weaknesses in his theory. However,

---

[39] Wislicenus, 'Anordnung', p. 76.
[40] Ibid., p. 74.

the papers published by Wilhelm Lossen (1887), Rudolf Fittig (1889–1892) and Arthur Michael (1888–1892) prompted Wislicenus to defend himself, and the resulting exchanges reveal much about Wislicenus' theoretical and methodological assumptions.

Anschütz, Skraup and Lossen

In 1889, Anschütz published a second installment of his 1887 paper on the 'History of Fumaric and Maleic Acid', in which he addressed directly Wislicenus' claims made in 'Spatial Arrangement'.[41] He noted that the 'treatise by J. Wislicenus was received with the greatest applause from many quarters', and had become largely convinced himself by Wislicenus' argument. Wislicenus had 'applied a unitary explanatory principle on the whole of previously unexplained isomeric properites of the unsaturated carbon compounds', as many of the cases of isomerism 'at the moment, as far as I know, have no better explanation'.[42] Anschütz could not accept completely Wislicenus' analysis of maleic and fumaric acid, however, specifically his explanation for the conversion of maleic to fumaric acid with hydrobromic acid (Figure 5.6). Anschütz called Wislicenus' explanation 'captivating' (*bestechend*) at first glance, but it was 'not difficult to show that it irreconcilably contradicts trenchantly noted facts and therefore cannot be correct'.[43] The explanation rested on the stability of the proposed bromo- or chlorosuccinic acid intermediate, both of which were known compounds. Indeed, it was known that chlorosuccinic acid decomposed at high temperature to give fumaric acid, lending support to Wislicenus' mechanism. But, as Anschütz pointed out, the conversion between maleic and fumaric acid with hydrochloric acid also took place at lower than room temperature, and at this temperature, chlorosuccinic acid was known to be stable, and did not decompose to fumaric acid.

In 1891, the Austrian chemist Zdenko Skraup published an extensive study of the precise conditions for the conversion of maleic to fumaric acid.[44] He considered changes in temperature, acid concentration and type of acid, and noted that acids such as sulfuric, nitric and oxalic acid were sufficient for the transformation, even though they were known not to add to double bonds 'in the sense of Wislicenus' theory'. He concluded that in the mechanism of the conversion, the assumption of additional intermediate compounds was unnecessary, and that 'at least in its current form,

---

[41] Richard Anschütz, 'Zur Geschichte der Isomerie der Fumarsäure und Maleinsäure. II. Abhandlung', *Annalen*, 1889, *254*, 168–82.

[42] Ibid., p. 169.

[43] Ibid., p. 172.

[44] Zdenko H. Skraup, 'Über die Umwandlung der Maleinsäure in Fumarsäure', *Monatshefte*, 1891, *12*, 107–45, and 'Zur Theorie der Doppelbindung', *Monatshefte*, 1891, *12*, 146–50.

Wislicenus' idea is certainly not correct'.[45] Skraup and Anschütz were not alone in their criticism of the proposed mechanism for the conversion of maleic to fumaric acid. Arthur Michael and Victor Meyer would also later recognize this problem with bromosuccinic acid, and regarded it as a true deficiency in the theory, which was certainly one of the weakest links in 'Spatial Arrangement'.

Lossen responded to 'Spatial Arrangement' in December 1887 with an explicit theoretical criticism that was discussed in Chapter 4.[46] Lossen's critique was short but penetrating, and we shall examine Wislicenus' equally illuminating response below.

Rudolf Fittig

In the course of his study of the chemistry of the angelic and tiglic acids (the dimethyl acrylic acids), Wislicenus had reinterpreted and corrected the experimental work of Rudolf Fittig. An exact contemporary of Wislicenus, as a student Fittig had been assistant to Heinrich Limpricht and Friedrich Wöhler at Göttingen, and in 1875 had succeeded Adolf von Baeyer as director of the chemical institute at the University of Strasbourg. Between 1876 and 1883, Fittig and his students had thoroughly examined the addition reactions of unsaturated acids, and in many cases had not reached any conclusions about their structure. In 1878, Fittig and one of his students, Alexander Pachenstecher, had published the results of experiments on tiglic and angelic acid, in which they found that both acids gave the same addition product with bromine. Wislicenus finished his article on tiglic acid and angelic acid in mid-November 1888, as the fourth and final article in the series of supporting work for 'Spatial Arrangement', and it appeared in the *Annalen* in the spring of 1889. Wislicenus and his student Maximilian Pückert found that the reaction of angelic acid with bromine gave a unique, previously unknown addition product, dibromoangelic acid; the existence of this product, coupled with its behavior in a sequence of related reactions lent credence to the theory that angelic and tiglic acid were spatial isomers, and to the assignment of their configurations. They had been unable to obtain this addition product in completely pure form, but its dramatic difference in properties, particularly its highly hygroscopic character, from Pachenstecher's dibromotiglic acid obtained from angelic acid convinced them they had obtained the true dibromide of angelic acid.

---

[45] Skraup, 'Über die Umwandlung', p. 108. Skraup followed the article on the analysis of maleic and fumaric acid with a short theoretical paper outlining a theory of the double bond and addition processes that was compatible with his conclusion that there could be no intermediates in the conversion process. Skraup, 'Zur Theorie der Doppelbindung'.

[46] Wilhelm Lossen, 'Über die Lage der Atome im Raume', *Berichte*, 1887, *20*, 3306–10. An analysis of Lossen's critique and Wislicenus' response is given in Peter J. Ramberg, 'Johannes Wislicenus, Atomism, and the Philosophy of Chemistry: A Translation and Commentary', *Bull. Hist. Chem.*, 1994, *15/16*, 45–53.

Fittig could not accept that he or one of his students could have missed the formation of dibromoangelic acid, and wrote to Wislicenus in April of 1889 to suggest a collaboration to remove the discrepancy. Wislicenus apparently agreed to Fittig's request, and sent Fittig a sample of Pückert's angelic acid dibromide. Their planned collaboration would not last. In the late spring of 1889, Wislicenus became ill with typhoid, and was unable to return to the lab until the following fall. By the winter of 1889–90, there had been no meaningful results; Fittig became impatient, and could no longer wait for Wislicenus. Their collaboration and correspondence ended in January 1890. Fittig began his own set of experiments and prepared his own publication that appeared in the *Annalen* in June of 1889, 'for if I wanted to remain silent longer, my silence could be interpreted as agreement, and the inaccuracies would even pass into textbooks'.[47] Fittig made clear at the beginning of his article that he thought Wislicenus and his students, in a rush to confirm their own theoretical preconceptions, had ignored the chemical facts generated in his own laboratory:

> With the aid of this conception about the geometric position of atoms, [Wislicenus] attempted to explain some isomeric relationships that had not been sufficiently explained by previous notions of atomic linkage. In the process, many of the facts observed by me and my students stood in his way and wouldn't fit into his theoretical picture, so he has seen himself fit to repeat our research and has found results that agree better with his theoretical ideas.[48]

It appeared initially that Fittig was justified. Crystallographic measurements of Fittig's two dibromides prepared from angelic acid and tiglic acid proved them identical (they had the same form and incident angles), and the sample dibromide Wislicenus sent to Fittig was 'missing all the properties of a pure chemical compound … I could certainly establish that it was not a homogenous compound, and that it contained a lot of tiglic acid dibromide contaminated with other solids, among them *calcium compounds*'.[49] Fittig therefore concluded that he could not have been mistaken: 'as I read Wislicenus' treatise, these differing observations were absolutely puzzling to me. An error on my part was inconceivable.'[50]

Fittig repeated the bromine addition to angelic acid under several different conditions, including his own and Pückert's conditions, and he isolated only tiglic acid dibromide in varying degrees of purity. Fittig concluded that the presence of light was necessary to complete the reaction. He determined the end of the reaction by observing the loss of the characteristic dark orange color of bromine, and under all the

---

[47] Rudolf Fittig, 'Über die Einwirkung von Brom auf die Angelicasäure und Maleinsäure', *Annalen*, 1890, *259*, 1–40, p. 4.
[48] Ibid., pp. 1–2.
[49] Ibid., pp. 4–5.
[50] Ibid., p. 3.

reaction conditions, the color did not completely go away until Fittig uncovered the reaction vessel or moved it towards the window. This explained Pückert's results:

> since [Pückert] used an excess of bromine, it escaped his notice that its influence had ended long ago. He then probably put the solution into light, and only during the removal of the carbon disulfide did the main reaction occur, during which a lot of tiglic acid dibromide must have formed. So he obtained a substance similar to that described in experiment 5 and since it possessed entirely different properties than the pure dibromotiglic acid, he was misled and thought the impure solid was pure or nearly pure dibromoangelic acid.[51]

With good reason, Fittig considered his earlier results completely vindicated: 'and I underwrite today word for word what I published concerning this subject twelve years ago. My results are absolutely correct and I need not retract one line from them.'[52]

Wislicenus regretted that their planned collaboration had not succeeded, and blamed the failure on Fittig's impatience. Unfortunately, he was not able to respond quickly to Fittig's article. Following his bout with typhoid, he suffered from influenza twice, and in the summer of 1890 his son Alwin died unexpectedly. Furthermore, his own comprehensive re-examination of the addition reaction required time to complete, and he also found it necessary to repeat Fittig's experiments. His response therefore did not appear until mid-1892. In a very personal introduction, Wislicenus explained his position, and described an increasingly unsteady relationship with Fittig:

> Unfortunately, as Fittig already related, it never came to agreement. Fittig's impatience to settle definitively the dibromoangelic acid dibromide and to see him forever denied the path into textbooks, was all too sorely tried at the beginning by the very slow progress of my work. As Fittig knew, in May I was ill with typhoid, and could energetically begin independent laboratory work again only in the fall. Besides, the first results had already made it impossible to agree with Fittig's conclusions, so that he notified me that he would cancel our scheduled meeting on January 27, 1890, if I would not answer yes or no in the next few days to the question: 'Do you want jointly under both our names to revoke and designate as erroneous the entire content of Pückert's work with all its conclusions?' The following reproach, 'Now you have treated the matter sluggishly nearly an entire year,' freed me of the necessity to continue further a correspondence that began with the judgment, 'You consider the facts through a veil that your theoretical views have pulled in front of your otherwise so clear eyes, and force the facts to something that matches these conceptions,' and had continued many times with similar grave accusations, placing my integrity as teacher and scholar in question, and I was entirely suited to end the personal friendly relationship which to my happiness I had with Fittig. Better an open duel before colleagues on the validity of facts than such correspondence that has ultimately also forced me to become increasingly bitter![53]

---

[51] Ibid., p. 27.
[52] Ibid., p. 17.
[53] Johannes Wislicenus, 'Über die Bromadditionsproducte der Angelica- und Tiglinsäure', *Annalen*, 1892, *272*, 1–99, p. 7.

In a detailed 90-page discussion, Wislicenus analyzed every aspect of the addition process. He developed an extraction technique to measure the amounts of angelic acid dibromide and tiglic acid dibromide formed, isolated crystalline angelic acid dibromide and characterized it, and agreed with Fittig that light had an influence on the course of the addition. He had known of this influence before the appearance of Fittig's critique, but had not determined its exact nature.[54] It was not, as Fittig claimed, simply to drive the reaction to completion – it caused the production of the abnormal tiglic acid dibromide addition product itself. The stronger the light, and the less bromine added, the greater the amount of tiglic acid dibromide that was formed from angelic acid. Fittig had produced precisely those conditions (bright light, insufficient bromine, warm conditions) that Wislicenus had found to give rise to the abnormal addition product. Both Fittig and Pückert had worked unaware of this influence of light and had therefore reached opposite experimental results. It was only through complete accident on both sides that Pückert had ever isolated the dibromoangelic acid:

> That Pückert first became aware of this isomer of tiglic acid dibromide, obtained in a pure state in different ways by Fittig and his pupils, and that this discovery had not been made in Fittig's laboratory years ago, is due to an accident. In the upper work rooms of the Leipzig laboratory, nearly all the hoods in the outer walls are fastened *between* the windows and are therefore dusky; on the other hand, in the Strassbourg laboratory they lie *in* the windows and are filled with the brightest daylight. So Pückert did the addition of bromine to angelic acid without intending to exclude bright light, and obtained a product consisting mostly of dibromoangelic acid, while Fittig and his pupils – just as unintentionally working in intensive daylight – obtained mainly dibromotiglic acid under otherwise completely similar conditions. Were the construction of the Leipzig laboratory the same as in Strassbourg, my necessary revision of Pachenstecher's experiments would of course as yet have had no different result. Meanwhile, the true dibromoangelic acid would probably already have been discovered.[55]

Discovery here of course, meant the recognition that this compound was the dibromide of angelic acid, a claim Fittig was not willing to make until Wislicenus demonstrated its existence in 1892. Fittig himself had actually noticed an impurity in the preparation of the angelic acid dibromide, both in 1878 and in 1890, and had also noted that its amount increased in indirect light. He was unable to isolate this substance or characterize it, but indicated it 'may be an isomer' of dibromotiglic acid:

---

[54] This only became clear after Fittig, in his second reply to Wislicenus, declared priority for the discovery of light's influence. In a second response to Fittig, Wislicenus cited dates in his laboratory notebook to show that he had discovered this influence in early 1890, after their correspondence had been cut off.

[55] Wislicenus, 'Angelica- und Tiglinsäure', pp. 97–8. Wislicenus' emphasis.

each experiment repeatedly indicates that the crude product contains completely formed dibromotiglic acid, but its properties are initially masked by another compound ... But what is this second compound? I have no certain answer, because I have always had such minimal success in purifying it. But I consider it more probable now than before that it is an isomer of dibromotiglic acid.[56]

In other words, Fittig could not suspect that this substance could actually be angelic acid dibromide. Since Fittig had not even suggested this possibility, Wislicenus could not resist the temptation to call attention to Fittig's myopia. He called Fittig's proposed 'masking substance' a 'legend', into which Fittig:

in the course of time has worked himself so tightly that he cannot free himself ... But what about this additional product formed 'in small amounts' from angelic acid in the addition of bromine dibromotiglic acid, which should have the 'strange' effect of completely changing or 'disguising' the properties of great amounts of dibromotiglic acid? Fittig confesses explicitly that he has not succeeded in isolating this special compound. He has no certain answer to the question of what it may be, but considers it 'more probable now than before that it is an isomer of dibromotiglic acid'. Meanwhile, he still cannot draw the conclusion that it is perhaps dibromoangelic acid, a claim, after all, that is the bridge he will finally reach to recognize the existence of the compound he has so persistently denied.[57]

Fittig, then, could very well have been the 'discoverer' of angelic acid dibromide, but by 1890, to do so would have been to admit his earlier results were wrong. To be fair, he may not have been able to isolate this impurity, and since it was an 'impurity', he had no reason to do so. Wislicenus, on the other hand, required from his theory the existence of angelic acid dibromide, and then looked for it. He felt satisfied that he had 'the right to subject the work of other chemists to a revision under the influence of my theoretical views', and that '*with the aid of advanced methods, under the influence of improved concepts and with greater skill more complete observations* could be made'.[58]

Fittig was quite correct: Wislicenus did see the 'facts' of the reaction through a 'veil of theory', and claimed the right and duty to revise the literature on the basis of that theory. Had he not been attempting to confirm his theory, he never would have attempted to correct Fittig's results, nor would Pückert have claimed the existence of angelic acid dibromide. And later, Wislicenus himself would not have isolated or even attempted to isolate the pure dibromoangelic acid as a response to Fittig. By no means was Pückert at first entirely successful. The impurity of Wislicenus' sample and the identity of Fittig's results with Pachenstecher's justified Fittig's complaint that Wislicenus and his students had ignored the facts generated in Fittig's laboratory.

---

[56] Rudolf Fittig, 'Einwirkung von Brom', p. 26.
[57] Wislicenus, 'Angelica- und Tiglinsäure', pp. 88 and 90.
[58] Ibid., p. 98. Emphasis added.

It was unfair, however, for Fittig to claim that Wislicenus had distorted the facts to force them into his theoretical scheme, because it was clear to Wislicenus and Pückert that they *had* isolated two different bromides. It was indicative of the nature of chemical research that theory can predict the existence of a compound, but it required skill and patience to find and isolate it. And for that, Fittig finally gave Wislicenus credit, although until Wislicenus' rebuttal he refused to acknowledge that Wislicenus and Pückert's results could actually be correct.

At this point, a description of the methodological differences between Fittig and Wislicenus will help to put the conflict in a broader light. As may already be evident from the above discussion, Fittig was wary of grand theoretical schemes like that in 'Spatial Arrangement' into which 'facts' were forced. 'As long as I can live and work,' Fittig wrote to his student Edvard Hjelt in 1883:

> I still insist that in a science like chemistry, observation and the exact determination of facts must precede speculation. But this truly exact establishment of facts is diminishing more and more now; in an age of steam and of electricity, chemical research is supposedly also to be done at a similar pace, and if you read closely all of what is published, which so far I still do, and continually see in so many cases how imprecise the observations are and how imprecise the establishment of facts are, and how far-reaching theoretical conclusions are drawn from such highly deficient observations, at times you cannot fend off the disgust and even temporarily lose your love for science. Contribute to your new sphere of activity by your example, acting against this overgrowth in scientific research, and always remember that in science one exact, completely established fact is of more value than ten imprecise observations.[59]

Fittig's empiricism is also evident in his papers that read like a textbook example of inductive methodology. He started with detailed descriptions of the reaction he or his students performed, and followed these with short theoretical interpretations, many times deferring a complete explanation. Fittig had struggled for years to explain the behavior of unsaturated acids in structural terms, and 'when from elsewhere a fortunate conception was applied to this case, then Fittig prepared himself with a deeper distrust'.[60] His commitment to an inductive methodology gave him an intuitive mistrust of modern science that relied too much on speculation and not enough on 'facts'.

Wislicenus, in contrast, jumped back and forth between theory and experiment, and revealed a greater play in his thought between the prediction of theory and experimental procedure. He was enthusiastic about the claims of 'Spatial Arrangement' as a broad theoretical scheme that would organize the known 'facts' and predict new ones. Well known to be a fair and even-tempered man, Wislicenus

---

[59] Rudolf Fittig to Edvard Hjelt, 1883, quoted in F. Fichter, 'Rudolph Fittig', *Berichte*, 1911, *44*, 1339–83, p. 1364.

[60] Ibid., p. 1372.

nevertheless became weary of the public battle with Fittig. In 1893, he wrote to Emil Fischer:

> Now I need to deal with Fittig again, and sent to Volhard [the editor of the *Annalen*] the day before yesterday my reply to his last, apparently so meaningful, but so absolutely insignificant submission. I hope the pitiful [*leidige*] affair is finally done and I can work again on what I want.[61]

Arthur Michael

The reactions to 'Spatial Arrangement' we have seen so far were for the most part minor. They dealt either with a specific theoretical problem in the model of double bonds or methodology (Lossen), alternative explanations for the unsaturated acids (Fittig and Anschütz), problems with a specific explanation Wislicenus offered for the transformation of maleic to fumaric acid (Skraup), or the existence of a particular compound (Fittig). None of these objections, even the extended discussion with Fittig, could be taken as presenting a serious danger to the overall structure of the argument in 'Spatial Arrangement'. Entirely different, however, was the strong and persistent criticism of Arthur Michael, an eclectic American chemist trained in Germany, towards Wislicenus' chemistry. The principal result of his extensive theoretical and experimental critique, that halogens add to multiple bonds in an *anti* manner (in nineteenth-century terms, 'axialsymmetric' addition), has become imbedded in modern chemical theory, and historical accounts of the history of stereochemistry have traditionally credited Michael with this fundamental discovery.[62] Like many episodes from the history of science, the story was not so simple. Michael never advocated axialsymmetric addition, even though his results pointed to that conclusion, because he refused to adopt the principles of stereochemistry. His results pointed to axialsymmetric addition only if one accepted the premises of stereochemical theory. The reasons for his rejection are complex, and here I can only present a brief overview of the major reasons for Michael's disagreement.[63]

When he first encountered Wislicenus' essay, Michael had been established at Tufts University for seven years. He was born in Buffalo in 1853, the second child

---

[61] Wislicenus to Emil Fischer, 16 January 1893, Emil Fischer Papers, BANC MSS 71/95 z, The Bancroft Library, University of California.

[62] At least two accounts of Michael's work have drawn this conclusion: Albert Costa, 'Arthur Michael (1853–1942) – the Meeting of Thermodynamics and Organic Chemistry', *J. Chem. Ed.*, 1971, *28*, 243–6, and Douglas Tarbell and Ann Tracy Tarbell, *Essays on the History of Organic Chemistry in the United States, 1875–1955*, Nashville, TN, Folio Publishers, 1986.

[63] A full analysis of Michael's critique and his own chemistry is recounted in Peter J. Ramberg, 'Arthur Michael's Critique of Stereochemistry, 1887–1900', *Hist. Stud. Phys. Biol. Sci.*, 1995, *26*, 89–138.

of the first-generation German immigrants John and Clara Michael.[64] John Michael was a prosperous real estate investor, and Arthur's rather eclectic, transient education reflected the fortunate circumstance of being born into a wealthy family. John Michael's business brought his family to Europe, and in 1871, Arthur entered A.W. Hofmann's laboratory in Berlin. After a year, he transferred to the University of Heidelberg and studied for two years under Robert Bunsen, returned to Berlin and Hofmann's laboratory for three years, and finally, spent an additional year in Paris under Adolph Wurtz. During this period, he managed to publish descriptions of small syntheses and new synthetic methods, but despite twelve publications, he never formally obtained a doctorate.

Upon his return to the US in 1880, he became Professor of Chemistry at Tufts College, where he remained until 1889. He married the plant chemist Helen Abbott in June 1888, and they took an extended world tour. They returned in late 1889, when Michael was to become Professor of Chemistry at the newly established Clark University. He resigned from Clark before assuming his duties, and set up a private laboratory at Bonchurch on the Isle of Wight, where he, his wife and a few students remained until 1894, when he returned once more to Tufts College.[65] In 1907, he retired from Tufts, and in 1912 became Professor of Chemistry at Harvard, where he remained until his death in 1942. Although officially at Harvard, he directed his own research from his personal laboratory in Newton, Massachusetts. Before 1900, he published nearly all his extensive research articles in German chemical journals; the *Journal für praktische Chemie* appeared to be his predominant choice, although a few of his articles also appeared in the *Berichte*.

Upon his return from Europe, Michael began to study the cinammic and crotonic acids, and concluded that both acids existed in two structurally identical forms. He referred to them as alloisomers, from the greek *allo* for 'difference', or 'deviation', but declined to offer a reason for this difference, and wanted his term to go no further than to 'unite properties with names', and to recognize the 'type of isomerism'.[66] In the midst of this study of crotonic acid, Michael encountered 'Spatial Arrangement', with its reinterpretation of the crotonic acids in spatial terms. Michael responded

---

[64] This biographical sketch is drawn from the following sources: Costa, 'Thermodynamics and Organic Chemistry', 'Arthur Michael', in Charles Gillispie, ed., *Dictionary of Scientific Biography*, New York, Scribner's, 1973, and 'Arthur Michael', in Edward T. James, ed., *Dictionary of American Biography*, Suppl. 3, New York, Scribner's, 1973. Louis F. Fieser, 'Arthur Michael', *Biog. Mem. Nat. Acad. Sci.*, 1975, *XLVI*, 331–66, draws heavily on Costa's 1971 article; W.T. Read, 'Arthur Michael', *Ind. Eng. Chem.*, 1930, *22*, 1137–8; anonymous, 'Arthur Michael', *American Cyclopedia of Biography*, 1916, *17*, 172.

[65] Michael resigned because the president of the new college refused to allow Michael's wife access to the laboratory. Paul R. Jones, 'The First Half Century of Chemistry at Clark University', *Bull. Hist. Chem.*, 1991, *9*, 15–19.

[66] Arthur Michael, 'Zur Alloisomerie in der Zimmtsäurereihe. III. Mittheilung', *J. prak. Chem.*, 1889, *40*, 63–8, p. 68.

immediately, and his first critique appeared as a 34-page article in the *Journal für praktische Chemie* in May 1888.[67] Michael pulled no punches – in the first sentence he labeled the essay 'superficial' and declared that the 'agreement between theory and fact is only apparent, since it is dependent in part on unjustified assumptions, in part from biased interpretations and from the oversight of a great number of facts'.[68] Many of the 'facts' were reactions of the crotonic and cinammic acids with which Michael had become familiar in the course of his own research. He proved himself extremely adept at using Wislicenus' own principles against him, deriving multiple configurations or even multiple structures for the same compound. He pointed out the improbability of migrations of atoms from valence to valence, the appearance of intermediates that were either unstable or unlikely to exist, or the assumption of intermediate reactions that did not or could not occur. Wislicenus' explanations could only be *ad hoc* rationalizations after the fact, driven by theoretical necessity and derived from assumptions such as planesymmetric addition that themselves were never proven.

Wislicenus replied to Michael's critique in 1888, in a section of his paper on crotonic acid. He thought Michael had falsely applied his theory to examples that were not yet thoroughly studied. He felt his work was incorrectly interpreted and improperly cited, and presented alternate explanations under his own theoretical framework for those reactions that Michael had analyzed. He could not agree with Michael's outright dismissal of his theory:

> Ready at any moment to drop my hypothesis as soon as perfectly good and exhaustive observed facts absolutely contradict it, because of my experiments, I also certainly do not find myself, even after Michael's critique, in the position of having to surrender. On the contrary, up to now not only its applicability, but also its often unsuspected curious successes, still surprise me again and again.[69]

Wislicenus admitted that some transformations remained unexplained under the new theory, and constituted legitimate criticisms. He was confident, however, that such gaps in theory would eventually be closed, and the existence of weak points in a theory did not justify its dismissal.[70] The many cases that *were* successfully interpreted under his principles argued for the validity of his theory, and Wislicenus gave great merit to a theory that found itself applicable in a great variety of examples. He also recognized the use of his principles to predict new chemical facts and indicate

---

[67] Arthur Michael, 'Zur Kritik der Abhandlung von J. Wislicenus. "Über die räumliche Anordnung der Atome in organische Molekülen"', *J. prak. Chem.*, 1888, *36*, 6–39.

[68] Ibid., p. 6.

[69] Wislicenus, 'Untersuchungen zur Bestimmung der räumlichen Atomlagerung. Dritte Abhandlung', pp. 354–5.

[70] Ibid., pp. 354–5.

directions for his research. The 1887 isolation of a new α-chlorocrotonic acid, he claimed, was a direct result of applying his new theory.

Michael did not attach much value to Wislicenus' claim of predictive power. He himself had isolated this new α-chlorocrotonic acid at the same time as Wislicenus, but there was no reason to suspect that Wislicenus' synthesis of the new acid confirmed any of Wislicenus' principles or supposed configurations, because the existence of such an acid could be inferred from the existence of alloisomeric acids in other groups of unsaturated acids. 'Wislicenus,' Michael complained, 'apparently out of unfamiliarity with my work, has overlooked that the formation of such acids already had an experimental basis; since moreover each of the chlorocrotonic acids must possess many configurations, we can ascribe no great value to Wislicenus' conjectures.'[71]

Wislicenus appeared bewildered at Michael's agnosticism, and found his analysis lacking precisely because he presented nothing to take the place of geometrical isomerism as the cause of the differences in isomers. Instead, Michael offered the term 'allo' only to label these differences and distinguish these compounds from one another. Michael's work on the chlorocrotonic acids was 'actually quite valuable ... but even the five general concluding statements contain nothing that would be in the position to shed light on the secrets of "allo"-isomerism'.[72] Since Michael had introduced 'neither ... new methods nor ... new theoretical points of view, but only ... a new name', Wislicenus felt entirely justified in studying the same group of compounds as Michael.[73]

Michael freely admitted that 'the phenomena that I have named alloisomerism still hides many secrets'.[74] Wislicenus was certainly entitled to study the reactions of chlorine with the crotonic acids, but he had not recognized 'that the correct interpretation of such experiments was already recognized before the appearance of his hypothesis'.[75] After such a response, it may not be surprising that Wislicenus never responded again directly to Michael, for they appeared to be talking at cross purposes. Michael found Wislicenus' adherence to an underlying causal scheme disruptive to the 'natural' generation of chemical theories, and Wislicenus did not recognize Michael's priority for empirical research independent of assumptions about an underlying causal structure.

In 1892, Michael published a long experimental critique to complement his

---

[71] Michael, 'Zur Kritik', p. 34.
[72] Wislicenus, 'Untersuchungen zur Bestimmung der räumlichen Atomlagerung. Dritte Abhandlung', p. 343.
[73] Ibid., p. 343.
[74] Arthur Michael, 'Bemerkung zu der Abhandlung von J. Wislicenus: "Zur geometrischen Constitution der Krotonsäuren und ihrer Halogensubstitutionsprodukte"', *J. prak. Chem.*, 1889, *40*, 29–44, p. 44.
[75] Ibid., p. 44.

theoretical analysis. In an extensive series of experiments, he showed that bromine added to acetylene dicarboxylic acid to form dibromofumaric acid as the major product, and reported that dibromofumaric acid, with bromine atoms in the supposedly *trans* position, reacted cleanly with zinc to give acetylene dicarboxylic acid. Michael noted that reaction after reaction gave addition products that were inconsistent with Wislicenus' principle of planesymmetric addition. Michael filled three pages of tables listing contradictory results, and accused Wislicenus of running reactions under conditions that led to erroneous results.

Michael's motivations became decidedly less murky in 1895, when he published a series of articles on the unsaturated acids in which he unveiled a number of empirical laws of alloisomerism. These empirical laws would constitute Michael's 'replacement' for Wislicenus' theoretical scheme. Michael used maleic and fumaric acids as the starting point, and classified the alloisomeric acids as either the 'maleinoid' or 'fumaroid' modification. Unlike Wislicenus' theoretical ordering under the headings of *cis* and *trans*, Michael's classification was entirely empirical. When Michael assumed a compound to be fumaroid, he meant only that it was the more stable of the two acids, and possessed a higher melting point and a lower solubility than the corresponding maleinoid modification.

Michael proposed 20 empirical laws that described the relationships between maleinoid and fumaroid modifications of alloisomers. They described the transformations that could occur, the predominant modification that formed in a given reaction, the predominant direction followed in a conversion between modifications, and so forth. Michael's objectives were now clear. From his first experiments with cinammic acid, he had intended to construct a series of statements that would describe the behavior of the unsaturated acids. In 1895, he had finally gathered enough information to present these laws, and declared that they reflected the natural order of the unsaturated acids.[76]

The greatest source of Michael's dissatisfaction with stereochemistry arose when he attempted to reconcile Wislicenus' theory with this natural order and classification of unsaturated acids. When he attempted to apply Wislicenus' principles consistently, he contradicted his own laws of alloisomerism, and when he attempted to apply his own system consistently, he was led to the improbable conclusion that maleic acid possessed the *trans* configuration. This contradicted the most secure of all stereochemical assumptions – that anhydride formation was a consequence of the spatial proximity of carboxyl groups. How could maleic acid possibly form an anhydride if the carboxyl groups were located *trans* from one another? Even Michael agreed this was the most probable, or at least the least dubious, of all assumptions in the stereochemist's repertoire.

---

[76] Wislicenus, 'Untersuchungen zur Bestimmung der räumlichen Atomlagerung. Dritte Abhandlung', p. 349.

Between the false assumption of planesymmetric addition and elimination, and the control of anhydride formation, Wislicenus' scheme led inevitably to inconsistency with natural law. It was impossible to reconcile Wislicenus' theoretical classification with Michael's empirical laws. One could not have theoretical unity without disarray in the natural order of the acids, nor have natural order with theoretical unity:

> In other words, Wislicenus had sacrificed unity in the grouping of unsaturated compounds to Van 't Hoff's hypothesis; of course he at least attempted a far-reaching stereochemical explanation for the behavior of such bodies, but then everything else is sacrificed. To keep Van 't Hoff's hypothesis intact [*aufrecht*], there would be neither unity in the grouping, nor in the reactions of unsaturated compounds.[77]

The methodological basis of Michael's dissatisfaction with stereochemistry now became extraordinarily clear. In the attempt to arrange the reactions of the unsaturated acids under a single underlying principle, Wislicenus had therefore sacrificed the 'natural order' of the unsaturated acids. Michael regarded the discovery of this 'natural order' as the first priority of a chemist, and therefore regarded Wislicenus' principles with great suspicion from the outset. Nor is it surprising that Michael found planesymmetric addition and configurational analysis suspect, because they had no empirical basis. Michael complained that:

> The natural coherence within the unsaturated compounds has become veiled by the attempts of J. Wislicenus to apply systematically Van 't Hoff's hypothesis to their relationships. A healthy development of theoretical chemistry in this area appears possible to me only if we abolish Wislicenus' purely theoretical conclusions and deductions based on inaccurate experimental trials.[78]

All of the efforts Wislicenus had made to explain anomalies in his theory, such as intramolecular rearrangements, atomic migrations and favored configurations, were in vain because he made them without a secure empirical foundation, and they were made for no other reason other than simply to force 'well-established facts' into his theoretical framework. Such explanations were therefore highly improbable, and did nothing to further the progress of chemistry.[79]

The power of Michael's results seems difficult to ignore, but chemists nevertheless appeared to have little more understanding of Michael's arguments than Wislicenus had. In a letter to his friend and former student Victor Meyer, Adolf von Baeyer

---

[77] Ibid., pp. 360–1.
[78] Ibid., p. 344.
[79] Another objection was the a stark, lifeless view of molecules governed by mechanics that stereochemistry offered. Michael was convinced that chemical phenomena were simply too complex and variable to be served adequately by mechanical forces, because this reduced the importance of their fundamental chemical qualities. See Ramberg, 'Arthur Michael's critique', p. 134.

described Michael's concept of alloisomerism as 'mysticism', and his response was typical.[80] While most chemists agreed that Michael had adequately shown that Wislicenus was overenthusiastic about his theory, no one rushed to abandon the basic premises of 'chemistry in space'. As late as 1930, Michael's experiments were given only a passing mention in most stereochemical monographs. The most apparent reason for the neglect of Michael's results appeared to be the 'mysticism' of alloisomerism that presented no alternate causal structure.

Michael's experimental results had shown the definite existence of axialsymmetric addition, but only if one accepted the models of stereochemistry. In fact, axial-symmetric addition is precisely the assumption needed for the reconciliation of the conflict Michael saw between natural law and theoretical unity. Yet, as Alfred Werner and others would point out, there was no acceptable model for conceiving axialsymmetric addition as a process. A mechanism involving axialsymmetric addition, even if the 'only way out', was difficult to comprehend. Wislicenus' ability to unite the behavior of the unsaturated acids in a 'unified manner' appeared to outweigh the validity of any criticisms Michael had to offer.

Michael's success in arguing against Wislicenus relied not so much on outright contradiction of stereochemical principles, but on a contradiction premised on an alternate methodology, in which the search for underlying causes was subordinated to an inductive procedure to uncover the 'laws' of alloisomerism. Because of this methodological commitment, his argument appeared weaker than he would admit, and the resulting reception of Michael's results reflect this difference. By the late 1890s, when he finally began to offer a theoretical alternative based on thermodynamic principles, stereochemistry had become fully entrenched into chemical theory. By that time, it was too little, too late. Michael did not appear to fit within a scientific atmosphere that demanded a unified causal account of isomerism that would guide experimental work. He was a classic example of a scientist striving to determine simple empirical rules that would organize facts independent of theory. He therefore offered only a non-committal term that indicated nothing but the existence of a form of isomerism that he refused to explain. The reluctance to adopt the full implications of Michael's results indicates the tenacity of chemists to adopt Van 't Hoff's theory as closer to the truth, despite its clear, acknowledged limitations. When given a choice between the 'mysticism' of alloisomerism and a theory that had admitted weaknesses, chemists chose the latter.

---

[80] Adolf von Baeyer to Victor Meyer, 8 November 1892, Deutsches Museum München, Archiv, HS 7073.

Atomism and Methodology in Wislicenus' Chemistry

Significantly, Wislicenus considered Van 't Hoff's theory an important milestone in the development of the atomic theory, which he felt to be fundamentally important for the success of chemistry. In his various public lectures and addresses, he repeatedly emphasized that the atomic theory was the most secure theoretical foundation of chemical theory. Without atomism, he said, 'the individual pieces of chemical knowledge would be a desolate heap of unrelated and incomprehensible observations, indeed, it would be less than that: to a great extent, it would not even exist'.[81] He was also convinced of the reality of the atomic theory:

> For modern chemistry, even if it remains absolutely conscious of the hypothetical nature of the atomistic view, elementary atoms are reality. Even though nobody has perceived them with their senses, or will ever perceive them, we do know certain properties they have, some precisely measured ...[82]

In other words, the reality of atoms was made evident by their measureable effects, and in his address to the Wiesbaden *Naturforscherversammlung* on the principal argument contained in 'Spatial Arrangement', Wislicenus related this work directly to the establishment of the atomic theory: 'My current task will be fulfilled if I have achieved in some way the proof that [the great question of chemical isomerism] can be answered only from the base of the atomistic view of nature, and to bear witness for the existence of atoms.'[83]

Wislicenus' ardent enthusiasm for Van 't Hoff's principles illustrated in 'Spatial Arrangement' represents an important shift in Wislicenus' own views on atoms and molecules that is also reflective of the overall shift from a chemical to physical atomism inherent in the emergence of 'chemistry in space'. As we saw in Chapter 2, Wislicenus initially maintained the traditional ontological and epistemological divide between considering atoms as physical unsplittable atoms, and atoms as empirically derived combining ratios. In 1859 he had ascribed no physical reality to chemical formulas whatsoever, and he considered them only as reaction formulas until 1875. It is unclear whether before 1875 Wislicenus believed in physical atoms, but after reading *La chimie dans l'espace*, he certainly saw that chemical theory could benefit by incorporating a physical atomism. By 1885 or 1886, and probably much earlier (indicated by his comments at the 1877 Munich *Naturforscherversammlung*), it

---

[81] Johannes Wislicenus, *Die Chemie und das Problem von der Materie*, Leipzig, Alexander Edelmann Verlag, 1893, p. 20.
[82] Ibid., p. 21.
[83] Johannes Wislicenus, 'Entwickelung der Lehre von der Isomerie chemischer Verbindungen', *Gesellschaft Deutscher Naturforscher und Ärtze*, 1887, *60*, 47–56, p. 56.

became clear to him *how* the physical aspects of molecules could be used for achieving the aims of chemistry beyond Van 't Hoff's original suggestions.

Wislicenus made his most direct statement about the nature and appearance of these physical atoms in 1888 as a reply to Lossen's 1887 critique mentioned above. Wislicenus answered Lossen with a short but penetrating justification for his chemistry that provides great insight into his views on atomism and chemical methodology. He answered Lossen quite directly – it was impossible, he said, *not* to conceive of atoms as 'spatial objects' with their affinities located in different areas of those objects. Atoms were *necessarily* material objects with a shape and form. Given his enthusiastic support of Van 't Hoff's theory, it is not surprising that Wislicenus considered carbon atoms to be truly tetrahedral:

> I do not consider it impossible that a carbon atom may be an object whose form more or less (perhaps quite closely) resembles a regular tetrahedron; further, it is not impossible that the causes of every effect that actually manifests itself in the affinity unit concentrate themselves in the corners of this tetrahedral object, and for analogous reasons, would possibly be similar to the electrical effect of an electrically charged metal tetrahedron. The actual carrier of this energy would ultimately be the primitive elementary atoms (*Uratome*), exactly like the chemical energy of compound radicals undoubtedly is a product of the inherent energy of the elementary atoms within them.[84]

Therefore, carbon atoms were tetrahedrally shaped carriers of chemical energy. But this statement must be considered carefully. It is clear elsewhere in his response that Wislicenus did not regard the *Berichte* as the place for such speculations, and he made them only out of courtesy to Lossen. At the most, it was perhaps a hunch or a feeling, based on his work so far. He was, on the other hand, absolutely committed to a broad conception of a tetrahedral carbon atom, since he was convinced that the experimental evidence confirmed this basic assumption. Perhaps he refused to commit to a specific interpretation of the nature of the tetrahedron: the ultimate stuff that composed them, the actual cause of chemical affinity, or to the precise nature of the affinity unit, as Lossen had requested. His model was therefore intentionally vague, and did not possess the 'value of a scientific conviction'. He clearly identified the sites of bonding as the corners and not the faces of the tetrahedron, but did not offer a physical explanation for the free rotation of single bonds or the restricted rotation of double bonds. For this reason, Wislicenus' 'space formulas' did not imply much more about the appearance of atoms than Van 't Hoff's original drawings.

Also revealed in his response to Lossen was Wislicenus' conviction that the spatial arrangement of atoms was 'accessible to experimental test', and that the establishment

---

[84] Johannes Wislicenus, 'Über die Lage der Atome in Raume: Antwort auf Lossen's Frage', *Berichte*, 1888, *21*, 581–5, p. 584. A full translation of Wislicenus' response is given in Peter J. Ramberg, 'Johannes Wislicenus, Atomism, and the Philosophy of Chemistry: A Translation and Commentary', *Bull. Hist. Chem.*, 1994, *15/16*, 45–53.

of configurational formulas led precisely to a deeper understanding of atoms and, furthermore, to subatomic structure:

> ... our considerations about the configuration of molecules exclude the assumption that atoms may be 'material points'. One cannot avoid imagining them as spatial objects, thereby transferring the location of the units of chemical effect [*Wirkungseinheiten*] on multi-valent elementary atoms into different regions of these spatial objects. In principle, this idea is in no way hindered by difficulties, provided we conceive the so-called elementary atoms not as atoms in a strict sense, but as composed of groups of still more fundamental elementary atoms [*Urelementaratome*] of a simpler sort – similar to the more compound radicals at more complex levels.[85]

As components of the recognized elementary atoms, the 'primitive atoms' (*Uratome*) were the carriers of chemical affinity. This was the closest he ever came to defining an affinity unit, but he fell short of actually producing a concrete definition, and he remained content merely to make an analogy between the primitive atoms and atoms of compound radicals at a higher level.

The idea that the known elements were possibly divisible was not a new idea with Wislicenus, as he himself was quick to point out. Adolph Wurtz, August Kekulé, and Crum Brown made similar speculations in the development of structure theory, and it permeated much of chemical thought in the nineteenth century.[86] In an address to the German Chemical Society, Wislicenus remarked that chemists held a 'deep conviction [that] the elementary atoms may not be coincidental things existing with and by one another, but members and phenomena of a higher unity'.[87] According to Wislicenus, the periodic table showed unequivocally that the current elements were not simple substances. Because valence rose along with an element's atomic weight, it appeared to be a function of atomic weight, and other periodic properties indicated that they could be caused by multiples of more elementary units.[88] In his 1893 address as Rector of the University of Leipzig, 'Chemistry and the Problem of Matter', Wislicenus indicated that the elements were not stable things. 'As modifications of a basic unit,' Wislicenus stated, 'the elementary atoms cannot be considered as eternal, but only thought of as transient [*Gewordene*].' He compared the formation and

---

[85] Ramberg, 'Johannes Wislicenus'.

[86] See Alan J. Rocke, 'Subatomic Speculations and the Origin of Structure Theory', *Ambix*, 1983, *30*, 1–18, W.V. Farrar, 'Nineteenth Century Speculations on the Composition of the Elements', *Brit. J. Hist. Sci.*, 1965, *2*, 297–323, and Britta Görs, *Chemischer Atomismus: Anwendung, Veränderung, Alternativen im deutschsprachigen Raum in der zweiten Hälfte des 19. Jahrhunderts*, Berlin, ERS Verlag, 1999.

[87] Johannes Wislicenus, 'Die wichtigsten Errungschaften der Chemie im letzen Vierteljahrhundert', *Berichte*, 1892, *25*, 3398–410, p. 3407. Wislicenus' suggestion that the elements evolved was a relatively popular idea during the late 1880s and 1890s, and was originally suggested by William Crookes. William H. Brock, *From Protyl to Proton: William Prout and the Nature of Matter, 1785–1985*, Bristol, Adam Hilger, 1985.

[88] Wislicenus, *Die Chemie und das Problem von der Materie*, p. 24.

existence of the elements to the evolutionary relationships between animals, 'for as all transients [*Gewordene*], the chemical elements are certainly also subject to the general universal laws of evolution'.[89]

Although Wislicenus presented his ontology in its most extended form in his 1893 *Rectorat*, his response to Lossen would contain his most detailed public statement about the physical nature of atoms. He never offered any concrete suggestions about the ultimate nature of matter, but in his *Rectorat*, he offered several intriguing speculations, including an allusion to Wilhelm Ostwald's program in energetics:

> How does chemistry imagine the last principle of matter is constituted? Can it be the luminiferous ether required by physics to explain certain classes of phenomena, above all the propagation of transverse waves? Is it perhaps beings of entirely different magnitude than the elementary atoms, perhaps extensionless, turbulent centers of force that act on one another, through which only their spatial aggregations form the simplest corpuscular units? Isn't matter a concept actually formed of experience, like all our external experience, only a product of the effects of the unchanging quantity of energy in space?[90]

But the science of chemistry, Wislicenus claimed, 'has no specific answer for these kind of questions, since all these ideas – like the materially conceived luminiferous ether – are of a purely hypothetical nature'. Chemical experiment may not be able to answer such questions, Wislicenus admitted, but new methods, for example the developing field of spectroscopic analysis, had made these properties accessible to observation.[91] Despite his certainty about his ability to ascertain the spatial properties of molecules and understand them as physical objects, Wislicenus recognized that chemical methodology by itself could not elucidate the ultimate nature of matter.

Wislicenus' response to Lossen is also revealing for what it says about Wislicenus' methodological assumptions. Whereas Lossen had suggested that speculation about the arrangement of atoms in a molecule should come *after* ascertaining an atomic form, Wislicenus addressed the problem in exactly the opposite way: 'only *after* the spatial arrangement of elementary atoms in molecules is determined, and not *before*, is it possible to consider Lossen's question [about affinity units] seriously.'[92] It was precisely the research into the spatial properties of *molecules* that had given him his ideas about the shape and form of the carbon *atom*. This approach, Wislicenus believed, was firmly imbedded into the general aims of chemical research. The study of chemical substances had led to the laws of atomism and chemical combination, which in turn led to consistent atomic weights, and eventually to the law of atomic linkage and structure. All of these accomplishments rested on the study of the

---

[89] Ibid., p. 25.
[90] Ibid., pp. 26–7.
[91] Ibid.
[92] Ramberg, 'Johannes Wislicenus', p. 582.

properties of molecules, and not atoms. It is only through the manipulation of molecules, said Wislicenus, that chemists have gained any knowledge about the nature of the constituent atoms. The same process, he was confident, would elucidate the nature of the parts of those atoms. This, Wislicenus claimed, put him squarely in the middle of traditional research in organic chemistry. He considered research on geometrical isomerism a natural outgrowth of that empirical, inductive tradition.

Wislicenus also argued that a methodological change had occurred in the middle of the century, coincident with the development of the structure theory:

> So chemistry lost more and more of its original purely inductive character and has partially converted to deductive procedures. The repeated prophecy of disadvantage and decay by the overconservative representatives of an earlier time has subsequently not occurred in our science, for its conclusions never do without the incorruptible test of experiment, whose development continues at an equal pace.[93]

He had already recognized this change as early as 1859, when he mentioned it in his essay 'Theory of Mixed Types'. Wislicenus did not regard it as a regressive step to postulate hypotheses – in fact, he defended the right to pursue 'inevitable' hypotheses about spatial arrangements. Because these speculations – which meant for Wislicenus the deductive predictions of inductive generalizations – were testable experimentally, they did not threaten 'the danger of losing the solid ground from under our feet'.[94]

The theory of geometrical isomerism, as a logical outgrowth of the structure theory, was exactly such a hypothesis. It was not a cautious hypothesis, built from the slow accumulation of facts and observations, but it did have certain empirical consequences that could be tested. This was precisely the criterion Wislicenus required for a good theory:

> [Van 't Hoff's theory] has fully achieved what can ever be achieved from an hypothesis; for it has made possible the explanation of previously incomprehensible facts that apparently are outside the basic chemical theories ... it has stimulated empirical research by creating new tasks, amassed great amounts of factual material, shown the path towards the invention of new observational methods, has subsequently become accessible to experimental test, and has simultaneously become a stimulus for a significant movement, in a sense even for a new epoch in our science.[95]

Wislicenus regarded theories as tools to generate new facts. It was precisely the testability of a hypothesis and its ability to generate new facts regardless of its absolute correctness that gave a theory its value to science. Because he could not test

---

[93] Wislicenus, *Die Chemie und das Problem von der Materie*, pp. 16–17.
[94] Johannes Wislicenus, 'Die wichtigsten Errungschaften', p. 3402.
[95] Johannes Wislicenus, 'Foreword', J.H. van 't Hoff, *Die Lagerung der Atome im Raume*, 2nd edn, Vieweg, Braunschweig, 1894, p. viii.

hypotheses about the ultimate nature of the primitive atoms, or the 'stuff' that made them up, he shied away from making concrete statements about them.

In a long 1877 letter to Hermann Kolbe, Wislicenus defended his support of Van 't Hoff's theory and the role of speculation and hypothesis in chemical theory. Speculation, Wislicenus admitted, was less certain than the cautious method of induction, but it none the less provided a quicker path to successful theories, provided such speculations can be supported by empirical investigation. They may be wrong, but that should not prevent us from making them and testing them. Even if tested hypotheses turn out to be wrong, the facts they generate could be reinterpreted by later generations of chemists to create a better theory. The future, and not the present, Wislicenus declared, was the best judge of scientific work:

> today we can err, but only when compared to a later stage in the development of scientific knowledge that no longer understands it as we do, and comprehends in a still more simple way than we are able to today with our relatively incomplete knowledge. So the future will judge the current conflict of views. I do not assume the present is a judge, although I have the solid conviction that in what divides us, it essentially stands on my side.[96]

Among organic chemists of his generation, he was perhaps the most explicit about the usefulness of hypotheses in chemistry. Like Van 't Hoff, he advocated vigorously the advantages of hypotheses and the imagination for the success of science. Wislicenus' scientific life therefore personified not only the incorporation of physical atomism into chemical theory, but also the appearance of a methodology in chemistry that explicitly advocated the role of theories as a predictive device.[97]

In the same letter to Kolbe, Wislicenus wrote: 'To what do we dedicate our science, if we do not have the serious aspiration to understand, that is seek to explain, all individual facts from a unified viewpoint?'[98] This search for a 'unified viewpoint' is also a central methodological principle that runs through Wislicenus' career, from his first paper on mixed types in 1859 to his last work on the geometrical isomerism of the bromobutenes in 1900. During the 1860s he looked for the proper means of representing lactic acid's 'individuality', and was forced to invent a new type of isomerism. The underlying processes of chemical attraction and molecular motion tied together the transformations of the unsaturated compounds at a common

---

[96] Johannes Wislicenus to Hermann Kolbe, 24 November 1877, Deutsches Museum München, Archiv, HS 3550.

[97] Alan J. Rocke cited this trend as a possible cause for the pathological nature of Hermann Kolbe's attacks, 'Kolbe vs. the "Transcendental Chemists": The Emergence of Classical Organic Chemistry', *Ambix*, 1987, *34*, 156; for a more general exploration of this trend earlier in the century, see also Rocke's 'Methodology and Its Rhetoric in Nineteenth Century Chemistry: Induction versus Hypothesis', pp. 137–55 in *Beyond History of Science: Essays in Honor of Robert E. Schofield*, Elizabeth Garber, ed. Bethlehem, PA, Lehigh University Press, 1990.

[98] Wislicenus to Kolbe, 24 November 1877.

theoretical level. Wislicenus could not accept science without the attempt to explain at a fundamental level, and a sufficient explanation involved an appeal to underlying causes, preferably to a few axioms. This view also explains why he found the periodic table and the concept of primitive atoms aesthetically pleasing ideas.

In the preceding analysis of Wislicenus' thought, I have tentatively assumed that we can rely on his own largely public statements as fully reflective of his views of science. Were these public statements about method actually followed in his laboratory work? This is one of the trickiest questions to answer in the history of science, for it is well known that published papers are notorious for disguising the actual methodology of the laboratory, and the rhetoric of such papers can often be crafted to make the argument more reflective of current fashion, rather than an accurate reflection of practice in the laboratory. A close reading of Wislicenus' papers does not, however, suggest any appreciable dichotomy between his public pronouncements about methodology and his private practice. Much of the experimental portions of the series of papers supporting the claims of 'Spatial Arrangement' revolve around the isolation of particular compounds predicted by the use of his theoretical principles. Wislicenus had no reason to attempt the isolation of these compounds without a theory to predict their existence. Furthermore, Wislicenus was never afraid to voice his opinions frankly and make them public – he was unable, for example, to renounce the radical religious views of his father, and he quickly became a strong public advocate for an uncertain theory that others had hesitated to support in public.[99] Although we must still be wary of equating Wislicenus' public statements with his private thoughts, there does seem to be some justifiable correlation between the two.

Wislicenus was extraordinarily confident that Van 't Hoff's theory was correct, to the point of being accused of being overenthusiastic by its other supporters. Yet it does not seem that his enthusiasm should be mistaken for a dogmatism. In the 1877 letter to Kolbe, Wislicenus wrote:

> I know that I can be mistaken, but I also know that I see no reason to have me removed from the list of exact scientists, for almost as much as you, I have a longing to serve the truth, and an awareness, now and in the past, to do this according to my capability and also with serious attention to the views that I oppose. My scientific opinions are certainly changeable and have already changed in many situations. But such changes occurred only under the force of facts – found by me or others. I have never attempted to interpret preconceived hypotheses by falsifying facts.[100]

---

[99] I could also note his active involvement in the German community in Zürich, when he was involved in a demonstration and uprising among German residents in Zürich following a celebration of German unification.

[100] Wislicenus to Kolbe, 24 November 1877.

Several times Wislicenus said he was ready to drop his theory should it not fit the facts. But not once during the course of the research did he think that drastic step was necessary. He reached a high point in his 1892 defense against Fittig. At about the same time, Michael's lengthy experimental critique appeared, in which he showed that the major product of bromine addition to ADCA was in fact dibromofumaric acid. Wislicenus' program lost impetus at that point, and although he continued to publish articles on geometrical isomerism, they mostly appeared in obscure journals and had none of the *tour de force* of his 1887–89 series of papers. He was most likely addressing Michael's arguments in 1894, when he wrote the foreword to the second edition of Van 't Hoff's *Die Lagerung*:

> For that matter, the previous categorical resistance has almost died out; where it does still rise, it turns to the last foundation: against the atomistic view itself, but does not deny that the doctrine of spatial atomic arrangements is a stage, perhaps the last, in the consistent and necessary development of chemical atomism. The opposition is primarily directed – indeed often justified – towards particular applications that explain individual, real relationships and processes, but that do not actually question the theory's basic content.[101]

The tetrahedral model, despite its weaknesses in the light of Michael's results, continued for Wislicenus to be the most plausible explanation.

---

[101] Wislicenus, 'Foreword', p. viii.

Chapter 6

Victor Meyer:
The New Science of Stereochemistry

> ... we may no longer count on atoms as material points, but are forced to consider their dimensions, and can already obtain ideas about their relative proportions, if only on a modest scale.
>
> Victor Meyer, 1890

Early in 1888, Victor Meyer joined Wislicenus in the defense of 'chemistry in space' with a brilliant series of articles describing the constitutional identity of the benzildioximes that cogently argued for the necessity of spatial explanations in chemical theory. Meyer's enthusiasm for the new science was unmistakable. The concept of geometrical isomerism took on a central role in his conception of organic chemistry, and his massive 1893 textbook (written with his assistant, Paul Jacobsen) was among the first organic chemistry texts to use spatial theories as a guiding principle, and the implications of Van 't Hoff's and Wislicenus' 'space formulas' also renewed Meyer's long-standing interest in questions about the nature of valence.[1] He cared deeply about how to represent a molecule's structure and configuration without also conveying misleading ideas about the nature of valence and bonding, but also carefully separated his speculations about the physical nature of valence from his chemical theories.

Like Wislicenus, Meyer was convinced that the spatial properties of molecules had become accessible to experimental test, and that those spatial properties would lead to a better conception of the atom itself, but Meyer offered a tempered excitement that balanced Wislicenus' occasional overenthusiasm for the merits of molecular dynamics and mechanico-chemical explanation. For example, Meyer did not adopt the overtly mechanistic models developed by Wislicenus to assign configurations. Instead, he employed a more cautious inductive approach to assign configurations to the benzildioximes. Meyer did adopt and rely on several of the mechanical

---

[1] Victor Meyer and Paul Jacobsen, *Lehrbuch der organische Chemie*, 3 vols, Leipzig, Veit, 1893–1923.

assumptions of stereochemical theory (one could not adopt its principles without making such an assumption), but he did not attempt to extend them or develop these mechanical aspects further. This general theoretical caution was related to his careful, methodical style, and is the only obvious characteristic that pervaded Meyer's enormously diverse interests in nearly all areas of chemistry.

Meyer's career was marked by meteoric brilliance and tragedy.[2] Aged 18, he was awarded a PhD in chemistry from the University of Heidelberg, and Robert Bunsen chose him as a personal assistant to carry out analyses of mineral waters. In 1868, he returned to his native Berlin to broaden his education in chemistry. He attended A.W. Hofmann's lectures on organic chemistry at the University of Berlin, and entered Adolf von Baeyer's laboratory at the Berlin Industrial Academy in Charlottenburg, where he formed close friendships with fellow students Carl Graebe and Carl Liebermann. Meyer's enthusiasm for all aspects of organic chemistry, as well as his theoretical caution, derived from his relationship with Baeyer, which evolved from mentor and student to colleague and close friend (in the mid-1880s, they began to address each other in correspondence with the informal '*Du*'). They carried on an active and extensive correspondence until Meyer, plagued by chronic insomnia and neuralgic pains, ended his life with cyanide in 1897.[3]

Meyer showed such promise as a chemist that he was allowed to skip the usual step of writing a formal *Habilitationsschrift*, and on Baeyer's strong recommendation, he was awarded a position at the Stuttgart Polytechnical Institute at the age of 23. Eighteen months later, he succeeded Wislicenus at the Zürich Polytechnical Institute, where he stayed for thirteen years, by most accounts the happiest years of his life. While in Zürich, Meyer built an outstanding reputation as an organic chemist by developing a new method for preparing nitroaliphatic compounds via aliphatic halogens and silver nitrite (Figure 6.1(a)). This group of compounds had been virtually unknown before Meyer's investigation, as nitro groups had previously only been known on aromatic compounds.[4] This chemistry was also part of Meyer's broader aim to determine the influence of negative substituents (such as nitro groups or halogens) on the substitution of hydrogen atoms. In 1883, his accidental discovery of the sulfur compound thiophene initiated a long investigation into its structure in

---

[2] The most comprehensive biography of Meyer was written by his brother Richard, who made extensive use of his own correspondence with Victor, and Meyer's correspondence with their parents and Adolf von Baeyer. Richard Meyer, *Victor Meyer: Leben und Wirken eines deutschen Chemikers und Naturforschers 1848–1897*, Leipzig, Akademische Verlagsgesellschaft, 1917.

[3] The friendship between Baeyer and Meyer has been emphasized by Richard Meyer and in the short biographies by Carl Liebermann and Georg Lunge. Carl Liebermann, 'Victor Meyer', *Berichte*, 1897, *30*, 2157–68, and G. Lunge, 'Nachruf auf Victor Meyer', *Vier. Zür. naturf. Ges.*, 1897, *42*, 347–61, p. 349.

[4] Gustav Schmidt, 'The Discovery of Nitroparaffins by Victor Meyer', *J. Chem. Ed.*, 1950, *27*, 557–9.

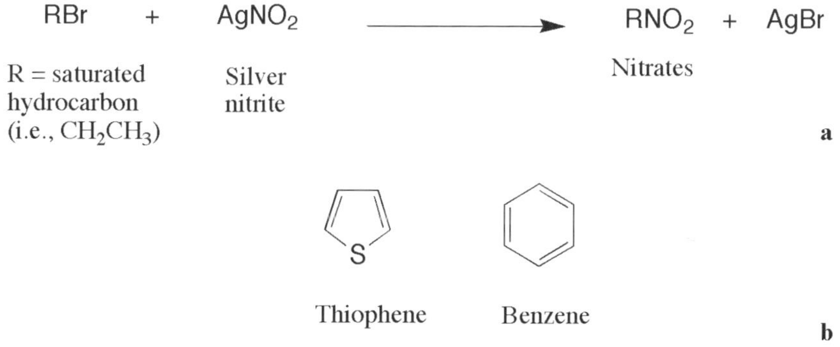

```
RBr      +    AgNO₂                    →        RNO₂   +   AgBr
R = saturated    Silver                          Nitrates
hydrocarbon      nitrite
(i.e., CH₂CH₃)
                                                                    a


              ⟨S⟩          ⟨⟩
            Thiophene    Benzene
                                                                    b
```

6.1 (a) Meyer's synthesis of nitro compounds. (b) Thiophene as an analog of benzene.

which Meyer showed it was a structural analog to benzene (Figure 6.1(b)) where a sulfur atom took the place of a C_2H_2 unit.[5]

In addition to classical organic chemistry, Meyer also became involved in the techniques and problems of physical chemistry. Prompted by the need to calculate molecular weights for many of his organic compounds, Meyer developed new techniques for determining molecular weight by means of vapor densities at high temperatures. Until his death in 1897, he was continually occupied with the design and construction of new ovens to reach higher temperatures for volatilizing substances with high melting points, and by the 1890s he had reached 3 000° C. As the research progressed, Meyer also attempted to reach temperatures high enough to detect the dissociation of the known elements into simpler components.[6]

Meyer and the Tetrahedral Carbon Atom, 1875–1887

In 1890, Meyer recalled the 1869 Innsbrück *Naturforscherversammlung* at which Wislicenus wrote 'on the blackboard of the Innsbrück University laboratory a formula that did not yet possess much similarity to today's stereochemical formulas, but

---

[5] Margaret Davis Cameron, 'Victor Meyer and the Thiophene Compounds', *J. Chem. Ed.*, 1949, **26**, 521–4.

[6] Albert B. Costa, 'Victor Meyer', pp. 354–8 in Charles Gillespie, ed., *Dictionary of Scientific Biography*; New York, Scribner's, 1973. Victor Meyer, 'Probleme der Atomistik', Winter, Heidelberg, 1896. The context of high temperature measurements of decomposition and dissociation of gases is given in Diana Barkan, *Walther Nernst and the Transition to Modern Physical Science*, Cambridge, Cambridge University Press, 1999.

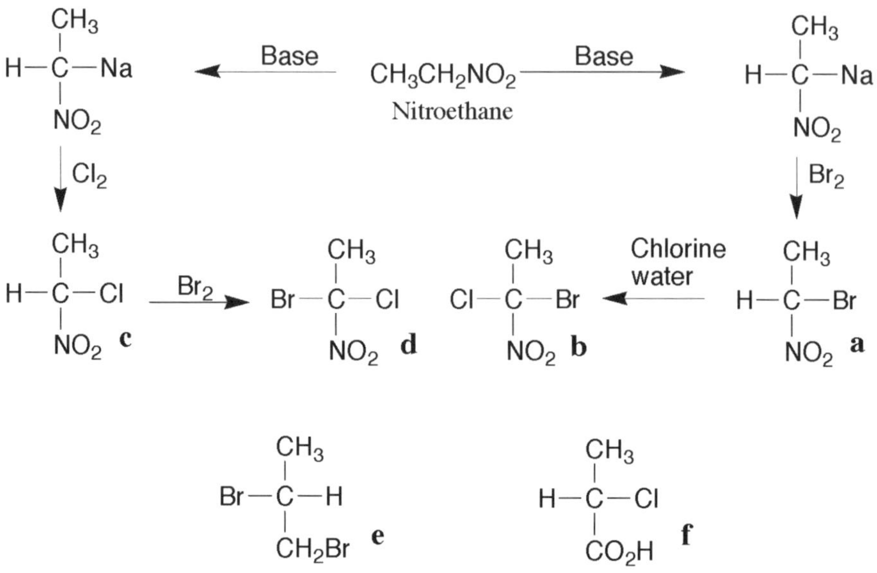

$$\text{H}-\underset{\underset{\text{NO}_2}{|}}{\overset{\overset{\text{CH}_3}{|}}{\text{C}}}-\text{Na} \xleftarrow{\text{Base}} \text{CH}_3\text{CH}_2\text{NO}_2 \xrightarrow{\text{Base}} \text{H}-\underset{\underset{\text{NO}_2}{|}}{\overset{\overset{\text{CH}_3}{|}}{\text{C}}}-\text{Na}$$

Nitroethane

$\downarrow \text{Cl}_2$ $\downarrow \text{Br}_2$

H–C(CH₃)(NO₂)–Cl **c** →[Br₂] Br–C(CH₃)(NO₂)–Cl **d**   Cl–C(CH₃)(NO₂)–Br **b** ←[Chlorine water] H–C(CH₃)(NO₂)–Br **a**

Br–C(CH₃)(H)–CH₂Br **e**   H–C(CH₃)(Cl)–CO₂H **f**

6.2 Meyer's test of Van 't Hoff's theory, proposed in a letter to Adolf von Baeyer.

nevertheless expressed the thought that our previous symbols that simply lent expression to the ideas of the structure theory did not suffice'.[7] Although Meyer said this well after the fact, it would indicate that in 1869 Wislicenus' plea had made a memorable impression on him. The earliest known mention of Van 't Hoff's theory by Meyer was in a July 1875 letter to Van 't Hoff requesting a copy of *La chimie dans l'espace* (mentioned in Chapter 4).[8] In an 1876 letter to Baeyer, Meyer proposed a test of Van 't Hoff's theory of optical activity by the preparation of bromochloronitroethanes by the two routes shown in Figure 6.2. 'If Van 't Hoff's view is correct,' Meyer noted:

> [2b] and [2d] must be isomers, but if it is incorrect, they must be identical. Moreover, I have also recently attempted the following: according to Van 't Hoff's view, propylene bromide [2e] must rotate optically. I have tested it, but found it absolutely inactive;

---

[7] The exact nature Wislicenus' symbols are unknown. Victor Meyer, 'Ergebnisse und Ziele der stereochemischen Forschung', *Berichte*, 1890, *23*, 567–619, p. 568.

[8] According to a later student, Paul Jacobsen, Meyer was among the earliest chemists to include Van 't Hoff's theory in his general lectures. Paul Jacobsen, 'Victor Meyer', *Naturwiss. Rund.*, 1897, *12*, 553–6 and 564–7, p. 565.

α-chloropropionic acid [2f] should also rotate, and a thousand other bodies that don't actually do it!⁹

These thoughts were never published, nor did Meyer indicate that he ever carried out the first experiment, but his ideas reveal two noteworthy points about his early conception of Van 't Hoff's hypothesis. First, according to Meyer, the *order* of substitution of hydrogen atoms determined the configuration. If chlorine substituted first, one obtained one isomer, if bromine were added first, the other isomer would obtain. This indicates that Meyer did not fully understand Van 't Hoff's hypothesis, or at least could not yet think clearly in three dimensions. According to Van 't Hoff, the two hydrogen atoms in nitroethane would be *completely equivalent*. It would not matter which halogen atom was added first, because it could take the place of either hydrogen atom without preference. Therefore, Meyer's conclusion that the order of addition would lead to different configurations indicated he still thought relatively unclearly in spatial terms.

Second, Meyer's statement that there were 'a thousand other bodies' that didn't rotate the plane of polarized light would indicate some skepticism on Meyer's part toward the validity of the tetrahedron, and exemplifies well the general early cautious attitude of chemists towards Van 't Hoff's theory of optical activity outlined in Chapter 4. It is difficult to tell whether Meyer was truly a skeptic, or simply reflecting on current experimental reality and not on any deeply held belief in the theory's validity. Whatever the motivation for these comments to Baeyer, they indicate that Meyer was reflecting on the implications of the tetrahedral carbon atom for his own chemistry, but like many other chemists, he did not deem it of sufficient importance to warrant a full investigation.¹⁰

In October 1885, Meyer wrote two postcards and a lengthy letter to his mentor in which he displayed his excitement about the recent publication of Baeyer's strain theory. 'The experimental part,' Meyer wrote, 'is enchanting, like experiencing a fairy tale.' He had no 'appreciable objections' to the theoretical portion, but proceeded to ask some pointed questions about Baeyer's concept of strain, specifically why benzene was stable with three double bonds, why rings of larger than three carbon atoms would be planar, and the apparent contradiction when Baeyer noted that formation of a double bond increased the stability, when single bonds would be the most stable and could only decrease in stability. In a later letter, Meyer offered an ingenious alternate geometrical conception of polymethylene compounds in which the carbon rings had a '*physical, not planar* form'. The four carbon atoms

---

⁹ Victor Meyer to Adolf von Baeyer, 11 May 1876, Deutsches Museum München, Archiv, HS 7092.
¹⁰ Jost Weyer presumed that Meyer fully accepted the theory on the basis of this letter. 'Die Aufnahme der van't Hoff'schen Hypothese vom aymmetrischen Kohlenstoffatomen (1874) in Deutschland', pp. 311–20 in Gunter Mann and Rolf Winau, eds, *Medizin, Naturwissenschaft, Technik und das Zweite Kaiserreich*, Göttingen, Vandenhoeck and Ruprecht, 1977.

 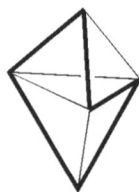  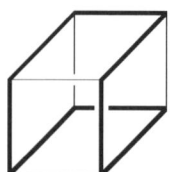

6.3 Meyer's alternative geometrical conception of cyclobutane, cyclopentane, cyclohexane and cyclooctane. The bold lines represent the ring of the molecule, while the thin lines fill in the remainder of the polyhedron. Meyer's original drawings are reproduced in the translation of his letter to Baeyer in Appendix 5.

in cyclobutane would occupy the corners of a tetrahedron, the five carbons in cyclopentane would occupy a double tetrahedron (*Doppeltetraëder*), the six in cyclohexane an octahedron, and so forth (Figure 6.3).[11] Meyer noted that according to his theory, the heat of combustion for cyclohexane would be exactly half that of cyclopropane, and all rings larger than three carbon atoms would have approximately equal strain energy. Meyer's conception of the three-dimensional arrangement of carbon atoms in rings is remarkably close to a modern conception of carbon rings, with the important difference that his model was not dynamic – that is, he did not postulate different conformations of varying energies, but a single geometric arrangement of carbon atoms in ring compounds. It illustrates that in the period since he first proposed his simple test of Van 't Hoff's theory to Baeyer in 1885, Meyer's thoughts on the geometrical properties of molecules had undergone significant changes.

The Benzildioximes

The Chemistry of Hydroxylamine

Meyer did not take an active interest in spatial properties of molecules until early 1888, when he reported the results of two projects. Meyer and his student Robert Demuth had found that different substitution products of benzylcyanide ($C_6H_5CH_2CN$) behaved differently towards further substitution, implying that the two

---

[11] Meyer to Baeyer, 5 October 1885, pp. 195–6 in Meyer, *Victor Meyer*, and Deutsches Museum München, Archiv, HS 7158. Full translations of the postcard and letter are given in Appendix 5. Meyer's emphasis.

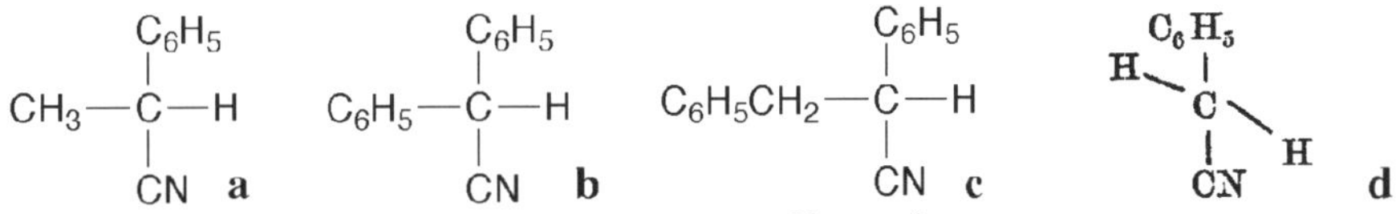

$$\underset{\underset{\text{H atom is easily substituted}}{\text{a}}}{\overset{\overset{C_6H_5}{|}}{\underset{|}{CH_3-C-H}}\atop{CN}} \qquad \underset{\text{b}}{\overset{\overset{C_6H_5}{|}}{\underset{|}{C_6H_5-C-H}}\atop{CN}} \qquad \underset{\underset{\text{casily substituted}}{\text{H atom is not}}}{\overset{\overset{C_6H_5}{|}}{\underset{|}{C_6H_5CH_2-C-H}}\atop{CN}} \qquad \underset{\text{d}}{\overset{\overset{C_6H_5}{|}}{\underset{|}{H\diagdown C\diagdown H}}\atop{CN}}$$

6.4 Derivatives of benzylcyanide in which the remaining hydrogens proved difficult to substitute. The structure on the far right is from the anonymous report of Meyer's lecture (see note 12).

hydrogen atoms were not equivalent.[12] For example, the methyl-substituted nitrile (Figure 6.4(a)) or diphenylacetonitrile (Figure 6.4(b)) could replace the remaining hydrogen atom easily with alkyl groups (for example, methyl). On the other hand, another similar compound (Figure 6.4(c)) could not be alkylated at all. Meyer concluded that the two hydrogen atoms were not equivalent and expressed this difference by using different bond angles in the formula by (Figure 6.4(d)). These results indicated to Meyer that the atoms were linked by an unequivalent attraction: 'one of the two hydrogen atoms is located nearer to the cyano group than the other.'[13] This conclusion remained unclear (it is difficult to see a tetrahedron in his drawing), and he never published a full report on this research.

Meyer's most extended foray into 'chemistry in space' would derive from the study of the reactivity of hydroxylamine with compounds containing carbonyl groups (a carbon–oxygen double bond). In 1882 and 1883, with his student Alois Janny, he reported that hydroxylamine uniformly reacted with organic compounds containing a carbonyl group to give a product with the atomic grouping CNOH. Through a painstaking series of experiments, they identified and named it the oximido, or oxime group ($R_2C=N-O-H$). In nearly all cases, the product oximes were highly crystalline substances isolated in high yields that Meyer or a student could easily purify and characterize.[14] In Meyer's hands, hydroxylamine became a definitive test of chemical structure that indicated unequivocally the existence of a ketone group in a compound's structure.

In 1887, Meyer used his new hydroxylamine chemistry to test Van 't Hoff's theory

---

[12] Anonymous report of Meyer's lecture to the Göttingen Chemical Society in *Chemiker-Zeitung*, 1888, *12*, 140.
[13] Ibid.
[14] Victor Meyer and Alois Janny, 'Über stickstoffhaltige Acetonderivate', *Berichte*, 1882, *15*, 1164–7, 'Über die Einwirkung von Hydroxylamin auf Aceton', *Berichte*, 1882, *15*, 1324–6, and 'Über die neue Bildungsweise der α-Nitrosopropionsäure und die Wirkungsweise des Hydroxylamins," *Berichte*, 1882, *15*, 1525–9.

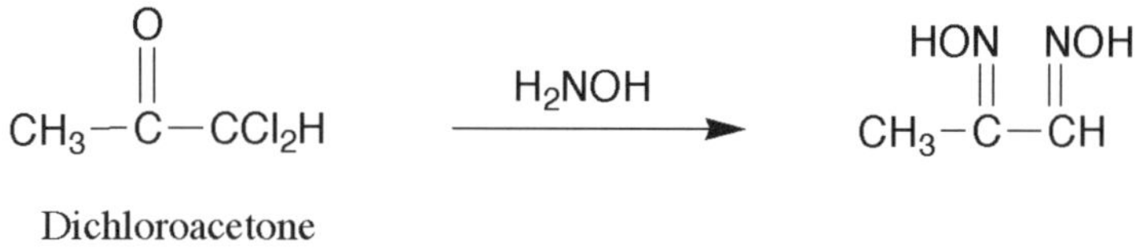

$$CH_3-\overset{\overset{O}{\|}}{C}-CCl_2H \quad \xrightarrow{H_2NOH} \quad CH_3-\overset{\overset{HON}{\|}}{C}-\overset{\overset{NOH}{\|}}{CH}$$

Dichloroacetone

6.5 The reaction of hydroxylamine with geminally disubstituted carbons to yield oximes.

of unsaturation directly. In the course of elucidating the structure of the oximido group, Janny had found that dihalogen substituted carbon atoms behaved like ketones with hydroxylamine (Figure 6.5).[15] Because of this reactivity, hydroxylamine provided a means of deciding between Van 't Hoff's and Fittig's explanation for the difference between maleic and fumaric acid *via* their bromine addition products. According to Fittig, maleic acid would add bromine unsymmetrically, resulting in an αα-dibromosuccinate, while fumaric acid would yield an αβ-dibromosuccinate. According to Van 't Hoff's theory of unsaturation, both maleic and fumaric acid would yield an αβ-dibromosuccinate (Figure 6.6).[16]

Hydroxylamine, Meyer asserted, would form an oxime with αα-dibromosuccinate, but not with αβ-dibromosuccinate. Meyer found that both dibromides (that is, fumaric acid dibromide and maleic acid dibromide) were completely indifferent to hydroxylamine, and therefore both compounds must contain the two bromine atoms at different carbon atoms.[17] Strictly speaking, this result did not prove decisively that the difference between maleic and fumaric must be spatial in character, as Meyer went on to assume, but it did indicate that both maleic and fumaric acid contained a double bond, and were therefore structurally identical. As Meyer proceeded to widen the scope of this test by attempting to prepare αα-dibromosuccinic acid, he learned of Wislicenus' essay on unsaturation and turned to the 'quite curious' (*höchst merkwürdige*) problem of isomerism in the benzildioximes that his student Heinrich Goldschmidt had found four years earlier, in 1883.

Using hydroxylamine to establish the structure of benzil, Goldschmidt had found that benzil and hydroxylamine yielded a monoxime, and with two equivalents of hydroxylamine, a dioxime. This had indicated that benzil contained a diketone

---

[15] Meyer and Janny, 'Über stickstoffhaltige Acetonderivate', and 'Über die Einwirkung von Hydroxylamin auf Acetone', *Berichte*, 1882, *15*, 1324–6. A summary of this work is depicted in F. Henrich, 'Victor Meyer', pp. 374–90 in Gunter Bugge, ed., *Das Buch der großen Chemiker*, Berlin, Verlag Chemie, 1930.

[16] Colin A. Russell, *The History of Valency*, New York, Humanities Press, 1971, pp. 228–30.

[17] Victor Meyer and R. Demuth, 'Zur Kenntniss der Isodibrombernsteinsäure', *Berichte*, 1888, *21*, 264–70.

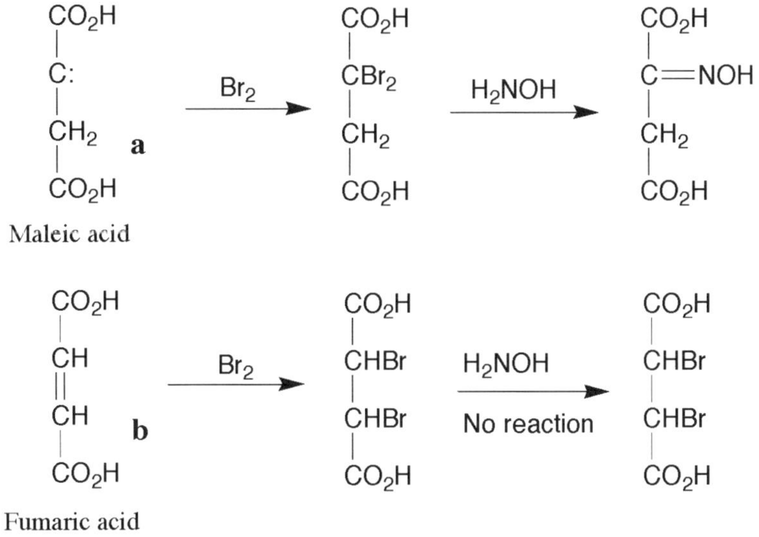

```
    CO₂H              CO₂H                  CO₂H
    |                 |                     |
    C:         Br₂    CBr₂     H₂NOH        C=NOH
    |         ──→     |        ──────→      |
    CH₂   a           CH₂                   CH₂
    |                 |                     |
    CO₂H              CO₂H                  CO₂H
 Maleic acid


    CO₂H              CO₂H                  CO₂H
    |                 |                     |
    CH         Br₂    CHBr     H₂NOH        CHBr
    ||        ──→     |        ──────→      |
    CH    b           CHBr    No reaction   CHBr
    |                 |                     |
    CO₂H              CO₂H                  CO₂H
 Fumaric acid
```

6.6 Meyer's proposed experiment, based on the reaction in Figure 6.5, to decide between the Kolbe/Fittig and Van 't Hoff theories of unsaturation.

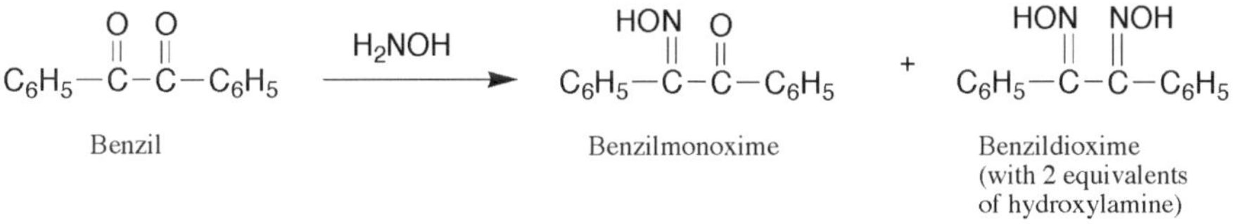

```
         O  O                              HON  O                        HON  NOH
         ||  ||           H₂NOH             ||  ||                        ||  ||
 C₆H₅—C—C—C₆H₅         ──────→      C₆H₅—C—C—C₆H₅          +      C₆H₅—C—C—C₆H₅
       Benzil                            Benzilmonoxime                Benzildioxime
                                                                    (with 2 equivalents
                                                                     of hydroxylamine)
```

6.7 The reaction of benzil with hydroxylamine.

unit (Figure 6.7). Shortly afterwards, while examining similar compounds for the presence of ketones, Goldschmidt had isolated a second benzildioxime, and named it β-diphenylglyoxime to indicate that it was different from the original, but had remarked that 'where the isomerism of both diphenylglyoximes is established is not easily explained'.[18] Privately, Meyer wrote to Baeyer that they had found similar 'puzzling' (*räthselhaft*) results for the product of furfural (C_4H_3OCHO) and

---

[18] Victor Meyer and Heinrich Goldschmidt, 'Über das Benzil', *Berichte*, 1883, *16*, 1616–17, and Heinrich Goldschmidt, 'Über die Einwirkung von Hydroxlamin auf Diketone', *Berichte*, 1883, *16*, 2176–80, p. 2178.

hydroxylamine. Meyer supposed that the difference between the compounds was 'either isomerism, or the lower melting [compound], in spite of its magnificent appearance, is not yet quite pure'.[19] Meyer gave no indication prior to 1888 that the difference between the two benzildioximes or the furfural derivatives could have a spatial cause, perhaps because he did not yet have full confidence that these pairs of isomers were actually structurally identical.

A number of impending projects had forced Meyer to drop any further investigation of the two benzildioximes until 1887. He had become deeply involved in writing a comprehensive review of his pyrochemistry, publishing a book in June 1884. In 1885, he moved from Zürich to succeed Hans Hübner as director of the chemical institute at Göttingen and began plans for a new laboratory building. Finally, the excitement surrounding his 1883 discovery of thiophene had not abated until January 1888, when he published the complete results.[20] When he returned to the benzildioximes, probably in mid-1887, he had a new assistant, Karl Auwers, and 'chemistry in space' had found new life at the hands of Wislicenus.

Meyer and 'Spatial Arrangement'

We can date quite precisely when Meyer first read 'Spatial Arrangement'. On 26 April 1887, amid the construction of his new chemical institute in Göttingen, Meyer wrote to his brother Richard, then assistant to Baeyer in Munich:

> Despite the great workload I have from construction, yesterday I read Wislicenus' brochure, filling 77 large octavo pages to the end in one sitting, and indeed with mixed feelings. The basic idea is quite attractive, but I fear that he has interpreted the matter a little too enthusiastically, and skips over great deficiencies. I am curious how the matter will develop further. What does Baeyer say about it?[21]

Meyer's feelings towards 'Spatial Arrangement' were indeed mixed. He clearly marveled at Wislicenus' skill in marshaling the evidence in support of Van 't Hoff's second and third hypotheses, because until then, only Van 't Hoff's first hypothesis was 'accepted as a solid pillar in the intellectual scaffolding (*Lehrgebäude*) of theoretical chemistry'. 'It is the merit of Johannes Wislicenus', Meyer claimed, 'to have grasped with full clarity the importance of both of the latter hypotheses, and to have placed their fundamental importance in the correct light.'[22] Meyer also would later promote the concept of planesymmetric addition, stating that this principle

---

[19] Victor Meyer to Adolf von Baeyer, 12 June 1883, Deutsches Museum München, Archiv, HS 7135.
[20] Victor Meyer, *Die Thiophengruppe*, Vieweg, Braunschweig, 1888.
[21] Victor Meyer to Richard Meyer, 26 April 1887, pp. 210–11 in Meyer, *Victor Meyer*.
[22] Victor Meyer and Karl von Auwers, 'Untersuchungen über die zweite van't Hoff'sche Hypothese', *Berichte*, 1888, *21*, 790–817, p. 785.

made possible the *a priori* assignment of configurations to addition or elimination products.[23]

Yet amid Meyer's praise was also a healthy skepticism. Meyer thought that Wislicenus had been overenthusiastic in the application of his basic principles, and that 'we must still emphasize that we are not satisfied with a few explanations he has given'.[24] For example, Meyer did not accept Wislicenus' proposed mechanism for the rearrangement of maleic to fumaric acid, because the intermediate bromosuccinate (see Chapter 5) was stable under the reaction conditions. Second, he regarded Wislicenus' space formulas as inaccurate representations of the molecule, because 'the four valences of the carbon are not even drawn, but rather the six tetrahedral edges that are of really no interest in this matter'.[25] He agreed with Lossen that the space formulas Van 't Hoff and Wislicenus used for portraying double and triple bonds were physically impossible because known attractive forces could not operate around corners. Finally, Meyer was skeptical about Wislicenus' methods for assigning configurations to the unsaturated acids. Unlike Wislicenus, Meyer did not develop his own axioms or principles, nor would he apply Wislicenus' principles in the comprehensive manner that Wislicenus did. He did not consider his own theory of the benzildioximes as a 'fruit' of Wislicenus' theory, because it was not simply an expansion or use of Wislicenus' ideas. If Meyer found it a 'high honor and joy' to see his work associated with that of Wislicenus, 'so indeed we must lodge a protest against identifying our efforts with those of Wislicenus'.[26]

Although Meyer clearly wanted to distance himself theoretically from Wislicenus, it is still difficult to tell whether Meyer would have resumed the study of the benzildioximes had Wislicenus' essay not appeared in 1887. According to Meyer himself, the benzildioximes led him 'from an entirely different side to areas which in part touch the area that Wislicenus recently treated so comprehensively'.[27] This would suggest that Meyer had an independent interest in the subject of geometrical isomerism, but the timing of Meyer's publications seems to diminish this self-declared independence. His first published research articles treating 'chemistry in space', those on the benzildioximes, on unsaturation and benzylcyanides mentioned above, appeared in early 1888, less than a year after he first read 'Spatial Arrangement'. Also, as we shall see below, much of Meyer's theory of the benzildioximes was couched in terms that were clearly adapted from 'Spatial Arrangement'. Meyer

---

[23] Meyer and Jacobsen, *Lehrbuch*.
[24] Victor Meyer and Karl Auwers, 'Weitere Untersuchungen über die Isomerie der Benzildioxime', *Berichte*, 1888, *21*, 3510–29, p. 3513.
[25] Meyer and Auwers, 'Untersuchungen', p. 786.
[26] Ibid., p. 3513.
[27] Victor Meyer and Robert Demuth, 'Zur Kenntniss der Isodibrombernsteinsäuren', *Berichte*, 1888, *21*, 264–70, p. 265.

certainly did not agree with all of what Wislicenus said, but he found his arguments as a whole compelling enough to adopt some of their fundamental ideas and to address issues that Wislicenus had neglected. He thus reopened the investigation into the benzildioximes he had begun in 1883.

Van 't Hoff's 'Second Hypothesis', 1887–1890

Although we cannot know with certainty when Meyer realized that the isomeric benzildioximes could be explained in spatial terms, he certainly did by the latter part of 1887. A spatial explanation would be only a last resort, however, and he would employ it only when and if he could show that the two benzildioximes possessed identical constitutions. Meyer and Auwers therefore subjected the two isomers to an exhaustive series of reactions to compare their products. The results of these reactions provided the ammunition for an exceptionally detailed argument for their constitutional identity. Unlike the case of lactic acid or the unsaturated acids, the reactions of the benzildioximes appeared decisive and unequivocal, largely because of the highly crystalline nature of oxime-containing compounds and their derivatives. Meyer's and Auwers' description provided one of the clearest arguments ever presented for the necessity of spatial explanations, and it is therefore worthwhile to recount briefly the chemistry of the benzildioximes that indicated structural identity.

The two isomers were the product of benzil with hydroxylamine. The first isomer, α-benzildioxime (melting point 237° C) crystallized directly from the reaction mixture and the second, β-benzildioxime (melting point 206–7° C) crystallized later during the isolation procedure. The isomers possessed identical chemical compositions ($C_{14}H_{12}O_2N_2$) but different crystal forms. The α-dioxime, when heated to 170–90° C in ethanol, transformed easily into the β isomer, and both dioximes reacted with hydrochloric acid to give benzil (Figure 6.8). Because Meyer had earlier shown that these conditions merely replaced the oxime group with the ketone (that is, these conditions performed the reverse reaction of hydroxylamine with ketones), the carbon skeleton remained unchanged, and both isomers must contain the group

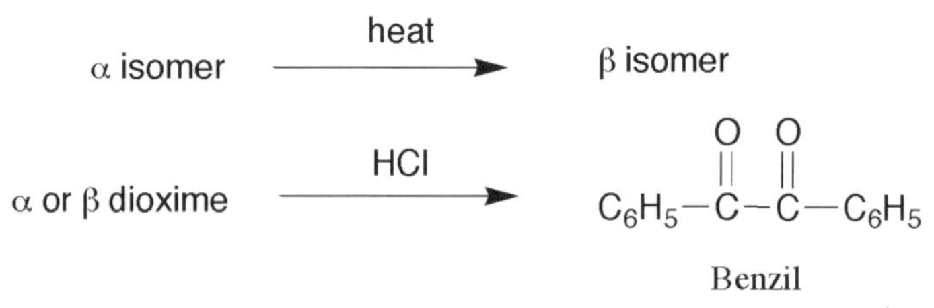

α isomer $\xrightarrow{\text{heat}}$ β isomer

α or β dioxime $\xrightarrow{\text{HCl}}$ $C_6H_5-\overset{\overset{O}{\|}}{C}-\overset{\overset{O}{\|}}{C}-C_6H_5$

Benzil

6.8   Reactions of the isomeric benzildioximes.

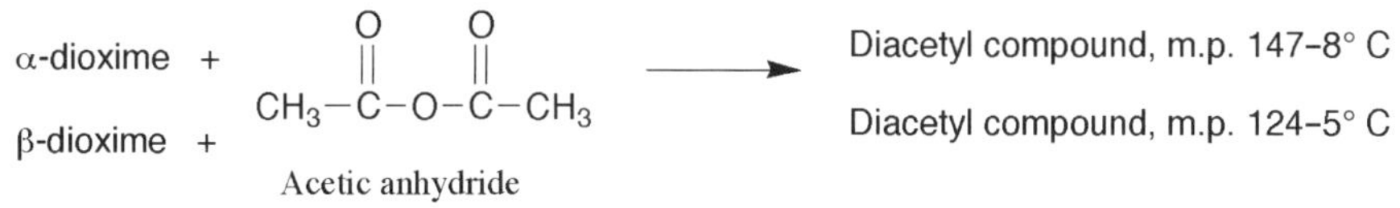

α-dioxime +
                    O   O
                    ‖   ‖
              CH₃—C—O—C—CH₃   →   Diacetyl compound, m.p. 147-8° C

β-dioxime +   Diacetyl compound, m.p. 124-5° C

                 Acetic anhydride

6.9 The reaction of α and β benzildioximes with acetic anhydride to give different compounds.

$C_6H_5-C-C-C_6H_5$. Any constitutional difference must therefore lie in the CNOH grouping of atoms, not within the main carbon skeleton. With this restriction, there were still several ways that carbon, nitrogen, oxygen and hydrogen could be connected in the second isomer. Each isomer reacted with two equivalents of organic acid anhydrides to yield isomeric esters that possessed different melting points (Figure 6.9). This indicated to Meyer that the benzildioximes were indeed chemical, not physical isomers, because true physical isomers (such as optical isomers) would yield absolutely identical products.[28] Both dioximes must therefore contain the group $C_6H_5-C-C-C_6H_5$ and two groups capable of reacting with anhydrides to form an ester: two hydroxyl groups (OH), two amine groups (NH), or one amine and one hydroxyl group. Under these restrictions, there were seven possible structural formulas for the second isomer (structures (b)–(h) in Figure 6.10).

The oxidation of both benzildioximes with potassium ferricyanide provided the conclusive evidence of structural identity. Both isomers gave the same cyclic product (Figure 6.11(a)). Formulas 6.10(c), 6.10(d) and 6.10(e) would yield oxidation products with totally different structures (6.11(b), 6.11(c), or 6.11(d)). Like Wislicenus, Meyer assumed the principle of least structural change: 10c, 10d, and 10e must undergo substantial connective changes to give the known oxidation product. If the two isomers possessed identical structures, only a change in configuration, not a drastic rearrangement of chemical connections, was necessary to form the anhydride from either isomer. Since both isomers unequivocally gave identical oxidation products, 'by this the structural equivalence is demonstrated, if the constitution of the oxidation products demonstrate, that is to say the presence of the unchanged group $C_6H_5-C-C-C_6H_5$ is established'. As additional support for this conclusion, Meyer and Auwers reported that the molecular weight of both isomers, determined by Raoult's new freezing point depression method, were identical.[29]

The chain of reasoning Meyer and Auwers followed here reflects perfectly the methodology of organic chemistry under the structure theory. Because they could

---

[28] 'Physical isomerism' most likely meant 'polymer' or 'dimer', since these would give identical esters with acid anhydrides.

[29] Meyer and Auwers, 'Untersuchungen', p. 805.

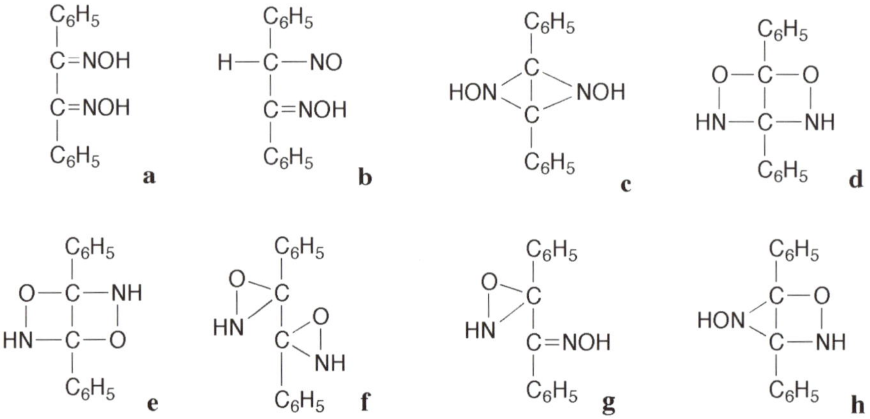

6.10 Possible structural isomers containing the group $C_6H_5-C-C-C_6H_5$.

One of the two isomers was assumed to have structure (a). The other isomer would have one of the other structures, except for (b), as Meyer had shown earlier that such nitro groups transformed spontaneously into an oxime group.

isolate benzil (whose structure was known beforehand) from benzildioxime in a simple manner, it argued that between the two transformations the connections between the carbon atoms remained unchanged. It remained, therefore, firstly to follow the rules of valence to connect the remaining nitrogen, oxygen and hydrogen atoms to the carbon atoms to determine the possible structures, and secondly to reason from these possible structures to an experiment (that is, a chemical reaction) that would show which of the structures was most plausible. If the two dioxime isomers possessed different structures, presumably one of the isomers possessed the dioxime, and the other isomer possessed one of the remaining options. The difference in atomic

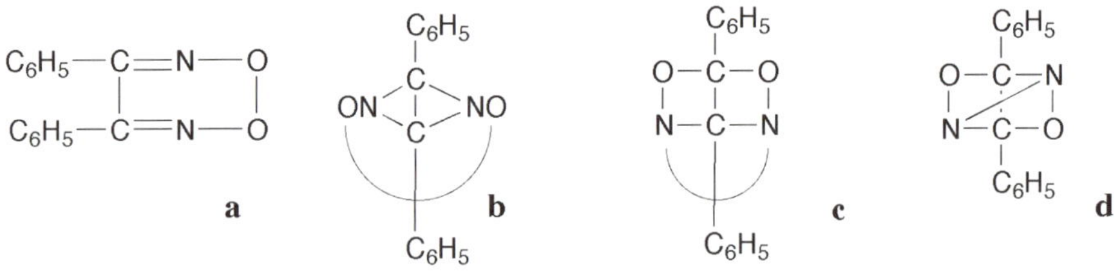

6.11 Possible reaction products of compounds (c), (d) and (e) in Figure 6.10 with potassium ferrocyanide.

connections between atoms (its structure) implied a difference in behavior towards a chemical reagent, specifically the oxidizing agent potassium ferricyanide. Because both isomers yielded the same oxidation product, they must contain the same sequential connections between atoms. Of course, Meyer and Auwers assumed in this case that in the reaction of one of the isomers with the oxidizing agent, a substantial rearrangement of connections between atoms did not occur. This assumption of least structural change (also behind the assumption that the $C_6H_5-C-C-C_6H_5$ group did not change between benzil and benzildioxime) stood behind all claims of chemical structure, and without it, assuming a structure would be impossible.

Meyer and Auwer's argument also reveals a novel, but central assumption in the understanding of 'structural identity', a crucial concept in the emergence of spatial theories of molecules. Structural identity did not mean that two compounds were identical in all respects. They differed in two ways. First, they possessed different gross physical properties such as melting point and crystal structure (characteristics that obviously indicated their differences). Second, the isomers possessed subtle differences in their characteristic reactions, for example each isomer gave an acetyl derivative that possessed a different melting points, or reacted at a slower or faster rate. Strictly speaking, then, this fact would argue *against* any claim for structural identity because their reactions were *not* absolutely identical. What mattered, however, was that both oximes yielded an acetate derivative: they reacted with acetic anhydride in the same way. The individual differences in the products were irrelevant. This fundamental assumption, that differences in degree of reaction were subordinate to differences in kind of reaction, underlay all claims for structural identity.

Constitutional identity assured, the only way for Meyer and Auwers to explain the difference between the benzildioxime isomers remained spatial; the benzildioximes must 'exist in two modifications with equivalent structure and molecular magnitude that differ in their derivatives, and so are not physical isomers whose differences can only be explained by deviating configurations'.[30] Theoretically, however, Meyer had no way to express a spatial difference. Benzildioxime contained no asymmetric carbon atom, and Van 't Hoff's second hypothesis precluded any lasting existence of different rotational configurations about the carbon–carbon axis, unless, in certain cases, this assumed free rotation did not exist. Meyer and Auwers chose this route. They saw no other conclusion but that:

> Van 't Hoff's second hypothesis must undergo a modification ... Hence it follows that for the nature of carbon's valence, two singly bound carbon atoms *can be bound in two different ways: one that permits opposing rotation, and another that does not allow such rotation*.

---

[30] Ibid., p. 815.

The conclusions which can be drawn from this unexpected result on the form of the carbon atom and the nature of valence shall be extensively discussed in another treatise.[31]

Although both Meyer and Auwers agreed that the rotation appeared restricted, they disagreed on the theoretical interpretation of that restricted rotation. They would publish separate accounts interpreting these results – Meyer in late 1888, and Auwers in his *Habilitationsschrift* in late 1889. Their decision to delay the publication of any theoretical explanation for their conclusion reveals a caution on their part, a separation between what they regarded as certain (the existence of restricted rotation in some single bonds) and what remained uncertain (the cause of that restricted rotation). This theoretical caution is also reflected in Meyer's hesitancy to publish their conclusions. In January 1888, he presented their conclusion about Van 't Hoff's second hypothesis to the Chemical Society of Göttingen, but felt apprehensive about immediately publishing the entire study. On his brother Richard's advice, he held back publication until March, when together with Auwers he published 'Investigations on Van 't Hoff's Second Hypothesis'.[32]

Three characteristics of the essay stand out. First, Meyer and Auwers devoted nearly three-quarters of its 33 pages to recounting the experimental trail that led to their conclusion of structural identity. In the remaining quarter of the paper, Meyer presented a careful theoretical analysis to complement these experimental results. As he suggested in the title of the paper, he wanted not to address the chemistry of the dioximes itself, but the theoretical issues that chemistry raised about the nature of spatial isomerism and carbon–carbon bonding. Meyer recognized that Van 't Hoff's assumption of free rotation was justified to account for the non-existence of isomerism in molecules like ethane. Wislicenus' conception of favored configurations was also justified, since 'the great success that he has achieved with this assumption allows *no doubt*, as it appears to us, *that one such type of carbon bond actually exists that allows free rotation, at least in a great number of cases*'.[33] But the existence of free rotation in *all* carbon–carbon single bonds had not been ruled out:

> But whether this bonding is always present, *whether it is the only way in which two simply bound carbon atoms can be connected, is for the moment still an open question.* If we are successful in detecting, contrary to the assumptions of Van 't Hoff and Wislicenus, that compounds constituted in this manner *exist in not just one, but two durable isomeric forms, it would subsequently be demonstrated that in addition to the 'free' rotation assumed by Van 't Hoff and Wislicenus, in which linkage of simply bound carbon atoms also allows opposed rotation, another, second type of single bond exists, in which that free mobility is*

---

[31] Ibid. Meyer's emphasis.
[32] Ibid., and Victor Meyer to Richard Meyer, 25 January 1888, p. 216 in Richard Meyer, *Meyer*.
[33] Meyer and Auwers, 'Untersuchungen', p. 790. Meyer's emphasis.

removed and a rotation of the two carbon atoms about a common axis in different senses is no longer possible.[34]

The benzildioximes, of course offered just the example Meyer sought. He carefully grounded his conclusion in the inherent limitations of Van 't Hoff's claim for free rotation as an inductive generalization applicable to *all* carbon–carbon single bonds. Because of the benzildioximes, Van 't Hoff's second hypothesis was not a universal statement and presented an obvious:

> departure from the axiom that isomerism in configurations with two singly bound carbon atoms is only possible if the atoms cannot be brought into identical positions by free rotation about the axis that lies in the direction of the binding affinity (Van 't Hoff) and that these types of compounds contain predominantly the 'favored' configuration (Wislicenus).[35]

Just as Wislicenus' essay had brought forth the overwhelming evidence (in Meyer's opinion) in favor of the spatial explanation of the unsaturated acids, the chemistry of the benzildioximes forced Meyer to similar results for single bonds. In some cases, the rotation about the carbon–carbon single bonds must be restricted.

The third principal characteristic of Meyer and Auwers' essay was its reliance on the chemistry of the unsaturated acids as an analogy to the benzildioximes. Although Meyer wished to distance himself, with some justification, from Wislicenus' own theoretical conclusions by disputing his universal claim of free rotation, Meyer's assumptions and conclusions were framed under the guidance of Wislicenus' theory. Wislicenus, following Van 't Hoff, had assumed that the chemical nature of geometrical isomers depended exclusively on the spatial properties of the carbon atom, and located the cause of isomerism and mechanism in the nature of the carbon–carbon bond. Meyer did not deviate significantly from this assumption; he assumed that the cause of configurational deviations in the benzildioximes also arose directly from nature of the carbon atom. He would later emphasize that the radicals attached to the carbon atom determined this type of bonding, but the primary cause still lay within the nature of the carbon atom's valence.[36]

Meyer's theoretical interpretation also relied heavily on the spatial configurations of maleic and fumaric acid and the concept of favored configurations that appeared in several places throughout 'Investigations'. First, Meyer applied them in the crucial point of his proof of structural identity – the oxidation of the benzildioximes with potassium ferricyanide. He noted that both maleic and fumaric acid, possessing different configurations, formed maleic anhydride. Since, like maleic anhydride, the cyclic oxidation product could exist only in the *cis* form, 'so it was expected, that

---

[34] Ibid., Meyer's emphasis.
[35] Ibid., p. 788. The original was emphasized in its entirety.
[36] Victor Meyer, 'Ergebnisse', p. 610.

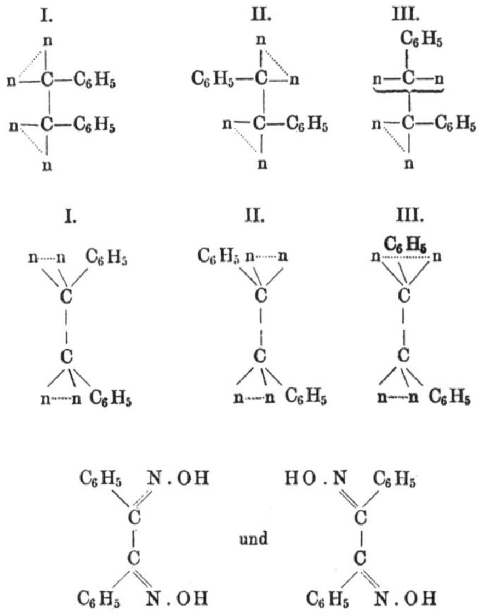

6.12 Meyer and Auwer's configurational analysis of the benzildioximes. The top row is drawn without perspective. The middle row is drawn with perspective. The bottom two formulas (the maleinoid and fumaroid forms) were also used by Meyer to emphasize the analogy between the unsaturated acids and the benzildioxime isomers.

both oximes, in case they are of the same structure, would give the same oxidation product – exactly like the anhydride of the isomeric maleic acid always forms from fumaric acid in anhydride formation'.[37] Meyer employed the analogy between the benzildioximes and maleic/fumaric acid several times throughout his essay.

Meyer also performed a configurational analysis of the benzildioximes, using diagrams similar to those in 'Spatial Arrangement'. Meyer's formulas for the benzildioximes showed three possible configurations (Figure 6.12, top two rows). If these three configurations were 'durable' (*dauernd*) and represented distinct substances, then (I) would be more stable than (II) or (III). (II) and (III) were not identical, but the difference between them was not of the same 'order' (*Ordnung*) as between either of them and (I), and Meyer considered the difference between them so slight as to make them practically identical. The non-existence of a third isomer seemed to justify this conclusion.

---

[37] Meyer and Auwers, 'Untersuchungen', p. 803.

Meyer abbreviated these cumbersome diagrams further by assuming again the analogy between the two benzildioximes and maleic and fumaric acid (Figure 6.12, bottom). Configuration (I) could be shortened to the formula on the left with a *cis* (maleinoid) configuration, and configuration (II) could be shortened to the formula on the right with a *trans* (fumaroid) configuration.[38] Because α-benzildioxime transformed into the β compound it must be the analogue to maleic acid and the less stable of the pair, and must possess the maleinoid form. Echoing Wislicenus, Meyer and Auwers also called this maleinoid form the 'less favored' configuration, and the more stable β-benzildioxime must possess the 'favored' fumaroid configuration.[39]

Previous historians who have examined Meyer and Auwers' theory have labeled it as *ad hoc*, largely because it appears that they imposed an arbitrary restriction on the bonding capability of the carbon atom that violated Van 't Hoff's second hypothesis.[40] Because Meyer could not unequivocally give any rules to predict absolutely when rotation would or would not be present, his theory did possess some *ad hoc* character. But if one considers the context in which he developed his theory, it seems clear that Meyer's hypothesis originated in a quite reasonable analogy to the chemistry of maleic and fumaric acid, demonstrated by Wislicenus to Meyer's satisfaction to be configurational isomers. Meyer's careful theoretical analysis indicated that a restricted rotation in a carbon–carbon single bond was *not* precluded by Van 't Hoff's theory. Perhaps where Meyer erred, and where he could be accused of being *ad hoc*, was his exclusive reliance on the spatial properties of carbon. As we will see in the next chapter, Meyer's theory of the benzildioximes was soon superseded by the Hantzsch/Werner theory that located the source of spatial isomerism in the double bond and the nitrogen atom.

During the next two years, amid Meyer's 1889 move to Heidelberg as Robert Bunsen's hand-picked successor, Meyer and Auwers continued their study, consolidating further evidence for the structural identity of the two benzildioximes.

---

[38] Meyer did not use the terms *cis* and *trans*, but referred to the benzildioximes as 'maleinoid' or 'fumaroid'. I have used these terms for consistency and clarity.

[39] It might be necessary to distinguish Meyer's explanation from a possible misinterpretation that derives from the modern understanding of restricted rotation in carbon–carbon single bonds. According to Meyer, the radicals attached to the carbon atoms did not simply 'bump' into one another and prevent rotation of the carbon atoms about their axis. Meyer did not even consider the spatial properties of the radicals attached to the carbon atom; in his theory, spatial properties of radicals had *no effect* on the rotation. The source of the ability or inability of carbon atoms to rotate about the carbon–carbon axis lay exclusively in the nature of the bond between carbon atoms, and no aspects of the modern concept of 'steric hindrance' were present in Meyer's theory.

[40] See W.V. Farrar, '"Chemistry in Space" and the Complex Atom', *Brit. J. Hist. Sci.*, 1968, *4*, 65–7, and George B. Kauffmann, 'Stereochemistry of Trivalent Nitrogen Compounds. Alfred Werner and the Controversy over the Structure of Oximes', *Ambix*, 1972, *19*, 129–44.

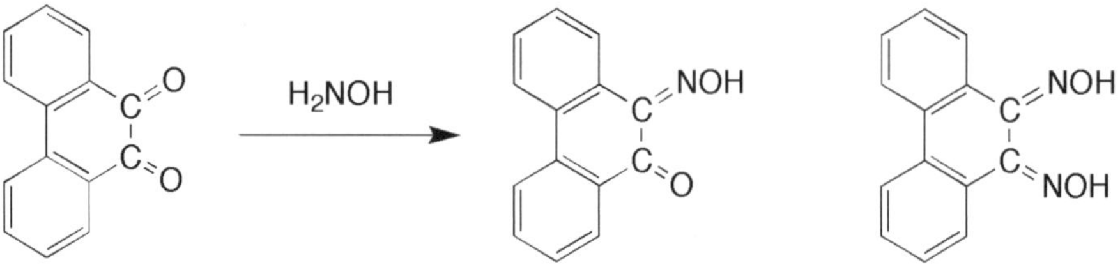

[Figure 6.13 structures: phenanthrenequinone + H₂NOH → monoxime → dioxime]

6.13 Meyer's preparation of phenanthrenequininonedioxime.

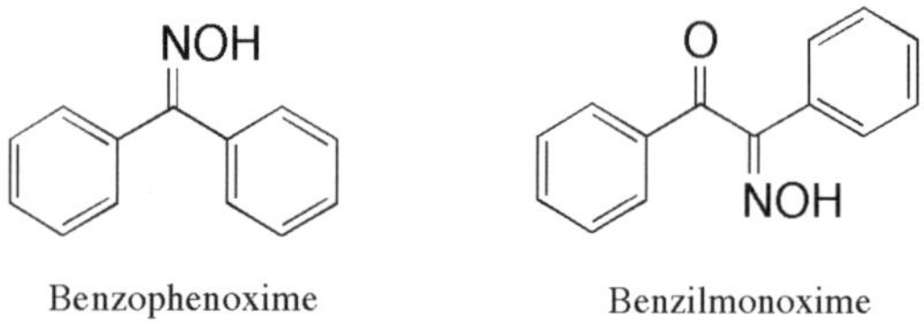

[Figure 6.14(a) structures: Benzophenoxime and Benzilmonoxime]

Benzophenoxime   Benzilmonoxime

a

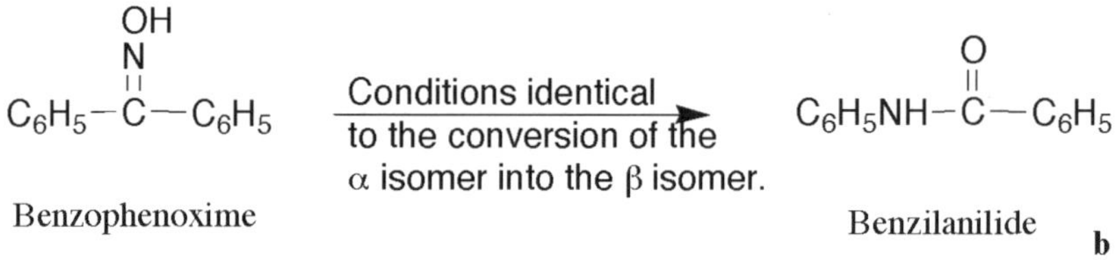

$$C_6H_5-\underset{\underset{N}{\overset{OH}{|}}}{C}-C_6H_5 \xrightarrow{\text{Conditions identical to the conversion of the } \alpha \text{ isomer into the } \beta \text{ isomer.}} C_6H_5NH-\underset{\overset{O}{\|}}{C}-C_6H_5$$

Benzophenoxime   Benzilanilide

b

6.14 (a) Benzophenoxime and benzilmonoxime. (b) The behavior of benzophenonoxime under heat.

Methylation of both dioximes gave identical products, and like the oximes, the α-methylation product converted easily into the β-methylation product. Zinc reduction of both dioximes gave identical products.[41] Because the structural analog of benzil, phenanthrenequinone, could not assume different rotational configurations, it gave, as expected, only one monoxime and only one dioxime (Figure 6.13).[42] In early 1889, Meyer and Auwers isolated two isomeric benzil*mon*oximes, and a single product (benzophenoxime) from the reaction of hydroxylamine and benzophenone, an analogue of benzil with only one carbonyl group (Figure 6.14(a)). They failed to

---

[41] Meyer and Auwers, 'Weitere Untersuchungen'.
[42] Victor Meyer and Karl Auwers, 'Über die Oxime des Phenanthrenchinons', *Berichte*, 1889, *22*, 1985–95. In modern terms, it is unclear why Meyer could not isolate a second or third stereoisomer. It may be that the aromatic group prevents a second isomer from forming.

$$\underset{\alpha}{\overset{C_6H_5}{\underset{C_6H_5}{\overset{|}{\underset{}{C}}}}\overset{NOH}{\underset{NOH}{\overset{}{C}}}} \qquad \underset{\beta}{\overset{C_6H_5}{\underset{C_6H_5}{\overset{|}{\underset{}{C}}}}\overset{NOH}{\underset{NOH}{\overset{}{C}}}} \qquad \underset{\gamma}{\overset{C_6H_5}{\underset{HON}{\overset{|}{\underset{}{C}}}}\overset{NOH}{\underset{C_6H_5}{\overset{}{C}}}}$$

6.15 Meyer's and Auwers' new conception of the three isomeric benzildioximes, arranged in order from least stable to most stable. In the middle isomer, the two groups are perpendicular to one another.

isolate any trace of an isomer of benzophenoxime, and under the same conditions conducive for the transformation of α- to β-benzilmonoxime, benzophenoxime gave a *structurally* different product (benzilanilide, Figure 6.14(b)). An oxime group must therefore exist in both benzilmonoxime isomers, because one would expect that any difference in atomic connections within the CNOH group of benzilmonoxime would also occur in benzophenoxime. By analogy, both groups in the benzil*di*oximes must also be true oximes.[43]

The strongest support for their theory appeared in March 1889, when Meyer and Auwers succeeded in isolating the third benzildioxime predicted by their earlier configurational analysis. The third isomer was extremely unstable, and immediately converted into the β isomer (the reason, Meyer noted, why it had not been isolated earlier). It also was extremely prone to forming an anhydride, and therefore more likely to be the true maleinoid compound. Meyer reassigned the three configurations – the new compound became α-benzildioxime with the maleinoid configuration, the old α-benzildioxime became β-benzildioxime, the third, most stable fumaroid isomer (the original β-dioxime) he renamed γ-benzildioxime (Figure 6.15). Although earlier Meyer argued that the difference between the configurations (II) and (III) in Figure 6.12 was so small as to make them practically the same, the ability to isolate this third isomer 'affords a strong argument for the correctness of our views. The new isomer is ... a true chemical isomer of both of the previously known benzildioximes an undoubtedly has the same chemical constitution as each of them'.[44]

In December 1888, Meyer noted their theory had provoked a considerable response, both public and private, from those in favor of their conclusion, and those that 'expressed apparently less in scientifically based opposition, than in a certain tenacious reluctance to recognize novelty'.[45] Soon after the appearance of

---

[43] Victor Meyer and Karl Auwers, 'Über zwei isomere Benzilmonoxime', *Berichte*, 1889, 22, 537–51.
[44] Victor Meyer and Karl Auwers, 'Über das dritte Benzildioxime', *Berichte*, 1889, 22, 705–20.
[45] Meyer and Auwers, 'Weitere Untersuchungen', p. 3511.

'Investigations', Bethmann, Baeyer, and Graebe offered independent confirmation of Meyer's conclusions by finding other apparent instances of restricted rotations.[46] Unfortunately, almost all of the private reaction remains unknown, although in a letter to Meyer, Baeyer described Meyer's 'wonderful work' on benzil, that he 'certainly found extremely interesting, as do all chemists'.[47]

There were relatively few extended negative critiques. Arthur Michael doubted the structure of the product in the key oxidation reaction in Meyer's proof of structural identity, but Meyer gave this criticism little attention.[48] Richard Anschütz questioned Meyer's claim that the two benzildioximes possessed identical molecular weights. Anschütz had found, also using Raoult's method, that racemic (optically inactive) tartaric acid and optically active tartaric acid had the same molecular weight, even though racemic acid should have double the molecular weight of one of its components.[49] He suggested that the two benzildioxime isomers had a similar composition, one active monomeric form and one dimeric racemic mixture with twice the molecular weight of the monomer that gave misleading results when tested for its molecular weight.

Meyer also gave little attention to Anschütz's specific criticism, but did use the opportunity to defend vigorously geometrical isomerism. He thought Anschütz was in a 'hopeless struggle' to explain maleic and fumaric acid on structural terms because Anschütz's lactone structure was only made possible by the presence of a divalent oxygen atom in maleic acid (Figure 4.1). Anschütz could not, for example, explain the same type of isomerism in tolane dichloride ($CH_3CCl=CClCH_3$) because chlorine was only monovalent:

> If despite the existence of these compounds – that nonetheless irrefutably shake our desire for causality, making the expansion of our theories necessary for any unbiased thinker – if many chemists scorn this expansion anyway and prefer to evade the emerging difficulties by the tried and true method of the ostrich by sticking their head in the sand, then the time has not yet come, we believe, to call out 'too much' to chemists, as expressed by Lothar Meyer.[50]

Meyer here referred to a short article by Lothar Meyer that appeared during the summer of 1888, in which he commented on Adolf von Baeyer's spatial formula

---

[46] T.E. Thorpe, 'Victor Meyer Memorial Lecture', *J. Chem. Soc.*, 1900, p. 194.

[47] Adolf von Baeyer to Victor Meyer, 6 May 1888, Deutsches Museum München, Archiv, HS 7056.

[48] Arthur Michael, 'Zur Kritik der Abhandlung von J. Wislicenus: "Über die räumliche Anordnung der Atome in organische Molekülen', *J. prak. Chem.*, 1888, *36*, 6–39, p. 36.

[49] In Anschütz's view, racemic inactive compounds consisted of a dimer (a complex of two identical molecules) of both enantiomers of the substance, and therefore possessed double the molecular weight. Richard Anschütz, 'Über die Bildung von Diacetyltraubensäuredimethyläther und die Bestimmung seiner Moleculargrosse nach der Methode von Raoult', *Annalen*, 1888, *247*, 111–22.

[50] Meyer and Auwers, 'Weitere Untersuchungen', p. 3512.

for benzene. Lothar Meyer had used the opportunity to comment on the field of geometrical isomerism in general: 'I would not like to suppress the question whether we now tread on the danger of laying too great a value on the spatial ideas so badly frowned upon earlier.'[51] He was generally sympathetic to the *idea* behind geometrical isomerism. The principles of spatial isomerism must be followed so far as possible, and those who did would not 'suffer damage to their souls'.[52] Their use, nevertheless, required great caution since, like the benzene ring, they gave rise to a danger that students may come to see them as 'certain established discoveries'. His remarks were not specifically addressed to Victor Meyer, but Meyer emphatically disagreed.

Only an enthusiastic adoption of new theories, Meyer said, brought them into the mainstream, and chemists had too long stayed within the comfortable boundaries of the structure theory. 'For many years,' Meyer remarked, 'chemists have stayed unshakably fast to the old structure theory, so that a quite considerable number of facts unexplained by it have remained completely unobserved for a long time.'[53] No, stereochemists did not have 'too much' trust in stereochemical principles, they did not have enough. Meyer admitted that when chemists hesitated to go beyond anything that 'could not be easily expressed on paper by the usual four valence lines', it spoke for the 'excellence' of the structure theory. But it was precisely the excellence of the theory that resulted in an intellectual 'inertia' that prevented them going beyond structure to the next step. Rather, critics 'cling fearfully to secure ground and recoil from every step into an unknown neighboring country'. Meyer's commentary was less restrained than Wislicenus, who in his reply to Lossen had not so explicitly ridiculed attempts to cling to the structure theory. Although Meyer may have entered the field hesitant, by late 1888 he was fervent in his promotion of 'chemistry in space', defending the correlation between chemical structure and spatial arrangement against critics he found too cautious and conservative.

Spokesman for the New Science of Stereochemistry, 1889–1890

Meyer's most enduring contribution to 'chemistry in space' was the introduction of the word 'stereochemistry' (from the Greek word *stereos*, or solid) to describe the new field of research. Although it is generally known that Meyer invented this term, what has not been noticed is that the first use of the prefix 'stereo-' was actually as an adjective. In a footnote to 'Investigation', Meyer and Auwers expressed the long phrase 'constitution under consideration of geometrical position' with the phrase 'stereochemical constitution' (*stereochemische Constitution*), whose meaning

---

[51] Lothar Meyer, 'Über die Constitution des Benzols', *Annalen*, 1888, *247*, 251–4, p. 254.
[52] Ibid.
[53] Meyer and Auwers, 'Weitere Untersuchungen', p. 3510.

was identical to Wunderlich's term 'Configuration'.[54] Meyer introduced the term 'stereochemistry' itself only later, in 1889.[55] This new label served the function of giving a general term to replace the rather vague or cumbersome ones such as 'chemistry in space', 'chemistry in terms of spatial considerations', or 'doctrine of geometrical isomerism'. The prefix 'stereo-' implied all these things in a simple way, making an inference to the central claim of the new field – molecules were solid objects, and stereochemists studied the gross external characteristics of those objects. The term came into common use almost immediately, even by opponents of stereochemistry.

In 1889, Wislicenus, then President of the German Chemical Society, invited Meyer to give the first of a series of comprehensive lectures to the society. At his mentor Baeyer's urging, Meyer chose to devote the lecture to stereochemistry, 'for other things, such as thiophene, vapor pressure and so on is old and worn out. And in a "guest performance", "impact" is still the important thing'.[56] Meyer was by all accounts a gifted lecturer (as a child, he had wanted to become an actor), and on 28 January 1890, he addressed the society for two hours on the 'Aims and Achievements of Stereochemistry', discussing the historical development and foundations of 'chemistry in space'. He recognized the importance of Wislicenus' lecture at the Innsbruck *Naturforscherversammlung* and recounted Van 't Hoff's initial arguments. The great initial success of the theory was due to the recognition that compounds that contain an asymmetric carbon atom were spatial isomers, and capable of optical rotation.[57] Wislicenus had expanded the original theory and 'opened a new phase of stereochemical research', by rekindling interest in Van 't Hoff's second hypothesis by showing how it could be applied with great success.[58]

One direct result of this 'new phase' of stereochemical research was, of course, Meyer's own study of the benzildioximes, to which Meyer devoted over half of his address. He recounted the development of his own theory, and how he was led to modify Van 't Hoff's second hypothesis. Using a new means of representation (Figure 6.16), drawing the molecule with the carbon–carbon bond perpendicular to the page (much like a modern Newman projection). Meyer predicted the existence of a fourth benzildioxime.

---

[54] Wislicenus used the term 'stereometric' in 'Spatial Arrangement', but Meyer was the first to suggest this prefix for spatial properties. Meyer and Auwers, 'Untersuchungen', p. 789.

[55] Victor Meyer, *Chemische Probleme der Gegenwart*, Heidelberg, Carl Winter, 1890, pp. 18 and 106.

[56] Victor Meyer to Adolf von Baeyer, 31 December 1889 and 6 January 1890, Deutsches Museum München, Archiv, HS 7180 and 7181. Emil Fischer gave the second lecture in the spring on his new research on the configuration of the sugars, and Otto Wallach lectured on his research on terpene derivatives. Wislicenus to Emil Fischer, 15 December 1889, Emil Fischer Papers, BANC MSS 71/95 z, The Bancroft Library, University of California, Berkeley.

[57] Meyer, 'Ergebnisse', p. 573.

[58] Ibid., p. 583.

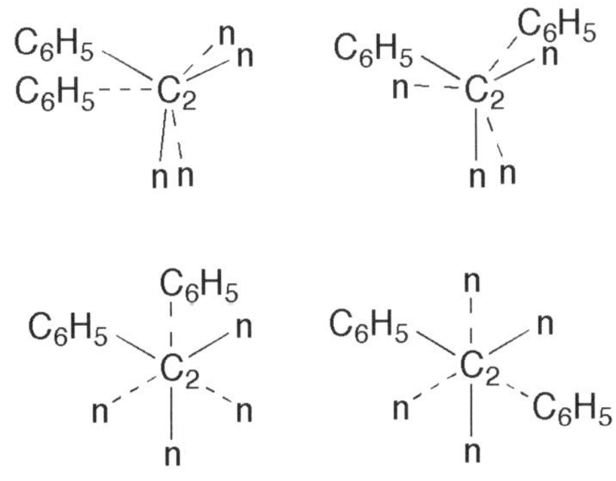

6.16 **Meyer's prediction of a fourth benzildioxime isomer in 1890. The configurations are drawn end-on, looking through the C−C axis (solid lines are radicals contained by the front carbon, dotted lines are held by the carbon atom behind). In the two upper drawings, the groups in front should cover completely the radicals in back. They are staggered slightly for clarity.**

This analysis revealed in Meyer a fluidity of thought and an ability to make comparisons between classes of compounds. The existence of three isomers indicated an analogy that he had already drawn between the benzildioximes and the spatial isomerism in the unsaturated acids or the tartaric acids. But the new representation rested on an analogy to benzene, in which a two-carbon unit in ethane could function as a six-carbon unit in benzene. 'The newer hypothesis' elaborated Meyer, 'offers the advantage of an analogy to the positional isomerism [*ortho*, *meta* and *para*] of the benzene derivatives, not dextro- and levo-tartaric acid; the first also has the advantage that the three predicted benzildioximes are already known.'[59]

The most pressing issue, however, was the recent challenge to Meyer's theory from his successor in Zürich, Arthur Hantzsch, and Hantzsch's student Alfred Werner. Meyer had learned of the Hantzsch/Werner theory of the benzildioximes in December 1889, a month before it was published as the first article of 1890 in the *Berichte*.[60] Meyer was not ready to abandon his theory and defended it against Hantzsch until 1893, when he finally admitted its superiority. However much he disagreed with Hantzsch and Werner in 1890, Meyer took great pride in the establishment of the constitutional identity of the benzildioximes. Whatever the ultimate cause of this isomerism:

---

[59] Ibid., p. 594.
[60] Alfred Werner and Arthur Hantzsch, 'Über die räumliche Anordnung der Atome in stickstoffhaltigen Molekülen', *Berichte*, 1890, *23*, 11–30.

I consider establishing the existence of structurally identical benzildioximes as the fundamental result of my endeavours in this field. Whether the hypothesis put forth by me or another theory in the long run promises to become more justified to the newer phenomena, it appears of little significance when compared to the evidence for the existence of a new class of isomeric phenomena that could be foreseen by none of the previously known theories.[61]

This statement reveals the nature of Meyer's enthusiasm towards the accomplishments of stereochemistry. Meyer did not defend single individual hypotheses as strongly as he defended the general *idea* of stereochemistry. According to Meyer, the structure theory had reached its limitations and required modification to include considering the possible physical form of molecules. The 'great and overall important achievement' of stereochemistry was not his or Hantzsch's theory of the benzildioximes, or Van 't Hoff's theory of the unsaturated acids, but the convincing evidence that spatial explanations had become necessary for chemical theory, that 'we may no longer reckon atoms as material points, but are forced to consider their dimensions, and can already obtain ideas of the relative proportions between them, if only on a modest scale'.[62]

The Nature of Valence and Bonding

Stereochemistry and Valence

In addition to considering the principles of stereochemistry as important in explaining new cases of isomerism, Meyer also found it important for the insights it brought into the general nature of valence, an issue which had concerned him as early as 1876 in a short theoretical article entitled 'On Carbon's Valence and its Bonding Ability'.[63] In this paper, Meyer had noticed that despite the success of the structure theory in ordering organic compounds and explaining isomerism, 'the nature of that what we call a valence or affinity, for the moment is still completely unclear'.[64] The more the theory of valence proved its value, said Meyer, the more one needed a 'certain, physically permissible conception' of valence. In this theoretical paper, Meyer attempted to form a vague idea of the nature of valence by studying the limitations on the carbon atom's bonding ability. For example, he pointed out that all reactions in which he expected cyclopropanes (three-membered rings) as the product gave only products with other structures. The fact that these compounds did not exist,

---

[61] Meyer, 'Ergebnisse', p. 596. Meyer's emphasis.
[62] Ibid., p. 618.
[63] Victor Meyer, 'Zur Valenz und Verbindungsfähigkeit des Kohlenstoffs', *Annalen*, 1876, *180*, 192–206.
[64] Ibid., p. 192.

coupled with the non-existence of C_2 (carbon with a 'quadruple' bond), despite the fact that these compounds were 'easily expressable by our usual formulas', indicated specific limits on the nature of valence.[65] This remained Meyer's only paper on pure theoretical organic chemistry until 1888.

This interest in the nature of valence reappeared when Meyer encountered and recognized the principles of stereochemistry. He wrote to his brother Richard in early 1888 concerning the publication of his theory of the benzildioximes, excited about the opportunity this chemistry offered for a more detailed picture of atoms and valence:

> I am so terrifically excited about all these things, dreaming of them, sometimes going through the day as if in a dream. For I have the feeling that this is a great step further in the understanding of nature, we are gaining concepts about atoms and valence. Above all, we are now enthusiastically working at making real isomers that unfortunately presents many technical difficulties and is not going quickly. It is, as Hofmann said in his time as he anticipated the phosphonium bases: 'Enough of theory! Experiment drags its feet and follows the flight of the exhilarated imagination only slowly and from a distance!'[66]

Furthermore, Meyer made it clear that the assumption of the tetrahedral carbon atom raised important questions about the physical nature of valence:

> As soon as one assumes Van 't Hoff's second hypothesis ... an hypothesis, whose fundamental meaning Wislicenus recognized first and which he has taken as the basis of his theoretical system and has executed with admirable consequence – certain conceptions about the form of atoms and the nature of valence must be introduced, if speculations should not dispense with a rigorous foundation.[67]

Wislicenus, of course, had avoided giving anything other than vague speculations on the nature of valence and affinity, and would not likely have published even that had Lossen not forced the issue. But according to Meyer, the very success of the tetrahedron in solving problems in isomerism – that is, his conviction in the *reality* of the tetrahedral arrangement of valences – compelled him to offer a more concrete physical model of valence and affinity.

Meyer's interest in the fundamental importance of valence was also reflected in his stereoformulas, in which he emphasized the distribution of valences, and not the shape of the carbon atom. The 'Newman' projections of 1890 also modeled the distribution of valence, and not the atom. 'Assuming that valences meet in certain sites in empty space free of atoms,' Meyer noted, 'cannot as such be made, and is only possible on paper or with a model, where lines and wires are used instead of forces

---

[65] Ibid., pp. 196–7. William H. Perkin Jr synthesized cyclopropane derivatives in 1885.
[66] Victor Meyer to Richard Meyer, 25 January 1888, p. 216 in Meyer, *Victor Meyer*.
[67] Victor Meyer and Robert Demuth, 'Zur Kenntniss der Isodibrombernsteinsäuren', *Berichte*, 1888, 21, 264–70, p. 265.

[valences].'⁶⁸ The meaning Meyer attributed to his stereoformulas was in fact ambiguous. Meyer agreed with Wislicenus on the reality of the double bond and the existence of free rotation about the carbon–carbon axis, and like Wislicenus, was convinced that access to the spatial properties of molecules also gave access to the spatial properties of atoms. As he stated in 1890, this was the principal success of stereochemical theory – atoms could no longer be considered material points, and drawings of molecules represented not only a molecule's chemical structure but also its physical form. But Meyer also agreed with Lossen, that the conception of bonding given in Wislicenus' drawings implied impossible physical situations, and that in double bonds valences met where there was no atom. Atoms could no longer be material points, said Meyer, but they could not be solid tetrahedra, at least not tetrahedra with the points of affinity at the corners, as Wislicenus assumed.

The Meyer/Riecke Theory of the Atom

Shortly after the appearance of his first paper with Auwers on the benzildioximes, Meyer presented a novel physical account of valence together with his colleague on the Göttingen physics faculty, Eduard Riecke.⁶⁹ Born in Stuttgart, Riecke entered the University of Tübingen in 1866 to study physics, occupied a teaching position for mathematics in Stuttgart, and in 1870, he entered the University of Göttingen, where he would remain until his death in 1915. He earned a PhD in 1870 with a study of the magnetization of iron, and in 1881 became Wilhelm Weber's successor at Göttingen. His wide-ranging interests covered the origins of ferromagnetism, elastic effects, thermodynamics and pyroelectricity.⁷⁰ Meyer, who met him after his 1885 move to Göttingen, said: 'He is the ideal colleague in physics for me, you can articulate any physical question to him, and be certain of understanding correctly; you always get a short answer free of scholarly detail, so he has become quite indispensable to me.'⁷¹

In 1886 and 1887, Riecke had published two articles on the pyroelectricity of tourmaline (a brittle crystalline mineral used for gems), in which he described the properties of its surface conductivity. In a series of experiments, Riecke found that the molecular unit in crystalline tourmaline possessed a permanent polarity, and concluded that 'each valence was caused by a certain combination of two opposed

---

⁶⁸ Ibid., p. 265.
⁶⁹ Auwers did not agree entirely with Meyer, and presented an alternate interpretation in his *Habilitationschrift*.
⁷⁰ Pyroelectricity is the ability of some minerals to develop electrical poles when heated. W. Voigt, 'Eduard Riecke als Physiker', *Phys. Zeit.*, 1915, *16*, 219–21.
⁷¹ The only complaint Meyer had was Riecke's strong Swabian dialect. Victor Meyer to Emil Fischer, 8 February 1888, Emil Fischer Papers, BANC MSS 71/95 z, The Bancroft Library, University of California, Berkeley.

electrical components'.[72] Meyer had drawn similar conclusions after his work with Demuth and Auwers. In his lecture to the Göttingen Chemical Society on the benzildioximes, he supposed that carbon atoms were composed of spheres surrounded by an ether shell that formed the seat of valence. Each valence he conceived as an 'electrule', composed of a positive and negative piece arranged in a straight line. Each electrule underwent isochronous oscillation, and would therefore attract other valences with different oscillations, and repel valences with the same oscillation. The four valences of the carbon atom must therefore repel each other to form a tetrahedron, and in the presence of other atoms, the valences arranged themselves in an irregular tetrahedron determined by the attached atoms.[73] Riecke was present at Meyer's lecture, and remarked in the discussion that the ether shell of pentavalent elements such as nitrogen or phosphorous could be ellipsoid, three valences occupying the corners of an equilateral triangle, and two occupying the poles.

Meyer and Riecke subsequently collaborated and co-authored a paper, containing 'A Few Remarks on the Carbon Atom and Valence'. Drawing from each of their results, they composed an ingenious account of carbon–carbon bonding that explained both the free and restricted rotation of single bonds, as well as the nature of multiple bonding. On the basis of chemical theory, three concepts had become evident. First, the four valences of the carbon atom were directed towards the corners of a tetrahedron. Second, in a single bond there were two types of linkage: one which allowed free rotation, and one that did not. Third, in double bonds, rotation did not exist. About the nature of the atom itself, however, chemical theory could say nothing further: 'with this, what we have to say about valence on the basis of certain chemical facts is exhausted.'[74]

Three general considerations of physical properties led Meyer and Riecke to conceive of atoms as composed of electric particles. First, the separation of 'electrical fluids' (via the generation of static electricity) suggested that the chemical elements were not bound simply by chemical affinity, but also by electricity, and because any body can become positive or negative, it must contain both kinds of electrical fluid. Second, the existence of pyroelectricity suggested that in some crystals, the molecules they contained were held together by a system of electric poles in fixed positions relative to each other and to the molecule. A mineral such as tourmaline then would be held together by two opposite electric poles that were arranged in the direction of the crystallographic main axes on opposite sides of the molecule. This also led to the assumption that atoms must also have certain systems of negative

---

[72] Victor Meyer and Eduard Riecke, 'Einige Bemerkungen über das Kohlenstoffatom und die Valenz', *Berichte*, 1888, *21*, 946–56, p. 946.
[73] Anonymous reviews of Meyer's lecture to the Göttingen Chemical Society in *Nature*, 1888, *37*, 327, and *Chemiker-Zeitung*, 1888, *12*, 140.
[74] Meyer and Riecke, 'Einige Bemerkungen', p. 950.

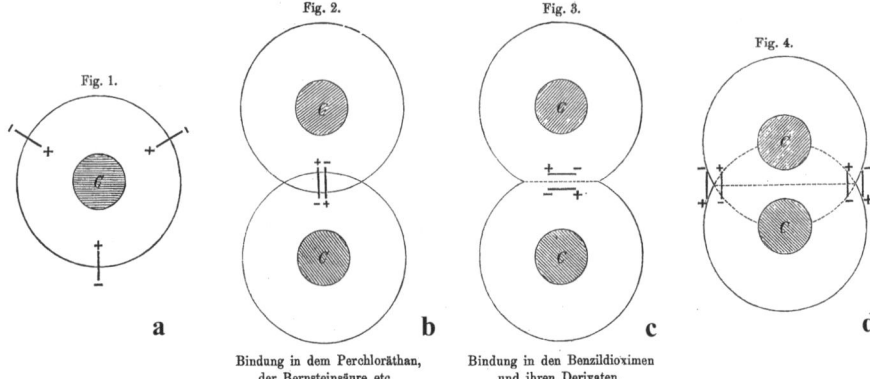

6.17 Meyer and Riecke's illustration of their model of the carbon atom and bonding. (a) This is the carbon atom itself. One valence is behind the atom pointing away from the viewer. (b) This illustrates a single bond capable of free rotation, while (c) illustrates a single bond incapable of rotation. (d) This illustrates the formation of a carbon–carbon double bond.

and positive poles. Third, the law of electrolysis suggested that each valence unit possessed the same amount of electricity, carried by ions, whose movement mediated the conduction of galvanic current.[75]

Based on these chemical and physical conceptions, Meyer and Riecke proposed the following model. The carbon atom:

> is surrounded by an ether shell, that in an isolated atom ... possesses a spherical form; the atom itself we consider as the carrier of the specific affinities, the surface of the shell as the seat of the valences. We consider each valence caused by the presence of two opposed electrical poles that are small straight lines in comparison to the diameter of the ether shell fastened at the endpoints. (Figure 6.17(a))[76]

The carbon atom possessed a stronger affinity for positive than for negative electricity, and the positive pole was stronger than the negative pole.[77] Each dipole would then position itself such that the positive end pointed toward the carbon atom. The result would be a natural tendency for the negative ends, pointing out away from

---

[75] The reference Meyer and Riecke offered was Hermann Helmholtz's 1881 Faraday lecture, *Chem. Soc. Trans.*, 1881, p. 277.

[76] Meyer and Riecke, 'Einige Bemerkungen', p. 951.

[77] This was an arbitrary choice. Meyer noted that he could assume the opposite conditions to achieve the same result.

the atom to repel each other and take up the positions of the tetrahedron. Furthermore, because they were freely movable within the ether envelope, they could be distorted from a perfect tetrahedron.

Meyer and Riecke then used this model to depict the nature of carbon–carbon bonding. In a normal single bond (Figure 6.17(b)), two dipoles, one from each atom, aligned by means of electrical attraction along the carbon–carbon axis. The two atoms remained free to rotate about the carbon–carbon axis. Meyer and Riecke also assumed that the dipoles were free to rotate within the ether shell, and this suggested a second way in which two dipoles would align to give a second form of a single bond (Figure 6.17(c)). In this case, rotation of both carbon atoms about the axis was not possible. The existence of these different kinds of single bonds were then determined by the nature of the radicals bound to carbon.[78] In double bonds, two pairs of dipoles aligned themselves, and in a triple bond, three pairs (Figure 6.17(d)). This model for multiple bonding also precluded the possibility of rotation about the carbon–carbon axis.

This model presented a strikingly different and more detailed view of the carbon atom than Wislicenus had offered to Lossen one month earlier. Unlike Wislicenus, Meyer and Riecke offered a specific depiction of the parts of the atom and how they worked together to produce the chemical effects that Meyer and Auwers had found. It offered a higher-level explanation for the conclusions of stereochemical theory. Why did some atoms combine to allow free rotation and others not? Why were the valences arranged in a tetrahedron? The plastic nature of the dipoles offered a clear picture of how the carbon atom could distort when it was attached to different radicals. An anonymous writer for *Nature* commented that the theory offered a 'strikingly natural explanation … of the nature of single double and triple linking of carbon atoms'.[79]

Striking as it was, this model had an extremely short life, and Meyer and Riecke hesitated to display much confidence in it, as it was not comprehensive. Meyer and Riecke were also careful to point out that this model did not adequately address the 'essence' of chemical affinity:

> No one feels clearer than ourselves how distant the opinion presented here still is from the path to a comprehensive physical theory of chemical phenomena. Our investigation may be seen only as modest progress on this still very dark path that makes a start towards a more clear understanding of some phenomena and presently is certain to prompt greater penetration in the subject of valence.[80]

Meyer would later say that this conception of atoms and valence had little to do with his actual work on the benzildioximes, and it seems striking that none of the experimental papers on the benzildioximes mentioned, even in passing, the Meyer/

---

[78] Meyer and Riecke, 'Einige Bemerkungen', p. 953–5.
[79] *Nature*, 1888, *37*, 327.
[80] Meyer and Riecke, 'Einige Bemerkungen', p. 956.

Riecke theory. Meyer emphasized that their model of the atom wasn't necessary for explaining the isomerism of the benzildioximes, but:

> originated simply from the necessity that both Riecke and I felt to suppose a comprehensible idea in place of the previously entirely indefinite concept of valence. That our hypothesis harmonizes certain pyroelectric phenomena as well as the observations on the benzildioximes can serve in a sense to support it; on its own, the inverse is incorrect, that any hypothesis of valence may be necessary for my explanation of the benzildioximes. Accordingly, I must protest against Hantzsch and Werner's statement that our conception of the isomeric benzildioximes *'demands a new principle of the essence of valence'*.[81]

Although Meyer found this explanation appealing on both a chemical and physical level, he did not regard it, or any other hypothesis, as necessary to reinforce his conclusions arrived at independently with Auwers.

By 1890, Meyer had already abandoned this model when he suggested that Wunderlich's conception of the carbon atom (see Chapter 4) was the best approximation of the nature of affinity, because it preserved the physically pleasing idea of linear attractive forces and best approximated the actual nature of chemical bonding.[82] But even Wunderlich's model was not definitive; in 1893, Meyer emphasized in his *Lehrbuch der organische Chemie* how little was known about the actual reality of valence:

> We still do not have a clear idea about the essence of valence – about the cause that determines the different saturation capacities of the elements. For now, we can only conclude from observations that a carbon atom possesses an atomic binding power four times greater than a hydrogen atom, just because it is in a condition to express its affinity to four other atoms.[83]

Meyer could only assert with confidence that all four affinity units must be of equivalent power and hold all radicals equally. Otherwise, compounds such as methylamine (CH_3NH_2) would exist in different isomeric forms.

The 'Infancy' and Maturity of Chemical Theory

This tension in Meyer's thought between the desire for a reductionist model of the atom and valence and his reluctance to endorse fully his own theory that met this desire was not atypical. Nineteenth-century chemists had created a largely autonomous and non-mathematical science with its own unique characteristics, yet they also had a lurking suspicion that chemistry would not become a 'true' science

---

[81] Meyer, 'Ergebnisse', p. 601.
[82] Ibid., p. 618.
[83] Meyer and Jacobsen, *Lehrbuch*, vol. 1, p. 58.

until it had been reduced to physical laws described by mathematics. In 1890, for example, Hermann Sachse employed extensively 'the language of mechanics, which ultimately shall just dissolve [*sich auflösen*] the language of our science'.[84] Meyer best exemplified this tension between autonomy and reduction in his 1889 lecture 'Contemporary Chemical Problems', given in a general address to the Heidelberg *Naturforscherversammlung*. At the same meeting where Heinrich Hertz spoke of the relationship between light and electricity, Emil Fischer spoke of his preliminary research on the sugars and a special exhibit of Edison's new phonograph took place, Meyer gave an address to the general session that put forth the best summary of his philosophy of chemistry and its status as a science.

Meyer oscillated between advocating the reduction of chemical theory to mathematical physics and then advocating its theoretical autonomy. The final goal of chemical theory, according to Meyer, was a complete reduction of chemical reactions to mathematical mechanics, because 'nature is not understood until we are able to reduce its phenomena to simple movements, mathematically traceable'.[85] He was certain that all chemical explanations would one day be completely understood in mathematical terms:

> The time will come, even for chemistry, when this highest kind of treatment will prevail. The epoch in which the foremost impulse of its research was a serenely creative phantasy will then have passed; the joys, but also the pangs and struggles, peculiar to youth, will have been overcome. Reunited to physics, her sister science, from whom her ways at present are separated, Chemistry will run her course with firm and unfaltering steps.[86]

If this were truly the goal of all science, declared Meyer, 'a science, which is so far distant from this aim as to look merely for the *path* that shall someday lead to it, must be considered in its infancy'.[87]

According to Meyer, the 'infancy' of chemistry was not a 'blemish', nor did it detract from the 'immense achievements [it had] registered on its own'.[88] The lack of rigorous mathematical principles, Meyer claimed, gave chemical thought a plasticity

---

[84] Hermann Sachse, 'Eine Deutung der Affinität', *Zeit. phys. Chem.*, 1893, *11*, 185–219, p. 185. See also Mary Jo Nye, *From Chemical Philosophy to Theoretical Chemistry: Dynamics of Matter and Dynamics of Disciplines, 1800–1950*, Berkeley, CA, University of California Press, 1993, and Mary Jo Nye, 'Physics and Chemistry: Commensurate or Incommensurate Sciences?', pp. 205–24 in *The Invention of Physical Science: Intersections of Mathematics, Theology and Natural Philosophy Since the Seventeenth Century*, Mary Jo Nye, Joan L. Richards and Roger H. Stuewer, eds, Dordrecht, Kluwer, 1992.

[85] Victor Meyer, *Chemische Probleme der Gegenwart*, Heidelberg, Carl Winter, 1890. L.H. Friedburg translated the address in 1890, 'The Chemical Problems of Today', *J. Am. Chem. Soc.*, 1889, *11*, 101–20. I have retranslated the title, but all subsequent translations are Friedburg's.

[86] Ibid., p. 120.
[87] Ibid., p. 103. Meyer's emphasis.
[88] Ibid., p. 102.

that the logical, mathematical sciences lacked. Chemists had a greater tendency for imaginative thought or speculation (*Phantasie*) that brought a creative enjoyment similar to that experienced by artists, and was aided by 'chemical feeling' or intuition that would disappear 'as soon as the progressive approach of chemistry to the mathematical physical basis shall have disclosed its meaning'.[89]

Two important points about Meyer's philosophy may be made here. First, chemistry could not be reduced to physics until it resembled physics in its methods and produced experimentally testable theories of valence, or until experimental chemistry resembled experimental physics. Chemistry would therefore continue on a separate path until such methods were developed. Second, when Meyer spoke of chemical methods, he remained fixed on methods that would provide answers to chemical, and not physical problems. The primary experimental task facing any chemist was the ability to isolate pure substances, and. to accomplish this Meyer felt limited to purely accidental properties – crystallization and distillation:

> Have not those thousands of amorphous substances, which cannot be characterized by any chemical property, and which the chemist is forced to lay aside because he is unable either to purify them or to transform them into volatile or crystallizable bodies; have they not the same claim upon our interest as their more beautiful and more manageable comrades? ... What we want are: *New methods for recognizing the individuality of substances.*[90]

Meyer therefore made a fundamentally chemical request. He did not ask for new methods to measure the intensity of physical properties of substances, but new ways to *identify* substances, or *recognize* new substances. Chemistry was the study of matter's transformations, and the elucidation of the units of those transformations, and it remained of urgent importance to recognize when a transformation had occurred.

'Contemporary Problems' epitomizes the ambiguity in nineteenth-century chemists' minds between autonomy and reduction, and a facet of Meyer's attitude towards chemistry that did not appear in his several other public lectures on the periodic table (*Die Umwalzung in der Atomlehre*, 1883), stereochemistry (*Ergebnisse*, 1890), and pyrochemistry and the problem of 'Uratoms' (*Probleme der Atomistik*, 1895). In 'Contemporary Problems', Meyer touched on the entire enterprise of chemical science and revealed an ambiguity about the reduction of chemistry to physics. Chemistry had proceeded successfully with methodologies, axioms and theories developed independently of physics, and should continue to do so. At the same time, the entirety of this enormously successful enterprise seemed 'trifling to the mind that looks down from its standpoint of mathematical mechanics when compared with the

---

[89] Ibid., p. 104.
[90] Ibid., pp. 116–17. Meyer's emphasis.

work of a promised Newton of chemistry, who some day will represent chemical reactions in the thought and in the language of mathematical physics'.[91]

Conclusion

A number of final observations can be drawn from this brief study of Meyer's reaction to 'chemistry in space'. Once convinced of the need for spatial explanations, Meyer found them appealing for two reasons. First, as he hinted, but did not fully explain in 'Contemporary Problems', he felt that only a complete reduction of chemistry to mathematical physics would allow a true understanding chemical phenomena. Stereochemistry appealed to Meyer because it did in part reduce chemical phenomena to mechanical parameters and offered a limited 'mathematical' or geometrical understanding of chemical phenomena. Second, stereochemical principles laid further ground work for the deeper understanding of the distinctly chemical concept of valence. According to Meyer, valence and affinity must be understood in an experimentally testable fashion (testable in the sense that physical theories are testable) before any complete reduction could occur. Whereas Wislicenus saw in stereochemistry a potential for addressing the chemical problem of isomerism, Meyer embraced stereochemistry because of its ability to clarify the chemical concept of valence.

Meyer's cautious theoretical conclusions about the benzildioximes were indicative of his scientific style in general. Meyer investigated problems exhaustively to a project's logical conclusion. In his biography of Meyer, Carl Liebermann wrote:

> one sees how he cleared a path stepwise towards a goal, following a thought, how he wrings out nature's secrets, ferreting towards the left and right, coping with or circumventing obstacles, preparing aid or striking out a new goal beyond that achieved.[92]

While the major assumptions of structure theory certainly guided Meyer's work, he did not, in contrast to Wislicenus, place great emphasis on using deductive conclusions to propagate stereochemical experiments. He thoroughly studied the properties of the benzildioximes and assigned configurations to the three spatial isomers in an inductive manner. He did not propose definite reaction intermediates, or perform configurational analyses to establish the configuration of a product. In 1893, Meyer used similar inductive reasoning when he proposed the concept of steric hindrance (Meyer's other major contribution to stereochemistry) from the reaction rates of aromatic acids.

---

[91] Ibid., p. 102.
[92] Carl Liebermann, 'Victor Meyer', *Berichte*, 1897, *30*, 2157–68, p. 2163.

To postulate the appearance of intermediates would have violated Meyer's sense of empiricism, where the only configurations (or structures) he could assign were to compounds he could *isolate*. This reflects the strong influence of the Baeyer school, and Meyer did not deviate from this tradition, which included, in addition to Baeyer himself, fellow Baeyer students Emil Fischer, Carl Graebe, Carl Liebermann and Eugen Bamberger. It should also be noted that when Baeyer and Fischer explored stereochemical questions (the strain theory and sugar chemistry respectively), their results were also derived in a more cautious manner, in which they postulated no intermediates or structures for compounds they had not isolated and characterized. Accepting stereochemical principles, then, was not necessarily completely tied to the adoption of a deductive methodology. Meyer accepted the new mechanical assumptions of stereochemistry – specifically the mechanical nature of the linkage between atoms and the possibility of different kinds of bonding – without assuming their truth as universal statements to guide extensive research programs.

Chapter 7

Arthur Hantzsch:
The Stereochemistry of Nitrogen

[Structural chemistry is] the predominantly valid kind of isomerism for carbon compounds; inversely, stereoisomerism will be recognized as the more widely distributed, more general form of isomerism.

Arthur Hantzsch, 1894

When Victor Meyer wrote to Adolf von Baeyer in 1889 to ask whether he should give a lecture on stereochemistry before the German Chemical Society, he worried that 'I am already no longer the last word in the subject, since [Arthur] Hantzsch has published a stereochemistry of nitrogen in the latest issue of the *Berichte*'.[1] Meyer had received advance word of a paper – the first article in the 1890 volume of the *Berichte* – by Hantzsch and his student Alfred Werner that challenged Meyer's theory of the benzildioximes by supposing that the presence of the carbon–nitrogen double bond, and therefore the stereochemistry of the nitrogen atom itself, produced the different isomers. Their theory was the first serious attempt to establish the stereochemistry of an element other than carbon, but as we shall see, it was not the first nor the only attempt. Meyer had good reason to worry, for in six months he would drop his and Auwers' theory in the face of Hantzsch's experimental results. Although Hantzsch and Werner had not (publicly, at least) expressed any previous interest or devotion to stereochemical theories, they instantly became two of the strongest supporters of 'chemistry in space', and never wavered in their confidence that spatial causes were necessary for the success of chemical theory. Hantzsch and Werner's theory proved itself applicable not only to the benzildioximes, but to other classes of compounds containing carbon–nitrogen double bonds, specifically the hydrazones and diazo compounds, and Hantzsch spent the 1890s attempting to establish the stereochemistry of these compounds and the configurations of their isomers.

---

[1] Victor Meyer to Adolf von Baeyer, 31 December 1889, Deutsches Museum München, Archiv, HS 7180. Also printed in Richard Meyer, *Victor Meyer: Leben und Wirken eines deutschen Chemikers und Naturforschers 1848–1897*, Leipzig, Akademische Verlagsgesellschaft, 1917, p. 250.

In his extensive study of Alfred Werner's career, George Kauffman has written the most detailed discussion of the history of nitrogen chemistry, including the Hantzsch/Werner theory and its differences from Meyer's theory of the benzildioximes. More recently, Joachim Stöcklöv has published a portrait of Arthur Hantzsch as an introduction to an edition of Hantzsch's correspondence with Wilhelm Ostwald.[2] I do not differ significantly from Kaufmann's or Stöcklöv's overall interpretations of Hantzsch's chemistry, but add three significant enhancements to their analyses. First, I offer a more detailed reconstruction of the theoretical discussion, only touched on by Kauffman and Stöcklöv, between Hantzsch and Meyer on the spatial isomerism of the oximes. This series of papers exchanged between Hantzsch and Meyer was the earliest controversy between rival stereochemical theories, and therefore is useful for understanding some of the principal assumptions behind different spatial explanations. Second, I also offer a reconstruction of the controversy between Hantzsch and Meyer as stereochemists, and Adolf Claus, who denied that spatial explanations were necessary for the benzildioximes and hydrazones. Claus' critique reveals the subtleties behind the concept of 'structural identity' – one of the crucial assumptions in stereochemistry. Finally, it will also become clear how Werner's conception of the spatial nitrogen atom, and Hantzsch's ensuing attempt to assign configurations to the benzildioximes, hydrazones and diazocompounds, was derived from the principles of stereochemistry laid out by Wislicenus in 'Spatial Arrangement'.

Hantzsch was born in 1857 in Dresden, where he enrolled at the local *Polytechnikum* and studied chemistry under Rudolf Schmitt, a former student of Kolbe. He began research on paraoxyphenetol and hydroquinone for his promotion, but because the *Polytechnikum* did not grant doctoral degrees, Hantzsch enrolled for a semester at the University of Würzburg, where in 1880 he received his doctoral degree. Although he formally received the doctorate under the direction of Wislicenus, who administered Hantzsch's oral exam, most of his formal training in chemistry came from Schmitt, to whom he remained close until Schmitt's death in 1898. After spending the summer of 1880 in A.W. Hofmann's laboratory in Berlin, Hantzsch became an assistant in Georg Wiedemann's laboratory in physical chemistry at the University of Leipzig.

---

[2] George B. Kauffman, 'Stereochemistry of Trivalent Nitrogen Compounds: Alfred Werner and the Controversy over the Structure of Oximes', *Ambix*, 1972, *19*, 129–44. In the title, Kauffman used the modern sense of the conception of 'structure' which includes spatial configuration. The *structure* of the oximes was never in doubt between Hantzsch and Meyer. Rather, it was the *configuration*, or specific spatial cause that was under dispute. Joachim Stöcklöv, *Arthur Rudolf Hantzsch im Briefwechsel mit Wilhelm Ostwald*, Berlin, ERS-Verlag, 1998. Other sources on Hantzsch include T.S. Moore, 'The Hantzsch Memorial Lecture', *J. Chem. Soc.*, 1936, 1051–66, and Arnold Weissberger, 'Arthur Hantzsch', pp. 1067–83 in Eduard Farber, ed., *Great Chemists*, New York, Interscience, 1961.

His primary experimental interests were in classical organic chemistry, and his *Habilitationsschrift* discussed a new efficient synthesis of pyridine derivatives from aldehydes and acetoacetic ester derivatives, today known as the Hantzsch pyridine synthesis. While in Leipzig, he also undoubtedly became acquainted with the use of physical methods in chemistry, and indeed, in his 1882 *Habilitation* lecture he discussed the relationship between physical properties and chemical constitution that foreshadowed some of his later interests. On the basis of his synthetic work, at the age of 28, Hantzsch succeeded Victor Meyer as full professor at the Zürich *Polytechnikum*, where he would stay until 1893, when he succeeded Emil Fischer in Würzburg. In 1903 he would succeed Wislicenus at Leipzig.

During Hantzsch's early period in Zürich, he could easily be characterized as a classical organic chemist. He and his students studied derivatives of quinones, the cleavage of phenol derivatives to aliphatic compounds, and the subsequent novel formation of five-membered rings. His most famous discovery during this period was thiazoline, an aromatic compound in which a sulfur atom 'mimicked' a C_2H_2 unit in pyridine. Thiazoline was a logical analogue to thiophene, discovered by Meyer in 1883, in which a sulfur atom took the place of a C_2H_2 unit in benzene, and Meyer's synthesis had spurred Hantzsch to contemplate the existence of a similar compound derived from pyridine. In two theoretical papers, Hantzsch discussed the recently proposed concept of tautomerism (which would play an important role in his later establishment of configurations), but his research during this period was the straightforward work of a 'structural workman' (*Strukturhandwerker*), as Hantzsch described it.[3] The only unusual characteristics were several conductivity measurements, carried out by Wilhelm Ostwald, on derivatives of the cleavage products of substituted phenols, to establish differences in structure.[4]

Hantzsch would become interested in the issues of stereochemistry in 1889, when Werner, who had recently completed his *Diplomarbeit* in chemistry at the University of Zürich and had begun working as an assistant in Hantzsch's laboratory, proposed to Hantzsch a new spatial model for the benzildioximes. He suggested to Hantzsch that the cause of isomerism in the benzildioximes might be due to the spatial properties of the nitrogen atom. After some deliberation, Hantzsch agreed, and they proceeded to work out the details of the theory.[5] Werner submitted his essay as the written portion of his *Doktorarbeit*, and the theoretical part appeared as the first article in the 1890 volume of the *Berichte* (before Werner had passed his doctoral exams), a 19-page article entitled 'On the Spatial Arrangement of Atoms in Molecules Containing

---

[3] Hantzsch to Ostwald, 30 October 1890, in Stocklöv, *Arthur Rudolf Hantzsch im Briefwechsel*, p. 84.
[4] Ibid., p. 33.
[5] George B. Kauffman, *Alfred Werner: Founder of Coordination Chemistry*, Berlin, Springer-Verlag, 1966, p. 17.

Nitrogen'.[6] Although the paper was co-authored by Hantzsch and Werner, Hantzsch acknowledged that in essence, its principle idea, that 'perhaps nitrogen could also give rise to geometric isomerism in a manner similar to carbon', was the 'intellectual property of Herr Werner'.[7] As Werner's interests shifted to inorganic chemistry (discussed in Chapter 9), Hantzsch himself would present the experimental and theoretical defense of the theory.

Hantzsch's study of the nitrogen atom's spatial properties took place in two distinct phases. Between 1890 and 1894, he defended Werner's theory of the oximes against Meyer's rival stereochemical theory, and against Adolf Claus' attempts to explain the oximes structurally. In 1894, Hantzsch shifted to the diazo compounds (those containing nitrogen–nitrogen double bonds). In both phases, Hantzsch pursued two separate but related goals. First, he wished simply to establish that the isomerism in oximes, hydrazones, and diazocompounds could be traced back to the spatial properties of the nitrogen atom. Second, like Wislicenus, Hantzsch attempted to ascribe configurations to these isomers. At first, Hantzsch pursued these two goals separately, but as he continued (and especially as he began the detailed studies of the diazo compounds), he concentrated exclusively on the second goal, perhaps because he had become convinced that he no longer needed to justify their spatial properties. Also during both phases, Hantzsch's interests in classical organic chemistry merged with a growing interest in physical methods for detecting the presence of new compounds and establishing configurations.

The Hantzsch/Werner Theory

An Alternate Theory for the Benzildioximes

Hantzsch and Werner's article began with an analysis of the Meyer/Auwers theory of the benzildioximes. They explicitly acknowledged Meyer's and Auwers' principal result, that the 'isomeric benzilmonoximes and dioximes are surely isomeric, but certainly not structurally isomeric'.[8] As was discussed in Chapter 6, Meyer and Auwers had explained this isomerism by supposing that Van 't Hoff's second hypothesis must be altered to allow for the possibility that this rotation could be removed in some cases. Werner and Hantzsch recognized that the suggested restricted

---

[6] Arthur Hantzsch and Alfred Werner, 'Über die räumliche Anordnung der Atome in stickstoffhaltigen Molekülen', *Berichte*, 1890, *23*, 11–30. An English translation is given in George B. Kauffman, 'Alfred Werner's Inaugural Dissertation', *J. Chem. Ed.*, 1966, *43*, 155–65. I have altered slightly his translation of the title, but all subsequent translations from this article are Kauffman's.

[7] Kauffman, 'Alfred Werner's Inaugural Dissertation', p. 165.

[8] Ibid., p. 157.

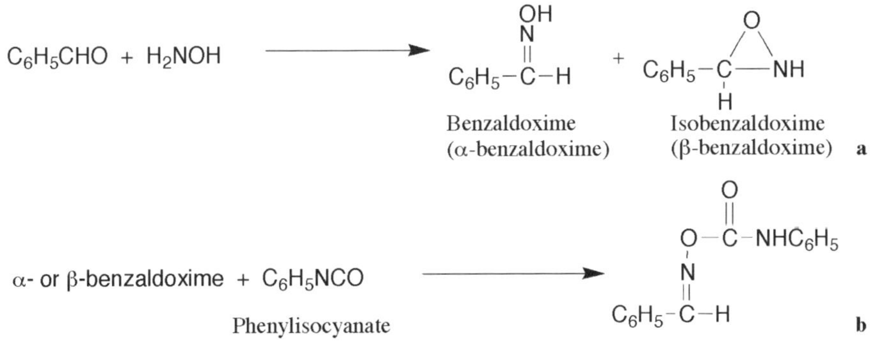

$C_6H_5CHO + H_2NOH \longrightarrow$

$\underset{\text{Benzaldoxime}}{\underset{(\alpha\text{-benzaldoxime})}{C_6H_5-\overset{\overset{\overset{OH}{|}}{\underset{||}{N}}}{C}-H}}$ + $\underset{\text{Isobenzaldoxime}}{\underset{(\beta\text{-benzaldoxime})}{C_6H_5-\overset{O}{\overset{|}{C}}\underset{H}{\overset{}{\diagup}}NH}}$ **a**

α- or β-benzaldoxime + C_6H_5NCO $\longrightarrow$ $\underset{}{C_6H_5-\overset{\overset{\overset{O-\overset{O}{\overset{||}{C}}-NHC_6H_5}{|}}{\underset{||}{N}}}{C}-H}$

Phenylisocyanate

b

7.1 (a) Ernst Beckmann's structures for the isomeric benzaldoximes.
(b) Heinrich Goldschmidt's demonstration of structural identity for the benzaldoximes.

rotation in carbon–carbon single bonds (and the Meyer/Riecke model of the atom) was possible, but the non-existence of isomers for certain compounds such as tetramethyl succinic acid ($HO_2CCCH_3CH_3CCH_3CH_3CO_2H$) and the identity of succinic acids made from the reduction of maleic or fumaric acid indicated that an alteration of Van 't Hoff's second hypothesis was unnecessary. The benziloximes 'constitute the sole exception' to Van 't Hoff's second hypothesis, Hantzsch and Werner noted, 'so the hypothesis of V. Meyer and E. Riecke is limited to this single case, but nevertheless, it may, of course, not only be justified, but may also be completely correct'.[9]

Hantzsch and Werner also noted other problems with the Meyer/Auwers theory. First, according to Meyer and Auwers, there should be three benzil*mon*oximes, while only two were known to exist. Second, only the oximes of benzil showed evidence of spatial isomerism, not benzil itself, despite the structural similarities between benzil and its oximes (the presence of a similar carbon–carbon bond). But Hantzsch and Werner's most crucial argument in support of their theory came from the known properties of the isomeric oximes of benzaldehyde (the benzaldoximes), isolated in 1888 by Ernst Beckmann, the first assistant in Wislicenus' laboratory in Leipzig. Beckmann had identified two isomeric oximes resulting from the reaction of hydroxylamine with benzaldehyde, and because isobenzaldoxime (also called β-benzaldoxime) formed a nitrogen ether and benzaldoxime (also called α-benzaldoxime) formed an oxygen ether (Figure 7.1(a)), Beckmann concluded that the two oximes were structural isomers. A year later, Heinrich Goldschmidt (the same

---

[9] Ibid., p. 158.

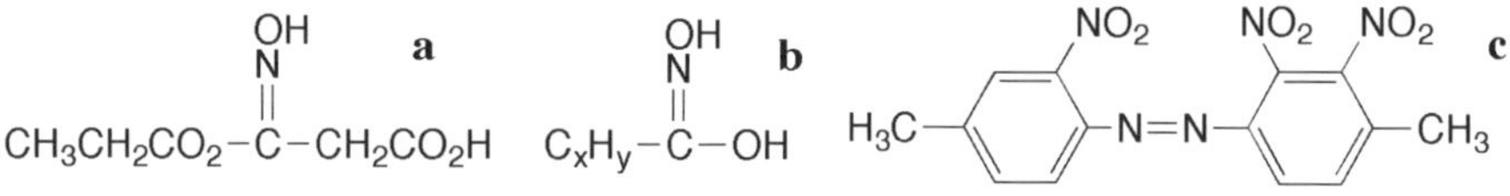

$$\text{CH}_3\text{CH}_2\text{CO}_2-\underset{\underset{\text{N}}{\overset{\text{OH}}{|}}}{\overset{\|}{\text{C}}}-\text{CH}_2\text{CO}_2\text{H} \quad \textbf{a}$$

$$\text{C}_x\text{H}_y-\underset{\underset{\text{N}}{\overset{\text{OH}}{|}}}{\overset{\|}{\text{C}}}-\text{OH} \quad \textbf{b}$$

$$\text{H}_3\text{C}-\underset{}{\text{C}_6\text{H}_3(\text{NO}_2)}-\text{N}=\text{N}-\underset{}{\text{C}_6\text{H}_2(\text{NO}_2)_2}-\text{CH}_3 \quad \textbf{c}$$

7.2 Structures for (a) oximidosuccinic acid, (b) hydroxamic acid and (c) trinitroazoxytoluenes.

student who had earlier discovered the two isomeric benzildioximes under Meyer's supervision, and had continued as an assistant in Hantzsch's laboratory) argued on the basis of the oximes' reaction with phenylisocyanate that the two oximes were actually structurally identical (Figure 7.1(b)).[10] Both the α and β aldoximes reacted with phenylisocyanate as if they contained a hydroxyl group, and in the presence of trace amounts of acid, the product derived from the β aldoxime converted into the product derived from the α aldoxime.[11] Goldschmidt himself did not interpret these results, and concluded only that 'perhaps it is actually a type of stereochemical isomerism which cannot be interpreted in terms of the hypotheses which are now known'.[12] But if Beckmann's structure for isobenzaldoxime were true, Hantzsch and Werner noted, it would require a significant amount of structural rearrangement to form the product amide. Both isomers must therefore contain an oxime group.

Hantzsch and Werner also noted other similar cases of unexplained isomerism in the oximidosuccinic acids, hydroxamic acid, and the trinitroazoxytoluenes (Figure 7.2). Given the number of examples, Hantzsch and Werner suggested that because all these compounds contained at least one nitrogen–nitrogen or nitrogen–carbon double bond, the difference between isomers could be located in that structural feature:

> Perhaps the cause of this isomerism could be sought in certain properties of the nitrogen atom. Indeed, since geometric isomerisms are undoubtedly present in many of these cases, it might be due to a different spatial arrangement of the groups bound to the nitrogen atom with respect to this atom itself. In other words, the task was to investigate whether the hypothesis developed by Van 't Hoff and Wislicenus for the carbon atom could not also be extended to the nitrogen atom and finally perhaps even to other polyvalent atoms.[13]

It also seemed clear that given the fact that the tetrahedral carbon atom must create bonds with nitrogen atoms, the valences of the nitrogen must 'not always lie in a plane

---

[10] Meyer and Auwers had used Beckmann's argument for the structural difference as a support of their own hypothesis, that the cause of the stereoisomerism could not be within the CHNOH group.
[11] Kauffman, 'Alfred Werner's Inaugural Dissertation', p. 158.
[12] Ibid.
[13] Ibid., p. 159.

I. Es können also den beiden stereochemisch isomeren Kohlenstoffverbindungen der Form XY:C = (CH)Z

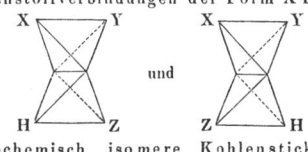

und

zwei stereochemisch isomere Kohlenstickstoffverbindungen von der Form XY:C ⇁ (N)Z entsprechen:

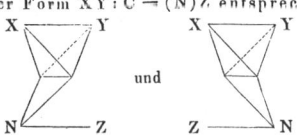

und

7.3 The Hantzsch/Werner model for nitrogen–carbon bonding. Note the explicit analogy drawn between the carbon–carbon and carbon–nitrogen double bonds.

with the nitrogen atom itself'. They used the example of the cyanide group to argue this point:

> in the cyanide compounds of the form RCN, and in all those nitrogen compounds in which the nitrogen atom N''' replaces a methyne group (CH'''), such as in pyridine, thiazole, quinoline, etc.) the valences of nitrogen must deviate from the plane of the nitrogen atom, because they are saturated by the three valences of the tetrahedral carbon atom.

Furthermore, if one considered Baeyer's proposal that a strain existed in small ring compounds and multiple bonds, and Wislicenus' claim that the valences were 'diverted from their original directions by the atoms that are attached to the atom', then it was 'inconceivable that the three carbon valences which are bound to nitrogen should lie in one plane in compounds like HCN'. This led to the basic hypothesis: 'in certain compounds the three valences of the nitrogen atom are directed toward the corners of a (in any case irregular) tetrahedron whose fourth corner is occupied by the nitrogen atom itself'.[14]

A carbon–nitrogen double bond would result when two of the affinities joined between a carbon and nitrogen atom, and Hantzsch and Werner proposed that the isomerism in carbon–carbon double bonds and carbon–nitrogen double bonds would be completely analogous. Carbon–nitrogen double bonds would possess restricted rotation and create spatial isomers analogous to *cis* and *trans* isomers in the unsaturated acids (Figure 7.3). This restricted rotation then easily explained the

---

[14] Ibid.

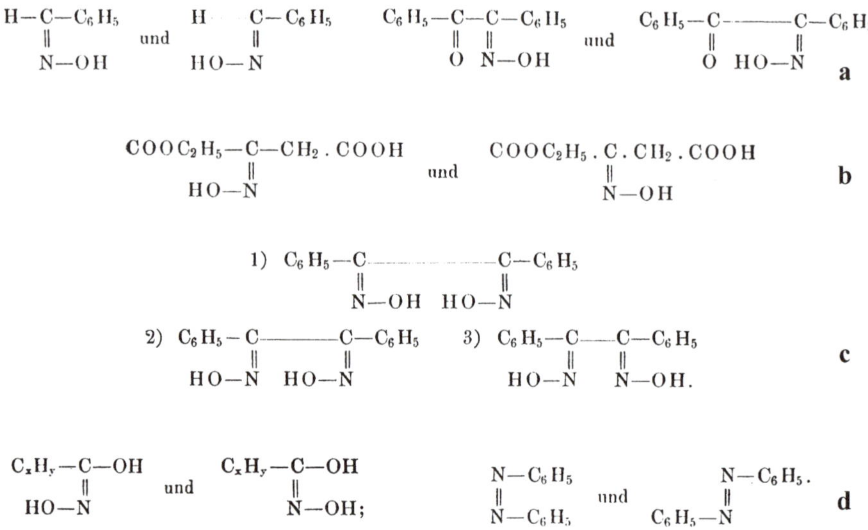

$$
\begin{array}{cccc}
\underset{N-OH}{\overset{\parallel}{H-C-C_6H_5}} \text{ und } \underset{HO-N}{\overset{\parallel}{HC-C_6H_5}} & \underset{ON-OH}{\overset{\parallel\parallel}{C_6H_5-C-C-C_6H_5}} \text{ und } \underset{OHO-N}{\overset{\parallel\parallel}{C_6H_5-C\text{---}C-C_6H_5}} & \text{a}
\end{array}
$$

$$
\underset{HO-N}{\overset{\parallel}{COOC_2H_5-C-CH_2.COOH}} \text{ und } \underset{N-OH}{\overset{\parallel}{COOC_2H_5.C.CH_2.COOH}} \quad \text{b}
$$

$$
\begin{array}{l}
1)\ \underset{N-OH\ \ HO-N}{\overset{\parallel\parallel}{C_6H_5-C\text{-----}C-C_6H_5}} \\[4pt]
2)\ \underset{HO-N\ \ HO-N}{\overset{\parallel\parallel}{C_6H_5-C\text{-----}C-C_6H_5}} \quad 3)\ \underset{HO-N\ \ N-OH.}{\overset{\parallel\parallel}{C_6H_5-C\text{-----}C-C_6H_5}}
\end{array} \quad \text{c}
$$

$$
\underset{HO-N}{\overset{\parallel}{C_xH_y-C-OH}} \text{ und } \underset{N-OH;}{\overset{\parallel}{C_xH_y-C-OH}} \qquad \underset{N-C_6H_5}{\overset{\parallel}{N-C_6H_5}} \text{ und } \underset{C_6H_5-N}{\overset{\parallel}{N-C_6H_5.}} \quad \text{d}
$$

7.4 Hantzsch and Werner's new formulas for various oxime isomers.

isomerism in the oximes and other compounds with multiple bonds to nitrogen. Benzaldoxime and benzilmonoxime could each exist in two spatial configurations, and the benzildioximes would exist in three different configurations. This form of restricted rotation would also be present in any compound with a double bond to nitrogen, and Hantzsch and Werner gave tentative spatial formulas for the isomeric hydroxamic acids and azobenzene (Figure 7.4(c) and (d)). Finally they suggested methods by which they could assign tentative configurations to the oximes.

Hantzsch and Werner were careful to make explicit the limitations of their hypothesis, calling it a 'first inadequate attempt to devise a correlation where one has hitherto not yet been recognized with certainty'.[15] They did not claim to postulate a true analog to the asymmetric carbon atom (an 'asymmetric nitrogen atom') and they recognized that the evidence did not warrant that assumption. Compounds of the type $CR_1R_2R_3R_4$ nearly always existed in two geometrically isomeric pairs, but pairs for the corresponding type $NR_1R_2R_3$ had not yet been observed. To account for this fact, Hantzsch and Werner made an important qualification to their hypothesis – the valences of nitrogen were directed out of planarity only in certain cases. They were forced to conclude that the original direction of the valences in the carbon and nitrogen atoms were different. Hantzsch and Werner concluded with a summary of the advantages of their hypothesis:

---

[15] Ibid., p. 165.

even in its present undeveloped state, our theory possesses the following advantages. It requires neither a new principle for the nature of valence nor the assumption of two different types of atomic bonding in order to explain the cases of isomerism in question. It proceeds from the concepts developed for geometrically isomeric carbon compounds, maintains completely van't Hoff's second proposition, and attributes a series of isomerism which have heretofore been considered partly geometric, partly structural, and partly insufficiently explained, to the same cause: the different spatial arrangement of the radicals bound to nitrogen in relation to the atom itself.[16]

We can make several inferences about Hantzsch's and Werner's motives from these passages. First, despite the fact that Werner, not Hantzsch, was the principal intellectual author of the theory, the idea that the nitrogen atom could 'mimic' the CH''' group reflected Hantzsch's own interest in the concept of chemical mimicry. If Hantzsch himself was not the author of the theory, in this sense its presentation strongly bears his stamp. Wislicenus also cast a long shadow on the essay's theoretical content. Hantzsch and Werner assumed from the outset that the isomeric differences were spatial, not structural, and drew the primary justification for their hypothesis from Wislicenus' essay on unsaturation: carbon–nitrogen double bonds were a physical analog to carbon–carbon double bonds. Like 'Spatial Arrangement', Hantzsch and Werner's essay was a theoretical reinterpretation of data, not a discussion of conclusions drawn from new experiments. Finally, Hantzsch and Werner showed a commitment to preserving Van 't Hoff's second hypothesis. Although they admitted that Meyer and Auwers might be correct, the many examples where free rotation did appear to exist did not seem to warrant the conclusion that another form of bonding existed.

At this point, it might also be useful to point out some unusual characteristics of the stereoformulas used by Hantzsch and Werner. Van 't Hoff's and Wislicenus' stereoformulas portrayed the edges of the tetrahedron, but as Meyer pointed out, these aspects of the drawing were physically and chemically meaningless, therefore they had not used them. Hantzsch and Werner's depiction of the tetrahedral form of the nitrogen atom, on the other hand, was physically more pleasing because the edges of their nitrogen polyhedra actually represented the bonds of the nitrogen atom. Yet the affinities pointed in a single direction, leaving one side of the atom without any affinity. Privately to Ostwald, Hantzsch referred to this asymmetry as 'crooked nitrogen' (*schiefe Stickstoff*). They also continued to use the curious polyhedral representation for the carbon atom, without making it clear whether they meant that the carbon atom was at the center with radiating valences towards the corners (as in Van 't Hoff's stereoformulas), or that the carbon atom itself was the tetrahedron (as in Wislicenus' stereoformulas). Hantzsch and Werner's stereoformulas were

---

[16] Ibid., p. 165.

$$C_6H_5-C\equiv N-\overset{H}{\overset{|}{O}}H \quad \longleftrightarrow \quad C_6H_5-\overset{H}{\overset{|}{C}}=N-OH$$

7.5 Meyer's explanation of the isomeric benzaldoximes by a 'mobile hydrogen atom' that was located on either the nitrogen or the carbon atom.

therefore a curious mixture of conventions that has largely gone unnoticed, even by their contemporaries.

Restricted Rotation (Meyer) versus the Spatial Nitrogen Atom (Hantzsch and Werner)

Meyer responded quickly to Hantzsch's theory on 28 January 1890 during his major lecture on stereochemistry to the German Chemical Society.[17] Meyer welcomed them to the discussion of benzildioxime stereochemistry, but disagreed with their hypothesis. He began with the supposed structural identity of the benzaldoximes. Meyer did not find Goldschmidt's argument for structural identity compelling. Beckmann's supposed structure for β-benzaldoxime (Figure 7.1) was the first substance given this kind of structure; its chemistry therefore remained relatively unknown, and its reactivity towards phenylisocyanate could not be determined *a priori*. And if α-benzaldoxime transformed easily into β-benzaldoxime, why shouldn't a simple derivative from phenylisocyanate also transform easily? 'Goldschmidt's argument – as interesting as his attempts seem – is devoid of a factual basis,' Meyer concluded, 'the opposite conclusions could be drawn from his observations with just as much or just as little justification.'[18] If the assumed structural identity of the benzaldoximes was in doubt, then Hantzsch's theory was unnecessary, and the isomerism could likely be caused by the aldehyde's 'mobile hydrogen atom' (Figure 7.5).[19]

Further, Meyer claimed the benzildioximes and the benzaldoximes were not analogous compounds, because the α-benzaldoxime converted easily into β-benzaldoxime and the β just as easily into the α. This relationship, Meyer noted, did not exist within the benzildioximes, where there was a definite stable form that could

---

[17] For simplicity, from this point on I will use only Hantzsch's name to refer to work by Hantzsch and Werner. As the discussion between Hantzsch and Werner on one side and Meyer on the other grew, Werner played less and less of a role, and by mid-1890 he had dropped out entirely. Werner's participation is indicated as an author of the cited papers.

[18] Victor Meyer, 'Ergebnisse und Ziele der stereochemischen Forschung', *Berichte*, 1890, *23*, 567–619, p. 597.

[19] Ibid., p. 598.

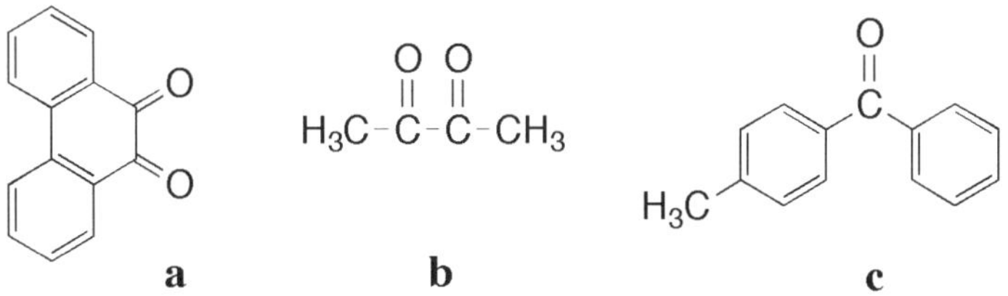

7.6 The structures of (a) phenanthrenequinone, (b) diacetyl and (c) phenyltolylketone.

not be converted into either of the other two isomers. Meyer concluded that the same kind of evidence used for concluding the structural equivalence of the benzildioximes could lead to the conclusion that the benzaldoximes were not structurally identical.[20] Meyer then asked his audience: 'How can the phenomena be produced [according to each theory], and how can it be removed?'[21] The answer to these questions favored his own theory. First, phenanthrenequinone (Figure 7.6(a)), whose structure excluded any rotation about the carbon–carbon axis, formed only one oxime, as expected by Meyer's theory. According to Hantzsch and Werner, the dioxime of phenanthrenequinone should have three isomers. Second, Meyer's explanation depended on the existence of negative groups such as phenyl to produce the restricted rotation. The isomerism should disappear when non-negative radicals took the place of the phenyl groups in benzildioxime, and in fact, diacetyl (Figure 7.6(b)), with 'neutral' methyl groups instead of negative phenyl groups, did form only one oxime.

The crucial point between the two theories, Meyer noted, was their different predictions. The isolation of a third isomer of benzil*mon*oxime would argue in favor of his theory of restricted rotation in single bonds, and the isolation of isomeric oximes from an unsymmetrical monoketone would support Hantzsch and Werner's theory. Auwers' unsuccessful attempt to isolate two isomeric oximes of phenyltolylketone (Figure 7.6(c)), therefore cast doubt on Hantzsch and Werner's theory.[22] Meyer dismissed the other cases where Hantzsch had applied the theory because so little was known about benzhydroxamic acid and the structural identity of the azotoluols had not been established. 'With these sorts of speculations', Meyer reflected, 'Hantzsch and Werner completely leave the base of established fact.'[23]

Finally, Meyer discussed what he considered the most significant problem with Hantzsch's theory. It applied only to the oximes, and not to all compounds containing

---

[20] Ibid., p. 600.
[21] Ibid., p. 604.
[22] Karl Auwers, 'Über das Oxim des p-Tolylphenylketon', *Berichte*, 1890, *23*, 399–403.
[23] Meyer, 'Ergebnisse', p. 608.

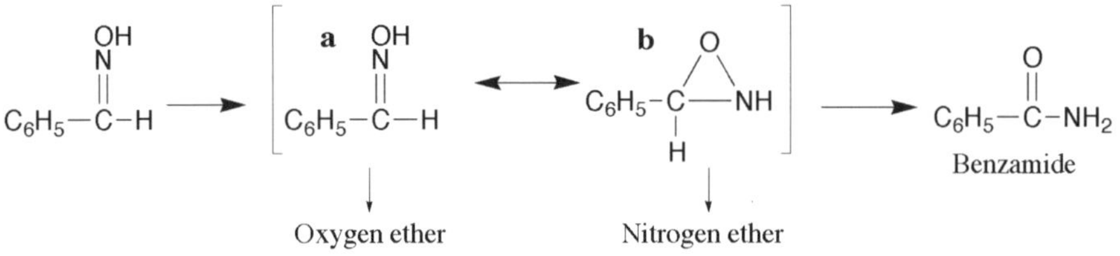

$$C_6H_5-\overset{\overset{OH}{\underset{\|}{N}}}{C}-H \longrightarrow \left[\underset{\text{Oxygen ether}}{\overset{a}{C_6H_5-\overset{\overset{OH}{\underset{\|}{N}}}{C}-H}} \longleftrightarrow \underset{\text{Nitrogen ether}}{\overset{b}{C_6H_5-\underset{H}{\overset{\triangle}{C}}\text{---NH}}} \right] \longrightarrow \underset{\text{Benzamide}}{C_6H_5-\overset{\overset{O}{\|}}{C}-NH_2}$$

7.7 The formation of benzamide from benzaldoxime via a tautomeric intermediate.

nitrogen. As Hantzsch and Werner themselves noted, they had not postulated an 'asymmetric nitrogen atom', but only a spatial model of the nitrogen atom when it existed in multiple bonds. Meyer noted that neither hydroxylamine nor its derivatives (R_1R_2NOH) had ever been observed in stereochemical forms, and concluded that there must be something present in the stereoisomeric oximes that diverted the hydroxyl group from its symmetrical position towards either of the two carbon groups. In the benzildioximes, the phenyl group was capable of such attraction. If this were the case, however, the configuration of β-benzildioxime (Figure 7.4 (c)) would be so stable as to preclude any other configurations.

Hantzsch replied in February to Meyer's lecture, defending Goldschmidt's results and the assumption of structural identity in the benzaldoximes.[24] Hantzsch suggested that both structural isomers (the true oxime, (a), and the cyclic imide form, (b), in Figure 7.7) were present when Beckmann attempted to isolate the nitrogen ether of β benzaldoxime. The isobenzaldoxime itself did not exist as the cyclic structure (7b) as such; it was an intermediate form between β-benzaldoxime and benzamide. Hantzsch assumed that the two proposed structural isomers of the benzaldoximes were tautomeric with one another; Beckmann's results were inconclusive because under the reaction conditions, the oxime group was tautomeric and could react in either form.[25] Hantzsch's proposed configuration for β benzaldoxime, in which the hydrogen atom and hydroxyl group were *cis* to one another, explained more satisfactorily its easy conversion to benzonitrile.[26] Meyer's proposed 'mobile

---

[24] Arthur Hantzsch and Friedrich Kraft, 'Über das Auftreten von Stereoisomerie bei nicht oximartigen Stickstoffverbindungen', *Berichte*, 1891, *24*, 3511–28.

[25] Conrad Laar introduced the term 'tautomerism' (*tauto*: the same) in 1885 to describe the peculiar observation that some organic compounds appeared not to have a unique structure, and behaved as if they possessed two different structures. The most famous example is keto-enol tautomerism. The subsequent development of tautomerism is too great to discuss in detail here, but it played a great role in Hantzsch's later theoretical interpretation of the diazocomounds. Arthur Hantzsch, 'Über Stereoisomerie bei Diazoverbindungen und die Natur der "Isodiazokörper"', *Berichte*, 1894, *27*, 1702–26, p. 1706.

[26] The term *cis* is used here only for convenience to prevent confusion between the isomerism here

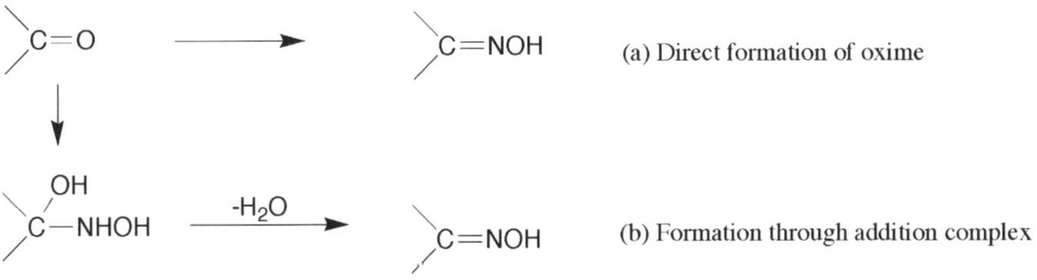

$\diagup$C=O $\longrightarrow$ $\diagup$C=NOH (a) Direct formation of oxime

$\downarrow$

$\diagup$C—NHOH with OH $\xrightarrow{-H_2O}$ $\diagup$C=NOH (b) Formation through addition complex

7.8 The formation of an oxime either directly (a) or indirectly (b) through an intermediate addition of hydroxylamine to the ketone.

hydrogen atom' was also unlikely because the conversion between isomers required 'a previously never observed and therefore improbable atomic movement'.[27]

The negative evidence in favor of Meyer's theory (the non-existence of unsymmetrical oximes), Hantzsch noted, could be countered by the negative evidence in favor of Hantzsch's theory. Why do the supposed structural isomeric differences of the benzaldoximes exist only in the benzaldoximes, and not in all oximes? Phenanthrenequinone did form only one oxime, but Hantzsch noted that the behavior of groups in rings can be entirely different than in open chains.

Hantzsch suggested that his hypothesis allowed both a less ambiguous assignment of configurations to the benzildioximes and an 'unforced' explanation for their stability, while Meyer and Auwers' explanation contained a 'certain capriciousness' (*gewissen Willkür*). Of Meyer's three configurations (see Figure 6.15), only the α isomer could be unambiguously assigned a configuration. Spatially, it was the most likely to form an anhydride, and the α benzildioxime formed an anhydride most easily. The β and γ isomers, on the other hand, could not be assigned configurations so readily. The two configurations could be interconverted by a rotation of the bottom carbon atom system, but Hantzsch noted that these simple manipulations of the formulas did not correspond to the behavior of the isomers: the β and γ isomers were not equally stable. Meyer's second means of representing the configurations of the benzildioximes (the 'staggered' and 'eclipsed' configurations in Figure 6.16) also resulted in an ambiguous assignment of configurations.

Hantzsch agreed with Meyer that in the absence of asymmetry in hydroxylamine itself, something must be present to divert the hydroxyl group from its symmetrical

---

and in ethylene compounds. *Syn* and *anti* are the modern terms to describe stereoisomers of oximes. The terms were introduced by Hantzsch in 1891. Arthur Hantzsch, 'Zur Nomenclatur stereoisomerer Stickstoffverbindungen und stickstoffhältiger Ringe', *Berichte*, 1891, *24*, 3479–88.

[27] Arthur Hantzsch and Alfred Werner, 'Bemerkungen über stereochemisch isomere Stickstoffverbindungen', *Berichte*, 1890, *23*, 1243–53, p. 1246 (hereafter referred to as 'Bemerkungen I').

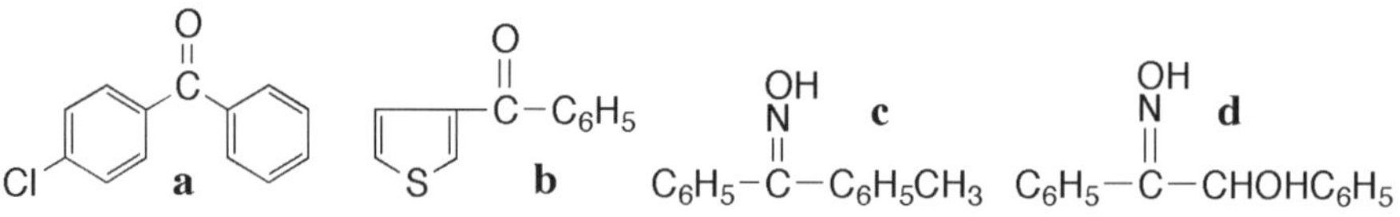

7.9 The structures of monoketones that formed oxime stereoisomers according to Hantzsch's theory.

position, but the phenyl group in the benzildioximes could not, because it would lead to only one isomer. Meyer had therefore assumed a direct exchange of oxygen for the oxime group. Such a process 'is not proven, and not even probable'. Instead, Hantzsch envisioned the hydroxylamine molecule forming an addition complex with the ketone that decomposed with the loss of water to form the oxime (Figure 7.8):

> But then we could also think that the hydroxyl formed in this first phase fills the spatial location that the original hydroxyl of hydroxylamine sought to occupy; that the latter will therefore be forced to another location, and also to remain there after removal of the original obstacle, that is after cleavage of water, until it is caused by direct encounter from outside to go into the newly free favorable position. With this sort of idea we really only want to suggest how labile forms can still form directly, and can only indirectly be converted into the stable form.[28]

The experimental evidence in favor of Hantzsch's theory appeared in July 1890, when Auwers isolated two oximes of chlorobenzophenone (Figure 7.9(a)). Meyer was not certain that the isomerism was due to the asymmetry, as it could have been caused by the presence of chlorine.[29] Meyer had informed Hantzsch privately of these results, and Hantzsch wrote a reply four days after Meyer had submitted the chlorobenzophenone paper, in which he sketched out the results of similar experiments that indicated that the difference in groups on the oximes, and not their nature, caused the stereoisomerism in the oximes.[30] In a subsequent series of papers, Hantzsch discussed the isolation of pairs of oximes from phenylthiophenyl-oxime and phenyltolyl ketone (Figure 7.9(b) and (c)).[31] The oximes of the latter, which

---

[28] Ibid., p. 1252.

[29] Victor Meyer and Karl Auwers, 'Über Oxime halogenirter Benzophenone', *Berichte*, 1890, *23*, 2063–4. Although this appeared to favor Hantzsch's theory, Meyer also published a short defense of his own theory by citing recent examples that indicated the existence of other restricted rotations. Victor Meyer and Karl Auwers, 'Zur Stereochemie der Aethenderivate', *Berichte*, 1890, *23*, 2079–83.

[30] Arthur Hantzsch and Alfred Werner, 'Über stereochemische Isomerie asymmetrischer Monoxime', *Berichte*, 1890, *23*, 2322–5.

[31] Arthur Hantzsch and Alfred Werner, 'Die stereochemisch-isomere Oxime des p-Tolyl-phenylketon', *Berichte*, 1890, *23*, 2325–32, and Arthur Hantzsch, 'Vorläufige Mittheilung über stereochemisch isomere Oxime des Phenylthienylketons und der Phenylglyoxyls', *Berichte*, 1890, *23*, 2332–3.

Auwers had been unable to isolate, required painstaking fractional crystallization from a different solvent system, and Hantzsch noted that the difficulty in preparing pure samples of the β isomer could be due to the great similarity between the phenyl and tolyl groups. The isomeric oximes of benzoin and desoxybenzoin (Figure 7.9(d)) appeared to indicate that the cause of the isomerism was located in the nitrogen atom, and not in the electrical nature of the groups attached to the oximes as Meyer's theory implied.[32]

The Conditions of Stereoisomerism in Nitrogen Compounds

In the light of these results, particularly Hantzsch's ability to isolate the second tolylphenyloxime, Meyer graciously admitted the defeat of his and Auwers' hypothesis, but he did not adopt Hantzsch's theory.[33] The cause of stereoisomerism could not be in the nitrogen atom, as Hantzsch maintained, since it was:

> difficult to understand why a similar isomerism has not been observed in derivatives of ammonia, or, for that matter, in azo- and azoxy-compounds, and so forth. As long as such isomers are missing, it seems appropriate to look for a cause not in the peculiarities of nitrogen, but in hydroxylamine.[34]

Meyer modeled the hydroxylamine molecule using the Kekulé ball and stick models (Figure 7.10(a)). By analogy to the carbon atom, where the different substituents attached to it distorted it from a regular tetrahedron, Meyer reasoned that the same distortions would occur in hydroxylamine:

> it is quite improbable that the hydrogen atom in the hydroxyl group of hydroxylamine, subjected to the attraction of both the nitrogen and the oxygen, remains in a position that is close only to the *oxygen* atom, but stays as far from the nitrogen atom as possible.[35]

Meyer argued that the hydrogen atom of the oxygen sought a position directly between oxygen and nitrogen where its attraction to both was as evenly allocated as possible. The molecule of hydroxylamine then took on the following form (the hydrogen–oxygen bond was actually coming out of the plane of the paper – Figure 7.10(b)). The oxime isomers would then appear as shown in Figure 7.10(c) and (d).

Meyer admitted, however, that the one weakness of this theory was the absence of optical isomers of hydroxylamine itself.

---

[32] Alfred Werner, 'Über ein zweites Benzoïnoxim', *Berichte*, 1890, *23*, 2333–6.
[33] Victor Meyer and Karl Auwers, 'Über die Isomeren Oxime unsymmetrische Ketone und die Configuration des Hydroxylamins', *Berichte*, 1890, *23*, 2403–9.
[34] Ibid., p. 2406.
[35] Ibid. Meyer's and Auwers' emphasis.

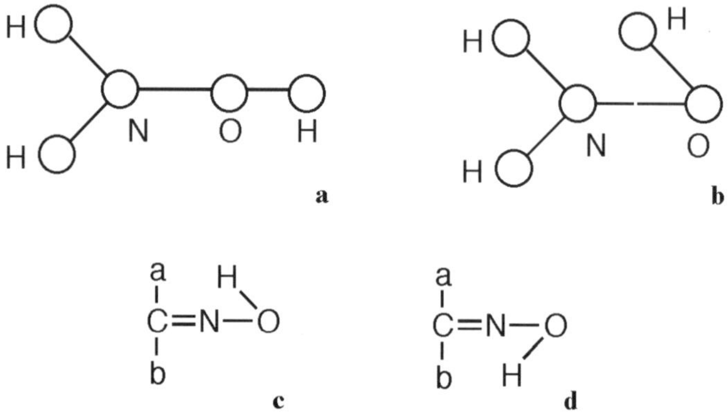

7.10 **(a) and (b)** Meyer's drawings of the ball and stick models for hydroxylamine. (a) shows the attraction of the hydroxyl hydrogen for the nitrogen atom, and is meant to be pointing up out of the plane of the page. **(c) and (d)** The two oxime isomers resulting from Meyer's model of hydroxylamine.

At this point, the discussion between Hantzsch and Meyer took on a different character. The dispute was no longer between the stereochemistry of the nitrogen atom and the nature of carbon–carbon bonds, but between the stereochemistry of nitrogen and the special properties of hydroxylamine. Nitrogen stereochemistry, Meyer noted, seemed confined to the oximes. Hantzsch largely agreed, noting that all attempts to isolate an asymmetric nitrogen atom had been fruitless, and that several analogs of carbon–carbon double bonds implied by his theory were unknown, such as the analog of maleic and fumaric acid or stilbene (Figure 7.11). Furthermore, simple imine compounds (Figure 7.11(c)) did not exhibit isomerism, nor did it seem possible to perform additions to the carbon–nitrogen triple bond in a manner similar to the addition to the carbon–carbon triple bond. Even the examples of isomerism in the oximes appeared limited, as some carbonyls gave no oximes while others yielded a single oxime (Figure 7.11), and still others (such as benzil) gave two stereoisomeric oximes. Hantzsch reasoned that the formation of oximes was hindered when an alkyl existed directly on the carbonyl, or if it existed at the ortho-position of a phenyl group, since if the two ortho-positions were occupied, no oxime formed at all.[36]

But Meyer's new theory also appeared inadequate to Hantzsch, who offered two

---

[36] Hantzsch attributed this to what is today called 'steric hindrance', and appears to pre-date Victor Meyer's discussion of steric hindrance by three years.

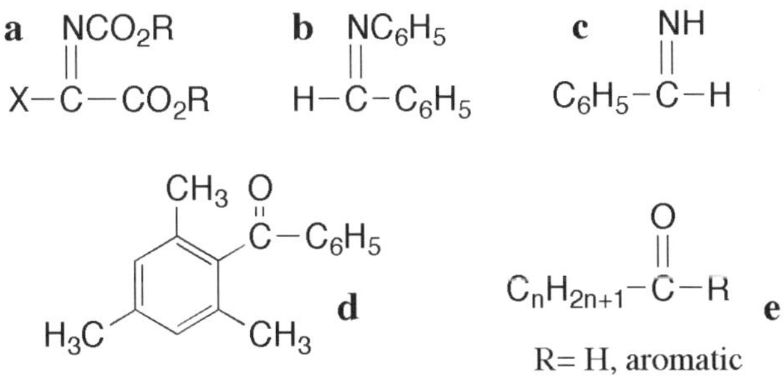

a $\underset{X-C-CO_2R}{NCO_2R}$ **b** $\underset{H-C-C_6H_5}{NC_6H_5}$ **c** $\underset{C_6H_5-C-H}{NH}$

d 2,4,6-trimethylphenyl-C(=O)-C$_6$H$_5$ (mesityl phenyl ketone)

e $C_nH_{2n+1}-\underset{\|}{\overset{O}{C}}-R$ R= H, aromatic

7.11 **(a), (b) and (c) Unknown nitrogen analogs of well-known compounds containing carbon–carbon double bonds. (d) and (e) Compound (d) did not react with hydroxylamine to give an oxime, and compound (e) formed only one oxime.**

principal objections to it.[37] First, although Hantzsch recognized that a compound with an asymmetric nitrogen atom had not yet been isolated, this problem was only of 'secondary importance' for his theory, while it actually endangered Meyer's theory.[38] In Hantzsch's theory, the carbon–nitrogen double bond, not the nitrogen atom itself, caused the isomerism, and Meyer's configurations consisted of 'more or less favored phases' of intramolecular rotation of the hydrogen about the nitrogen–oxygen bond. Why shouldn't this hydrogen rotate to a favored configuration determined by the affinity of the constituent atoms? Third, Hantzsch used Meyer's hypothesis to support his own: if the nitrogen could attract the hydroxyl hydrogen atom, then why couldn't the oxygen atom attract the amine hydrogen atoms? If this were the case, the result would be a tetrahedron, and hydroxylamine would then simply be a special case of Hantzsch's own hypothesis. Hantzsch repeated that:

> it may be possible that nitrogen actually forms stereochemical isomers only in oximes, that is, in union with oxygen – it may be, as we have always suggested, that perhaps the double bond between carbon and nitrogen in this case is more essential – but all the eventual further development in this area will still be based on the stereochemistry of nitrogen.[39]

---

[37] Arthur Hantzsch and Alfred Werner, 'Bermerkungen über stereochemisch isomere Stickstoffverbindungen', *Berichte*, 1890, *23*, 2764–9 (hereafter referred to as 'Bemerkungen II').

[38] Friedrich Kraft's attempts to isolate possible optical isomers of various amines, hydrazines, and hydroxyl amine compounds by means of fractional crystallization were all unsuccessful. Arthur Hantzsch, 'Die stereochemische-isomere Oxime des p-Tolylphenylketons. II. Mittheilung', *Berichte*, 1890, *23*, 2776–80, and Arthur Hantzsch and Friedrich Kraft, 'Zur Frage des asymmetrischen Stickstoffatoms', *Berichte*, 1890, *23*, 2780–4.

[39] Hantzsch and Werner, 'Bemerkungen II', p. 2767.

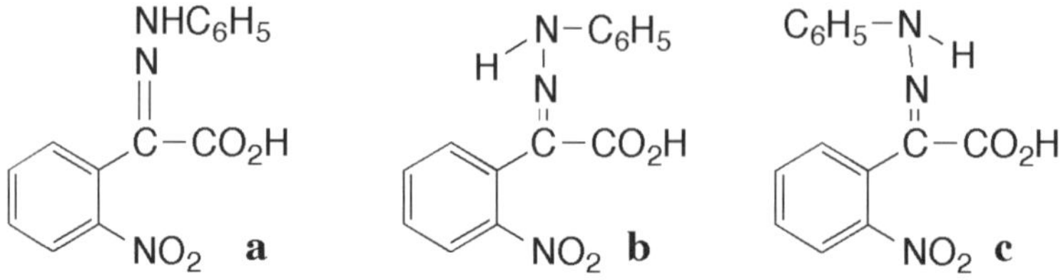

[Structures a, b, c showing hydrazone isomers with NHC₆H₅/N=, C-CO₂H, and NO₂ substituents]

7.12 Hydrazones isolated reported by Meyer from the reaction of ketones with phenylhydrazine.

Meyer continued to assert that if stereoisomers of other nitrogen compounds such as anilines or amine derivatives did not exist, the nature of the atom attached to the nitrogen (such as oxygen) seemed a better indicator of isomerism than the nitrogen atom itself. When a compound with a monovalent group (such as a hydrogen atom, or phenyl group) attached to nitrogen was found to exist in different isomeric forms, Hantzsch's theory would gain more credibility.

Meyer's and Hantzsch's independent isolation of isomeric hydrazones appeared to confirm both stereoisomeric theories. In late 1890, Meyer isolated two apparent isomeric hydrazones of o-nitrophenylglyoxylic acid (Figure 7.12(a)).[40] Because the nitrogen atom was connected to another multivalent atom, it appeared to bolster Meyer's assumption that the nature of hydroxylamine, not nitrogen, caused the isomerism. It was 'not improbable' that his revised theory could be modified from hydroxylamine to hydrazine, to give the stereoformulas 7.12(b) and 7.12(c), but Meyer did not assume that the two isomers were structurally identical.[41]

Nearly a year later, in November 1891, Hantzsch and Kraft announced the isolation of isomeric hydrazones prepared from various 'ketone chlorides' (Figure 7.13).[42] Anisylphenyl hydrazone (Figure 7.13(a)) appeared to have two isomeric forms, and the two hydrazones 7.13(b) and 7.13(c) also appeared to be spatial isomers, although Hantzsch admitted that they had not yet completely separated and isolated them. He based their existence on a consistently imprecise melting point of 7.13(b), and the presence of an isomeric oil left after recrystallization of 7.13(c). Hantzsch immediately assumed that they had isolated the long-awaited stereoisomers analogous to the oximes that possessed the two space formulas 7.13(d) and 7.13(e).

---

[40] R. Demuth and M. Dittrich, 'Über Oxime halogenirter Benzophenon', *Berichte*, 1890, *23*, 3609–17.

[41] Albert Krause, 'Über die isomeren Formen des Hydrazons der Orthonitrophenylglyoxylsäure', *Berichte*, 1890, *23*, 3617–22.

[42] Hantzsch had found earlier that normal ketones did not react with hydrazine to produce isomeric hydrazones. Arthur Hantzsch and Friedrich Kraft, 'Über das Auftreten von Stereoisomerie bei nicht oximartigen Stickstoffverbindungen', *Berichte*, 1891, *24*, 3511–28.

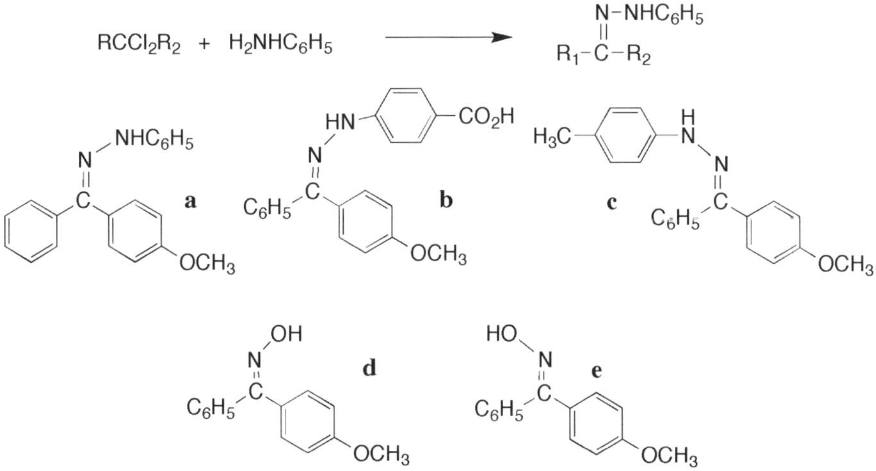

7.13 **The formation of hydrazones from various ketone chlorides by the general reaction given at the top.**

Meyer repeated that true chemical isomerism in these compounds had not been proved, nor if true isomerism existed, that it was caused by spatial properties of the molecule.[43] Like the oximes, in the hydrazones there was no monovalent group attached to the nitrogen atom, and 'the existence of this type of isomerism is demanded equally by our theory and by Hantzsch and Werner'.[44] The analogy between hydroxylamine and hydrazine explained 'the fact that until now the isomeric phenomena in question have only been observed in those nitrogen derivatives in which the unique and analogous character of hydroxylamine and phenylhydrazine is expressed'.[45] Meyer admitted that if Hantzsch should succeed in isolating the 'presumed labile' modification of hydrazones 7.13(b) or 7.13(c), and attribute the difference between them to a stereochemical cause, then 'the special position of oximes and hydrazones would end'.[46]

Hantzsch continued to maintain that the properties of these compounds vindicated his analogy drawn between the isomeric hydrazones and the isomeric oximes.[47] The melting points and stability of the isomeric hydrazones corresponded exactly to those

---

[43] Victor Meyer and Karl Auwers, 'Bemerkungen zu der Abhandlung von A. Hantzsch und Friedrich Kraft: Über das Auftreten von Stereoisomerie bei nicht oximartigen Stickstoffverbindungen', *Berichte*, 1891, *24*, 4225–30.
[44] Ibid., p. 4229.
[45] Ibid.
[46] Ibid., p. 4230.
[47] Arthur Hantzsch, 'Über Stereoisomerie bei asymmetrischer Hydrazon', *Berichte*, 1893, *26*, 9–17.

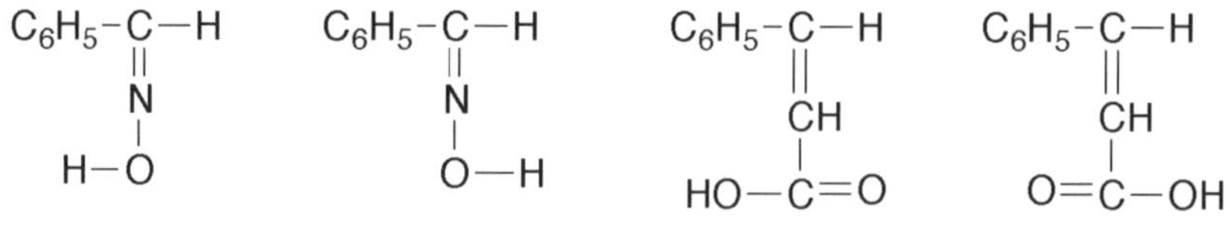

$$
\begin{array}{cccc}
C_6H_5-\underset{\underset{H-O}{\underset{|}{N}}}{\overset{\overset{H}{|}}{\overset{||}{C}}} & C_6H_5-\underset{\underset{O-H}{\underset{|}{N}}}{\overset{\overset{H}{|}}{\overset{||}{C}}} & C_6H_5-\underset{\underset{HO-C=O}{\underset{|}{CH}}}{\overset{\overset{H}{|}}{\overset{||}{C}}} & C_6H_5-\underset{\underset{O=C-OH}{\underset{|}{CH}}}{\overset{\overset{H}{|}}{\overset{||}{C}}}
\end{array}
$$

7.14 Hantzsch's comparison of the oximes and cinammic acids as an example of extending Meyer's theory of the oximes to other analogous structures.

of the oximes. The major product possessed the higher melting point and was less soluble, and the minor product possessed the lower melting point and had a higher solubility. The isomer with the lower melting point was more labile, and transformed into the isomer with the higher melting point. As 'proof' (*Beweis*) of structural identity, Hantzsch noted that diphenylhydrazine formed two isomeric hydrazones that lacked the hydrogen atom necessary to produce the isomerism in Meyer's theory.

Hantzsch also claimed that Meyer's analogy between hydroxylamine and phenylhydrazine was not complete or unified as Meyer claimed.[48] Meyer's configurations for the oximes and hydrazones were not analogous, because only the hydrazones could be 'geometrically isomeric nitrogen compounds', while the oximes would be 'geometrically isomeric oxygen compounds'.[49] Hantzsch again emphasized that his analogy to the carbon–carbon multiple bond gave a more consistent theory: 'Our theory rests directly on the parallel between stereoisomeric bodies with a double bond between two carbon atoms and stereoisomeric bodies with a double bond between a carbon and nitrogen atom.'[50] Hantzsch drove home his point by drawing an analogy between Meyer's stereoformulas for the oximes and cinnamic acid (Figure 7.14). The formulas for cinnamic acid predicted isomers of cinnamic acid that were 'just as improbable'.[51] Analogous phenomena for the unsaturated acids were found in ethylene derivatives, not in carboxylic acids. Similar analogous compounds for the oximes would be found in additional compounds with a nitrogen–carbon double bond without oxygen (such as $R_1R_2C=NC_6H_5$). But if the non-existence of this compound:

> is subsequently used as an objection to our formulas, it naturally has the same weight, for example, as if before the discovery of the isomeric tolanedibromides, cinammic or isocinammic acids, or fumaric and maleic acid, one had wanted to interpret them not as geometrically isomeric carbon, but as geometric oxygen compounds.[52]

---

[48] Ibid., p. 12.
[49] Ibid., p. 13.
[50] Ibid., pp. 12 and 14.
[51] Ibid., p. 14.
[52] Ibid.

Meyer versus Hantzsch: The Role of Theoretical Analogy

A number of points can be drawn from the discussion between Hantzsch and Meyer. The most obvious characteristic of their discussion was its civility. Although they clearly disagreed on the cause of the isomerism in the benzildioximes, they almost appeared to be in collaboration, each discussing their reasonable objections to and reflections on the other's theory. None of their papers contains an extended polemical argument or misdirected personal attack. This civility perhaps arose from the theoretical level of the discussion, for there was never a doubt on either side that the explanation would involve the spatial characteristics of molecules. Both Hantzsch and Meyer assumed the benzildioximes to be structurally identical, and the cause of the differences between them was stereochemical, not structural. Therefore they were not talking past one another, as Wislicenus appeared to do with Michael, but discussing the merits and demerits of various alternative spatial explanations. And as we shall see below, both Meyer and Hantzsch displayed obvious impatience with objections to either of their spatial explanations.

The focus of the controversy was Hantzsch's assumed stereochemistry of the nitrogen atom itself. Hantzsch was willing to overlook the non-existence of an 'asymmetric nitrogen atom' for the existence of nitrogen stereoisomers in other groups of compounds. What drove Hantzsch's conviction was the strong analogy he saw between the nitrogen atom and the methyne group, or the ability of the nitrogen atom to mimic this group. 'Once again,' Hantzsch noted, 'we wish to emphasize that our view follows completely the principles of Van 't Hoff and Wislicenus, in that it assumes the ability of N''' to replace CH'''; that in these cases, in which Auwers and V. Meyer concur, the three nitrogen valences cannot lie in a plane.'[53]

Meyer, on the other hand, could not see the validity of Hantzsch's theory because Hantzsch had not developed a general spatial model of the nitrogen *atom*. In part, this was due to the fact that many nitrogen compounds were pentavalent, and did not exhibit the properties of an asymmetric nitrogen atom fully analogous to the asymmetric carbon atom. He was thus led to suppose that the cause was contained within the hydroxylamine (or hydrazine) unit, rather than the nitrogen atom. The difference between Meyer and Hantzsch can also be summarized in terms of their use of analogy. Meyer found the analogy to Van 't Hoff's first hypothesis necessary, while Hantzsch found the analogy to Van 't Hoff's third hypothesis compelling. He was willing, as Meyer was not, to overlook the absence of analogy to the asymmetric nitrogen atom, and consider the analogy to the third hypothesis independently of the first.[54] It may be that Meyer found Van 't Hoff's first hypothesis more fundamental,

---

[53] Hantzsch and Werner, 'Bermerkungen II', pp. 2768–9.

[54] It is unclear whether Meyer ever actually accepted Hantzsch's theory. The second volume of Meyer and Jacobsen's *Lehrbuch der organische Chemie* (1897) discussed Hantzsch's theory, but this volume

from which the others derived, whereas Hantzsch saw all three hypotheses on more or less equal theoretical level.

The Configuration of the Benzildioximes

Chemical Methods for Establishing Configurations

In a series of several articles beginning in the first issue of the 1891 volume of the *Berichte*, Hantzsch reported the results of his attempts to establish specific configurations for the isomeric oximes. Not only did the nitrogen atom (or some molecules containing it) have spatial properties, but the absolute configuration of the molecules could be ascertained by means of their chemical properties. In their original 1890 article, Hantzsch and Werner had indicated how they could make such assignments, and in later articles, Hantzsch simply elaborated on their applicability. The method primarily relied on the assumption of planesymmetric elimination introduced by Wislicenus in 'Spatial Arrangement'.[55] First Hantzsch assigned configurations to the aldoximes. Since β-benzaldoxime cleaved into benzonitrile and water in the presence of base (Figure 7.15), Hantzsch reasoned that it must proceed by a simple elimination reaction in which the hydrogen and hydroxyl left from the same side of the carbon–nitrogen double bond to form a carbon–nitrogen triple bond. Beckmann had assumed that the nitrile was only formed from an intermediate compound, a sequence that Hantzsch called 'incomprehensible', because β-benzaldoxime decomposed exclusively to benzonitrile at room temperature, and under identical conditions the α isomer gave no trace of benzonitrile.[56] Hantzsch also noted that the corresponding β-acetyl derivative decomposed into the nitrile under milder conditions (with carbonate at room temperature), while the α compound still reacted only slowly:

> Of course, this proof is really the same kind as all conclusions that can ever be drawn about the spatial arrangement of atoms from chemical reactions; in this sense, nitrile formation in β-oximes corresponds roughly to anhydride formation in maleinoid or planesymmetric [*cis*] dicarbonic acids.[57]

---

appeared after Meyer's death in 1897, and it is not clear whether Meyer or Jacobsen (who took over the project after Meyer's death) wrote the section. Kauffman ('Stereochemistry of Trivalent Nitrogen Compounds') assumed that Meyer was the principal author of volume 2. Richard Meyer noted that the second volume appeared posthumously and may have been changed. In journal articles, Meyer never mentioned, to my knowledge, that he finally adopted Hantzsch's theory.

[55] Arthur Hantzsch, 'Die Bestimmung der räumlichen Configuration stereoisomerer Oxime', *Berichte*, 1891, *24*, 13–30.
[56] Ibid., p. 18.
[57] Ibid., p. 19.

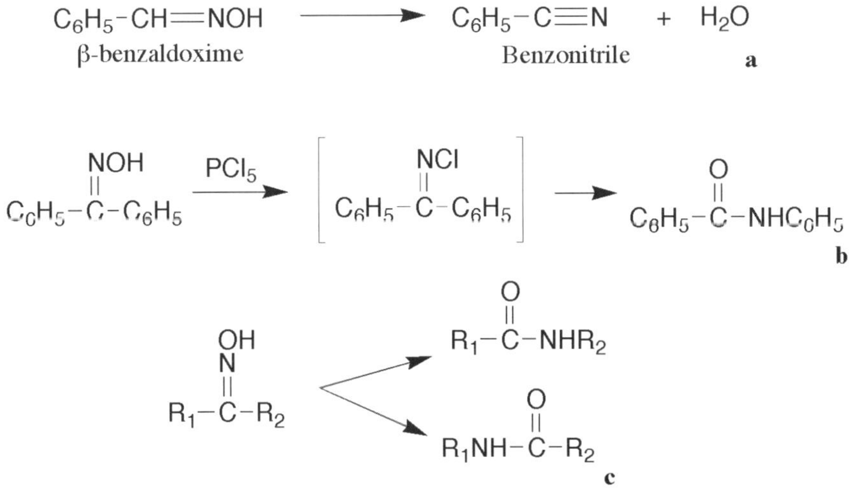

$$C_6H_5-CH=NOH \longrightarrow C_6H_5-C\equiv N + H_2O$$
$$\text{β-benzaldoxime} \qquad\qquad \text{Benzonitrile} \qquad \textbf{a}$$

$$\underset{C_6H_5-\overset{\overset{NOH}{\|}}{C}-C_6H_5}{} \xrightarrow{PCl_5} \left[\underset{C_6H_5-\overset{\overset{NCl}{\|}}{C}-C_6H_5}{} \right] \longrightarrow \underset{C_6H_5-\overset{\overset{O}{\|}}{C}-NHC_6H_5}{}$$
$$\textbf{b}$$

$$\underset{R_1-\overset{\overset{N}{\|}}{\underset{}{C}}-R_2}{\overset{OH}{}} \begin{matrix} \nearrow \\ \searrow \end{matrix} \begin{matrix} R_1-\overset{\overset{O}{\|}}{C}-NHR_2 \\ \\ R_1NH-\overset{\overset{O}{\|}}{C}-R_2 \end{matrix}$$
$$\textbf{c}$$

7.15 Hantzsch's methods for determining the configuration of the oximes. (a) Elimination of water to form a nitrile. (b) The Beckmann rearrangement to form an amide. (c) The two possible isomers resulting from the Beckmann rearrangement and formed by the migration of either R_1 or R_2.

Hantzsch also assigned configurations to the ketoximes using a slightly more complex reaction discovered by Beckmann in 1887, in which oximes reacted with phosphorous pentachloride to form amides by the migration of a carbon group to the nitrogen atom (Figure 7.15). In 1888, Victor Meyer named this reaction the 'Beckmann rearrangement' and had used it in his demonstration of the structural identity of the benzildioximes. In their 1890 paper, Hantzsch and Werner had indicated that according to their theory, stereoisomeric oximes would have two different carbon groups capable of migration, and should therefore form structurally isomeric amides under the conditions of the rearrangement (Figure 7.15). Because the migration also involved an elimination of the hydroxyl group from the nitrogen atom, Hantzsch reasonably inferred that this process must occur by a planesymmetric elimination. In 1891, Hantzsch's student Arturo Miolati determined that isomeric oximes did display differences in behavior compatible with this principle: 'In stereochemical oximes the radical that is located closer, or which in the sense of Werner's and my space formula lay on the same side of the binding axis between carbon and nitrogen, will shift with the hydroxyl.'[58] The structure of the product

---

[58] Ibid., p. 23.

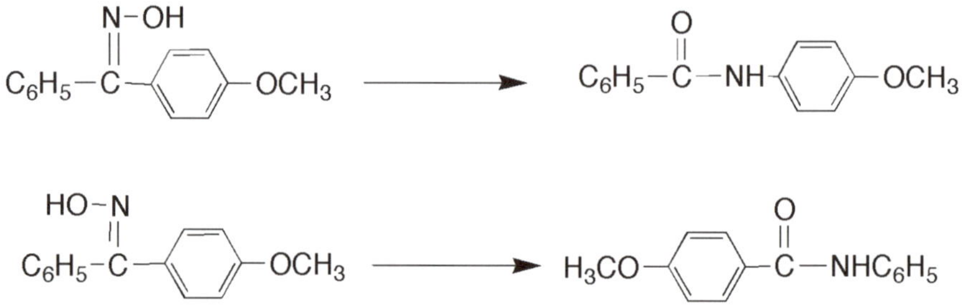

$$C_6H_5-\underset{\underset{\displaystyle \text{N-OH}}{\|}}{C}-\!\!\left\langle\;\right\rangle\!\!-OCH_3 \longrightarrow C_6H_5-\underset{\underset{\displaystyle O}{\|}}{C}-NH-\!\!\left\langle\;\right\rangle\!\!-OCH_3$$

$$C_6H_5-\underset{\underset{\displaystyle \text{HO-N}}{\|}}{C}-\!\!\left\langle\;\right\rangle\!\!-OCH_3 \longrightarrow H_3CO-\!\!\left\langle\;\right\rangle\!\!-\underset{\underset{\displaystyle O}{\|}}{C}-NHC_6H_5$$

7.16 An example of using the Beckmann rearrangement to determine configuration. In each case, the aromatic group *cis* to the hydroxyl group migrates to the nitrogen atom. Different stereoisomers therefore result in different product amides.

amide therefore provided a clue to the original configuration of the oxime. For example, the two isomers of p-methoxybenzophenonoxime gave two different amides (Figure 7.16). In the first product the phenyl group was attached to the nitrogen in the amide, and therefore it had moved from carbon to nitrogen during the reaction.[59]

Although Hantzsch recognized that these reactions were not always simple, his means of assigning configurations effectively introduced Wislicenus' principle of planesymmetric elimination to nitrogen stereochemistry. Like Wislicenus, Hantzsch assumed that only functional groups in close spatial proximity to one another would be capable of reacting. This remained an essential *a priori* principle necessary for stereochemistry to make legitimate claims about chemical reactions:

> It is solely because of this axiom that determining configuration is possible; an axiom that obviously has the same meaning for stereochemistry, but then again only the same meaning, as for example, the basic axiom in ascertaining constitutions in the structure theory, that in substitutions the entering radical takes the place of the leaving radical.[60]

Hantzsch felt comfortable using the Beckmann reaction in this manner, even though he admitted that it was the chloride, and not actually the oxime, that underwent the reaction, and that 'the true process of this strange rearrangement is not yet elucidated'.[61] This was, in fact, Meyer's principle criticism of Hantzsch's use of the Beckmann rearrangement:

---

[59] Ibid., p. 25.
[60] Ibid., p. 24.
[61] Ibid., p. 23.

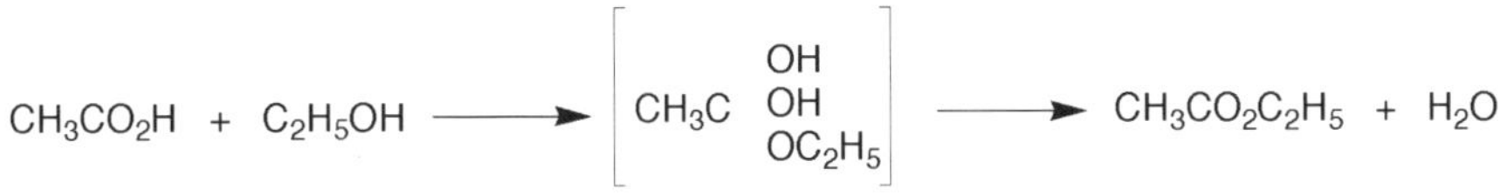

$$CH_3CO_2H + C_2H_5OH \longrightarrow \left[CH_3C \begin{array}{l} OH \\ OH \\ OC_2H_5 \end{array} \right] \longrightarrow CH_3CO_2C_2H_5 + H_2O$$

7.17 The formation of esters via an intermediate addition product, as speculated by Meyer.

giving this importance to the Beckmann rearrangement appears dubious, because the course of this curious reaction is still completely shrouded in darkness. As far as cleaving water from the oximes is concerned, perhaps Hantzsch's formulas express this more obviously, yet according to our formulas the explanation of this process appears to be no more difficult, because both hydrogen atoms that participate in the formation of water are also very close to one another in our formulas.[62]

Hantzsch replied that Meyer's caution was unnecessary because 'in the overwhelming majority of all chemical processes the dynamics of the process, the nature of the intermediate phases, and so forth, are entirely unknown; one knows the beginning and ending state and formulates on the basis of both'. For example, the process of esterification could be assumed to proceed through the 'highly probable' intermediate addition product in Figure 7.17, although the *actual* process remained 'completely obscure'. 'The rearrangements of the oximes into acid anilides are entirely similar,' Hantzsch continued:

no matter whether the dynamics of rearrangements or the nature of the intermediate products are completely known or not, or if the molecule's initial and final state is exactly known; this reaction can therefore be fully justified and exploited to determine configurations from the perspective of stereochemistry, because in doing so, observation and theory are in complete agreement.[63]

Although Hantzsch explicitly committed himself to the use of the Beckmann rearrangement and the principle of planesymmetric addition to determine the configuration of molecules, his use of them remained, in a sense, empirical. Unlike Wislicenus, Hantzsch did not commit himself to describing elaborate reaction mechanisms to explain the appearance of anomalous results. He thought it completely sufficient simply to know the starting and ending conditions of the reaction, because: 'We must be even more aware in a general sense that structural rearrangements cannot

---

[62] Victor Meyer and Karl Auwers, 'Bemerkungen zu der Abhandlung von A. Hantzsch und Friedrich Kraft: Über das Auftreten von Stereoisomerie bei nicht oximartigen Stickstoffverbindungen', *Berichte*, 1891, *24*, 4225–30, p. 4229.
[63] Arthur Hantzsch, 'Über Stereoisomerie bei asymmetrischer Hydrazon', *Berichte*, 1893, *26*, 9–17, p. 15.

ever be explained, and that only the initial and final state of a system can be expressed with structural formulas.'[64] As we have seen, Hantzsch did propose various intermediates, but he did not suppose the actual mechanism for their production or decomposition. The idea of a favored configuration did not enter into his thought, for it did not prove useful for elaborating the isomerism of the oximes and hydrazones or for assigning configurations.

In the second article in the series, Hantzsch extended these methods to ascertaining 'The Configuration of Asymmetric Oximes Without Stereoisomerism'. A number of oximes, a 'great majority', according to Hantzsch, were known to exist in only one isomeric form. For example, oximes containing both an aromatic group and an aliphatic group existed only in one configuration, established by the structure of its Beckmann rearrangement product. Aliphatic aldoximes also existed only in configurations that could be established by its elimination of water to form nitriles. Only one stereoisomer of the glyoximes, and particularly β-keto acids, was known, and the other possible stereoisomer of the oxime of β-keto acids 'disappear[ed] entirely' (*überhaupt verschwindet*) because the oxime formed a stable cyclic isoxazolone.[65] Other examples were still unclear, but Hantzsch assumed that stereoisomers of certain oximes were too labile to exist, and the stability of each stereoisomer was determined by the relative attraction of the negative hydroxyl group to one of the two groups attached to the carbon of the oxime.

In June of 1892, Hantzsch published a comprehensive article in which he attempted to establish further these general principles for the relationship between the constitution, configuration and chemical properties of the oximes. His principal motivation was the apparent confusion over the ability to predict the stability of a given configuration by the attraction and repulsion of the positive or negative character of the radicals in the molecule. For example, fumaric acid (more stable than maleic acid) had the negative carboxyl groups attracted to the positive hydrogen atoms, and *cis* dichloroethylene was unknown because of the strong repulsion between the negative chlorine atoms, and would spontaneously convert into the *trans* isomer. Yet the use of attraction to explain the presence of certain configurations was not universally recognized. Meyer had supposed that the configurations of the benzildioximes were established by a special kind of bonding, and Carl Bischoff, in

---

[64] Arthur Hantzsch, 'Über stereoisomerische Salze des Benzoldiazosulfonsäure', *Berichte*, 1894, *27*, 1726–9, p. 1731.

[65] Arthur Hantzsch, 'Die Configuration asymmetrischer Oxime ohne Stereoisomerie', *Berichte*, 1891, *24*, 31–6; 'Über die Einwirkung des Hydroxylamins auf β-Ketosäuren und β-Diketone', *Berichte*, 1891, *24*, 495–506.

his 'dynamic hypothesis', supposed that the rotation was hindered by the groups attached to the carbon atoms.

The conventional conception of positive and negative radicals, Hantzsch also noted, led to inconsistent assignment of configurations. For example maleic acid was unfavored because the two positive hydrogen atoms wanted to occupy a position near the negative carboxyl group, but when the hydrogen atoms were changed to more positive methyl groups, the maleinoid form was the only one known to exist. Simple substitution of hydrogen for methyl could change the stability of the stereoisomers such that only one configuration was stable, and the other became more labile, or even incapable of existence. Therefore, stability or instability could not be explained by a simple interaction of positive and negative groups, and Hantzsch concluded that consideration of the positive or negative character of radicals 'veiled' (*verschleiert*) the true influence of radicals on the stereochemical configurations. The chemistry of the oximes, however, offered a path into the relationships between various radicals, because it was this group of compounds in which the relationship between the nature of the radical and the 'existence, stability, and reactions', of stereoisomers had been most intensively studied.

The relative stabilities of the oximes would then provide a means of sorting radicals according to their influence on configuration. For example, the CH_2COOH group had the strongest attraction for the oxime group because it formed a cyclic isoxazalone, preventing entirely the formation of its stereoisomer. Similarly, because the aliphatic aldoximes existed in only one form in which the hydrocarbon chain and the oxime were *trans* to one another, the hydroxyl group must repel the hydrocarbon chain, or at least its attraction for the hydrogen atom was greater than for the carbon chain.[66] Methyl groups (CH_3) appeared to exert the strongest repulsion of all groups. Reasoning in this manner, Hantzsch ordered the radicals in a distinct sequence from the strongest to weakest attraction for the hydroxyl group, with number 1 showing the strongest attraction and number 10, the weakest:[67]

1 $COOH \cdot CH_2$ 5 C_6H_4X (*meta* or *para*)   8 $C_4H_3S(C_4H_3O)$
2 $COOHCH_2CH_2$ 6 $C_6H_5 \cdot CO$ 9 C_nH_{2n+1}
3 $COOH$ 7 C_6H_4X (ortho)   10 CH_3
4 C_6H_5

---

[66] Arthur Hantzsch, 'Über Beziehungen zwischen Constitution, Configuration, und chemischen Verhalten der Oxime', *Berichte*, 1892, 25, 2164–85, p. 2167.
[67] Ibid., p. 2168.

220   *Chemical Structure, Spatial Arrangement*

The sequence, Hantzsch noted, was independent of the radical's degree of positive or negative character.

Why then did the labile configuration, if it existed, form at all? Hantzsch's answer was quite simple: the assigned stability and lability applied only to the free oximes, and substituting the hydrogen on the oxygen affected the stability of the configurations. In basic solution, for example, ketones would form the sodium salt of the oxime, not the free oxime, and the two configurations of the salt had a stability opposite to the free oxime.[68]

Using Physical Properties to Establish Configurations

With his assumption of labile isomers and detectable but not isolable compounds, Hantzsch's interest in questions of pure organic chemistry – explanation of isomerism by the 'arrangement of atoms' – converged with his increasing interest in the use of physical methods for establishing the configuration of isomers. During the 1890s, Hantzsch's research program transformed from classical organic chemistry to a new physical organic chemistry dependent on physical measurements for the detection of new compounds. Stocklöv has argued that Hantzsch was the first organic chemist of any prominence to practice the methods of the new physical chemistry and join the 'wild army of ionists'.[69] Hantzsch's interest in physical measurements dates at least to 1888 (and probably earlier to his association with Wiedemann), when he wrote to Ostwald, 'although unfortunately a quite ordinary structural chemist, I still have at least an active interest in the new paths you are leading our science, and just between us, if I could study once more, I would go to school with you'.[70] Hantzsch had asked Ostwald to carry out conductivity measurements on thiazoline derivatives, and Ostwald continued to perform conductivity measurements for Hantzsch until April 1891, when Hantzsch acquired his own apparatus in Zürich by means of Ostwald.[71]

The earliest indication that Hantzsch was interested in ascertaining the relationship between the configuration of the oximes and their electrical conductivity was in April 1890, when he sent Ostwald a sample of the one known oxime of phenylglyoxylic acid to measure its conductivity. Hantzsch hoped to determine the influence of C=NOH group on the acid strength of a carboxyl group bound to it because, according to Meyer, the oxime and phenyl groups were of nearly equivalent

---

[68] Ibid., p. 2175.

[69] The phrase 'wild army of ionists' was coined by August Horstmann to refer to the proponents of the new physical chemistry in the 1880s. Elisabeth Crawford, *Arrhenius: From the Ionic Theory to the Greenhouse Effect*, Canton, MA, Science History Publications, 1995, p. 96.

[70] Hantzsch to Ostwald, 26 November 1888, in Stocklöv, *Arthur Rudolf Hantzsch im Briefwechsel*, p. 54.

[71] Stocklöv, *Arthur Rudolf Hantzsch im Briefwechsel*, p. 36.

negativity, and should enable the formation of three benzildioxime stereoisomers. Hantzsch suspected that the NOH group must be 'much less' negative than oxygen because the isomers of the benziloximes did not form for benzil itself. In June 1890, he sent a sample of the recently isolated isomer of the known oxime to Ostwald in order to determine whether, as expected, it was a stronger acid. The conductivity measurements confirmed Hantzsch's predictions, and he suggested configurations for the two oximes.[72] These results remained unpublished until 1892, when Hantzsch briefly mentioned, in the article on the relationship between configuration, constitution and chemical properties, that electrical conductivity measurements indicate 'that, of the components of the oxime group NOH, the nitrogen is more negative than the water fragment'.[73]

At nearly the same time, almost exactly a year after receiving his own conductivity apparatus, Hantzsch and his assistant Arturo Miolati published a landmark article, 'On the Connection Between Configuration and the Magnitude of Affinity in Stereoisomeric Nitrogen Compounds', Hantzsch's first paper in the *Zeitschrift für physikalische Chemie*. In this paper, Hantzsch and Miolati reported the results of conductivity measurements on various oximes, and in the introduction, Hantzsch made the clearest statement about the relationship between conductivity and configuration:

> If Ostwald has shown that affinity constants depend not only on the nature and composition of substances, but also very much on their constitution, so that they display qualities of an eminent constitutive nature, so the word 'constitution' must also contain the concept of 'configuration'; being declared and proved by example that the intramolecular influence of atoms on affinity constants are functions of their spatial distances.[74]

Stereoisomers should therefore display measurable differences in conductivity, and Hantzsch's results substantiated this assumption; the labile modifications of oximes were found to have consistently higher conductivities than the more stable isomer. Furthermore, these relative conductivities were similar to the conductivities observed for maleic and fumaric acid. Maleic acid, the more labile of the two isomers, also possessed a higher conductivity. This paper was not only Hantzsch's first paper in the *Zeitschrift*, but the first paper from a chemical institute run by an organic chemist.[75] Hantzsch would increasingly rely on physical methods when in 1894 he began an attempt to establish the configurations of the diazo compounds, many of which had

---

[72] Hantzsch to Ostwald, 29 April 1890 and 14 June 1890, in Stocklöv, *Arthur Rudolf Hantzsch im Briefwechsel*, pp. 77–81.
[73] Hantzsch, 'Über Beziehungen', p. 2166.
[74] Arthur Hantzsch and Arturo Miolati, 'Über die Beziehung zwischen der Configuration und den Affinitätsgrössen stereoisomerer Stickstoffverbindungen', *Zeit. phys. Chem.*, 1892, *10*, 1–33, p. 5.
[75] Stocklöv, *Arthur Rudolf Hantzsch im Briefwechsel*, p. 36.

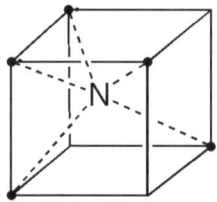

7.18 Van 't Hoff's model for the pentavalent nitrogen atom.

labile stereoisomers, and whose existence could only be detected by the use of physical measurements such as conductivity and cryoscopy (freezing point depression).

Other Spatial Models of the Nitrogen Atom, 1888–1893

Although Hantzsch and Werner's spatial model for the nitrogen atom would offer the most comprehensive, and in the long run most successful, spatial model for the nitrogen atom, they were not the first to speculate on the nitrogen atom's spatial properties, nor were they the last. In 1890, Werner was only one of many chemists to propose a novel spatial model of the nitrogen atom. Between 1888 and 1891, there were at least seven publications (including Werner's), all by lesser-known chemists (again including Werner) which suggested a spatial model of the nitrogen atom. Most were published in 1890 as alternatives to the Hantzsch/Werner theory, and accounted for the principal deficiency Meyer saw in Werner's theory: it was not a stereochemical model of the nitrogen *atom*, but only accounted for nitrogen in multiple bonds. Of the spatial models of nitrogen, only Werner's did not account for the known fact that nitrogen often appeared to be pentavalent. As such, Werner's model, at least in the form he presented it in 1890 was quite inadequate, because it did not address this apparent dual character in the valence of nitrogen: at times it appeared trivalent (in the alkyl amines), and at others it appeared pentavalent (in the ammonium salts). The chemistry of nitrogen had always remained somewhat problematic because of this puzzling dual valency.[76]

As early as 1877, Van 't Hoff had proposed a spatial model of nitrogen based on a cube, in which five corners were arranged in a quasi-trigonal bipyramidal arrangement defined by the corners of a cube (Figure 7.18). Such an arrangement had

---

[76] George B. Kauffman, 'Quinquevalent Nitrogen and the Structure of Ammonium Salts: Contributions of Werner and Others', *Isis*, 1972, *62*, 78–95.

several sets of equivalent valences. As this model was published in a Dutch journal at a time at which his theories were still relatively unknown, it subsequently had little notoriety or influence.[77] As we saw in Chapter 5, in 1877 Wislicenus suggested that a spatial model of nitrogen might be necessary, but there is no information about the precise nature of Wislicenus' thoughts. The first spatial model of the nitrogen atom of any consequence was published by Conrad Willgerodt in 1888 (Figure 7.19). An assistant in Adolf Claus' laboratory in Freiburg, Willgerodt proposed a true trigonal bipyramidal arrangement of valences in the nitrogen atom in connection with his study of substituted hydrazines.[78] Willgerodt found that reacting dinitrochlorobenzene with phenylhydrazine resulted in 'two different bodies with respect to their physical properties that lead to the same derivatives'.[79] One of these isomers was 'certainly' the symmetric dinitrophenyl-phenylhydrazine, but the structure of the other isomer remained ambiguous. Willgerodt was inclined to assume that it too was symmetrical, that is structurally identical to the other hydrazine, and that the two isomers should therefore be distinguished on the basis of their spatial properties, requiring an extension of nitrogen chemistry 'in a way similar to the hypotheses of Le Bel and Van 't Hoff that have now just become the common property of chemists, especially in the outstanding studies by J. Wislicenus, and Victor Meyer's subsequent publications'. Willgerodt's spatial model was straightforward:

> Assuming that the *nitrogen atom rests in the middle of a double tetrahedron, and that its three principle affinities* [Hauptaffinitäten] *always used in compounds are directed towards the corners of an equilateral triangle where the two tetrahedra come together*, while the two auxiliary affinities [Nebenaffinitäten], *that serve, for example to form the salts of ammonia and its derivatives, work at the two remaining corners*, the indicated cases of phenylhydrazine isomerism can be easily explained.[80]

Willgerodt's model was the prototype of nearly all the alternative spatial models of nitrogen, in that it contained three equivalent primary valences for amines, and two 'auxiliary' valences for the ammonium salts. The nitrogen–nitrogen single bond in the hydrazines was formed when two nitrogen atoms shared a vertex. This resulted in two possible isomers, analogous to the *cis-trans* relationship in carbon–carbon

---

[77] J.H. van 't Hoff, 'Over die bindingsrichtingen van het stickstoffstoom', *Maandblad voor Natuurwetenschappen*, 1877, 7, 109.

[78] C. Willgerodt and M. Ferko, 'Beiträge zur Kenntniss des Phenylhydrazins', *J. prak. Chem.*, 1888, 37, 345–58, C. Willgerodt, 'Vorläufige Mittheilungen zur Kenntniss der Hydrazin', *J. prak. Chem.*, 1888, 37, 449–54, 'Beitrag zur Kenntniss der Stereochemie von Verbindungen der Elemente der Stickstoffgruppe', *J. prak. Chem.*, 1890, 41, 526–8, 'Zur Kenntniss der räumlichen Anordnung der Atome in stickstoffhältige Verbindungen', *J. prak. Chem.*, 1890, 41, 291–300, 'Zur Kenntniss der Stereochemie isomerer Stickstoffverbindungen', *J. prak. Chem.*, 1890, 42, 63–4.

[79] Willgerodt, 'Vorläufige Mittheilungen', p. 449.

[80] Willgerodt, 'Vorläufige Mittheilungen', pp. 450–1.

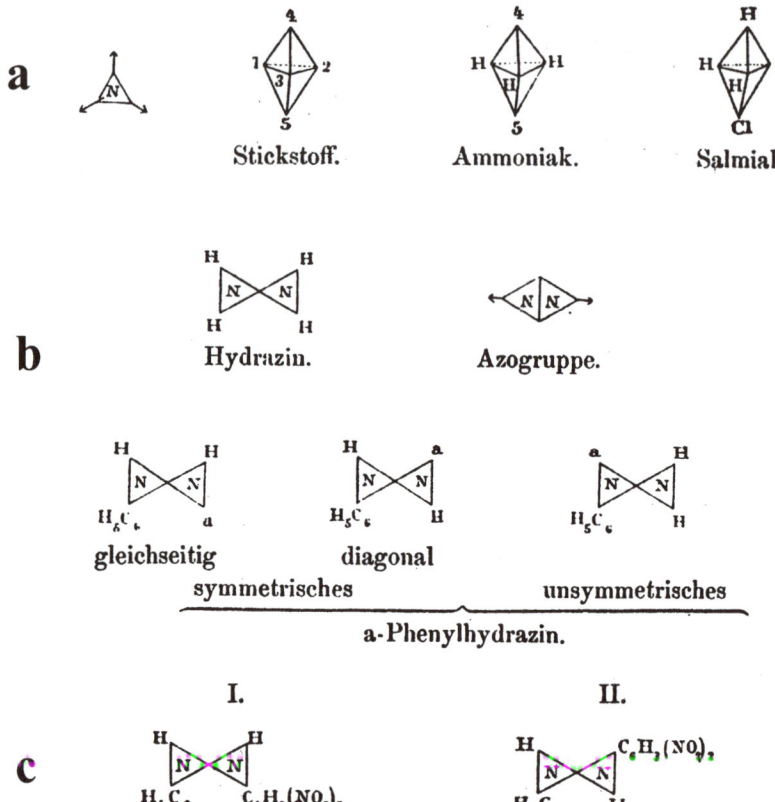

7.19 Willgerodt's spatial model for the nitrogen atom (1888). (a) The nitrogen atom's three principal coplanar valences (left) and stereoformulas for the pentavalent nitrogen atom, ammonia and ammonium chloride; (b) Willgerodt's stereoformula for hydrazine and azo compounds (top) and the possible spatial isomers resulting from disubstitution (bottom); (c) Spatial isomers for the phenylhydrazines.

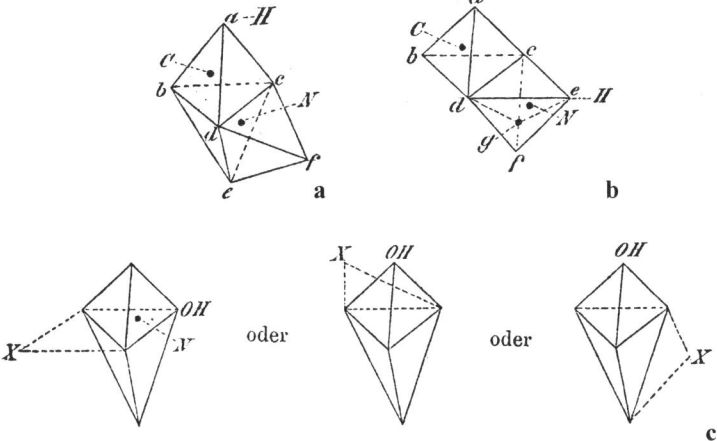

7.20 Examples of Willgerodt's 1890 space formulas for the pentavalent nitrogen atom. (a) Hydrocyanic acid. (b) An imine group with a carbon–nitrogen double bond. (c) Various possible formulas for the oximes.

double bonds. Willgerodt noted that compound I (Figure 7.19(c)), the '*cis*' isomer with the hydrogens in a planesymmetric relationship, could be oxidized more easily to azo compounds, whereas the '*trans*' isomer must rotate for oxidation to the diazo compound to take place. He remained unclear about the precise conditions for rotation, but presumably assumed there was a single bond between nitrogen atoms that had restricted rotation of some sort. In an article that appeared as a response to Hantzsch and Werner's paper, he provided detailed stereoformulas for various nitrogen-containing compounds (Figure 7.20).

In their 1890 article, Hantzsch and Werner mentioned Willgerodt's 1888 articles in passing, calling his theory only 'remotely similar' to theirs. In his response to Hantzsch and Werner, Willgerodt was clearly irked that Hantzsch and Werner thought their own theory was the 'first' spatial model of an atom of an element other than carbon. In a footnote to a paper following the initial article, they thought it 'unnecessary' (*überflüssig*) to respond to Willgerodt's 'polemical' article, or to address a theory 'which is solely based on the conjectural existence of two different structurally identical hydrazine derivatives'.[81] In a final response, Willgerodt remarked that if Hantzsch and Werner were correct, his own diagrams must not

---

[81] Hantzsch and Werner, 'Bemerkungen I', p. 1244.

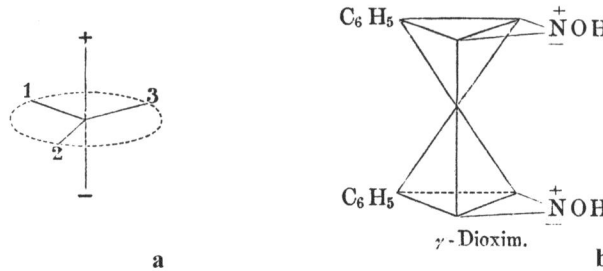

7.21 Robert Behrend's model for the nitrogen atom. (a) The model for the nitrogen atom itself, showing the three principal valences and the two auxiliary valences denoted with plus and minus signs. (b) A condensed version of the formula applied to the benzildioximes, showing how the position of the remaining charges would influence the favored configuration of the isomeric benzildioximes.

be 'stereometrical', yet the tetrahedral nitrogen atom was, 'with the marvelous [*wunderbar*] latent corner from which the nitrogen atom directs its three valences towards the remaining corners of the tetrahedron such that the three valences cannot be arranged in a plane'.[82]

Willgerodt's pentavalent model was copied in numerous contexts, and seems to be the most prevalent alternative model to the Hantzsch/Werner theory. In 1889, the English chemists Burch and Marsh proposed a similar geometry for the nitrogen atom, and presented vapor density measurements of amines that appeared to support it.[83] In 1890, spurred by Hantzsch and Werner's article, Robert Behrend, assistant to Wislicenus in Leipzig, argued that certain derivatives of hydroxylamine were isomeric and led to a conception of nitrogen 'following the conceptions of Van 't Hoff and Wislicenus'.[84] Behrend's model was identical to Willgerodt's, as Willgerodt himself later noted, but emphasized the lines of valence and presumed the presence of negative and positive poles at the two 'auxiliary' valences that would bind with, for example, positive hydrogen and negative chloride to form an ammonium salt (Figure 7.21). Under this theory, ammonium salts could exist in two enantiomeric forms. Another possible configuration could exist when one pole was occupied by a group that resulted in a planar arrangement of the remaining valences.

With this model, Behrend was able to provide an explanation that was 'in complete agreement with the views that have no doubt become generally common by the

---

[82] Willgerodt, 'Beitrag zur Kenntniss', p. 63.
[83] G.J. Burch and J.E. Marsh, 'The Dissociation of Amine Vapours', *J. Chem. Soc.*, 1899, 55, 656–64.
[84] Robert Behrend, 'Zur Stereochemie stickstoffhältiger Körper', *Berichte*, 1890, 23, 454–8, p. 454.

publications of Wislicenus on the forces with which indirectly bound atoms attract one another'.[85] The configurations of the benzildioximes would be caused by the various attractive and repulsive forces between the poles of the nitrogen atoms in the oximes and the radicals attached to the carbon atoms (Figure 7.23). The isomers were therefore caused by the 'the directive effect of those bound radicals', not by a hindered rotation of the carbon–carbon single bond, or the character of the carbon–nitrogen double bond.[86] Willgerodt rejected this model because analogous compounds in the nitrogen group of the periodic table such as PCl_5 would then have two positive poles to bind with negative chlorine atoms, even though, according to Behrend's model, the phosphorous in PH_3 should have one positive and one negative pole.[87]

Carl Bischoff, formerly an assistant to Wislicenus in Würzburg and successor to Wilhelm Ostwald at the Riga *Polytechnikum*, suggested that the 'solution to this problem [of the spatial properties of nitrogen] will come closer, if it is made clear that here there are three large groups of compounds that must be treated separately', namely the alkylamines, in which isomerism was not ruled out, compounds containing double or triple bonds with carbon, and most important, derivatives of hydroxylamine and nitrous acid.[88] In compounds where the nitrogen was pentavalent, the valences of nitrogen must be directed towards the corners of a quadratic pyramid, in which one valence would occupy a non-equivalent position to the other four. In conversion to compounds with a trivalent nitrogen, the pyramid converted into a trigonal bipyramid to achieve equivalent angle between the three remaining groups on the nitrogen. Bischoff did not provide a model of the carbon–nitrogen double bond, likely because 'too much unclarity dominates the concept of the double bond'.[89]

In 1893, shortly after the appearance of Hantzsch's monograph *Grundriß der Stereochemie*, the English chemist S.W.U. Pickering offered the same trigonal bipyramidal model after rejecting the Hantzsch/Werner theory as physically and chemically 'untenable'.[90] Hantzsch's model of the nitrogen atom was clearly inadequate, Pickering argued. The stereoformulas Hantzsch employed for the nitrogen atom showed it to be asymmetric, in which one side of the tetrahedron was incapable of bonding. In addition, why shouldn't deviation of the valences from planarity that Hantzsch presumed in multiple bonds not occur in C–N single bonds?

---

[85] Ibid.
[86] Ibid., pp. 454 and 457.
[87] Willgerodt, 'Beitrag zur Kenntniss'.
[88] Carl Bischoff, 'Beiträge zur Stereochemie des Stickstoffs', *Berichte*, 1890, *23*, 1967–72.
[89] Ibid., p. 1968. Bischoff considered the portrayal of double bonds with a two lines 'only as a heuristic aid'.
[90] S.W.U. Pickering, 'Note on the Stereoisomerism of Nitrogen Compounds', *J. Chem. Soc.*, 1893, *63*, 1069–75.

228   *Chemical Structure, Spatial Arrangement*

Pickering argued that line formulas expressed the spatial relationships just as well as the misleading 'tetrahedral' stereoformulas, and noted that Hantzsch himself had largely abandoned the stereoformulas for simple bond-line formulas. 'Indeed, in his *Grundriß*,' Pickering complained '… his views as to the tetrahedron are expressed in such an ambiguous manner that it is impossible to make out in what light he expects them to be received.' Hantzsch formed pentavalent nitrogen compounds by putting the nitrogen at the center of the tetrahedron (similar to the carbon tetrahedron) and then 'cramming' the remaining hydrogen in the middle. '[S]uch playing fast and loose with the atoms … is scarcely a scientific mode of dealing with the question.'[91] Pickering noted that Hantzsch's theory was also different from traditional attempts to explain cases of isomerism:

> Every case of isomerism which has hitherto been met with has been satisfactorily explained by the various atoms or groups being joined to different bonds in the different cases (to use a graphic, though unscientific, mode of expression), but here the isomerism is explained by the difference in the position occupied by the one and only bond concerned.[92]

The only acceptable way of deriving a spatial model of the nitrogen atom, Pickering noted, was to do it as Van 't Hoff had for carbon: he considered the 'simplest possible arrangement of the atoms round the central one with which they are combined'. This led to the same trigonal bipyramidal model as Willgerodt, Behrend, and Burch and Marsh. The three main valences, or 'triad', was highly stable because it was 'perfectly symmetrical', and the pentad less so because it was 'imperfectly symmetrical'. It would therefore split off the two equivalent groups with relative ease. Pickering then offered an explanation for the the isomeric oximes similar to Behrend's: two valences of the nitrogen would match with two on the carbon, and the hydroxyl would occupy the third planar position (Figure 7.22). In the presence of external attraction by radicals at a and b, the hydroxyl group would then occupy position 4 or 5, depending on which had the strongest attraction.

The most unusual geometry for the nitrogen atom was proposed in an eight-page pamphlet, *Das Stickstoffatom*, written by Wilhelm Vaubel in 1891 as a footnote to his earlier publications on the stereochemistry of benzene.[93] Vaubel did not mention Hantzsch and Werner (there are no references in the entire pamphlet), but like the other models of the nitrogen atom, it seemed directed towards it. Vaubel made two

---

[91] Ibid., p. 1071.
[92] Ibid.
[93] Wilhelm Vaubel, *Das Stickstoffatom*, Ottmann, Giessen, 1891. Vaubel's stereochemical model of benzene is discussed in Tonja A. Koeppel, *Benzene Structure Controversies 1865–1920*, PhD, University of Pennsylvania, 1973, and 'Significance and Limitation of Stereochemical Benzene Models', pp. 97–113 in O. Bertrand Ramsey, ed., *Van't Hoff-Le Bel Centennial*, Washington, DC, American Chemical Society, 1975.

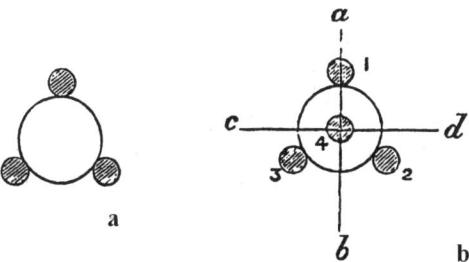

7.22 **S.W.U. Pickering's model of the nitrogen atom, showing only the three principal valences (a), and a view from one of the two auxiliary valences (b).**

assumptions. First, if the carbon atom was a tetrahedron with a specific volume, Vaubel argued, then the volume of the nitrogen must also have specific volume in proportion to its atomic weight, and calculation using the relative weights of carbon and nitrogen gave the volume of the nitrogen atom. Second, he started from his own stereochemical model of benzene, in which a nitrogen must be capable of substituting for carbon, but without any remaining free valences. In pyrrole, the nitrogen must substitute and leave one valence free. Under these constraints, Vaubel derived the form of the nitrogen atom as a curious multifaceted solid resembling an arrowhead with a cleft in the middle (Figure 7.23). The five valences were located on the vertices of this solid, and in a carbon–nitrogen triple bond the carbon atom fit neatly into the cleft, at which position three valences were satisfied. Vaubel's pamphlet included a large plate of stereoformulas that demonstrated the utility of this solid for modeling the compounds of nitrogen. Although ingenious, Vaubel's model drew little commentary from his contemporaries.

I have recounted each of these theories somewhat in depth because it is important to remember, as previous accounts have not, that Hantzsch and Werner's paper was only one among many that suggested a stereochemistry for the nitrogen atom. Furthermore, it would also seem that their theory, although ingenious, essentially ignored a fundamental characteristic of nitrogen agreed upon by nearly all chemists: that the nitrogen atom could be trivalent or pentavalent. Second, as Pickering noted, it was also derived in a way that differed from that for carbon. Hantzsch and Werner did not choose the simplest arrangement of valences around a central atom, but proposed a physically and chemically curious 'asymmetric' or 'crooked' arrangement of valences around it. For Hantzsch and Werner, the analogy to the carbon–carbon double bond must have been extraordinarily compelling, enough so that the obvious shortcomings of their theory could be ignored.

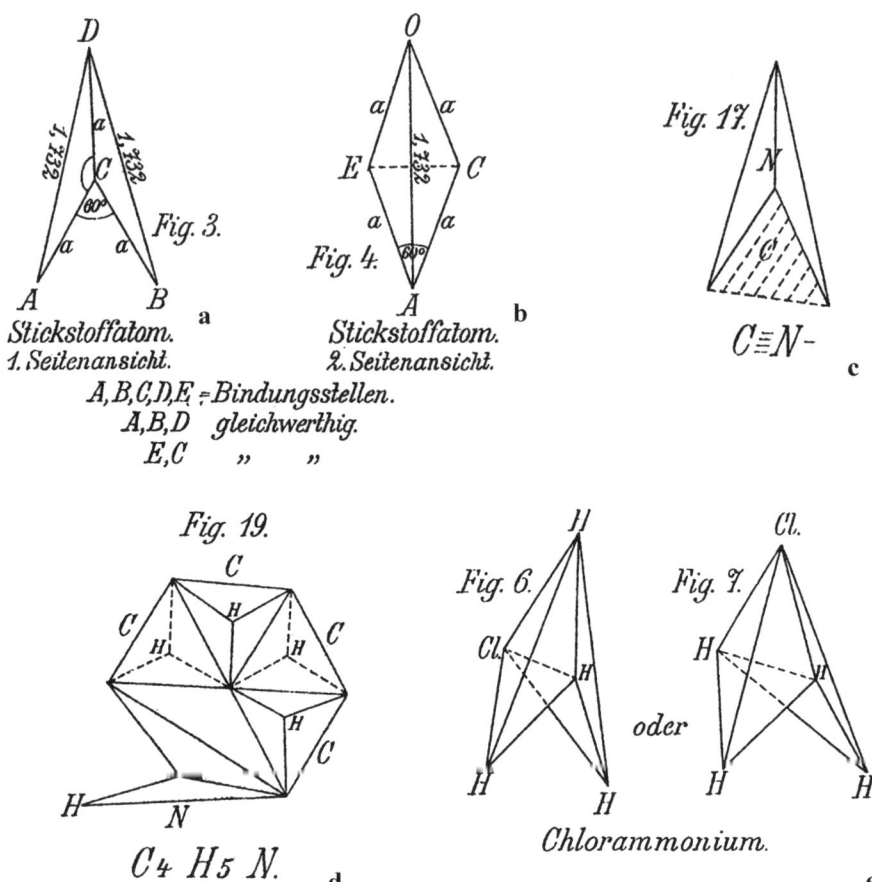

7.23 Wilhelm Vaubel's stereoformulas for nitrogen-containing molecules. (a) and (b) Two views of the arrowhead-shaped nitrogen atom. (c) The cyano radical, showing how the carbon atom fits into the cleft of the nitrogen atom, satisfying all four valences, leaving one valence on the pentavalent nitrogen atom. (d) and (e) Two representative stereoformulas with Vaubel's model: (d) Pyrrole; (e) Ammonium chloride.

Adolf Claus versus Meyer and Hantzsch: Is Stereochemistry Actually *Chemistry*?

The most extended and explicit critique of Hantzsch's and Meyer's stereochemical theories of the oximes – that is, the critique that both Hantzsch and Meyer spent a significant amount of time addressing – came from Adolf Claus, who in 1891 was at the University of Freiburg. Claus matriculated in 1859 at the University of Marburg as a medical student, but was soon swayed to chemistry by Hermann Kolbe's lectures. He earned a doctorate in chemistry at the University of Göttingen in 1862, and then served as an assistant to Lambert Babo at the University of Freiburg, where in 1866 he completed a *Habilitationsschrift* entitled *Theoretische Betrachtungen und deren Anwendung zur Systematik der organischen Chemie*. In 1867, he became extraordinary professor, and in 1875 ordinary professor at Freiburg, and on the retirement of Babo in 1883, was appointed director of the chemical institute. Most of Claus' research centered on the chemistry of condensed aromatic ring systems and quinones, but he is most famous for the 'centric formula' of benzene, a principal rival to Kekulé's cyclohexatriene formula. Claus was the first to use the word *Valenz*, and in 1866, he was the first chemist to ascribe a precise structural interpretation to Kekulé's hexagon formula for benzene.[94]

There are only two short biographical sources about Claus, and there is little information about his philosophy of chemistry outside his published works. He appears to have been a devoted teacher who emphasized the importance of extensive practical laboratory training over lectures. In this sense, he remained strongly influenced by Kolbe. Claus was deeply committed to the power of structural chemistry as an explanatory device, and as we shall see, he was insistent that *all* chemical differences between isomers should be explicable by a difference in structure.

In the summer of 1891, Claus began a series of articles critical of the Meyer and Hantzsch theories of the benzildioximes and hydrazones, in which he principally objected to the assumption that these isomers were structurally identical. Hantzsch's and Meyer's stereoformulas explained the increasing stability of the oximes in an 'extraordinarily captivating way', but they did not meet his criteria for the 'natural consequences' of structural identity.[95] A few examples of Claus' argument will suffice to demonstrate his point. Consider first the benzil*mono*ximes, first reported by Meyer

---

[94] Georg W.A. Kahlbaum, *Claus, Adolf Karl Ludwig*, Berlin, Reimer, 1906. G.N. Vis, 'Adolf Claus', *J. prak. Chem.*, 1900, *62*, 127–33. Kekulé's initial publications on the hexagon were schematic and suggestive, and he avoided using any specific structural formulas for benzene. Alan J. Rocke, 'Hypothesis and Experiment in the Early Development of Benzene Theory', *Ann. Sci.*, 1985, *42*, 355–81, p. 371.

[95] Adolf Claus, 'Zur Kenntniss der Oxime und der sogenannten Stereochemie', *J. prak. Chem.*, 1891, *44*, 312–35, p. 321 (hereafter referred to as 'Zur Kenntniss I').

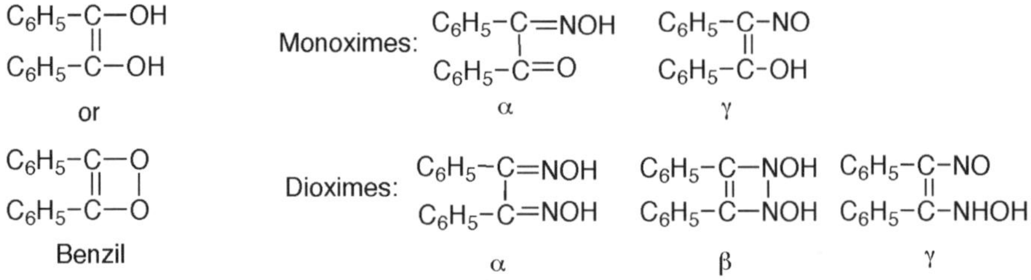

232   Chemical Structure, Spatial Arrangement

C_6H_5-C-OH
$\|$
C_6H_5-C-OH

or

C_6H_5-C-O
$\|\ |$
C_6H_5-C-O

Benzil

Monoximes:

$C_6H_5-C=NOH$
$|$
$C_6H_5-C=O$

α

C_6H_5-C-NO
$\|$
C_6H_5-C-OH

γ

Dioximes:

$C_6H_5-C=NOH$
$|$
$C_6H_5-C=NOH$

α

$C_6H_5-C-NOH$ C_6H_5-C-NO
$\|\ |$ $\|$
$C_6H_5-C-NOH$ $C_6H_5-C-NHOH$

β   γ

7.24   Claus' proposed structures for α-, β- and γ-benzildioximes.

and Auwers. The conversion of the α benzilmonoxime into the γ isomer occurred easily, but the reverse transformation did not take place. Second, the two isomers reacted with acetic anhydride to form acetate derivatives with different melting points. Finally, the α and γ monoximes did not react at an equal rate with hydroxylamine to produce benzildioxime, and the stability of the product dioximes was inverted relative to the stability of the monoximes (the α-dioxime was more stable than the γ-dioxime, but the γ-*mon*oxime was more stable than the α-monoxime).[96]

Consider also the behavior of the benzil*di*oximes. When the γ isomer was transformed into the β isomer, it went directly to the β, and not through the α isomer of intermediate stability. Since the β isomer could not be transformed into the α or γ isomer (that is, the transformation of α or γ into β was irreversible), the cause of the difference between them must require more than a simple difference in geometrical form. Finally, the formation of a single anhydride from all three isomers (one of Meyer's central claims for their structural identity) was 'of no importance at all', since the α isomer, Claus noted, converted first into the β isomer, and *then* into the anhydride, and not directly.[97] According to Claus, this meant that the α isomer itself did not convert into the anhydride, and was therefore not chemically identical. The stereochemical configurations gave 'an apparently very simple and natural appearing explanation' for the fact that the γ isomer has greatest tendency to form an anhydride, but one would expect the α isomer, on the basis of its presumed configuration, to form an anhydride at a rate *between* the β and γ isomers.[98] Given these inconsistencies between the assigned configurations and the chemical behavior of the oximes, Claus did not find it 'indispensable' to explain these isomeric differences by a spatial cause, and proposed new structural formulas to explain the difference (Figure 7.24).

In his initial reply to Claus, Meyer admitted that the chemical relationships Claus

---

[96] Meyer had offered this same sort of critique to defend the structural differences in the isomeric benzaldoximes until he recognized their structural identity in 1891.
[97] Claus, 'Zur Kenntniss I', p. 325.
[98] Ibid., p. 324.

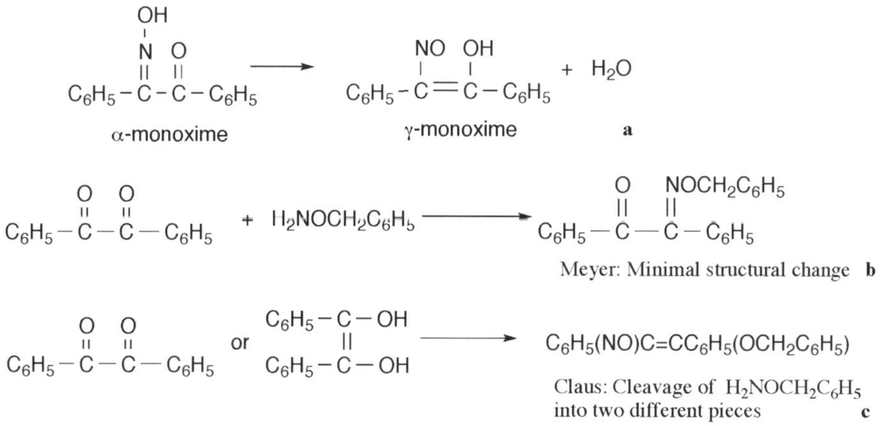

$$\underset{\alpha\text{-monoxime}}{\underset{|}{C_6H_5}-\overset{\overset{OH}{|}}{\underset{||}{N}}-\overset{\overset{O}{||}}{C}-C_6H_5} \longrightarrow \underset{\gamma\text{-monoxime}}{C_6H_5-\overset{\overset{NO}{|}}{C}=\overset{\overset{OH}{|}}{C}-C_6H_5} + H_2O \qquad \text{a}$$

$$C_6H_5-\overset{\overset{O}{||}}{C}-\overset{\overset{O}{||}}{C}-C_6H_5 + H_2NOCH_2C_6H_5 \longrightarrow C_6H_5-\overset{\overset{O}{||}}{C}-\overset{\overset{NOCH_2C_6H_5}{||}}{C}-C_6H_5$$

Meyer: Minimal structural change **b**

$$C_6H_5-\overset{\overset{O}{||}}{C}-\overset{\overset{O}{||}}{C}-C_6H_5 \quad \text{or} \quad \underset{C_6H_5-\overset{||}{C}-OH}{C_6H_5-\overset{||}{C}-OH} \longrightarrow C_6H_5(NO)C=CC_6H_5(OCH_2C_6H_5)$$

Claus: Cleavage of $H_2NOCH_2C_6H_5$ into two different pieces **c**

7.25 Meyer's analysis of the reaction of α-benzylhydroxylamine with benzil using his own configurational formulas and Claus' structural formulas. (a) The conversion of α- to γ-monoximes using Claus' structural formulas, a process that Meyer called 'extremely improbable'. Meyer's structure for the product of benzil with α-benzylhydroxylamine (b) appeared more probable than Claus' structure for the product (c).

mentioned were problematic, and that he was not yet able 'to explain all the genetic relationships of the substances in question in a completely satisfactory way'.[99] Claus' proposed structures, however, were not applicable to isomeric oximes derived from monoketones, such as benzophenone, or for aldehydes. More importantly, Claus' claim that the isomeric benzilmonoximes were structurally different rested on the difference in their reaction rates towards hydroxylamine. Because the α benzilmonoxime reacted more quickly with hydroxylamine than the γ isomer, they possessed different chemical properties, and according to Claus, required different structures. According to Meyer, however, this difference in reaction rates was 'only a *difference of degree, and not kind*, that very probably can be caused by the difference in spatial configurations'.[100] Furthermore, the conversion of α to γ monoximes appeared 'extremely improbable' using Claus' formulas, and the ease with which α-benzylhydroxylamine reacted with benzil to form a benzyl ether of benzilmonoxime made Claus' formulas equally suspect (Figure 7.25). Both reactions would require unusually complex structural rearrangements.

---

[99] Victor Meyer and Karl Auwers, 'Über die Claus'sche Theorie der Benziloxime', *Berichte*, 1891, *24*, 3267–71, p. 3268.
[100] Ibid., p. 3269. Emphasis added.

Claus had never intended that his formulas could or should be applied to the oximes derived from monoketones, and Meyer's admission that he could not:

> yet explain all the genetic relationships of the substances in question in a completely satisfactory way ... represents in any case the most benevolent and most favorable light which at the moment can ever be yielded to a criticism without directly harming the truth of this hypothesis about the stereochemistry of nitrogen.[101]

Because Meyer had dismissed the difference in reactivity between α and γ monoxime with hydroxylamine as a difference in degree and not kind, he had unjustifiably differentiated between these two kinds of chemical properties. According to Claus, a difference in the degree of reaction constituted a chemical difference as much as a difference in kind. The two isomers could not be structurally identical because they were not chemically identical.

Another argument for structural identity, the 'complete equivalence of the means of formation of both substances', also appeared insufficient. Claus pointed out that dozens, even countless reactions (such as sulfonation and nitration) proceeded under identical conditions and produced two or more isomers that undoubtedly possessed different structures.[102] Claus reacted to Hantzsch's recent paper on the discovery of stereoisomeric hydrazones in a similar way. Hantzsch had mentioned that crystals of the two hydrazone isomers were identical, a fact that would seem to contradict the assumptions of stereochemical theory. Claus also wondered why Hantzsch mentioned a 'considerable chemical difference' (*erhebliche Unterschied in chemischer Hinsicht*) between the two isomers – an alcohol solution of one hydrazone quickly turned brown and precipitated as a tarry oil. Hantzsch apparently did not regard this resinification as a significant chemical property, since he later claimed that the two hydrazones possessed identical reactions.

Hantzsch had used the identity of chemical reactions between the isomeric hydrazones to argue that the hydrazones were spatial isomers, but his claim was in 'stark contrast' to the principal assumption of stereochemistry, the 'unavoidable consequence' that two compounds displaying differences in chemical properties could not exist as spatial isomers. Chemical differences required a *structural* explanation. But that meant, according to Claus, that:

> *equal chemical connection* of the individual elementary components of a molecule by the different arrangement of one or the other constituents in *space* can cause *only physical and not chemical differences*. But recognizing this principle, as Hantzsch has explicitly

---

[101] Adolf Claus, 'Zur Kenntniss der Oxime und der sogenannten Stereochemie', *J. prak. Chem.*, 1892, *45*, 1–20, pp. 4–5 (hereafter referred to as 'Zur Kenntniss II').

[102] Ibid., 10–11.

done here, then for *the doctrine of those isomeric phenomena* that can be traced back to *the stereometric relationships within structurally identical molecules*, the term *'Stereo**chemistry**'* must appear inaccurate and be dropped, since these relationships *are without influence* on the *chemical reactions* of these molecules.[103]

Claus conceived of 'identical chemical behavior' here in a literal sense. In order for any two compounds to be considered structurally *identical*, they must share *all* reactions. This, as Hantzsch pointed out later, was practically an impossible task, since in nearly all cases stereoisomeric oximes, hydrazones and unsaturated acids *did* exhibit subtle differences in chemical properties, if only in degree and not in kind. Satisfactory criteria for 'structural identity' for Claus would be absolute identity in chemical behavior; substances could differ only by their differences in physical properties such as optical rotation.

Although Claus questioned the claim of structural identity, he did not doubt the 'elegant' means of explaining optical activity – a physical property – by the tetrahedral carbon atom. Yet Claus warned that current stereochemical theory was 'rushed and premature', when:

> differences of an undoubtedly chemical nature – and only in this case does the designation 'stereochemistry' have any justification in a logical sense – are supposed to be derived from a completely identical structural relationship of the atomic components solely from their spatial relationships.[104]

According to Claus, therefore, stereochemistry was not 'chemistry' at all, but 'a break with the entire concept of the chemical activity of matter as it has slowly developed in the last three decades'.[105]

Hantzsch considered Claus' alternative structural explanations weak because he had not dealt with the oximes or hydrazones as a whole, rather each new isomeric oxime received 'special new formulas' (*neue Specialformeln*). For example, Claus gave structures for the oximes of acetophenone or benzophenone based on his centric formula for naphthalene and the quinones (Figure 7.26). Why, Hantzsch asked, would Claus' supposed structures for these oximes not appear in the benziloximes? Hantzsch thought this structure 'at the least a highly peculiar and unusual conception', because quinones behaved like benzene derivatives: that is, they did not act as if only 'half' of the second phenyl ring were aromatic.[106]

Hantzsch also offered other objections, but in brief, Hantzsch accused Claus of

---

[103] Ibid., p. 397. Claus' emphasis.
[104] Ibid., pp. 12 and 13.
[105] Ibid., pp. 12–13.
[106] Arthur Hantzsch, 'Über die Claus'sche Auffassung der isomeren Oxime und Hydrazone', *Berichte*, 1892, 25, 1692–700.

| Claus' napthalene and quinone nucleus | Oxime of acetophenone | Oxime of benzophenone |

7.26 Claus' alternate formulas for the oximes.

employing *ad hoc* structural explanations for each group of oxime isomers, rather than attempting to base all of them in one cause. Claus completely ignored the isomers of ketoximecarbonic acids that lacked a phenyl group necessary for Claus' explanation:

> So again Herr Claus will put forward a new class of oximes with a completely different structure. Thus, according to Herr Claus, the isomeric oximes of benzil would already possess entirely different structural formulas as, for example, benzophenone; the aromatic ketoximes would not correspond at all to the aliphatic ketoximes, and the aromatic aldoximes would again by no means be analogous to the aromatic ketoximes ... Most of Claus' formulas for the group of oximes are chemical curiosities and possess only the commonality that they contradict the behavior of the relevant substances and their isomeric relationships.[107]

Hantzsch therefore found the presence of a common functional group in each set of isomers essential for deriving the most plausible explanation. He was frustrated that Claus refused to recognize the common characteristic – the carbon–nitrogen double bond – among the oximes and hydrazones. He was impatient with criticisms of stereoformulas based simply on alternate structural formulas.

> In reality [Claus] shows once again what no one doubts, that a host of different 'structural formulas' can be constructed by different arbitrary distributions of 'bonding lines', but that structural chemistry is not capable of explaining the questionable isomerism and its behavior.[108]

Claus' complaint that Hantzsch had been imprecise about the 'identity of chemical reactions' in the hydrazones also left Hantzsch rather impatient. Because the two radicals of anisyl-phenyl ketone were similar, Hantzsch argued, the crystals of their

---

[107] Ibid., p. 1697.
[108] Ibid.

oximes or hydrazones should also appear similar. When Hantzsch had noted that the crystals of the two hydrazones appeared identical, it was because the two radicals in the two hydrazones also has a similar external appearance, and would not give rise to significantly different crystal forms. Structural identity to Hantzsch *did not* mean identity in each and every way:

> Herr Claus makes such an absurd accusation in the literal conception of the words 'identity of chemical reactions of both hydrazones', that I must decisively protest against it. 'Identity of chemical reactions of both hydrazones' actually means what I, like any chemist, obviously have not especially emphasized, the 'chemical identity of these bodies in their properties as hydrazones' according to the structural formula $C_6H_5CN_2HC_6H_5 \cdot C_6H_4OCH_3$. I have never wanted to claim, and will never want to claim, that other still finer special distinctions of a chemical nature should not exist independent of these structural formulas.[109]

For example, Hantzsch noted that according to their structural formula, the two crotonic acids were identical 'methylated unsaturated carbonic acids'. But they were *not* identical in *all* their chemical reactions – those differences were expressed in the different stereoformulas. Therefore Claus made a logical error when he assumed that two substances *must* be structural isomers if they possessed different chemical properties, of *any* kind or degree.

Although Claus continued to attack the assumed stereochemistry of the oximes and hydrazones in several subsequent articles, Hantzsch declined to reply to them. The primary vehicle for Claus' arguments, the *Journal für praktische Chemie*, had been a consistent outlet for opponents of nearly all innovations in chemical theory – the structure theory, the tetrahedral carbon atom, the stereochemistry of the oximes and hydrazones, and finally the stereochemistry of the diazo compounds. This trend was not lost on Hantzsch, who wrote in 1897 that it was not surprising, that 'if the stereochemistry of double nitrogen compounds now befalls the same fate, this phenomena is only normal'.[110] The early objections to the structure theory and against the asymmetric carbon atom came primarily from the pen of the *Journal*'s editor, Hermann Kolbe. Given his reaction to both that we have seen in Chapter 4, this was understandable. Hantzsch did not understand, however, how Ernst von Meyer, Kolbe's successor as editor (and Kolbe's son-in-law) could allow the publication of Claus' polemical arguments. Hantzsch wrote to his mentor Rudolf Schmitt in 1892, shortly after Claus had replied to Hantzsch's response, and wondered how 'the editor of the "Journal", whom I highly regard personally, lets his journal be used as a repository for this kind of product. Isn't it truly regrettable that this journal now

---

[109] Ibid., p. 1699.
[110] Arthur Hantzsch and M. Schmiedel, 'Weiteres über Diazosulfonate und über freie Diazosulfonsäuren', *Berichte*, 1897, *30*, 71–88, pp. 88–9.

threatens to become a salon for every disgruntlement?'.[111] Five days later, he wrote to Schmitt again, apparently to justify his opinion concerning the *Journal*, 'I will dispute just as little as any reasonable person that the stereochemistry of nitrogen in its particulars does not need to be modified. But Claus' objections, to make use of V. Meyer's expression in his letters to me, are utter nonsense'.[112]

The conservative *Journal für praktische Chemie* may have been the only outlet for Claus. In a footnote to his response to Meyer's critique, Claus complained that he had submitted it to the *Berichte*, but it had been rejected because it 'did not contain new facts'. He was 'not surprised', however that Meyer's critique *was* accepted by the *Berichte*, even though Meyer had not included 'any new facts'. Tired of the 'eternal moaning on the part of the editorship and publication commission', Claus had quit the German Chemical Society several years earlier.[113]

Conclusion

What can we conclude from this discussion of Hantzsch's chemistry? Two broad points can be made. First, he appeared to lie somewhere between Meyer and Wislicenus with respect to his use of mechanism. He was not hesitant to postulate reaction intermediates or tautomeric forms, but was reluctant to postulate outright reaction mechanisms based on principles such as a 'favored' configuration. Hantzsch did, however, presume that oximes and hydrazones existed in 'labile' and 'stabile' forms. In this sense, Hantzsch could be accused of the same criticism as Wislicenus, that he proposed an isomerism, and then conveniently explained away the anomalies by assuming that one of the isomers spontaneously transformed into the other, or was simply too 'labile' to exist.

This reliance on intermediates that he had not isolated made his chemistry at best innovative, and at worst suspect. As he moved to the chemistry of the diazocompounds, he relied more and more on the assumption of tautomeric intermediates, and on the use of physical methods such as cryoscopy, conductivity, and spectroscopy to establish the existence of these intermediates. In a profession where compounds must be isolated to be labeled as existent, Hantzsch straddled the fence between innovative and traditional methods in organic chemistry. In the late 1890s, he had several articles rejected or returned for revision by the *Berichte*, and he complained to Emil Fischer that he was treated poorly by its publications

---

[111] Arthur Hantzsch to Rudolf Schmitt, 14 July 1892, Sammlung Darmstadter G1 1883, Staatsbibliothek zu Berlin, Preußischer Kulturbesitz.

[112] Arthur Hantzsch to Rudolf Schmitt, 19 July 1892, Sammlung Darmstadter G1 1883, Staatsbibliothek zu Berlin, Preußischer Kulturbesitz.

[113] Claus, 'Zur Kenntniss II', p. 1.

commission.[114] In frustration, he referred to the Deutsche Chemischen Gesellschaft in a letter to Werner as the 'Deutsche *Struktur*chemischen Gesellschaft'.[115]

The skepticism towards Hantzsch's theory of nitrogen and his innovative methods was also likely reinforced by the bold style of his papers, which presume the correctness of his theory from the outset. All three of his major papers – on the theory of oximes on methods for determining configurations of oximes, and on the stereoisomerism of the diazo compounds did not present a cautious presentation of theory from the facts, but simply outlined the points of his new theory. Only later did Hantzsch fill in the experimental details.

Hantzsch resembled Wislicenus closely in the use of broad hypotheses to generate his chemical research program. But as his study of the relationship between constitution and configuration indicates, he also retained a certain amount of empiricism, making general conclusions about the attraction and repulsion of radicals in molecules without prior assumptions about their electronegativity. His study of this relationship also reflects his use of chemical synthesis as a means towards an end, and not an end in itself.

The second major observation in Hantzsch's approach to stereochemistry is that Hantzsch did not go as far as either Wislicenus or Meyer to develop a concept of valence, or of the atom itself. In fact, he did not address the issue at all. In his 1893 monograph, *Grundriß der Stereochemie* (the first comprehensive monograph on stereochemistry that was not written by Van 't Hoff), Hantzsch explicitly denied that any kind of theory of valence was necessary for, or followed from, the principles of stereochemistry. On the first page of the *Grundriß*, Hantzsch remarked that:

> at least in its present stage of development, stereochemistry requires no specific idea about the type and cause of intramolecular cohesion of atoms, (the nature of chemical affinity), or concerning the type and cause of the ratios in which different atoms combine (the nature of valence); at present it requires only the idea, proved by the existence of isomerism itself, that the atoms are not situated within the molecule in a chaotic state, but in a stable equilibrium position within certain limits.[116]

Like Van 't Hoff, Wislicenus and Meyer, Hantzsch did not doubt the existence of molecular motions. He adopted the fundamental concept that the molecule was a 'static system of material points whose dynamics are considered only under certain

---

[114] Emil Fischer to Paul Jacobsen, 11 March 1903, Emil Fischer Papers, Bancroft Library, California. Jeffrey Johnson has noted this characteristic of Hantzsch's science, 'Hierarchy and Creativity in Chemistry, 1871–1914', *Osiris*, 1989, 5, 214–40, p. 234.

[115] Apparently, Werner also had papers suffer the same fate as Hantzsch, and Hantzsch had written to his former student in complete sympathy. The emphasis on 'Struktur' was Hantzsch's. Hantzsch to Werner, 8 July 1902, Sammlung Darmstadter G1 1883, Staatsbibliothek zu Berlin, Preußischer Kulturbesitz.

[116] Arthur Hantzsch, *Grundriss der Stereochemie*, Breslau, Trewendt, 1893, p. 1.

conditions'.[117] This did not, however, imply any specific theoretical concept of valence. When they began their work on the oximes, Hantzsch and Werner explicitly followed this assumption:

> The form in which we have developed and clothed our hypothesis ... is to be comprehended only as a graphic means of representation and expression in the garb of the usual chemical ideas. The hypothesis can be developed, how no doubt need not be argued, just as, only more extensively and also entirely independently, of the varying concepts of valence and the still more varying ideas on the 'direction', 'diversion', and 'bond' of valence and rather, ascribe it to general symmetrical relationships, just like the fundamental discoveries of Van 't Hoff and Wislicenus.[118]

Hantzsch rarely depicted the individual valences of the nitrogen atom, or the spatial properties of the atom itself, as Wislicenus and Meyer had done. Nowhere did he express a specific model to explain the appearance of the tetrahedron or to explain the phenomenon of valence. Unlike Meyer, he saw no urgent need to do so.

Yet privately, Hantzsch admitted in a revealing exchange of letters with Wilhelm Ostwald that his stereochemical theory of nitrogen required modifying current conceptions of valence. In an 1890 letter to Ostwald, Hantzsch had noted that Ostwald had not published a review of the theory of 'crooked nitrogen' in the *Zeitschrift für physikalische Chemie*, to which Ostwald replied: 'I don't have any special passion for crooked nitrogen, not any more than for little carbon windmills [*Kohlenstoffmühlchen*]', and offered to publish a review.[119] Hantzsch noted that he did not actually want a review, but had simply noticed that one had not appeared, despite 'the indubitable *fact* that an isomerism exists, [that] must also have some interest for the subject of physical [chemistry] – quite independently of the attempts to explain these facts'.[120]

Ostwald replied that Hantzsch's theory had 'difficulties' because of the resulting 'one-sided' nature of the three affinities of the nitrogen atom. Like most chemists, Ostwald assumed that the affinities must be evenly distributed around a central atom, so the resultant force (*Resultierende*) is zero, otherwise the nitrogen would not be in 'equilibrium' with other atoms in the molecule. Hantzsch agreed with Ostwald that 'crooked nitrogen' presented a problem in understanding the nature of valence, but that simply meant:

---

[117] Ibid., p. 2.
[118] Hantzsch and Werner, 'Bermerkungen II', p. 2769.
[119] Hantzsch to Ostwald, 30 October 1890, Ostwald to Hantzsch, 1 November 1890, in Stocklöv, *Arthur Rudolf Hantzsch im Briefwechsel*, pp. 84 and 85.
[120] Hantzsch to Ostwald, 4 November 1890, in Stocklöv, *Arthur Rudolf Hantzsch im Briefwechsel*, p. 86.

Arthur Hantzsch: The Stereochemistry of Nitrogen 241

In my opinion, it concerns modifying our conception of valence and affinity. Regarding the N-atom, we also really just assume – expressed in the image of the current valence theory – that the 3 valences of the N are symmetrically distributed, in that, what in C crudely and incorrectly is designated as a diversion of valences (or how else?) can possibly also occur in N. Yet then the resultant collective forces of the N-containing molecule must also equal zero? I fully agree with you that the 'little carbon windmills' have not only some, but quite a few imperfections. A reform of our ideas appears urgently necessary in both cases. Perhaps you will be somewhat interested in the 'Contributions to a Theory of Valence' that Herr Werner will publish shortly.[121]

Finally, Hantzsch made some interesting comments on the place of stereochemistry within chemical theory in a 1896 lecture 'On the Statics and Dynamics of Nitrogen Compounds', given at the celebration of the 150th anniversary of the Zürich Society of Natural Science. He noted that in organic chemistry proper, stereochemistry played a subordinate role as a branch of the structure theory (most accounts of stereochemistry would agree). But according to Hantzsch, these roles were *inverted* for nitrogen compounds. Nitrogen showed precious few instances of structural isomerism, but exhibited a great number of cases of stereoisomerism. As Werner's new coordination theory showed, in which stereoisomerism played a far greater role than structural isomerism, the same was true for inorganic compounds.

For elements other than carbon, 'molecules constructable by the different linkage of the "valence units", do not correspond at all to real forms', because they actually existed in only one structural form that took up different spatial arrangements.[122] Because it was limited to carbon, structural chemistry, not stereochemistry, was the special case. Despite its importance in the development of chemical theory, the structure theory appeared as:

> the predominantly valid kind of isomerism for carbon compounds; inversely, stereoisomerism will be recognized as the more widely distributed, more general form of isomerism. This also appears natural to me. For structural chemistry rests at least in part on more or less special fictions of valence and atomic linkage; in essence, stereochemistry makes only the necessary assumption, especially if valence is considered only as a number, that like all other individuals, the chemical individuals, molecules, are three-dimensional.[123]

Like Meyer and Wislicenus, Hantzsch did not doubt the physical nature of molecules or atoms, and assumed that their three-dimensional properties had become accessible to experimental test.

---

[121] Hantzsch to Ostwald, 18 November 1890, in Stocklöv, *Arthur Rudolf Hantzsch im Briefwechsel*, p. 87. Emphasis added.
[122] Arthur Hantzsch, 'Zur Statik und Dynamik der Stickstoffverbindungen', *Festschrift der Naturforschenden Gesellschaft in Zürich 1746–1896*, Zürich, Zürcher and Furrer, 1896, 186–202, p. 202.
[123] Ibid., p. 202.

Chapter 8

Emil Fischer and Carbohydrate Chemistry, 1884–1891

> To the initiated, modern configurational formulas proclaim, with the clarity and brevity of a mathematical expression, both the already established metamorphoses of the simple sugars and the many still to be expected, and one may indeed now say that the morphology and systematics of the monosaccharides has been achieved.
>
> Emil Fischer, 1894

The last three chapters have recounted the efforts of Wislicenus, Meyer and Hantzsch during the late 1880s and 1890s to adopt and modify Van 't Hoff's second and third hypotheses related to the nature of chemical bonds. At about the same time, Emil Fischer became occupied with methods for the identification and characterization of the known sugars. During the first phase of what would become a twenty-year project, Fischer realized that Van 't Hoff's theory of the asymmetric carbon atom could be employed to classify the increasing number of sugars and their derivatives isolated in the laboratory. Because the sugars had multiple asymmetric carbon atoms and were optically active, Fischer relied on the implications of Van 't Hoff's first hypothesis, and used it primarily as an isomer-counting device. Using the chemical relationships between the sugars and their derivatives, Fischer was also able to assign specific three-dimensional configurations, much as Wislicenus had done for the unsaturated acids. Fischer's configurational assignments were accompanied by a strong sense of empiricism and a distaste for speculation that was most likely inherited from his mentor, Baeyer. Nevertheless, like other stereochemists, Fischer was convinced that the static spatial properties of molecules had become accessible to experimental test.

Fischer was born in 1852 in Euskirchen, a small town near Bonn on the west bank of the River Rhine. He was the eighth and last child and only surviving son (of three) of Julie and Laurenz Fischer, a local businessman and entrepreneur. Fischer began to study chemistry at the University of Bonn and attended the lectures of August

Kekulé, but was disappointed with the practical instruction of analytical chemistry. During the summer of 1872, while he and his cousin Otto Fischer (who had recently studied chemistry at the Berlin *Gewerbeschule*) were weighing their options on where to continue studying chemistry, they read the lecture schedule for the University of Strasbourg published in the *Kölnische Zeitung*. As part of the Alsace region of France, Strasbourg and its university had been acquired a year earlier by Germany in the Franco–Prussian War, and in an effort to consolidate the German nature of the region, the university had installed German professors, among whom was Adolf von Baeyer, who had moved to Strasbourg from the Berlin *Gewerbeschule* in the spring of 1872. The cousins were aware of Baeyer's growing reputation in chemistry, and decided to move to Strasbourg. Although upon arriving the two cousins did not find the city itself inviting to Germans, they were received at the laboratory 'with a graciousness that exceeded our expectations', and they chose to stay.[1]

In the summer semester of 1873, Fischer formally began the study of organic chemistry under Baeyer, and earned his doctoral degree in 1874 with a dissertation on the two artificial phenolic dyes flourescine and phtaleine-arocine. Baeyer chose Fischer to be a private assistant in his research laboratory, and retained him when he moved to the University of Munich in 1875 as Liebig's successor. While in Strasbourg, Fischer began his investigation into the aromatic hydrazines that was published in six parts between 1875 and 1877 and comprised the work required for his *Habilitation*. In this work, he elucidated the structure of phenylhydrazine, the compound that would later both spectacularly advance his professional career and, because of its toxicity, cause him severe long-term health problems. While in Munich, he collaborated with his cousin Otto on the structure of the rosaniline dyes, first isolated by A.W. Hofmann, and together they established the presence of a triphenyl methane group in rosaniline and its derivatives. These publications on aromatic hydrazines and rosaniline quickly earned Fischer a reputation as an excellent organic chemist. In 1879, when Jakob Volhard, the long-time director of the inorganic division of the institute, left Munich for Erlangen, Baeyer quickly recommended that Fischer be promoted to the faculty as *ausserordentlicher* professor as Volhard's replacement. Fischer would succeed Volhard again in 1882, when Volhard left Erlangen to succeed Wilhelm Heintz at the University of Halle.

The early research conducted by Fischer in Baeyer's laboratory was by and large of intellectual rather than practical interest, but it was also strongly influenced by Baeyer's own close ties to the dye industry, and BASF in particular. This influence is seen in the compounds Fischer studied at the time – aromatic hydrazines, triphenylmethane derivatives, phenol dyes – which are all dyes or structurally related

---

[1] Emil Fischer, 'Errinerungen aus der Straßburger Studienzeit: 1872 bis 1875', pp. xx–xxvii in Adolf von Baeyer, *Gesammelte Werke*, Vieweg, Braunschweig, 1905, pp. xx–xxi.

compounds.² When Fischer arrived in Erlangen, he immediately moved away from Baeyer's emphasis on artificial and largely aromatic organic compounds, and began an entirely new research program on the chemistry of naturally occurring biological compounds that would occupy him the remainder of his life. His research can be broadly characterized as the application of classical chemical methods for establishing the structure and configuration of, and relationships between, the compounds in various classes of biological compounds. He began with the purines in 1882, moved to carbohydrates in 1884, and focussed on proteins and enzymes beginning in 1901. Fischer was trained as a classical organic chemist, and his projects retained the character of straightforward organic chemistry, but they were also closely related to, or even part of, the nineteenth-century tradition of physiological chemistry, in which the principal goal was the separation and analysis of materials and compounds found in plants and animals.³ Although he continued to have close ties with industry, his choice of personal research projects did not seem to have been driven by the needs of the chemical industry, as many them did not have any immediate practical value.

The Carbohydrates, 1884–1891

Fischer's cumulative research on the carbohydrates was immense, covering a period of 24 years between 1884 and 1908. Fischer himself collected his papers on carbohydrates in 1908, a volume that would eventually comprise one of eight large volumes in his *Gesammelte Werke*, published between 1907 and 1924.⁴ The volume on carbohydrates includes 109 articles published between 1884 and 1907, the overwhelming majority in the *Berichte*, and spans 912 pages, including the index. Although Fischer guided the overall research program and carried out some of the research himself, it clearly required a large group of assistants and students that were

---

² For more on Fischer's early work and the close relationship between Baeyer's research and industry, see Carsten Reinhardt, *Forschung in der chemischen Industrie: Die Entwickelung synthetischer Farbstoffe bei BASF und Hoechst, 1863 bis 1914*, Freiberg, TU Bergakademie, 1997.

³ This is opposed to biochemistry, which is concerning with understanding the dynamic chemical processes that take place at the cellular level. Robert Kohler, 'The Enzyme Theory and the Origin of Biochemistry', *Isis*, 1973, *64*, 181–96.

⁴ Several of the volumes appeared in Fischer's lifetime, and Fischer himself initiated the project. After Fischer's death, Max Bergmann continued supervising the project. The volumes included the papers on purines (1907), two volumes on amino acids (1906 and 1923), two volumes on carbohydrates (1908 and 1922), one on depsides and tannic acid (1919), one on Triphenylmethanes and hydrazines (1924), and a volume containing miscellaneous papers and lectures (1924). Fischer's autobiography was included as volume 1 of the series, making it nine volumes in total. Emil Fischer, *Gesammelte Werke*, 9 vols, Berlin, Springer, 1906–1924.

either included as co-authors or acknowledged at the end of the article.[5] A complete examination of Fischer's work on carbohydrates is not necessary or possible here, but we can focus on the initial period between 1884, the year he first reported the reaction of sugars with phenylhydrazine, and 1891, the year he first proposed a specific configuration of glucose and other monosaccharides. It was during this crucial period that he established the overall methodology for characterizing, organizing and representing the growing number of sugars characterized in his laboratory.

Fischer's work on the sugars has remained one of the most famous accomplishments of organic chemistry to this day, and nearly all organic chemistry textbooks contain a small section devoted to Fischer's carbohydrate chemistry. The existing studies of Fischer's work on carbohydrates tend to make several incorrect assumptions that I hope to correct here.[6] The first is the tendency to describe Fischer's work as either the 'discovery' or the 'proof' of the configuration of glucose. Fischer neither discovered nor proved the configuration of glucose, but *suggested* it, based on the chemistry of the known carbohydrates and Van 't Hoff's theory. Fischer's synthesis of glucose is also sometimes portrayed as the 'proof' of its configuration, which is not the case.[7] Accounts of Fischer's chemistry have also tended to assume that Fischer's original intention was to establish the spatial configuration of glucose, or that Fischer's establishment of the configuration was the first application of Van 't Hoff's theory of the tetrahedral carbon atom. This latter assumption is certainly not true, as Baeyer, Wunderlich, Wislicenus and Meyer all attempted to apply Van 't Hoff's fundamental theory to other cases.

As we shall see, examining Fischer's papers published between 1884 and 1891 in sequence reveals that Fischer did not begin with the goal of ascertaining configurations of the sugars. Fischer's research program for the sugars had no particular final goal other than to develop a general method for characterizing and classifying the known sugars. Van 't Hoff's theory did not make its appearance in his published work until 1889, when Fischer reported that mannose and glucose possessed the same structure, a result he described in 1890 as 'surprising'.[8] Rather,

---

[5] Out of the approximately 600 papers listed in Fischer's *Gesammelte Werke*, 185 had Fischer as sole author, and 295 were joint publications. The remaining 120 were published by his students as extracts of dissertations. Joseph S. Fruton, *Contrasts in Scientific Style: Research Groups in Chemical and Biological Sciences*, Philadelphia, PA, American Philosophical Society, 1990, p. 169.

[6] C.S. Hudson, 'Emil Fischer's Discovery of the Configuration of Glucose', *J. Chem. Ed.*, 1941, *18*, 353–7. Frieder W. Lichtenthaler, 'Emil Fischer's Proof of the Configuration of Sugars: A Centennial Tribute', *Ang. Chem. Int. Ed. Eng.*, 1992, *31*, 1541–56. Michael Engel, 'A Projection on Fischer', *Chemistry in Britain*, 1992, *28*, 1106–9.

[7] Bert Fraser-Reid, letter to *Chemical and Engineering News*, 27 September 1999.

[8] Emil Fischer, 'Synthesen in der Zuckergruppe', pp. 1–29 in Fischer, *Gesammelte Werke*, vol. 5, p. 10.

```
      CH₂                CH₂OH              CH₂OH                  CH₂           CH₂OH
     /CHOH               CHOH              /CH                    /CHOH          /CH
  O< CHOH      oder     /CH              O< CHOH                O< CHOH         O< CHOH
     CHOH             O< CHOH              CHOH                   CHOH            CHOH
     CHOH               CHOH               COH                    CHOH            C
     CHOH               CHOH               CH₂OH                  CH          ----O/ CH₂OH
        Dextrose                           Laevulose
```

Tollens

```
                                                       CH². OH                OH H
                                                       CH. OH             OH   C   OH
  {CH². OH      {CH². OH       {CH². OH                CH. OH             HC   CH
  {(CH.OH)⁴     {(CH.OH)⁴      {(CH.OH)⁴               CH. OH              |   |
  {CH². OH      {CH<OH/OH      {CH.O                   CH. OH             OHC   COH
     Mannit       hypoth. Alkohol   Glycose.           CH O                H    H
                                                                              C
                                                                            OH . H
```

Schiff   **Fittig**   **Baeyer**

8.1 Structures of the monosaccharides proposed by Baeyer (1870), Schiff (1870), Fittig (1871) and Tollens (1883).

Fischer initially appears to be one of many chemists of the mid-1880s who was interested in ascertaining the structure of the known sugars using contemporary chemical methods. The reports from Fischer's laboratory during this period consist of several separate but closely related streams of research that intertwined and eventually converged by 1890 into a coherent and effective methodology for characterizing natural and artificial sugars. These streams included (1) his initial use of phenylhydrazine to identify, isolate and characterize the various sugars, (2) the formation and identification of α- and β-acrose as synthetic sugars from acrolein bromide and glycerin, (3) the isolation of mannose and its derivatives, and (4) the oxidation and reduction of the monosaccharides to give, respectively, acids and alcohols. The chemistry of phenylhydrazine enabled the success of much of the remaining chemistry, as Fischer relied on phenylhydrazine's ability to react with sugars to form highly crystalline derivatives.

In 1888, as Fischer's research program was just under way, Bernhard Tollens summarized the state of sugar chemistry in his *Handbuch der Kohlenhydrate*, which provides a convenient entry point into Fischer's work. In the *Handbuch*, Tollens described only four well-characterized monosaccharides, or sugars with the formula $C_6H_{12}O_6$: grape sugar (also called dextrose, later renamed glucose by Fischer), levulose (fructose), galactose, and sorbose. Tollens also mentioned phlorose, crocose, lokaose, eucalyn, wood sugar (*Holzzucker*, later named xylose) and cerebrose as monosaccharides, but their exact nature remained unknown. Closely related to these monosaccharides were gluconic acid (the oxidized form of glucose), saccharic acid (*Zuckersäure*), a six-carbon diacid, and mannitol, a fully reduced carbon chain of

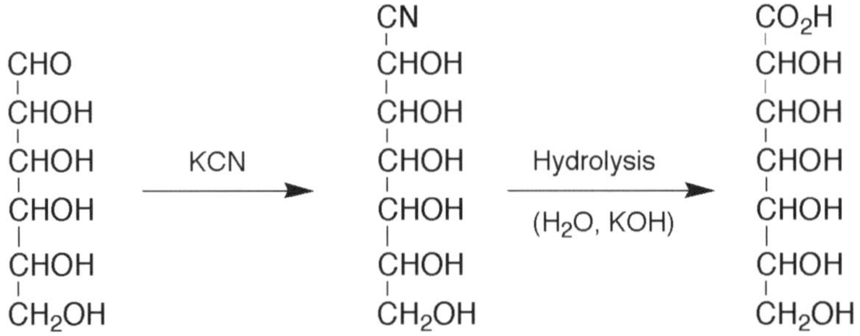

```
CHO              CN              CO₂H
|                |                |
CHOH             CHOH             CHOH
|                |                |
CHOH      KCN    CHOH   Hydrolysis CHOH
|         ──→    |       ──────→   |
CHOH             CHOH   (H₂O, KOH) CHOH
|                |                |
CHOH             CHOH             CHOH
|                |                |
CH₂OH            CH₂OH            CH₂OH
```

8.2 **Kiliani's two-step procedure for extending the carbon chain of monosaccharides. In this case a six-carbon sugar is lengthened to a seven-carbon sugar acid.**

six carbon atoms, each with a hydroxyl group.[9] In addition to these sugars were arabinose, shown by Heinrich Kiliani in 1887 to be a five-carbon sugar, and xylose, isolated by Koch in 1886 and shown to be an isomer of arabinose by Tollens and Wheeler in 1889. All of these compounds showed optical activity: glucose was dextrorotatory (hence the alternative name dextrose), as were gluconic and saccharic acid. Levulose and mannitol were levorotatory.

As a group, sugars were difficult to manipulate and identify, as they tended to form impure syrups rather than crystalline compounds, and therefore any molecular or structural formula for them remained somewhat uncertain. In 1870, for example, Baeyer proposed a six-membered ring, and Hugo Schiff suggested that glucose was the 'first aldehyde of mannitol'. In 1871, Rudolf Fittig suggested that glucose was the 'saturated sixfold alcohol of hexane', the structure that eventually proved most enduring, but in 1883, Bernhard Tollens suggested a cyclic pyranose ring (today called a hemiacetal) for glucose, levulose and sucrose (Figure 8.1).[10] In his *Handbuch*, Tollens recognized the probable existence of four asymmetric carbon atoms in the monosaccharides, and suggested that 'possibly the observed differences in the relationships of the glycoses rests' on the presence of multiple asymmetric carbon atoms.[11] As evidenced by Fischer's surprise in 1889 that glucose and mannose

---

[9] Also known were the disaccharides sucrose, maltose and lactose.

[10] Adolf Baeyer, 'Über die Wasserentziehung und ihre Bedeutung für das Pflanzleben und die Gährung', *Berichte*, 1870, *3*, 63–78. Hugo Schiff, 'Über die Consitution des Amygdalins und der Amygdalinsäure', *Annalen*, 1870, *154*, 337–53. Rudolf Fittig, *Über die Constitution der sogenannten Kohlenhydrate*, Tübingen, Fues, 1871. Bernhard Tollens, 'Über das Verhalten der Dextrose zu ammoniakalischer Silberlösung', *Berichte*, 1883, *16*, 921–4.

[11] Bernhard Tollens, *Kurzes Handbuch der Kohlenhydrate*, Leipzig, J.A. Barth, 1888, vol. I, pp. 12–13.

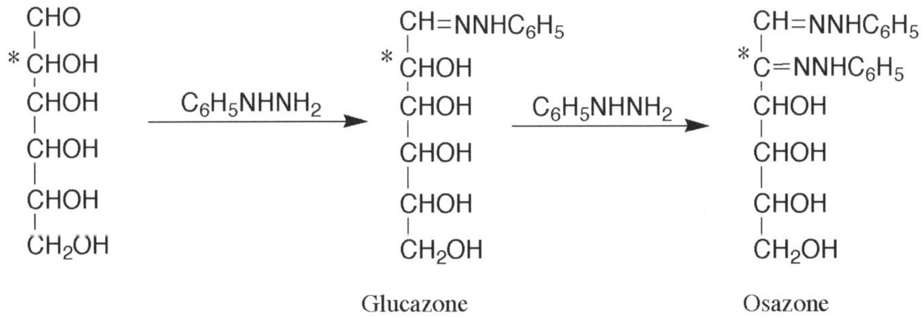

```
CHO                        CH=NNHC₆H₅              CH=NNHC₆H₅
*CHOH                      *CHOH                    *C=NNHC₆H₅
CHOH    C₆H₅NHNH₂    →     CHOH    C₆H₅NHNH₂  →    CHOH
CHOH                       CHOH                     CHOH
CHOH                       CHOH                     CHOH
CH₂OH                      CH₂OH                    CH₂OH
                           Glucazone                Osazone
```

8.3 The formation of phenylhydrazones and glucosazones with phenylhydrazine.

were structurally identical, it was not known with any certainty whether the known sugars were in fact the same or different structurally.

When Fischer began working on carbohydrates, he entered a field that had been occupied by Tollens, at the Göttingen Agrictultural Institute, and Heinrich Kiliani, a former student of Emil Erlenmeyer at the Munich Technische Hochschule. Like Fischer, Tollens and Kiliani had been occupied in the 1880s with understanding the chemical nature of the sugars. In 1885, Kiliani had reported a novel two-step procedure in which monosaccharides reacted with cyanide to give a cyano intermediate that could then be hydrolyzed to a carboxylic acid (Figure 8.2). The net result was the extension of the carbon chain by one carbon atom. Because the carboxylic acid could be reduced to the corresponding aldehyde, it provided a method to synthesize longer-chain monosaccharides. Kiliani had used the process in 1887 to demonstrate that arabinose was a five-carbon sugar.[12] Fischer and his assistants found they could repeat the process three times to create a nine-carbon sugar (a nonose). Fischer exploited this reaction repeatedly in his own research, and in 1890 referred to it as 'the greatest advance in the study of the sugar group in recent decades'.[13]

Phenylhydrazine and the Sugars

Fischer's research on the sugars began in 1884, when he published a short paper, the first of a series of five papers published between 1884 and 1889, on the reaction of phenylhydrazine with sugars. In this 1884 paper, he described the easily formed,

---

[12] Heinrich Kiliani, 'Über die Zusammensetzung und Constitution der Arabinosecarbonsäure bezw. der Arabinose', *Berichte*, 1887, *20*, 339–46.

[13] Fischer, 'Synthesen in der Zuckergruppe', p. 3.

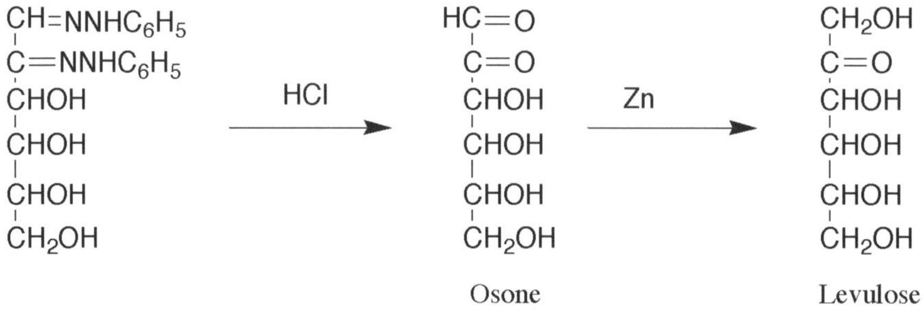

$$\begin{array}{c}\text{CH=NNHC}_6\text{H}_5\\|\\\text{C=NNHC}_6\text{H}_5\\|\\\text{CHOH}\\|\\\text{CHOH}\\|\\\text{CHOH}\\|\\\text{CH}_2\text{OH}\end{array} \xrightarrow{\text{HCl}} \begin{array}{c}\text{HC=O}\\|\\\text{C=O}\\|\\\text{CHOH}\\|\\\text{CHOH}\\|\\\text{CHOH}\\|\\\text{CH}_2\text{OH}\\\text{Osone}\end{array} \xrightarrow{\text{Zn}} \begin{array}{c}\text{CH}_2\text{OH}\\|\\\text{C=O}\\|\\\text{CHOH}\\|\\\text{CHOH}\\|\\\text{CHOH}\\|\\\text{CH}_2\text{OH}\\\text{Levulose}\end{array}$$

8.4 The removal of the osazone with hydrochloric acid and reduction of the osone to give levulose.

highly crystalline product of five known sugars (glucose, levulose, galactose, sorbose and cane sugar) with phenylhydrazine, and advocated the use of phenylhydrazine as a reagent for identifying and characterizing sugars as their phenylglucazone derivatives, including the detection of sugars in the urine of diabetics.[14] The second paper of the series did not appear until 1887, two years after he had moved from Erlangen to Würzburg, when he suggested a structure for the phenylglucazone (later termed an osazone by Fischer) and noted that the reaction with hydrazine took place in two phases. In the first phase, the aldehyde of the sugar molecule combined with the phenylhydrazine to form a phenylhydrazone. On heating, this phenylhydrazone combined with another molecule of phenylhydrazine in an oxidation reaction to give the osazone (Figure 8.3).

He also noted that glucose and levulose gave the same osazone, and also (without isolating it) that the intermediate phenylhydrazone of fructose must be isomeric with the phenylhydrazone created with glucose. In the third and fourth papers in the series that appeared in 1888, Fischer described a method by which he could relatively easily remove the osazone with hydrochloric acid to regenerate the dicarbonyl compound (called an osone), which in turn could be reduced with zinc metal to yield levulose (Figure 8.4).[15] This method offered a new and 'promising way to regenerate various sugars (*Zuckerarten*) from osazones'.[16]

---

[14] Emil Fischer, 'Verbindungen des Phenylhydrazins mit den Zuckerarten I', pp. 138–43 in Fischer, *Gesammelte Werke*, vol. 5.

[15] Emil Fischer, 'Verbindungen des Phenylhydrazins mit den Zuckerarten II', pp. 144–57, 'Über die Verbindungen des Phenylhydrazins mit den Zuckerarten III', pp. 158–61, and 'Über die Verbindungen des Phenylhydrazins mit den Zuckerarten IV', pp. 162–5, in Fischer, *Gesammelte Werke*, vol. 5.

[16] Emil Fischer, 'Über die Verbindungen des Phenylhydrazins mit den Zuckerarten V', pp. 166–76, in Fischer, *Gesammelte Werke*, vol. 5, p. 166.

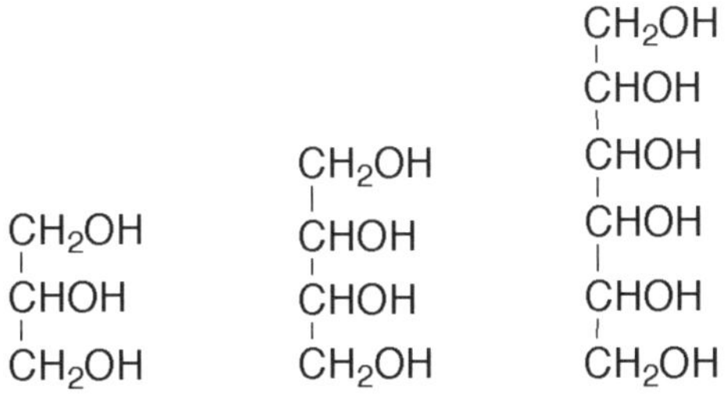

```
   CH₂OH
   |
   CHOH
   |
      CH₂OH      CHOH
      |          |
CH₂OH  CHOH      CHOH
|      |         |
CHOH   CHOH      CHOH
|      |         |
CH₂OH  CH₂OH     CH₂OH

Glycerin  Erythritol  Dulcitol
```

8.5   The structures of glycerin, erythritol and dulcitol.

Because each known sugar formed a unique highly crystalline osazone, the reaction of sugars with phenylhydrazine proved to be an extremely useful test for their presence in a reaction, and to unambiguously identify carbohydrates by the melting point of its osazone. 'As the previous experiments show anew', Fischer noted:

> Phenylhydrazine provides a convenient means for the isolation and recognition of products whose separation by older methods is extremely arduous or entirely impossible. It seems obvious to exploit such an aid in investigating the oxidation products of other polyalcohols.[17]

Fischer then offered a new definition of sugars as those compounds that were similar to glucose – those aldehydic and ketonic alcohols that reduced copper solutions and formed an osazone with phenylhydrazine. According to such criteria, the only true known sugars were glucose, levulose, galactose and sorbose, and Fischer established that phlorose, crocose and formose were either identical to one of the known sugars, or a mixture of two or more sugars.

The Synthesis of α- and β-acrose

In a second related project, published in three parts as 'Synthetic Experiments in the Sugar Group' with his assistant Julius Tafel, Fischer oxidized polyalcohols to aldehydes, and used phenylhydrazine to detect the presence of aldehydes or ketones in the mixture of products. Such oxidations of alcohols with nitrous acid had been done before, but the only isolated products had been acids, and Fischer reasoned that if aldehydes were produced as an intermediate compound, they would react with the phenylhydrazine to form derivatives. The oxidation of three alcohols – glycerin,

---

[17] Emil Fischer, ' Verbindungen des Phenylhydrazins II', p. 156.

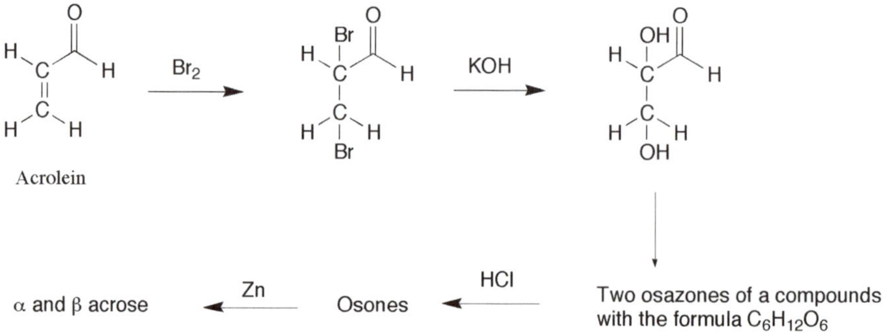

8.6 The synthesis of α- and β-acrose from acrolein.

erythritol and dulcitol (Figure 8.5) – followed by the addition of phenylhydrazine resulted in the formation of phenylosazones corresponding 'completely in composition, mode of formation and properties' to those osazones formed by the reaction of phenylhydrazine with the sugars.[18]

The structure of the oxidation product of glycerin was uncertain, however, and they therefore attempted to make it by another route from acrolein via its dibromo intermediate (Figure 8.6). Adding phenylhydrazine to the reaction mixture resulted in the formation of an osazone of a six-carbon compound that 'shows the greatest similarity to phenylglucazone, it melts at exactly the same temperature and can scarcely be distinguished from it externally'.[19] The only difference was the absence of optical activity in the new osazone. In a second paper, Fischer and Tafel confirmed this suspicion. It was 'undoubtedly' the osazone of a sugar, and differed from phenylglucosazone only in its optical inactivity. They also isolated a second isomeric osazone, and named the two sugar-like compounds that formed the osazones α- and β-acrose after acrolein, the original starting material. Furthermore, the osazone could be transformed into levulose by a two-step procedure. The importance of this osazone formation and their subsequent transformation into a sugar-like substance was not lost on Fischer, who noted that:

> With the isolation of the osazone the artificial formation of sugars from acrolein bromide is no doubt proven ... *This result does not only mean the reconversion of the osazone into a sugar, but at the same time is an easily comprehensible conversion of dextrose to levulose.*[20]

---

[18] Emil Fischer and Julius Tafel, 'Oxydation der mehrwertigen Alkohole', pp. 242–8 in Fischer, *Gesammelte Werke*, vol. 5, p. 242.
[19] Ibid., p. 247.
[20] Emil Fischer and Julius Tafel, 'Synthetische Versuche in der Zuckergruppe I', pp. 249–58 in Fischer, *Gesammelte Werke*, vol. 5, pp. 249–50. Fischer's emphasis.

I.	II.	III.
CH₂OH	CH₂OH	CH₂OH
ĊHOH	ĊHOH	ĊHOH
ĊHOH	ĊHOH	ĊHOH
ĊHOH	ĊOH.CH₂OH	ĊO
ĊHOH	ĊHO	ĊHOH
ĊHO		ĊH₂OH.

8.7 Possible structures for α-acrose, derived by determining the possible aldol products.

Fischer reasoned that the acrose isomers must form in an aldol-like reaction that resulted in three possible products with differing structures (Figure 8.7), but the similarity in the properties of the α-phenylacrosazone to glucosazone (derived from glucose) suggested that structure 8.7(I) was the most probable. Acrose and glucose would therefore be related to each other as tartaric acid (optically active) and racemic acid (optically inactive). In a third paper, Fischer and Tafel reported a more efficient alternate synthesis of α- and β-acrose by the direct oxidation of glycerin with nitrous acid and isolation with phenylhydrazine.

In the same article, Fischer and Tafel reported that α-acrosazone could be transformed into the acrosone with hydrochloric acid, and the resulting acrosone reduced to give α-acrose itself. The acrosone showed all the properties associated with the osones of the natural sugars, including reduction with zinc to give a sugar that fermented with beer yeast. Fischer concluded that:

> α-acrose is the first synthetic sugar of the hexane series that ferments with yeast. As the investigation of acrosone shows, it also yields all of the characteristic reactions of the natural sugars dextrose, levulose, and galactose. It can be distinguished from them only by its optical activity.[21]

In 1890, Fischer remarked that the discovery of the isomeric acroses was of key importance and 'moved my entire project in a particular direction'.[22]

---

[21] Emil Fischer and Julius Tafel, 'Synthetische Versuche in der Zuckergruppe III', pp. 267–70, in Fischer, *Gesammelte Werke*, vol. 5, p. 270.

[22] Fischer, 'Synthesen in der Zuckergruppe', p. 15.

Mannose

In 1888 and 1889, Fischer reported the results of another project involving a new sugar, mannose, in a four-part series co-authored by Josef Hirschberger. In the first paper, Fischer and Hirschberger announced the synthesis of a new sugar called mannose derived by oxidizing mannitol with nitrous acid. This sugar behaved in all respects like the other known sugars, both in its chemical reactions and in its ability to ferment yeast, so Fischer was confident that it would eventually be found in nature. In the subsequent papers, they noted that mannose gave the same osazone as glucose, could be oxidized to mannonic acid in a reaction analogous to the oxidation of glucose, and in adopting Heinrich Kiliani's method of chain extension in the sugars, found that it formed a seven-carbon sugar and a lactone that was very similar to the lactone Kiliani had produced from glucose. They also managed to isolate mannose from ivory nuts (*Steinnüße*), showing that it was also a naturally occurring sugar. In the second and most important paper of the series, appearing in February 1889, they suggested that the similarity in reactions between mannose and glucose indicated that they were structurally identical isomers, and 'as a consequence is to be considered a geometrical isomer of it'.[23] In the section on the 'constitution of mannose', Fischer and Hirschberger concluded that 'dextrose and mannose are the first example in the sugar group of two isomers that possess the same structure and can be transformed into each other'.

At this point, Fischer invoked Van 't Hoff's theory of the asymmetric carbon atom for the first time: 'We are explaining this isomerism entirely on the basis of the Le Bel-Van 't Hoff theory. The formula CHO·CH(OH)·CH(OH)·CH(OH)·CH(OH)·CH$_2$OH contains four asymmetric carbon atoms' which Fischer designated as as$_1$, as$_2$, as$_3$, and as$_4$ (beginning with the aldehyde group at the left), and he noted for the first time that the presence of four asymmetric carbon atoms predicted 'no less' than 16 possible geometric isomers. Using this formula, Fischer concluded that:

> From the previous experimental evidence [*Material*] it is easy to demonstrate that the isomerism of dextrose and mannose is limited to the carbon atom as$_1$. The *phenylhydrazones* of both sugars are completely different, but they change with the greatest facility into the same *osazone*. The latter posesses the structural formula:
>
> HC(N$_2$HC$_6$H$_5$)·CH(N$_2$HC$_6$H$_5$)·CH(OH)·CH(OH)·CH(OH)·CH$_2$OH
>
>   as$_1$ as$_2$ as$_3$ as$_4$
>
> in which the asymmetry of carbon as$_1$ is lost [*eingebüßt*].[24]

---

[23] Emil Fischer and Josef Hirschberger, 'Über Mannose II', pp. 294–305, in Fischer, *Gesammelte Werke*, vol. 5, p. 294.

[24] Ibid., p. 305.

As it seemed 'extremely improbable' that the other three asymmetric carbon atoms would change their configuration in the course of osazone formation, the difference between glucose and mannose must lay in the asymmetry of carbon atom as$_1$. Mannose was therefore the 'true' aldehyde of mannitol, and glucose had a corresponding hexitol that was geometrically isomeric to mannitol.[25]

Creating a Network of Compounds and Reactions

During the nearly two-year period between the fall of 1889 and the summer of 1891, Fischer consolidated his earlier results by extending both the number of monosaccharides and their reactions, creating what can conveniently be called a complex network of compounds linked by the reactions between them. In September of 1889, Fischer presented a lecture at the Heidelberg *Naturforscherversammlung* that closed the circle between the synthetic α-acrose and naturally occuring levulose and mannose. The contents of the lecture appeared in the *Berichte* in February 1890 as 'The Synthesis of Mannose and Levulose'. Except for the sign of its optical rotation, the dextrorotatory lactone of mannonic acid was identical to a levorotatory lactone made by Kiliani after reacting the five-carbon sugar arabinose with cyanic acid to form a six-carbon sugar acid (*Arabinoscarbonsäure*). A mixture containing equal amounts of both lactones was, as expected, optically inactive, and reduction of the lactone with sodium amalgam (using a process reported earlier by Fischer) yielded optically inactive mannose as a mixture of two enantiomers.[26] The discovery of dl-mannose was 'of special importance' because it shed light on the identity of the synthetic, optically inactive α-acrose. Reducing α-acrose gave α-acritol, a compound identical to the racemic mannitol created from the mixture of Fischer's and Kiliani's lactones. Furthermore, Fischer argued, the sugar derived from the reduction of the osone made from α-acrose was inactive levulose, because the sugar derived from the reduction of the osone from mannose was also active levulose. 'With these results', Fischer noted, 'the total synthesis of optically active natural sugars was finally made possible.'[27]

Fischer's argument was extremely compressed, and it is helpful to expand it somewhat. First, racemic mannitol (made from the totally synthetic α-acrose) could be oxidized to dl-mannose and dl-mannonic acid, which could in turn be resolved into its enantiomers with naturally occurring optically active strychnine. The resulting optically active acid could then be reduced to mannose, which in turn could be

---

[25] Ibid., pp. 304–5.
[26] Emil Fischer, 'Reduktion von Säuren der Zuckergruppe I', pp. 315–16, and 'Reduktion der Säuren der Zuckergruppe II', pp. 317–26 in Fischer, *Gesammelte Werke*, vol. 5.
[27] Emil Fischer, 'Synthese der Mannose und Lävulose', pp. 330–54 in Fischer, *Gesammelte Werke*, vol. 5, p. 332. This passage was emphasized by Fischer.

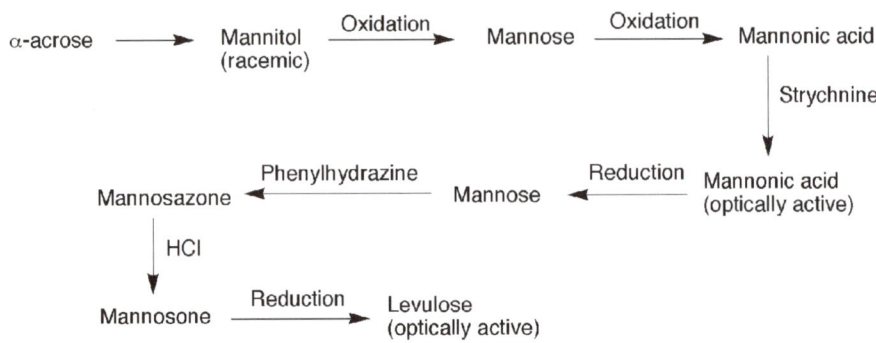

8.8 The total synthesis of optically active levulose from the synthetic α-acrose.

converted into its osazone and then the osone, which, finally, could be reduced to optically active levulose (Figure 8.8).

One month later, in March 1890, Fischer reported the synthesis of glucose itself by treating d-mannonic acid with the base quinoline.[28] His route to this reaction lay in his consideration of the stereochemical implications of two reactions: the formation of mannose from arabinose by the Kiliani chain extension, and the reduction of levulose to mannitol. In both cases, it was apparent that an asymmetric carbon was created at as_1. 'In mannose and mannonic acid,' Fischer noted, 'it is highly probable that the carbon atom previously designated as as_1 has no optical effect, but plays a role similar to the two asymmetric carbon atoms in racemic acid.' If carbon atom as_1 was 'optically ineffective', Fischer reasoned, then mannonic acid would be related to gluconic acid either in the same way that racemic acid was related to one of its optically active components, or in the way that racemic acid was related to inactive meso tartaric acid. If the first case were true, it would then be possible to 'resolve' the synthetic mannonic acid into gluconic acid and an isomer using an optically active base. If the second case were true, one acid could be converted into the other at high temperature.[29]

Attempts to perform a resolution with cinchoninic acid were unsuccessful, but treating d-mannonic acid with heat and quinoline did partially convert it into d-gluconic acid. This in turn could be reduced to the corresponding aldehyde to give

---

[28] The motivation for this conversion was the false belief that d-mannose was optically inert (optische unwirksam) at carbon 2, because both the reduction of levulose and the chain extension of arabinose produced a mixture of asymmetric carbon atoms at position 1. Therefore, Fischer attempted to separate d-mannose into isomers with quinoline.

[29] Emil Fischer, 'Synthese des Traubenzuckers', pp. 355–61 in Fischer, *Gesammelte Werke*, vol. 5.

glucose. Because Fischer had shown that the synthetic and optically inactive α-acrose was dl-mannose, which could be converted into the optically active mannonic acid by oxidation, he had accomplished, in effect, the synthesis of glucose from glycerine or even formaldehyde. The key to the synthesis in this case was his use of the lactones of mannonic and gluconic acid, which like the osazones, were highly crystalline and could be separated by means of fractional crystallization.

Soon afterwards, Fischer announced the synthesis of the unnatural (levorotatory) enantiomer of glucose when he reported that applying Kiliani's chain extension to arabinose resulted not in one six-carbon sugar acid, but two isomeric sugar acids, l-mannonic acid and l-gluconic acid. The levorotatory gluconic acid could then be reduced to l-glucose. Fischer found this new sugar to be inactive towards yeast, and described the preparation and properties of racemic gluconic acid, racemic glucose and racemic saccharic acid. Because the formation of these sugar acids involved the addition of cyanide to the aldehyde group of arabinose, Fischer discussed extensively the implications of this addition for the creation of asymmetric carbon atoms in the two products. In a section on the constitution of gluconic and mannonic acid, Fischer remarked that an asymmetric carbon atom produced by synthesis had always, according to 'previous experience' resulted in an inactive compound, either racemic or internally symmetric. This did not appear to be the case for the formation of an asymmetric center by the addition of cyanide to arabinose, because it resulted in two optically active compounds. His argument is worth repeating at length:

> Because the molecule of arabinose is unsymmetrical, a compound that would be comparable to meso-tartaric acid cannot form according to the theory of Le Bel and Van 't Hoff. Assuming that no stereometric rearrangements occur in the reaction, there consequently remain only two cases to consider. Two acids, with regards to the asymmetric carbon atom marked with a star in the formula
>
> $CH_2(OH) \cdot CH(OH) \cdot CH(OH) \cdot CH(OH) \cdot C^*H(OH) \cdot COOH$
>
> possess the opposite arrangement and a third resolvable compound that would be conceived as a combination of the two. Two of these acids are present in l-mannonic and l-gluconic acid. One of them could be resolvable. In the d-mannonic acid and d-gluconic acid, I have long endeavored to bring about such a resolution; I have not succeeded with either, but the mutual interconversion with quinoline does occur. So I consider it probable that gluconic and mannonic acid can be considered as right and left forms with regard to that carbon atom. But these two acids cannot be combined with one another. On the contrary, mannonic acid crystallizes from the mixture as a lactone. This observation appears to indicate that such isomeric substances by no means always combine with one another, as we are previously accustomed to assume.

Fischer was then forced to drop the conclusion he had drawn earlier in the reported synthesis of glucose:

So the opinion that I expressed earlier about the configuration of mannonic acid, the mannoses and mannitol regarding that asymmetric carbon atom [as$_1$] would become invalid. In fact, all previous experience in the sugar group confirms the view that for each optically active substance there exists an optically opposed isomer that combines with it to form an inactive compound; but the latter seems to be true for the asymmetry of the entire molecule, not for those of individual carbon atoms.[30]

The production of equal amounts of l-mannonic and l-gluconic acid from arabinose led Fischer to suspect that a similar process would occur in the reduction of levulose, during which the carbonyl carbon would become asymmetric – the production of mannitol from levulose did not create, as he had thought earlier, 'an analog of the resolvable inactive compounds'. The reduction, 'according to theory', should form two stereoisomeric products, that could 'further combine to a third product, which with regard to that one carbon atom would be considered as a combination of a right and left form'.[31] In the reduction of levulose, there should be a second isomer, which Fischer expected and found to be the known compound sorbitol. This was a second example of a reaction in the sugars that formed two stereoisomeric products that would not combine to form a racemic compound. 'All of these observations', Fischer concluded:

> show that stereoisomeric products can result during the formation of asymmetric carbon atoms that cannot be combined, but can be separated by a simple crystallization. And according to the results from the sugars it appears that under no circumstances do such isomers always form in equal quantities, such isomers do not form by any means always in equal quantities, rather one location can be favored.[32]

In 1891, Fischer reported the existence of yet another new sugar obtained from the reduction of the lactone of d-saccharic acid. Reduction of the lactone group formed a sugar acid identical to gluconic acid, except that the carboxyl group was at the opposite end of the carbon chain (Figure 8.9). Reduction of this acid gave a new sugar that Fischer named d-gulose (by transposing the 'l' and 'u' and dropping the 'c' in glucose). Fischer speculated that the reduction of the sugar diacids could be a general method for making new kinds of sugars.[33] In a paper that immediately followed, Fischer reported that the reaction of the five-carbon sugar xylose with cyanic acid formed the lactone of l-gulonic acid, the enantiomer of d-gulonic acid lactone formed

---

[30] Emil Fischer, 'Über die optischen Isomeren des Traubenzuckers, der Gluconsäure und der Zuckersäure', pp. 362–6 in Fischer, *Gesammelte Werke*, vol. 5, pp. 375–6.

[31] Emil Fischer, 'Reduktion der Fruchtzucker', pp. 377–80 in Fischer, *Gesammelte Werke*, vol. 5, p. 377.

[32] Ibid., p. 380.

[33] Fischer's claim that gulose was glucose with the aldehyde at the other end was quite novel and highly significant, but it is also mysterious, as Fischer makes no comment about how he knew this. It is not obvious from the text that this claim was true. Emil Fischer and Oscar Piloty, 'Reduktion der Zuckersäure', pp. 381–8 in Fischer, *Gesammelte Werke*, vol. 5.

d-saccharic acid lactone

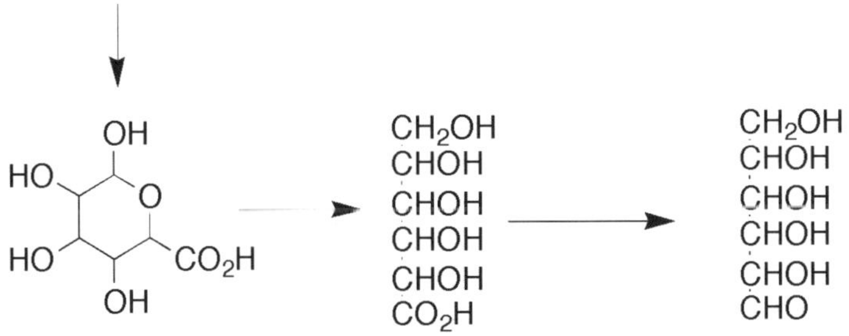

↓

[structure: pyranose with OH, HO, HO, OH, CO₂H] →

```
CH₂OH
CHOH
CHOH
CHOH
CHOH
CO₂H
```
→
```
CH₂OH
CHOH
CHOH
CHOH
CHOH
CHO
```

Gluconic acid with carboxyl group at the opposite end

Gulose

8.9 The synthesis of gulose from saccharic acid.

from d-saccharic acid. Oxidation of this lactone formed l-saccharic acid, and its reduction formed l-gulose.

By the spring of 1891, it had become increasingly clear that the relationships among the sugars were stereochemical, and Fischer repeatedly mentioned the implications of Van 't Hoff's theory for the configurations of various reaction products. In the discussion of gluconic and gulonic acid, Fischer mentioned again the number of possible stereoisomers:

> Gluconic and gulonic acid constitute the first example of stereoisomeric substances that lead to an identical product, when the molecule becomes symmetrical by transformation of the existing alcohol group into a carboxyl group. This observation appears an important confirmation of the theory of the asymmetric carbon atom. According to this theory, 16 isomers exist with the structure of gluconic acid. This number reduces to 10 for the related dibasic acids. Among the latter there are only 6 that can form monobasic acids. Among these must be d- and l- saccharic acid.
>
> From this we can conclude that soon we will be in a position to establish from true observations the configuration of the members of the sugar group in the sense of the Le Bel-Van 't Hoff theory.[34]

Finally, Fischer speculated that xylose and arabinose might be related as glucose and gulose, with the aldehydes located on opposite ends of the same carbon chain. Reduction of both to the pentitol would in both cases give a symmetrical (optically

---

[34] Emil Fischer and Rudolf Stahel, 'Zur Kenntnis der Xylose', pp. 389–99 in Fischer, *Gesammelte Werke*, vol. 5, p. 397.

	11	12	13	14	15	16
	+	+	+	+	+	−
	+	+	+	−	−	+
	+	−	−	+	−	−
	−	+	−	−	−	−
	−	+	−	−	−	−
	+	−	−	+	−	−
	+	+	+	−	−	+
	+	+	+	+	+	−
	5	6	7	8	9	10

	1	2	3	4
	+	+	−	−
	+	−	+	−
	+	−	+	−
	+	+	−	−

8.10 The possible configurations of the isomeric hexoses using Van 't Hoff's notation. The plus and minus signs are determined relative to the point between carbon atoms 3 and 4. Aldehyde groups are on the top. For a translation of this notation into Fischer's projection formulas, see Appendix 6.

inactive) compound, which did not prove to be not the case, as arabinol was optically active, and xylitol was inactive.[35]

The Configuration of Glucose

In August of 1891, Fischer collected all of this information together to propose configurations for glucose and its isomers in a paper entitled 'Über die Konfiguration des Traubenzuckers und seiner Isomeren' (hereafter referred to as 'Configuration'). Among his papers on the carbohydrates, 'Configuration' remains the most famous, and is the principal reason why Fischer's entire research program has often been characterized as a single search for the configuration of glucose. As we have seen, this was not Fischer's initial objective, nor was it his final one. Curiously, the two most comprehensive studies of Fischer's carbohydrate chemistry took this paper, quite rightly, as the basis for Fischer's argument, but did not actually reconstruct Fischer's verbal argument given in the paper nor show how Fischer employed Van 't Hoff's (admittedly confusing) configurational notation. For that reason, and because it represents Fischer's most extensive use of Van 't Hoff's theory of the asymmetric carbon atom, reconstruction of his argument is worth presenting here.[36]

Fischer began the article by reiterating that:

---

[35] Technically, both were inactive, but only arabinol showed optical activity on addition of borax to amplify the rotation. Xylitol showed no such optical activity on addition of borax.

[36] The two major articles discussing Fischer's carbohydrate chemistry do not follow the reasoning in this paper. Claude Hudson was explicit in his discussion that he was using another form of the argument developed later. Lichtentaler, in what is otherwise a very thorough treatment of Fischer's chemistry, also uses an argument that is not in Fischer's 1891 paper. Both Hudson and Lichtentaler also use Fischer projections, which Fischer introduced only in a second article (see note 42 below).

All previous observations in the sugars agree so completely with the theory of the asymmetric carbon atom that we may now venture an attempt to use this theory as the basis for the classification of these substances. The theory allows 16 isomers with the structure of grape sugar. This number reduces to 10 for the derivatives that are symmetrical.[37]

In order to make configurational assignments, Fischer adopted Van 't Hoff's convention for compounds with multiple asymmetric carbon atoms as originally published in *Die Lagerung* (Figure 8.10). In this system, each asymmetric carbon atom was assigned a configuration with either a + or − sign, determined from the viewpoint of the midpoint between carbon atoms 3 and 4. The resulting 16 stereoisomers could then be arranged in a table according to their spatial relationships. If the two groups at the ends of the chain were identical (as is the case for the saccharic acids and hexitols), the number of isomers was reduced to 10.[38]

Fischer began with saccharic acid (*Zuckersäure*), because it could be derived from both gluconic and gulonic acid, and made the following argument:

1. Saccharic acid must be among the sugars with configurations 5–10, 'because only these are each able to form from two stereoisomeric sugars'.[39]
2. Configurations 7 and 8 are optically inactive (meso compounds), and therefore cannot be correct, because both forms of saccharic acid were optically active.
3. Configurations 6 and 10 can also be eliminated, because glucose and mannose differ only at carbon atom 2. That is, if saccharic acid possessed configuration 6 or 10, then the saccharic acid derived from mannose and mannitol must have configurations 7 or 8, both of which are symmetric, and therefore optically inactive. Mannitol and the saccharic acid derived from mannose were both optically active.
4. Saccharic acid must then be either configuration 5 or 9. Because it was 'irrelevant, which was designated as + or −,' Fischer chose configuration 5 for d-saccharic acid.
5. d-glucose and d-gulose therefore each have one of the two configurations (the aldehyde is to the left):

$CHO \cdot CH(OH) \cdot CH(OH) \cdot CH(OH) \cdot CH(OH) \cdot CH_2OH$

```
   −       +       +       +
   +       +       +       −
```

---

[37] Emil Fischer, 'Über die Konfiguration des Traubenzuckers und seiner Isomeren I', pp. 96–115 and 417–27 in Fischer, *Gesammelte Werke*, vol. 5.

[38] For chemists familiar with the Fischer projections, these designations can be very confusing. See Appendix 7 for a translation of the +/− notation into the later Fischer conventions.

[39] Fischer, 'Über die Konfiguration des Traubenzuckers', p. 418.

6 Deciding between these two configurations rested on the reactions of arabinose and xylose. The asymmetry of carbon 2 in glucose and gulose is formed on their synthesis by the addition of cyanide to the aldehyde of l-xylitol and l-arabinose. Based on the two possible configurations of l-glucose and l-gulose (derived by changing the signs from + to − and vice versa), arabinose and xylose must therefore have the two configurations (the aldehyde is to the left, derived from configurations 5 and 9 by removing the first carbon):

CHO·CH(OH)·CH(OH)·CH(OH)·CH$_2$OH

− − −

− − +

7 When the first of these possible configurations of pentose is oxidized or reduced, the resulting compound is symmetrical and therefore optically inactive. When the second pentose is oxidized or reduced, the resulting compound will be optically active. When l-arabinose was reduced the resulting arabitol was optically active. When xylose was reduced, the resulting xylitol was inactive. Arabinose must correspond to the first configuration (+++), and xylose to the second (−−+).[40]

8 Finally, because the Kiliani chain extension of arabinose led to glucose, and that for xylose led to gulose, glucose must have configuration 5, and gulose configuration 9.

Fischer was then able to assign configurations to the following sugars (the aldehyde is to the left):

aldose:	CHO.CH(OH).CH(OH).CH(OH).CH(OH).CH$_2$OH			
d-glucose	−	+	+	+
l-glucose	+	−	−	−
d-gulose	+	+	+	−
l-gulose	−	−	−	+
d-mannose[41]	+	+	+	+
l-mannose	−	−	−	−

These configurations were followed by the configurations of d- and l-fructose, the sugar acids, the hexitols and the sugar diacids.

---

[40] Arabitol was active only on the addition of borax.
[41] Mannose differed from glucose only in the configuration of carbon 2.

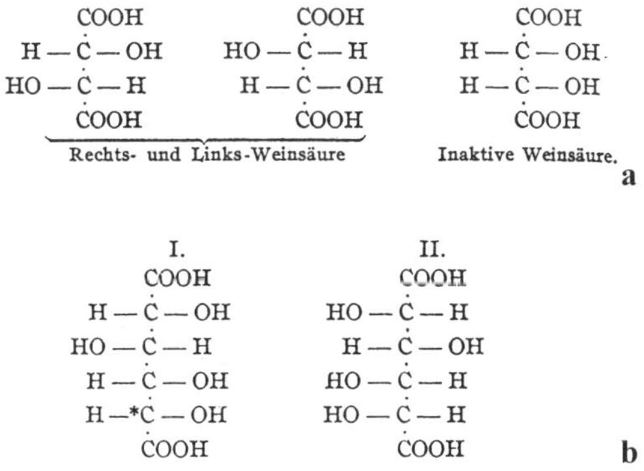

```
              COOH              COOH              COOH
          H — C — OH        HO — C — H        H — C — OH
          HO — C — H        H — C — OH        H — C — OH
              COOH              COOH              COOH
          Rechts- und Links-Weinsäure       Inaktive Weinsäure.
                                                            a

                     I.                II.
                    COOH              COOH
                H — C — OH        HO — C — H
                HO — C — H        H — C — OH
                H — C — OH        HO — C — H
                H —*C — OH        HO — C — H
                    COOH              COOH              b
```

8.11 **Fischer's first use in 1891 of the new projections for the tartaric acid isomers (a) and the saccharic acids (b).**

Two months later, Fischer abandoned the +/− notation because of the inherent ambiguities in assigning configurations to the carbon atoms. In his previous paper, he had used Van 't Hoff's convention and assigned + and − configurations from the viewpoint of the center between the second and third asymmetric carbon atoms. Looking towards the aldehyde, the alcohol could therefore be pointing towards the right or the left, and looking in the other direction towards the terminal alchohol, the hydroxyl group could also be pointing to the right or the left. But these configurational assignments could easily be misunderstood if the point of reference was taken to be the center between the first and second or the third and fourth carbon atoms. In that case, the reader could easily create a different stereoisomer than Fischer intended.[42]

Fischer therefore chose to expand his simple linear formulas to show explicitly the configuration of each asymmetric carbon atom:

Constructing molecules of dextro, levo and inactive tartaric acid with the help of the convenient Friedlander rubber models, they can be laid on the plane of the paper such that the four carbon atoms are located in a straight line and the relevant hydrogen and hydroxyl groups are over the plane of the paper. By projections, we obtain the following drawing:[43]

He then gave the formulas for d- and l-saccharic acid and glucose and mannose, choosing 'arbitrarily' formula I for the d-saccharic acid (Figure 8.11). These famous

---

[42] Emil Fischer, 'Über die Konfiguration des Traubenzuckers und seiner Isomeren II', pp. 427–31 in Fischer, *Gesammelte Werke*, vol. 5. A more detailed discussion of the inherent problems in Van 't Hoff's notation is given in C.S. Hudson, 'Historical Aspects of Emil Fischer's Fundamental Conventions for Writing Stereo-formulas in a Plane', *Adv. Carb. Chem.*, 1948, 3, 1–22.

[43] Emil Fischer, 'Über die Konfiguration des Traubenzuckers II', p. 428.

'Fischer conventions' did not depict tetrahedra; rather, he represented the three-dimensionality of each asymmetric carbon atom by means of a two-dimensional convention, derived by literally squashing the flexible carbon–carbon bonds of rubber models onto the plane of the paper.

With the publication of 'Configuration' in 1891, Fischer had in effect established the principal methodology for organizing the monosaccharides. Although he had assigned configurations to only 6 of the 16 possible sugars, in the next three years he continued to apply the same general methods, and by 1894 he had expanded the number of known isomers to 11.[44] In a period of ten years, Fischer had produced a remarkable body of work by almost any standard. A thorough study of the reception of Fischer's work is beyond the scope of this chapter, but it seems clear that Fischer's method for classification of the sugars was accepted by chemists extraordinarily quickly and without any controversy. In June 1889, Victor Meyer, obviously very excited, wrote to Fischer: 'You've really made my mouth water with your reports on the way you verbally and experimentally treat the sugars in your treatises.'[45] Meyer would also recount Fischer's argument and use Fischer projections in the first volume of his 1893 organic chemistry textbook.[46] In the second volume of his *Handbuch der Kohlenhydrate* (1895), Tollens recognized Fischer's new definition for the sugars, and referred repeatedly to Fischer's work using the new configurational assignments and projections for organizing the sugars. Fischer's Nobel Prize in 1902, the first awarded to an organic chemist, and before his mentor Baeyer received the honor, was for his studies of purines and sugars, and it was largely for this work that Fischer received the call to the University of Berlin in 1892 to succeed A.W. Hofmann.

There is little wonder that chemists found Fischer's reports so exciting and unobjectionable. Fischer drew his conclusions very carefully, drawing on aspects of Van 't Hoff's theory when necessary, but without reverting to speculation. When read in sequence, Fischer's papers read like a detective novel in which the identity of compounds is slowly revealed by their chemical relationships to the others. These chemical relationships were then given additional meaning by a configurational interpretation. Some synthetic sugars revealed themselves as identical or enantiomorphic to natural sugars. New sugars appeared from natural compounds and under artificial syntheses. Other sugars became stereoisomeric with existing sugars but not as enantiomorphs. Oxidations and reductions of sugars revealed synthetic routes to

---

[44] Emil Fischer, 'Synthesen in der Zuckergruppe II', pp. 30–75 in Fischer, *Gesammelte Werke*, vol. 5.

[45] Victor Meyer to Emil Fischer, 17 June 1889, Emil Fischer Papers, BANC MSS 71/95 z, The Bancroft Library, University of California, Berkeley.

[46] Victor Meyer and Paul Jacobsen, *Lehrbuch der organischen Chemie*, vol. I, Leipzig, Veit, 1893, pp. 896–914.

other isomeric sugars, while the presence of optical activity gave clues to molecular symmetry or asymmetry. Fischer could also suppress or regenerate the asymmetry of specific carbon atoms within sugar molecules.

Another reason chemists found little to object to in Fischer's research was the overall conventional nature of his methodology, which differed little from that already established under the structural theory. Classification of the sugars was made possible by considering their chemical relationships to each other and to their oxidized and reduced derivatives. Although he did not (to my knowledge) use the specific term, Fischer's methodology was identical to how Wislicenus used the 'genetic relationships' of the unsaturated acids to establish their configurations. In 1890, for example, Fischer proposed reorganizing the nomenclature of the monosaccharides by suggesting that all compounds derived from d-mannose should be labeled as d-compounds: that is, organized on the basis of their chemical relationships, regardless of their actual sign of rotation.[47] The centrality of genetic relationships between compounds in Fischer's work is illustrated further in his two comprehensive reviews of sugar chemistry, in which Fischer employed visual devices for showing the relationships between the sugars that closely resembled the phylogenetic trees found in biology (Figure 8.12).[48]

Fischer's research program between 1884 and 1891 lends itself to a variety of metaphors. In a rare moment of methodological reflection during his 1890 lecture on sugar chemistry to the German Chemical Society, Fischer compared his research program to the construction of a tunnel:

> I would like to compare these chemical investigations in which the material questions become more difficult with every step to the construction of a tunnel. If the mountain is not too broad, one can dig through in one direction. In other cases the engineer must also start working from the opposite side. But the engineer is fortunate in that he can determine the point of attack by exact measurements and has the certainty of bringing both parts together within the interior.
>
> Our science is unfortunately still far from being deductive enough to allow such calculations.
>
> The chemist can therefore count himself lucky if he digs his way through the material from the opposite points and makes the interior connection by several zig-zag paths.
>
> To show you how such a lucky accident led me to my goal, I must return to the natural sugars.[49]

By his own account, Fischer saw analysis and synthesis working from opposite directions towards the same end, but not always in a straight line.

---

[47] Fischer, 'Synthese der Mannose und Lävulose'.
[48] Fischer, 'Synthesen in der Zuckergruppe', and 'Synthesen in der Zuckergruppe II', note 44, p. 34.
[49] Fischer, 'Synthesen in der Zuckergruppe'. Translation from Karl Freudenberger, 'Emil Fischer and his Contribution to Carbohydrate Chemistry', *Adv. Carb. Chem.*, 1966, *21*, 1–38, p. 13.

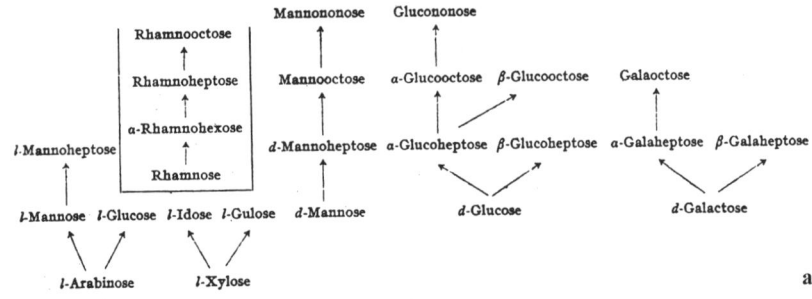

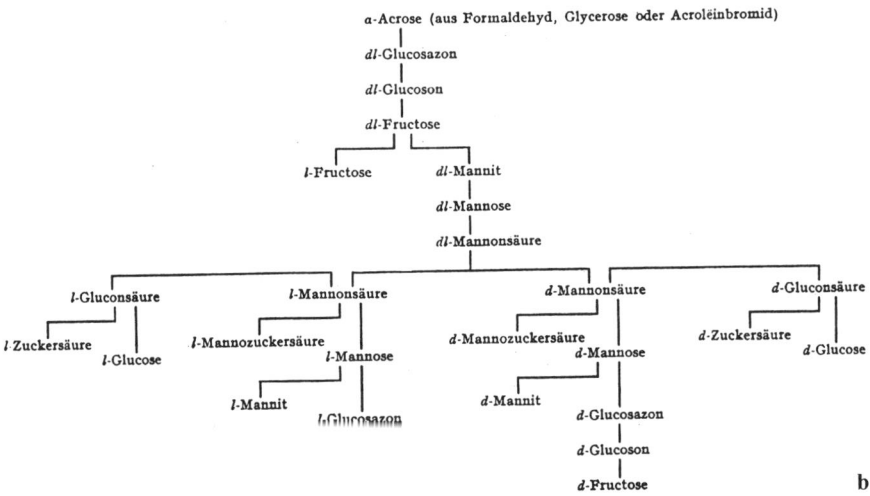

8.12 Fischer's 'family trees' for the synthetic and natural sugars that emphasise genetic relationships among the monosaccharides. (a) This shows the relationships between sugars as the number of carbon atoms increases from the bottom of the diagram towards the top. (b) This shows the route by which various naturally occurring sugars (and their alcohol and acid derivatives) can be derived from synthetic α-acrose.

Fischer's project also lends itself to analysis using more recent studies in history of science by Frederic Holmes and Hans-Jörg Rheinberger. In numerous books and articles, Holmes has elaborated on the concept of an 'investigative stream' created in the ordinary day-to-day work of the scientist.[50] While Holmes has developed this concept primarily in the course of reconstructing the activity of scientists from surviving laboratory notebooks, the sequence of Fischer's published papers can also be treated as describing such an investigative stream, or more properly, multiple streams that converged into a larger coherent methodology. Each of his papers were relatively short, encapsulating a small portion of his current research, yet also contained experimental details about the preparation of compounds. Most papers also had a short description of how the work derived from the results of previous papers and the possible or likely directions the project would take. But as is common in laboratory notebooks, Fischer's publications often gave no *specific* indication of the direction his research. The notebooks of Fischer and his students, if they were available, would no doubt tell a somewhat different, more detailed story about the specific chronology of the process and the mechanics of crafting and optimizing each individual chemical reaction. There is no doubt that each of the published papers is a polished version of the detailed day-to-day work of Fischer's students, but it does seem possible that we can reconstruct, in some cases, investigative streams of research from the published record, streams that occupy a different level than those found in laboratory notebooks.

Fischer's creation of a complex network of reactions also fits the description of Rheinberger's concept of an 'experimental system'. In *Toward a History of Epistemic Things*, Rheinberger describes experimental systems as a complex of assumptions in which no single experiment serves as decisive in affecting the outcome of the final theory. Rheinberger's systems are especially suited to biochemistry and molecular biology, sciences that develop through the study of particular physiological systems. He compares the developing system to the construction of a 'labyrinth, whose walls, in the course of being erected, in one and the same movement, blind and guide the experimenter'.[51] Fischer's network of reactions can also be seen in this light. No single experiment, but rather a cluster of syntheses, analyses and chance discoveries, coalesced into a single experimental system that eventually resulted in a coherent classification of the sugars. Following the results in Fischer's papers is indeed

---

[50] Frederic Lawrence Holmes, *Claude Bernard and Animal Chemistry*, Cambridge, MA, Harvard University Press, 1974. Holmes, *Lavoisier and the Chemistry of Life: An Exploration of Scientific Creativity*, Madison, University of Wisconsin Press, 1985. Holmes, *Hans Krebs: The Formation of a Scientific Life, 1900–1933*, New York, Oxford University Press, 1991. Holmes, *Hans Krebs: Architect of Intermediary Metabolism, 1933–1937*, New York, Oxford University Press, 1993.

[51] Hans-Jörg Rheinberger, *Towards a History of Epistemic Things: Synthesizing Proteins in the Test Tube*, Palo Alto, CA, Stanford University Press, 1997, p. 74–5.

reminiscent of constructing a labyrinth, in which the results of various reactions constrained and directed Fischer's subsequent moves without revealing the final exit. Rheinberger's metaphor complements nicely Holmes' concept of an 'investigative stream'. The former emphasizes the construction of a complex epistemic system whose final form is unknown to the investigators, while the latter emphasizes the temporal dimension of scientific work.

Sugars and Enzymes, 1894

Throughout his study of natural and synthetic sugars, Fischer had routinely used the behavior of beer yeast towards these sugars as a diagnostic tool – a characteristic 'reaction', as it were, for the sugars. When he began his study of the sugars in 1884, Fischer had already become familiar with the handling and identification of yeasts during a three-month stay in Strasbourg during the winter of 1876/1877. At that time, he had become acquainted with Pasteur's *Etudes sur la bière*, in which Pasteur had shown how beer yeasts can be contaminated by other microorganisms. By Fischer's own account, he was one of the first chemists in Germany to use a microscope to determine the quality of brewer's yeast during a stay in a brewery in Dortmund.[52] As Lichtentaler has noted, Fischer was very interested in the behavior of yeasts, and he 'certainly would have done my own research in this field had I stayed longer in Strasbourg'.[53] Therefore, it should not be surprising that after the successful synthesis of several artificial sugars, he should investigate further their behavior with yeast. The result was a collaborative paper with Hans Theirfelder, a physiological chemist at the hygiene institute in Berlin, with the title 'Behavior of Various Sugars with Pure Yeasts'. Fischer and others had already established that the natural sugars – glucose, galactose, fructose – were easily fermented, but arabinose was not. Recent research on brewer's yeast had shown that it was actually a mixture of various cultivated (*gezüchtete*) yeasts that fermented different sugars to different degrees. These yeasts had been separated into three different classes, depending on their ability to ferment certain sugars. Fischer himself, of course, had already reported in his earlier articles that of his synthetically derived sugars, the 'optical antipodes' (*optischen Antipoden*) of glucose, fructose, mannose, galactose, both d and l-gulose, and the synthetic heptoses and octoses, remained unchanged in the presence of yeast:

---

[52] Lichtentaler, 'Centennial Tribute', p. 2366.
[53] Frieder W. Lichtenthaler, '100 Years "Schlüssel-Schloss-Prinzip": What Made Emil Fischer Use This Analogy?', *Ang. Chem. Int. Ed. Eng.*, 1994, *33*, 2364–74, p. 2366. Translation in Lichtentaler, 'Centennial Tribute'.

As the compilation shows, the observed evidence [*Beobachtungsmaterial*] is by far the greatest for the hexoses and leads to the conclusion that fermentability has a close relationship to the geometrical construction of the molecule, and as a result may simply be characterized as a stereochemical matter.[54]

This apparent correlation between fermentation and configuration prompted Fischer and Thierfelder to use 12 'pure cultivated yeasts' with 14 sugars and methyl glucosides. They compiled the results in a table that illustrated the relative ability of yeasts to digest the sugars (Figure 8.13). The results showed a 'complete corroboration of our results and the earlier observations'. Of the nine known hexoses, d-glucose and d-mannose were easily digested, while galactose showed more difficulty with specific yeasts. Of the ketoses, only d-fructose was fermented. Fischer and Thierfelder drew their conclusions carefully. Because the configurations of the sugars were established, the relationship between configuration and fermentability could be established. If the configuration of the hydroxyl groups were changed only slightly from those found in galactose, then 'the fermenting power disappears'.[55] For example, the difference between galactose and talose was very minor (the configuration of only one carbon atom needed to be changed), but this:

> small additional geometrical shift is sufficient to remove entirely the fermenting power of talose. As a result, the yeasts are very choosy with regard to the configuration of the molecule; but the majority of them show the same preference, only a few are especially sensitive, such as *S. apiculatus*.[56]

Fischer and Thierfelder also noted an important new observation about the relationship between configuration and fermentability. The preference of microorganisms for only one of a pair of enantiomeric compounds was well known – it was first described by Pasteur, and Fischer himself had used this method to prepare pure samples of sugars with opposite optical rotations. Yet the difference in fermentation rates shown by the sugars involved:

> not only the contrast between optical antipodes, but a great number of geometrical forms, only a few of which meet the needs of the cell ... perhaps many chemical processes that occur in organisms are influenced by the geometry of the molecule. Under these circumstances it is certainly profitable to investigate the cause of that phenomena, and to seek the explanation in mainly stereochemical terms.[57]

---

[54] Emil Fischer and Hans Thierfelder, 'Verhalten der verschiedenen Zucker gegen reine Hefen', pp. 829–35 in Fischer, *Gesammelte Werke*, vol. 5, p. 830.
[55] Ibid., p. 834.
[56] Ibid.
[57] Ibid.

Tabelle[1]).

	d-Mannose	d-Fructose	d-Galactose	d-Talose	l-Mannose	l-Gulose	Sorbose	l-Arabinose	Rhamnose	α-Glucoheptose	α-Glucooctose	Rohrzucker	Maltose	Milchzucker	Methyl-glucosid[2]	Äthyl-glucosid[2]	Glucose-resorcin	Glucose-pyrogallol	Glucoseäthyl-mercaptal
S. Pastorianus I	+++	+++	+++	—	—	—	—	—	—	—	—	+++	+++	—	—	—	—	—	—
S. Pastorianus II	+++	+++	+++[3]	—	—	—	—	—	—	—	—	+++	+++?	—	—	—	—	—	—
S. Pastorianus III	+++	+++	+++	—	—	—	—	—	—	—	—	+++	+++	—	—	—	—	—	—
S. cerevisiae I	+++	+++	+++	—	—	—	—	—	—	—	—	+++	+++	—	—	—	—	—	—
S. ellipsoideus I	+++	+++	++	—	—	—	—	—	—	—	—	+++	+++	—	—	—	—	—	—
S. ellipsoideus II	+++	+++	†[3]	—	—	—	—	—	—	—	—	+++	+++	—	—	—	—	—	—
S. Marxianus	+++	+++	+++	—	—	—	—	—	—	—	—	+++	+++*)	—	—	—	—	—	—
S. membranaefaciens	—	—	—	—	—	—	—	—	—	—	—	—	—	—	—	—	—	—	—
Brauereihefe	+++	+++	+++	—	—	—	—	—	—	—	—	+++	+++	—	†	†	—	—	—
Brennereihefe	+++	+++	†	—	—	—	—	—	—	—	—	+++	+++	—	††	††	—	—	—
S. productivus	+++	+++	—	—	—	—	—	—	—	—	—	†	—	—	—	—	—	—	—
Milchzuckerhefe	††	+++	†	—	—	—	—	—	—	—	—	+++	—	+++	—	—	—	—	—

1) Erklärung der Zeichen:
††† bedeutet keine Reduktion der Fehlingschen Lösung nach 8 Tagen, also vollständige Vergärung,
†† „ eine ganz schwache Reduktion nach 8 Tagen, also fast vollständige Vergärung,
† „ deutliche Reduktion nach 8 Tagen, aber unzweifelhafte Gärung,
— „ keine Gärung.

2) Die Prüfung auf völlige Vergärung wurde hier unterlassen, da der Nachweis der Glucoside, welche erst nach der Hydrolyse reduzieren, durch die Anwesenheit des Hefeglycogens zu sehr erschwert wird.

3) Nach 14 Tagen war der Zucker ganz verschwunden.

*) siehe Berichte d. d. chem. Gesellsch. 28, 985 [1895] (S. 861), wo diese Angabe als unrichtig bezeichnet werden mußte.

8.13 Fischer's chart showing the relative fermen=ability of various sugars with various species of yeast. Three daggers indicates complete fermentation after eight days, and a dash indicates the absence of fermentation.

Then they reached their principal conclusion, 'that the yeast cells with their asymmetrically formed agents can only mesh and cause fermentation with those sugars whose geometry is not too far removed from that of grape sugar'.[58]

In a later paper, Fischer took the relationship between fermentation and configuration one step further when he tested the reaction of sugars with 'ferments that are separable from the organism, the so-called enzymes'.[59] Like the yeasts, the enzymes invertin and emulsin showed preference for sugars of a particular configuration. Fischer noted that he intended to try additional enzymes, but:

> the observations are sufficient to demonstrate in principle that the enzymes with regard to the configuration of their substrates [*Angriffsobjekte*] are as choosy as yeast and other microorganisms. In this respect, the analogy between both phenomena appears so total that we may assume the same cause for both, and so I return to the hypothesis mentioned earlier by Thierfelder and me. Invertin and emulsin have many similarities to the proteins and undoubtedly possess an asymmetrically constructed molecule. Their restricted action on the glucosides may then also be explained by assuming that only molecules with similar geometrical structure can approach each other closely and thus inititate the chemical reaction. To use a picture, I would say that that enzyme and the glucoside must fit each other like a lock and a key, in order to effect a chemical action on each other.[60]

The metaphor of a 'lock and key'(*Schloß und Schlüssel*) was one of Fischer's most enduring contributions to biochemistry, yet as Lichtentaler points out, Fischer used the metaphor casually (it is not emphasized in the text), and mentioned it only four times subsequent to its first appearance. This reluctance was no doubt related to his empiricism, which prevented him 'from placing this hypothesis side by side to the established theories of our science, and [I] readily admit that it can only be thoroughly tested, when we are able to isolate the enzymes in a pure state and thus investigate their configuration'.[61] Nevertheless, Fischer noted that this model of enzyme action 'shifted [the explanation] from the biological to the purely chemical domain', and that the 'geometric construction' (*geometrische Bau*) of the molecule played as great a role in its reactivity as chemical affinity:

> The realization that the effectiveness of the enzyme is confined so precisely to the molecular geometry may be of use to physiological research. But still more important for physiology, it seems to me, is the demonstration that the earlier commonly assumed distinction between chemical activity of the living cell and the effect of chemical agents with regard to molecular asymmetry does not actually exist. In particular, the analogy between 'living and

---

[58] Ibid., p. 835. This sentence was emphasized in the original.
[59] Emil Fischer, 'Einfluss der Konfiguration auf die Wirkung der Enzyme', pp. 836–44 in Fischer, *Gesammelte Werke*, vol. 5, p. 836.
[60] Ibid., p. 843.
[61] Emil Fischer, 'Die Bedeutung der Stereochemie für die Physiologie', pp. 116–37 in Fischer, *Gesammelte Werke*, vol. 5. Translation in Lichtentaler, '100 Years "Schlüssel-Schloss-Prinzip"', p. 2371.

lifeless ferments', so frequently emphasized by Berzelius, Liebig and others, is thereby significantly revived.[62]

Even though Fischer himself downplayed the significance of his metaphor, it was quickly adopted. In 1897, for example, Paul Ehrlich based his side chain theory of immunology on Fischer's lock and key metaphor. Fischer's concept of complementarity between enzyme and substrate remains one the most famous theories in the history of biochemistry and enzymology, and is still used today in introductory biochemistry textbooks.

The Significance of Sugar Chemistry for Fischer

Ascertaining Fischer's motives in his chemistry quickly becomes one of the significant challenges for anyone undertaking a study of his chemistry, for Fischer himself was quite reticent about making his motives public. He made no explicit statements about why he chose certain projects over others, and his autobiography is completely silent about his ultimate scientific goals, motivations or philosophy of science. His published work is also nearly free of any theoretical discussion about any of the fundamental theoretical problems in chemistry, making clear that Fischer was not greatly interested in chemical 'theory'. In a 1911 letter to the American chemist T.W. Richards, he confirmed this by writing: 'I do not derive much pleasure from theoretical things. The occupation with the Walden Inversion was rather a digression and recuperation from the extensive work on proteins. Moreover, so many limited heads have now jumped on this question, that the delight thereon is spoiled completely'.[63]

What Fischer meant by 'theory', however, must be considered carefully, because throughout all his projects, he was actually continually occupied with theoretical issues, from the classification of sugars and purines based on structure theory or the theory of the asymmetric carbon atom, to speculation on the nature of the interaction between sugar and enzyme molecules, or establishing the presence of an amide bond between amino acids in proteins. In these three projects, Fischer applied *established* theory to new groups of compounds. Fischer did not, on the other hand, develop or propose *novel* theoretical ideas. The two most prevalent examples of novel theory – the lock and key hypothesis and the attempted explanation of the Walden inversion – were cautious conclusions drawn from extensive study. In his Faraday lecture of

---

[62] Emil Fischer, 'Einfluß der Configuration', pp. 843–4.
[63] Emil Fischer to Theodore Richards, 11 November 1911, Fischer Papers. Translation in Lichtentaler, '100 Years "Schlüssel-Schloss-Prinzip"', p. 2372.

1907, he consciously invoked the memory of Faraday himself to illustrate his methodology:

> We cannot do better than strive to imitate the great example of Faraday, who always, with rare acumen, directed his attention to actual phenomena without allowing himself to be influenced by preconceived opinion, and *who in his theoretical conceptions gave expression only to observed facts.*[64]

The distinction between the use of established theory and the development of novel theory therefore gives us a good sense of what Fischer meant when he said that he did not like 'theory'.

Fischer's preference for the application of established theory is reflected in his extremely pragmatic use of structural and stereoformulas in his papers. He employed stereoformulas only sparingly, and then only when he saw a compelling need for them – indeed, his papers contained very few *structural* formulas. For Fischer, stereoformulas were isomer-counting devices for prediction and organization of the sugars. Unlike the stereochemists we have considered so far, Fischer neither addressed the nature of valence nor discussed any potential theoretical problems implied by the tetrahedral arrangement of valences. Reading his publications on the sugars in isolation, we would never even be aware of the potential theoretical problems with Van 't Hoff's model. Fischer's pragmatism also seems evident in his description of Van 't Hoff's theory. Although he repeatedly referred to the presence of 'asymmetric carbon atoms' in carbohydrates, there is no mention in his papers of the carbon atom as specifically *tetrahedral* until he introduced the famous projections derived from the rubber Friedländer models. Fischer obviously thought of the asymmetric carbon atom as tetrahedral and used tetrahedral models, but to a reader unfamiliar with Van 't Hoff's theory, an important detail of Van 't Hoff's theory was left out.

Finally, we must ask why Fischer turned to the chemistry of biological compounds. It seems likely that the fact that despite being significant for animal nutrition, little was known about their specific chemical nature. He repeated several times in review articles that the carbohydrates, next to proteins, were the most important nutrients (*Nährmaterial*) for animals, and in his 1890 lecture on the sugars, he made his motives more explicit:

> If I as the third [in the series of general speakers] venture to present you the modest [*schlichten*] results of an experimental investigation, so it happens in the conviction that the further development of hypotheses is prepared by the discovery of new facts, and, further, that organic chemistry grows by its hundredfold relationships to physiology, industry

---

[64] Emil Fischer, 'Synthetical Chemistry in its Relation to Biology', *J. Chem. Soc.*, 1907, *91*, 1747–65. Emphasis added.

and the requirements of everyday life and many other tasks, than by the development [*Ausbildung*] of its theory.[65]

This would complement the analysis given above, and would emphasize the application of established theory to unknown compounds.

Yet it also seems clear that Fischer chose certain compounds because of the chemical challenge they offered – how does one apply classical chemical methods to establish the structure and relationships between the known sugars, and eventually between the natural and synthetic sugars? The sugars, Fischer found, could not be completely classified by structure, but required the application of the theory of the asymmetric carbon atom for their classification. Establishing their configurations was the end result of a long, complex study of their genetic relationships:

> To the initiated, modern configurational formulas proclaim with the clarity and brevity of a mathematical expression both the already established metamorphoses of the simple sugars and the many still to be expected, and one may indeed say that at the moment, the morphology and systematics of the monosaccharides has been achieved.[66]

Earlier studies of Fischer's work have implicitly treated Fischer's work on the sugars as a 'confirmation' of Van 't Hoff's theory, which it certainly was. It was not, however, Fischer's explicit purpose in his sugar chemistry to confirm a specific hypothesis. Rather, he was looking for some way of establishing the 'morphology' and 'systematics' of the monosaccharides, and Van 't Hoff's theory happened to allow him to meet these requirements.

Chemically, Fischer was quite excited about the role that synthesis played in his study of the sugars, to the extent that he published the first artificial synthesis of a natural sugar. Beginning in 1889, when he and Tafel reported that α-acrose was indeed a sugar, Fischer mentioned repeatedly each instance of a synthetic sugar, and emphasized that even though many of his own syntheses were from other natural sugars, the relationships between the sugars and simpler starting materials (in particular the synthesis of α- and β-acrose from glycerin) allowed him to conclude that the completely artificial preparation of sugars was possible. In his 1890 lecture, Fischer placed his accomplishment firmly within the traditional dream of nineteenth-century organic chemistry to prepare compounds found in nature by synthesis, and that all of his previous work was only 'preparation for the synthesis of natural sugar', a 'thought [that] may be just as old as organic synthesis itself'.[67]

As Fischer's synthetic work proceeded, its biological significance also took on

---

[65] Emil Fischer, 'Synthesen in der Zuckergruppe', p. 1.
[66] Emil Fischer, 'Die Chemie der Kohlenhydrate und ihre Bedeutung für die Physiologie', pp. 96–115 in Fischer, *Gesammelte Werke*, vol. 5, p. 106.
[67] Fischer, "Synthesen in der Zuckergruppe', p. 13.

greater importance. The most famous result was the lock and key metaphor, but Fischer also speculated (despite his distaste for theory) numerous times on how plants form asymmetric sugar molecules. The earliest mention of biological mode of synthesis of sugars was in the last pages of his 1889 article on the synthesis of mannose and levulose, where he raised the question of asymmetric synthesis:

> A living organism's capability to prepare optically active substances that can never be obtained directly by chemical sytheses is such a curious fact that discovering its cause certainly constitutes a delightful problem for physiological research.[68]

He would mention this problem in various forms in later articles and lectures, but his most detailed conception of the solution to the problem of asymmetric synthesis in nature appeared in 1894, in his second comprehensive review article on the sugars:

> According to the plant physiologists, carbohydrate formation takes place in the chlorophyll granule, which itself consists entirely of optically active substances ... and that then, due to the already existing asymmetry of the entire molecule, the condensation to the sugars also takes place in an asymmetric fashion.
>
> Their asymmetry can thus be readily explained by the nature of the material from which they were produced. Of course, they also provide the material for new chlorophyll granules which in turn produce active sugar. In this manner, the optical activity propagates from molecule to molecule, as life itself does from cell to cell. Hence, it is not necessary to attribute the formation of optically active substances in the plant to asymmetric forces outside the organism, as Pasteur had supposed. The origin rather lies in the structure of the chlorophyll granule that generates the sugar, and with this conception, the difference between natural and artificial synthesis is completely eliminated.[69]

Because Fischer was so reticent about describing his motives, it is difficult to say with absolute certainty whether this was a deliberate contribution to the ongoing nineteenth-century debate between the chemical and biological theories of fermentation.[70] Yet, as we have seen, he was familiar with, and very interested in, the process of fermentation and with the work of Pasteur, at the least his book on beer fermentation. It seems unlikely that Fischer was totally unaware of the stakes involved, and even though he did not directly address proponents of the biological theory, his own statements support an unequivocal chemical and mechanistic view of fermentation and enzyme action, free of vital or asymmetric forces. In this light, we

---

[68] Emil Fischer, 'Synthese Mannose und Lävulose', pp. 352–3.
[69] Emil Fischer, "Synthesen in der Zuckergruppe II'. Translation in Lichtentaler, '100 Years "Schlüssel-Schloss-Prinzip" ', p. 2371.
[70] Joseph Fruton, *Molecules and Life: Historical Essays on the Interplay of Chemistry and Biology*, New York, Wiley, 1972, pp. 42–66. Fruton, *Proteins, Enzymes, Genes: the Interplay of Chemistry and Biology*, New Haven, CT, Yale University Press, 1999. Robert Kohler, 'The Enzyme Theory and the Origin of Biochemistry', *Isis*, 1973, 64, 181–96, Gerald Geison, 'Pasteur on Vital Versus Chemical Ferments: A Previously Unpublished Paper on the Inversion of Sugar', *Isis*, 1981, 72, 425–45.

can perhaps recast Fischer's ultimate motive in pursuing the chemistry of the sugars. Was Fischer interested in the chemical nature of the sugars, and therefore came across a chemical theory of enzyme action? Or was Fischer initially primarily interested in understanding the process of fermentation, studying the chemistry of the carbohydrates, the principal foodstuffs of yeasts, in order to reveal the nature of fermentation? Only a closer examination of Fischer's life can provide the answer.

Chapter 9

Alfred Werner and Coordination Chemistry, 1893–1914

> Chemistry must become the astronomy of the molecular world.
> Alfred Werner, 1905[1]

In the fall of 1896, Alfred Werner wrote two letters in which he gave his impression of a lecture given by Hantzsch at the recent Frankfurt *Naturforscherversammlung*. To his friend Arturo Miolati, Werner reported that 'Hantzsch lectured on nitrogen stereochemistry and also demonstrated with wire models and valences that a structural literalist certainly wouldn't be able to imitate him'.[2] Two weeks earlier, he wrote to Hantzsch about a specific discussion at the meeting:

> I am convinced that the gentleman concerned was lacking the necessary awareness of the relevant facts, but then he is such a structural literalist that endless discussions are required to make it clear to him that atoms are not balls equipped with wires.[3]

As we shall see in this chapter, these two passages reflect well Werner's own personal understanding of the meaning of chemical structures and molecular models. Werner used the term 'structural literalist' (*Strichchemiker*) privately to assess candidly those chemists who simply manipulated bond lines on paper without considering the reality of the atoms and bonds they represented.[4] Werner was emphatic that atoms were not really 'balls equipped with wires' that hooked onto other elements.

---

[1] Autograph quotation reproduced in George B. Kauffman, *Alfred Werner: Founder of Coordination Chemistry*, Berlin, Springer-Verlag, 1966.
[2] Alfred Werner to Arturo Miolati, 21 November 1896, copybook 1896–1900, Werner Archives, Handschriftenabteilung, Zentralbibliothek Zürich.
[3] Alfred Werner to Arthur Hantzsch, 6 November 1896, copybook 1896–1900, Werner Archives, Handschriftenabteilung, Zentralbibliothek Zürich.
[4] I have loosely translated Werner's word *Strichchemiker* as 'structural literalist' because the literal meaning, 'line chemist' or 'stroke chemist' is not meaningful in English and does not convey the pejorative aspect of Werner's usage. Another intriguing, although unsubstantiated, interpretation is that Werner might

Standard molecular models were misleading, according to Werner, because they implied the existence of valence as a directed attractive force, and because they assumed the constancy of the valence number. The known attractive forces were incapable of division, and the assumption of constant valence – so useful for organic compounds – led to difficulties in constructing formulas for inorganic compounds. He would make this clear in the opening sentence of the foreword to his *Neuere Anschauungen der anorganischen Chemie* (1905):

> The facts of inorganic chemistry speak ever more convincingly that the conceptions of valence developed from the constitutional relationships of carbon compounds do not allow a sufficient picture of the molecular construction of inorganic compounds.[5]

As he hoped to demonstrate, Werner's own coordination theory offered a consistent conception of valence that could be applied to both organic and inorganic compounds. Through the coordination theory, Werner would eventually re-create inorganic chemistry, and almost single-handedly create inorganic stereochemistry.

Werner supposed that groups could arrange themselves in an octahedron around a central metal atom and that these groups could be freely replaced by other groups to create whole series of compounds. He saw immediately that such an arrangement would lead to spatial isomers, and shortly after the initial publication of his theory he realized that groups could be arranged around the metal atom asymmetrically to create inorganic compounds capable of optical activity. Werner was relentless in the promotion and defense of the octahedron, from its initial inception in 1893 to the final isolation of optically active metal ammines between 1911 and 1914. In a spontaneous 1913 speech given on the occasion of receiving the Nobel Prize, Werner recalled that an unidentified 'northern colleague' referred to his coordination theory as an 'ingenious impudence' (*geniale Frechheit*).[6] Whether he was truly impudent or simply bold and self-confident (which Werner certainly was), this characterization succinctly described both Werner's coordination theory and his scientific persona.

Werner was born in 1866 in the strongly independent city of Mulhouse in the Alsace region of France, to Jean-Adam Werner, a foundry worker, and Salomé Jeanette Tesché.[7] From the time he was five, Werner's homeland was annexed by the new

---

have been making a pun by comparing *Strichchemiker* to *Strichdamen* (prostitutes). Werner's use of the word would seem to fit his personality, but the analogy does not seem quite appropriate.

[5] Alfred Werner, *Neuere Anschauungen auf der anorganischen Chemie*, Braunschweig, Vieweg, 1905, p. vi.

[6] George B. Kauffman, 'An Ingenious Impudence: Alfred Werner's Coordination Theory', *J. Chem. Ed.*, 1976, 53, 445–6.

[7] The summaries of Werner's life given in this chapter are taken from Kauffman, *Alfred Werner*.

German Empire as the result of the Franco–Prussian War, and anti-German sentiment in the region ran high throughout Werner's youth, a factor that may have influenced him to study chemistry in Switzerland at the Zürich Polytechnical Institute after his compulsory year of service in the German Army in Karlsruhe. In Zürich Werner studied chemistry with Georg Lunge and Arthur Hantzsch, did his *Diplomarbeit* on simple preparations of carbon disulfide, phenylhydrazine and other organic compounds, and became an assistant in Lunge's laboratory for technical chemistry.[8] At the same time, he began research for the doctorate in Hantzsch's laboratory, which he completed in late 1889. He passed his doctoral examinations in July 1890, although the first part of the dissertation on the stereochemistry of nitrogen (discussed in Chapter 7) had appeared in the *Berichte* six months earlier.

In October 1891, Werner submitted an innovative treatise on the nature of affinity, *Beiträge zur Theorie der Affinität und Valenz*, as a *Habilitationsschrift* to the faculty of the *Polytechnikum*. In the *Beiträge*, Werner agreed with his mentor Hantzsch that valence was simply a number, and with Lossen that valence units could not be a single directed force. Indeed, the motive behind Werner's treatise was the construction of a model of affinity that would avoid the implication that valence was a physically curious concept. Werner proposed a non-mechanical conception of the atom in which chemical affinity was an indivisible 'attractive force acting equally from the center of the atom toward all parts of its spherical surface', much like light from the sun.[9] The four radicals attached to the carbon atom 'arrange themselves in the mutual position of the corners of a regular tetrahedron, because in this way the greatest exchange of affinity between them and the carbon atom, i.e., the greatest bonding ability occurs'.[10] In the analogy to sunlight, they each received the greatest amount of light. In other words, the affinity of the carbon atom was distributed equally when the radicals occupied a symmetrical position. The net result was an apparent division of chemical affinity into units of valence. Valence was not a unit of directed force, but simply the effect of sharing affinity between the central atom and the surrounding radicals. Werner's model was one of the soundest reconciliations of the apparent division of affinity into separate units implied by chemical and stereochemical theory with the physical requirement of an indivisible attractive force. The model was also comprehensive, providing explanations not only for the tetrahedral arrangement of valences in carbon, but also for the process of racemization, the interconversion of unsaturated molecules, and the stereochemistry of nitrogen.

Before the *Beiträge* was accepted, Werner studied with Marcellin Berthelot in

---

[8] Kauffman, *Alfred Werner*, p. 14.
[9] Alfred Werner, 'Beiträge zur Theorie der Affinität und Valenz', *Vier. Zür. naturf. Ges.*, 1891, 36, 129–69, translation in George B. Kauffman, 'Alfred Werner's Habilitationsschrift', *Chymia*, 1967, 12, 183–216, p. 191.
[10] Kauffman, 'Alfred Werner's Habilitationsschrift', p. 191.

Paris during the winter semester of 1891/1892, a period in Werner's life about which little is known. On the successful acceptance of his *Habilitationsschrifft* in January 1892, Werner returned to the Zürich Polytechnic, where he gave an inaugural lecture on the benzene theory that employed his novel theory of valence and gave his first course, on the 'Atomic Theory', during the summer semester of 1892.[11]

Within three years of beginning his doctoral dissertation research, Werner had established himself as an extraordinarily original thinker who was interested in constructing a consistent theoretical but qualitative foundation for the concepts of valence and affinity. Judging by the originality of Werner's work completed at a very young age, he displayed a natural aptitude for both theoretical and practical chemistry. He might have done well working under nearly any chemist, but he was also fortunate to have chosen Hantzsch as his mentor, as Hantzsch actively encouraged and promoted independent and original thought among the staff of his laboratory. As revealed in an 1891 letter to his parents, Werner was an extremely ambitious, self-confident young man, already entirely devoted to a life in chemistry:

> I am beginning to take my place among the chemists of the time, and if heaven preserves my health, I intend to surpass them one by one, for glory is not an empty word; it is the personal satisfaction of a man who needs it as a stimulant in his moments of weakness ... My illusions have never left me, but man needs them. For without illusions there is no ideal. I am becoming an enthusiast with age; I often remain for long moments in ecstasy before the beauties of my science, for the more I advance into its mysteries, the more it seems to me grand, sublime, almost too beautiful for a mere mortal.[12]

Accompanying this ambition and devotion to chemistry was a tendency towards extreme overwork. Although he married and had hobbies, he rarely took extended vacations, and was almost invariably in the laboratory seven days a week – he was usually the first to arrive in the morning and the last to leave at night. This physically demanding regimen, along with heavy drinking and cigar smoking, led to a period of ill health between 1900 and 1904, and also likely led to the rapid deterioration of his health beginning in 1914, and his premature death in 1919 at the age of 53.

The Coordination Theory, 1893–1907

The Metal-ammonia Salts

The metal-ammonia salts were brilliantly colored compounds known since the late eighteenth century that had attracted a great deal of attention from chemists. The

---

[11] Kauffman, *Alfred Werner*, pp. 24–5.
[12] Werner to his parents, 11 May 1891, ibid., p. 19.

number of these compounds had increased dramatically in the first half of the nineteenth century, prompting the Swedish chemist Christian Wilhelm Blomstrand (1826–1897) to write in 1869:

> There is hardly any other class of bodies about which so much has been written, as about the metal-containing ammonias. We have entire books ... full of theoretical debates that should serve to interpret them, in addition to the rich journal literature that has obligingly supplied material for constructing formulas.[13]

These salts were ordinary chlorine or nitro salts of transition metals such as cobalt, platinum or rhodium whose molecular formula also included a distinct number of individual molecules of ammonia. In 1852, Fremy proposed a nomenclature for the cobalt series based on the color of the parent compound.[14] The series for the cobalt ammine chlorides (given in formulas valid after 1890) had the following members:

Luteo	Yellow	$Co(NH_3)_6Cl_3$
Purpureo	Purple	$Co(NH_3)_5Cl_3$
Praseo	Green	$Co(NH_3)_4Cl_3$
Hexammine salts		$Co(NH_3)_6Cl_3$
Roseo	Rose-red	$Co(NH_3)_5(H_2O)Cl_3$
Violeo	Violet	$Co(NH_3)_4Cl_3$

There were two peculiar chemical properties of these compounds that were difficult to explain. First, the ammonia in these compounds had lost all properties of a base and could not be neutralized by treatment with acid. Second, while in the luteo series all three chlorine atoms could be precipitated as silver chloride by treatment with silver nitrate, in the purpureo and praseo series not all of the chlorines showed this property. These peculiar chemical properties were accompanied by the theoretical question of their constitution. Was cobalt trivalent in cobalt chloride, nonavalent in the luteo series, and octavalent in the purpureo series? Or were these molecules examples of 'molecular compounds', which consisted of separate molecules (that is, ordinary cobalt chloride and six molecules of ammonia in the luteo series) held together by intermolecular forces? Answering 'yes' to either question was largely unsatisfactory. It seemed unwarranted, even if valence was assumed to be variable, to assume that cobalt could have a valence of 3, 6, 7, 8, or 9. Kekulé's concept of molecular compounds seemed equally unlikely as a desperate attempt to maintain the doctrine of constant valence that had been employed so usefully for organic compounds.

---

[13] Christian W. Blomstrand, *Die Chemie der Jetztzeit*, Heidelberg, Winter, 1869, p. 280.
[14] Colin A. Russell, *The History of Valency*, New York, Humanities Press, 1971, George B. Kauffman, 'Early Experimental Studies of Cobalt-ammines', *Isis*, 1977, 68, 393–403.

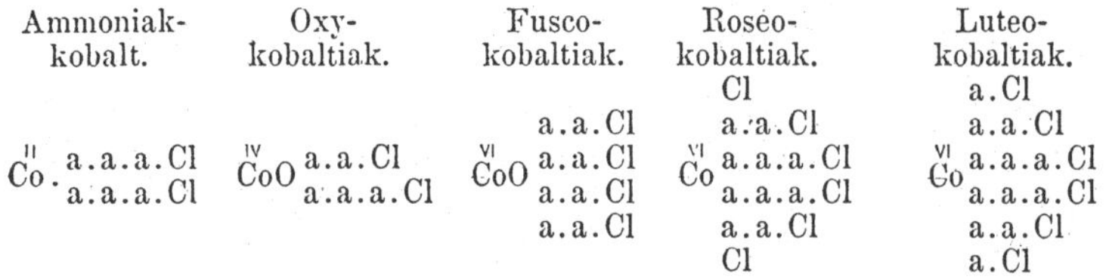

Ammoniak-kobalt.	Oxy-kobaltiak.	Fusco-kobaltiak.	Roseo-kobaltiak.	Luteo-kobaltiak.
$\overset{II}{Co} \cdot \begin{matrix} a.a.a.Cl \\ a.a.a.Cl \end{matrix}$	$\overset{IV}{Co}O \begin{matrix} a.a.Cl \\ a.a.a.Cl \end{matrix}$	$\overset{VI}{Co}O \begin{matrix} a.a.Cl \\ a.a.Cl \\ a.a.Cl \\ a.a.Cl \end{matrix}$	$\overset{VI}{Co} \begin{matrix} a.a.Cl \\ a.a.a.Cl \\ a.a.a.Cl \\ a.a.Cl \\ Cl \end{matrix}$	$\overset{VI}{Co} \begin{matrix} a.Cl \\ a.a.Cl \\ a.a.a.Cl \\ a.a.Cl \\ a.a.Cl \\ a.Cl \end{matrix}$

9.1 Examples of Blomstrand's formulas for the cobalt ammine complexes containing cobalt of varying valence. The luteo salt is on the far right. Blomstrand used 'a' as an abbreviation for the ammonia (NH$_3$) radical.

A reasonably satisfactory solution to this dilemma was offered by Blomstrand in *Die Chemie der Jetztzeit* (1869), an extensive monograph in which he attempted to reconcile Berzelian dualism with the newer unitary theory that had been so successful in organic chemistry. Throughout his *Chemie*, Blomstrand strongly opposed Kekulé's doctrine of constant valence and assumed that 'variable valence is a basic property of atoms'. He declared that: 'The main task of modern chemistry has become the explanation on atomistic grounds of the compounds previously considered, more or less resolutely, as molecular by the saturation capacities of their basic constituents.'[15] The concept of molecular compounds was unacceptable because it did not formulate structures based on the principles of valence or show all bonds in the molecule.

Blomstrand applied these principles to the metal-ammonia salts to construct rational formulas that would account for all chemical bonds in the molecule. Borrowing terminology from Berzelius, he supposed that the metal atom and the accompanying ammonia molecules were 'copulated' (*gepaarte*) and therefore the equivalent of radicals in organic compounds. Ammonia was therefore combined with the cobalt atom in a way that suppressed its basicity, and the resulting complex was then combined with chlorine atoms. The chlorine atom could either be bound directly to the cobalt atom, or bound indirectly through a chain of pentavalent nitrogen (Figure 9.1).[16] Blomstrand also noted that his formulas would differentiate between two chemically different chlorine atoms in the metal ammonia salts. Those bound directly to the metal atom would not precipitate with silver nitrate, whereas those

---

[15] Blomstrand, *Chemie der Jetztzeit*, note 13, pp. 185 and 127. Blomstrand's conception of the maximum valence for the elements apparently also considered their spatial arrangement. George B. Kauffman, 'Christian Wilhelm Blomstrand (1826–1897): Swedish Chemist and Mineralogist', *Ann. Sci.*, 1975, *32*, 13–37, p. 32.

[16] Kauffman, 'Christian Wilhelm Blomstrand', L. Tansjö, 'While waiting for Werner: Chemistry in Chains', pp. 35–40 in George B. Kauffman, ed., *Coordination Chemistry: A Century of Progress*, Washington, DC, American Chemical Society, 1994. P.S. Cohen, 'Effect of the Fixity of Ideas on the Werner-Jørgensen Controversy', *Advances in Chemistry*, 1967, *62*, 8–40.

bound to the end of the ammonia chain would precipitate because of their more distant proximity to the metal atom.

Blomstrand's chain theory would become the starting point for the Danish chemist Sophus Mads Jørgensen (1837–1914), who in 41 papers authored between 1878 and 1899 prepared and studied hundreds of the metal-ammonia salts of cobalt, chromium and rhodium. Jørgensen's education and his entire career in chemistry took place at the University of Copenhagen, where he was professor of chemistry from 1887 to 1908. Although he was eleven years younger than Blomstrand, the two became close friends, exchanging 78 letters between 1870 and 1897.[17] This friendship seems based on their similar attitudes towards chemistry, which included an empirically based methodology opposed to excessive speculation.

Jørgensen's work on the metal-ammines before 1893 consisted of three principal contributions. First, in 1890, Jørgensen used conductivity and freezing point depressions to establish a consistent molecular weight for the cobalt series, giving each member one cobalt atom and a maximum of six ammonia molecules. The formula for the luteo salt therefore became $Co(NH_3)_6Cl_3$.[18] Second, he showed that not only ammonia, but organic amines could participate in chain-formation, demonstrating that the complexes did not involve replacement of hydrogens in the ammonium (NH_3) radical.[19] Third, Jørgensen modified Blomstrand's initial theory by supposing that the chains of nitrogen atoms could vary in length from one to four ammonia molecules. The luteo and purpureo salts therefore were given the structures shown in Figure 9.2.

At heart an empiricist, Jørgensen did not use his formulas as predictive tools or create an overall classificatory scheme for the metal-ammines. Rather, he appeared to treat each compound individually, assigning a structure to it based on the overall principles of Blomstrand's chain theory, making slight modifications where he thought it necessary. This overall approach could not be more different from Werner's

---

[17] George B. Kauffman, 'Christian Wilhelm Blomstrand (1826–1897) and Sophus Mads Jørgensen (1837–1914): Their Correspondence from 1870 to 1897', *Centaurus*, 1977, *21*, 44–63. Kauffman, 'Sophus Mads Jørgensen (1837–1914): A Chapter in Coordination Chemistry History', *J. Chem. Ed.*, 1959, *36*, 521–7. Kauffman, 'Sophus Mads Jørgensen and the Werner-Jørgensen Controversy', *Chymia*, 1960, *6*, 180–204. Kauffman, 'Jørgensen, Sophus Mads', in Charles Gillispie, ed., *Dictionary of Scientific Biography*, New York, Scribner's, 1973.

[18] Kauffman writes that Jørgensen established simply that the molecular weight of the cobalt ammines should be halved. Examining Blomstrand's original text in *Chemie der Jetztzeit*, however, indicates that Blomstrand did not give a consistent atomic weight to the cobalt atom and gave it a valence of three or six, depending on the class of salt. Not all of Blomstrand's formulas, in other words, show two cobalt atoms. Kauffman, 'Sophus Mads Jørgensen and the Werner-Jørgensen Controversy', p. 186.

[19] Alfred Werner, 'Beitrag zur Konstitution anorganischer Verbindungen', *Zeit. anorg. Chem.*, 1893, *3*, 267–330. Translation in George Kauffman, *Classics in Coordination Chemistry, Part 1: The Selected Papers of Alfred Werner*, New York, Dover, 1968, pp. 12–13.

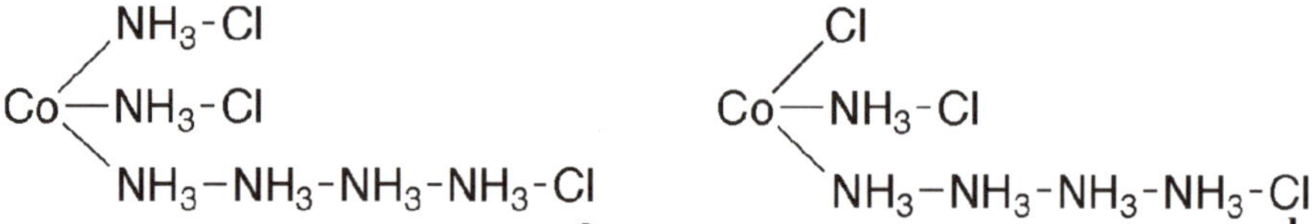

$$\text{Co} \begin{matrix} \diagup \text{NH}_3\text{-Cl} \\ - \text{NH}_3\text{-Cl} \\ \diagdown \text{NH}_3-\text{NH}_3-\text{NH}_3-\text{NH}_3\text{-Cl} \end{matrix}$$
a

$$\text{Co} \begin{matrix} \diagup \text{Cl} \\ - \text{NH}_3\text{-Cl} \\ \diagdown \text{NH}_3-\text{NH}_3-\text{NH}_3-\text{NH}_3\text{-Cl} \end{matrix}$$
b

9.2 The structures for the luteo (a) and purpureo (b) compounds, according to Jørgensen.

use of formulas as predictive tools and tendency towards speculation, against which Jørgensen found himself directing his energies beginning in 1893.

Werner's 'Beitrag zur Konstitution der anorganischen Verbindungen' (1893)

It is against this background that Werner entered the world of inorganic chemistry with a 63-page article in the third volume of the newly established *Zeitschrift für anorganische Chemie* entitled 'Contribution to the Constitution of Inorganic Compounds'. The appearance of a paper by Werner on purely inorganic chemistry was sudden and unexpected, given Werner's training as an organic chemist, and the precise origins of Werner's move from organic to inorganic chemistry are unclear. Although preserved notes show that Werner was aware of the existence of metal-ammonia complexes as early as February 1890, there is no record of Werner having any interest in inorganic chemistry much before the publication of this 1893 paper. Werner's assistant Paul Pfeiffer related that Werner became acquainted with the literature on the metal-ammine salts while writing a lecture, when he realized that conventional valence theory was inadequate to explain their constitution. Pfeiffer also noted that Werner's solution came in a 'flash' at two in the morning. Werner immediately sat down to write, and by five the next afternoon, 'the essential points of the coordination theory were achieved'.[20] Given the known reputation of Werner's speed as an author and devotion to work, this story does not seem unlikely. As Kauffman has noted, the lecture mentioned by Pfeiffer was prepared either for Werner's first course on Atomic Theory given in the summer semester of 1892, or perhaps more likely, for his second course, '*Ausgewählte Kapitel der anorganischen Chemie*', which began in the fall of 1892. Because Werner submitted the final article to the *Zeitschrift* in December 1892, he had evidently developed his theory and prepared it for publication during an extremely short time period of either two or six months.

---

[20] Kauffman, *Alfred Werner*, pp. 31 and 30–3.

If Werner was unfamiliar with the metal-ammine complexes before mid-1892, or even later, the resulting paper demonstrates that he had acquired a good, if not thorough, familiarity with the literature surrounding the known metal-ammine complexes. The article began with a five-page general overview of the types of known metal-ammonia salts, which Werner sorted into two groups: those that contained at most six ammonia molecules to one metal atom, and those that contained at most four ammonia molecules to one metal atom.[21] These two classes could further be subdivided into those of various metals with differing valences, such as the trivalent cobalt, tetravalent platinum or divalent nickel. In these complexes, individual ammonia molecules could be substituted by other molecules 'that behave in a manner similar to ammonia', such as water, but in a point Werner would continue to stress, '*the total number of such molecules added to the metal salt, however, remains constantly six*'.[22] After reviewing the possible constitutions for these compounds that had previously appeared in the literature, Werner concluded the introduction in his typically self-confident style:

> Chemists today, in the case of the metal-ammonia salts, therefore find themselves forced either to assume, with Blomstrand-Jørgensen, nitrogen–hydrogen chains copied from hydrocarbon chains or to conceive of these compounds as molecular compounds, that is to substitute a beautiful word for a confused concept.[23]

The bulk of the paper was divided into three sections. Part A considered the metals that contained a maximum of six ammonia molecules. Part B considered those metals that could contain a maximum of four ammonia molecules, and Part C contained a discussion on the 'Cause of the Peculiar Constitution of Inorganic Compounds', in which he introduced the coordination theory for the first time. Lack of space prevents us form considering all of Werner's lengthy argument, so I shall here focus specifically on two aspects of Werner's paper: (1) his discussion of the series of metal-ammines derived from the generic complex $M(NH_3)_6X_3$ and the introduction of the octahedron, both given in Part A, and (2) the introduction of the coordination theory given in Part C.

In a subsection of Part A, Werner described the 'Formation of compounds poorer in ammonia from metal-ammonia salts of the general formula $M(NH_3)_6X_3$'. He described the well-known property of the luteo salts ($M(NH_3)_6X_3$) to lose ammonia on heating to give purpureo salts ($M(NH_3)_5X_3$), and the well-known 'change in function of one acid residue X', upon the loss of ammonia, specifically that it lost the property of acting like an ion: that is one chlorine atom could no longer be precipitated

---

[21] Werner also included a third group, those 'in which the ratio of hydrogen to a nitrogen is less than that in ammonia', which were not discussed in the paper.
[22] Werner, 'Beitrag zur Konstitution anorganischer Verbindungen', in Kauffman, *Classics*, p. 10.
[23] Ibid., p. 13.

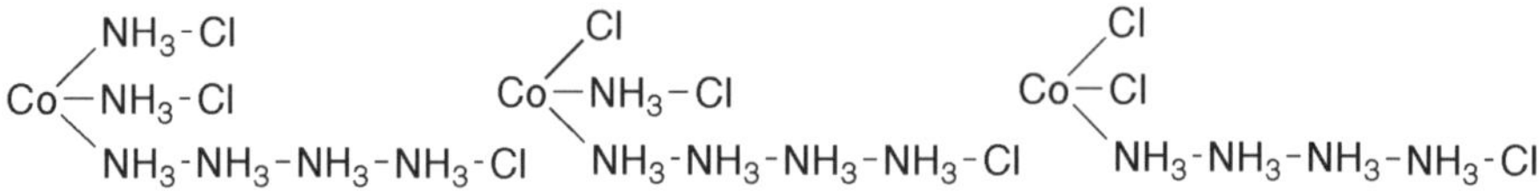

```
        NH₃-Cl                    Cl                        Cl
       /                         /                         /
    Co—NH₃-Cl                 Co—NH₃-Cl                 Co—Cl
       \                         \                         \
        NH₃-NH₃-NH₃-NH₃-Cl        NH₃-NH₃-NH₃-NH₃-Cl        NH₃-NH₃-NH₃-NH₃-Cl

           Luteo                    Purpureo                    Praseo          a
```

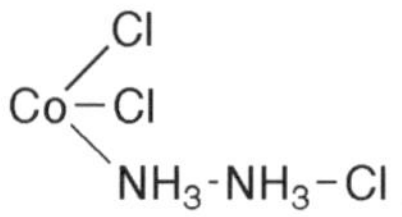

```
                                    Cl
                                   /
                                Co—Cl
                                   \
                                    NH₃-NH₃-Cl
                                                b
```

9.3 (a) Jørgensen's structures for the cobalt series in which the number of ammonia molecules decreases. (b) The structure (according to Jørgensen's theory) that results from Werner's thought experiment of removing four ammonia molecules from the luteo compound. Werner thought this structure to be 'highly peculiar and unlikely'.

as silver chloride by treating the compound with silver nitrate. It was also well known that the purpureo salts could also lose a molecule of ammonia to yield the praseo salts ($M(NH_3)_4X_3$), in which only one of the three chloride ions could be precipitated. According to the Jørgensen-Blomstrand theory, the formulas for these three compounds would then be as shown in Figure 9.3(a).

Werner continued this analysis by a thought experiment: what happens if ammonia molecules are continually removed from the metal, and how does the resulting sequence of formulas match the number of known compounds? Loss of ammonia from the praseo salt should result in a compound with the complex $M(NH_3)_3X_3$, which according to Werner should entirely lose the properties of a salt. Werner cited the iridum compound $Ir(NH_3)_3Cl_3$, recently reported by W. Palmaer, as an example. Unlike the praseo and luteo compounds, none of the chlorine atoms in this complex were exchanged for sulfate ions on boiling with sulfuric acid. This particular complex with three ammonia molecules, Werner noted, could no longer be explained by the Blomstrand-Jørgensen chain theory because removal of an ammonia molecule from the chain in the praseo formula did not change the position of the chlorine atom. According to the theory, it should remain 'distant' from the metal atom and therefore be capable of exchange.

Loss of a fourth ammonia molecule would yield a compound, which according to the Blomstrand-Jørgensen theory, would have the structure shown in Figure 9.3(b). Werner found this formula peculiar and unlikely: 'A compound of formula $M(NH_3)_2X_3$ does not exist at all; instead, with the loss of a further ammonia molecule, a simultaneous replacement of this molecule by an acid radical occurs: a complex

[M(NH$_3$)$_2$X$_4$] is formed.'[24] Removing the remaining two ammonia molecules would then result in further replacement by ions to give the two complexes M(NH$_3$)X$_5$ and MX$_6$R$_3$. Examples of the former were still unknown, but for the latter, Werner listed 18 examples with 6 metals and 5 different ions.

Werner's thought experiment for the M(NH$_3$)$_6$ complex therefore led to six possible series of metal-ammonia salts for each metal atom that was capable of binding to six ammonia molecules. In many, but not all, of the members of the series, Werner could point to specific known compounds as examples of each group. The replacement of ammonia with ions in the last three steps of the process convinced Werner that:

> in all the transitions caused by loss of ammonia we are dealing not merely with a loss of ammonia molecules, but rather with an actual substitution of ammonia molecules by acid residues, for at the moment when the acid residues found in the molecule have been substituted, a further acid residue enters the molecular complex from the outside.[25]

Werner continued his analysis with the tetravalent metal-ammine compounds, M(NH$_3$)$_6$X$_4$, and the divalent metal-ammines M(NH$_3$)$_6$Cl$_2$, comparing the existing and nonexisting members of the known metal-ammines to the predicting classes of complexes based on the sequential removal of ammonia molecules. Another entire section was devoted to looking at a similar sequence in which ammonia was sequentially replaced by water to give a series of six complexes, the last of which were the metal hydrates M(H$_2$O)$_6$X$_3$, in which all six ammonia molecules were replaced by six water molecules. These compounds were none other than the well known hydrates of the metal salts, which had been previously unrelated to the metal ammines. Werner was again attracted to the recurrence of the number six in the hydrates:

> The ever-recurring number of six molecules of water of crystallization in so many metal salts and of six molecules of ammonia in the corresponding ammonia compounds, which appears to be quite independent of the nature of the acids bound to the metal, cannot be a coincidence but must have a deeper cause. Since in double salts these metals also very frequently contain six definite radicals directly bound to them, a fact which may be shown again by the following examples:
>
> Fe(OH)$_6$Cl$_3$ Fe(CN)$_6$K$_3$ FeFl$_6$K$_3$ FeCl$_6$K$_3$
>
> we conclude that *the property of metal salts to bind six water molecules or six ammonia molecules has the same cause, which is inherent in the metal atom.*[26]

---

[24] Ibid., pp. 19–20.
[25] Ibid., p. 21.
[26] Ibid., p. 34.

Furthermore, Werner noted, water showed a complete analogy in its 'function', that is, in the complex $Co(H_2O)_5Cl_3$, only two of the chlorine atoms were ionic.

In Section VI of Part A, 32 pages into the article, Werner offered 'A Concept of the Constitution and Configuration of Radicals MR_6', which began with the statement:

> If, in the hydrates, metal-ammonia salts, and so on, we are to assume radicals which are formed in such a manner that six water molecules, six ammonia molecules, or six monovalent groups are arranged around the metal atom, then we may ask with what spatial configuration we can represent the entire molecular complex.
>
> If we think of the metal atom as the center of the whole system, then we can most simply place the six molecules which are bound to it at the corners of an octahedron.[27]

It is curious that Werner did not suggest, consider or reject any alternative spatial arrangement, as there are at least two other arrangements of six atoms around a central metal atom.[28] Nor did Werner discuss the position of any of the acid residues not attached to the octahedron. Rather, he immediately discussed the implications of the octahedral formula for the praseo compounds with the complex $M(NH_3)_4X_2$. Assuming an octahedral arrangement of groups, this complex would be capable of existing in two spatial forms (shown in Figure 9.4(a)), with the acid residues situated either axially (*trans*) or along the same edge (*cis*).

This predicted isomerism matched a pair of known isomers reported by Jørgensen, the isomeric dichloro bis ethylenediaminecobalt (III) complexes. These two compounds, Werner noted:

> behave exactly the same chemically; of the three acid residues only one functions as an ion. However, the two series differ very characteristically in their colors. One series is green, the so-called praseo salts; the others are violet and are called violeo salts.
>
> This interesting isomerism is the first confirmation of the conclusions resulting from the octahedral formula.[29]

Jørgensen had explained this isomerism by invoking a difference in the three cobalt valences – an explanation, Werner pointed out, that involves 'a somewhat obscure

---

[27] Ibid., p. 47.

[28] As Kauffman has noted in several places, Werner did offer an argument based on counting the isomers predicted by alternate geometries. Kauffman did not note, however, that this method appeared in 1895, in Werner's fourth article on metal ammines. He did not use it in his initial paper. Werner, 'Beitrag zur Konstitution anorganischer Verbindungen. IV', *Zeit. anorg. Chem.*, 1895, *9*, 382–417, pp. 391–2.

[29] Werner, 'Beitrag zur Konstitution anorganischer Verbindungen', in Kauffman, *Classics*, p. 49.

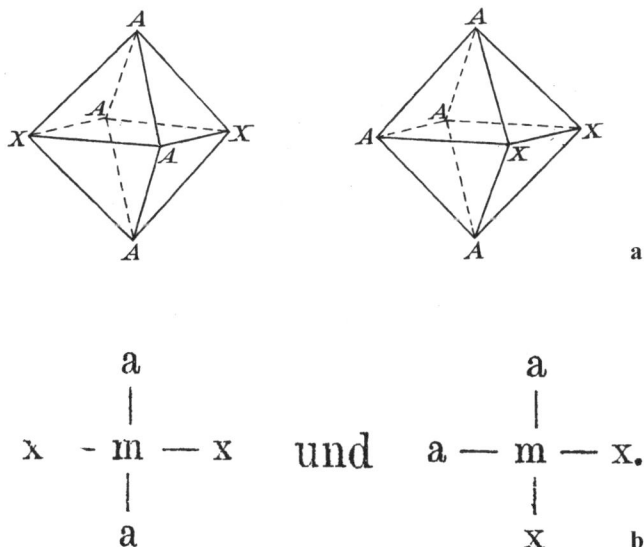

9.4 (a) Werner's depiction of the resulting *cis–trans* isomerism in a disubstituted octahedron. (b) The *cis–trans* isomerism resulting from the square planar arrangement of atoms in the MA₄ type.

concept, because valence itself is not a clear concept'.[30] Despite the complexity of the molecules and his novel explanation, Werner did not draw stereoformulas for the two possible isomers of this complex. Werner introduced a similar isomerism in the platinum series $Pt(NH_3)_2X_4$, and concluded that 'a stereochemistry of cobalt compounds and platinum compounds appears alongside the stereochemistry of carbon compounds and nitrogen compounds'.[31]

In Part B of the essay, Werner submitted metal-ammonia compounds containing the complex MA₄ to the same analysis, successively removing molecules of ammonia and comparing the resulting series of formulas with known metal-ammonia compounds. He considered these compounds as 'fragments of the MA₆ type compounds, and in many cases their modes of formation confirm this view'.[32] In a section on 'A concept of the Configuration of the Radicals MA₄', Werner repeated that the MA₄ radicals were formed:

---

[30] Ibid., p. 49.
[31] Ibid., p. 50. Werner also attempted to explain the existence of isomeric roseo salts with the complex $M(NH_3)_5H_2O$ as spatial isomers in which the ions lay in one of two possible planes in the octahedron.
[32] Ibid., p. 54.

by the loss of two A groups [from MA_6], and indeed in such a manner that no change in function of the acid residues occurs. Thus the release of the two A groups must occur in such a manner that the metal atom in the MA_4X_2 molecules remains in that plane in which the acid residues lie, and is protected from direct bonding with the acid residues by the four A radicals.[33]

The simplest way for two residues to leave the octahedron and leave the acid residues intact, would be 'through the departure of the two axially located residues a_1 and a_2'. The result was a square planar configuration of groups around the metal. If the metal were disubstituted, such a configuration would predict two spatial isomers (Figure 9.4(b)). This prediction was identical to Van 't Hoff's prediction in 1874 of the possible spatial isomers of disubstituted carbon atoms.

In Part C, Werner provided some additional thoughts on the 'Cause of the Peculiar Constitution of Inorganic Compounds'. Strictly speaking, Werner had already given a cause – the octahedral or square planar arrangement of groups around the metal atom. A specific metal-ammonia compound could be differentiated from others not only by the number and kind of groups surrounding the metal atom, but also in some cases by the specific three-dimensional configuration – *cis* or *trans* – of the groups around the metal atom. But Werner did not seem to regard this claim as causal or theoretical, because the 'developments presented in this paper on the constitution of inorganic compounds have been developed wholly from the facts', and the view that either four or six ammonia molecules are directly bound to the metal atom 'forces itself upon us so prominently when we consider comparatively the compounds corresponding to these radicals that its correctness cannot be doubted'.[34]

Werner had therefore already come to regard the octahedron and square planar arrangements as the 'correct' picture of the complexes in the metal-ammonia salts and therefore sought a deeper underlying reason for why and how groups would arrange themselves around the metal atoms in this fashion. Werner initially made clear that valence theory, and its implied directive forces, was entirely inadequate for explaining how a di- or trivalent atom could form bonds to six atoms or groups. Werner then recast the problem into the valence of the metal complex:

> [The] metal atom is a sphere, with the six complexes bound to it defining a sphere around it, and everything else in a second sphere. The valence of the complex is therefore the difference between the valence of the metal atom and the number of monovalent groups in the first sphere.[35]

---

[33] Ibid.
[34] Ibid., p. 79.
[35] Ibid., p. 81.

Water and ammonia, according to Werner, had the 'quite peculiar' function of 'shifting ... the force of affinity of the metal atom from the first sphere, in which it generally achieves activity, to a sphere further removed from the metal atom'. Werner drew an analogy to electricity:

> If we imagine the metal atoms as a sphere charged with positive electricity and the water molecules as a neutral shell covering this sphere, then in this neutral shell negative electricity will accumulate on the inner shell surface, while on the outer surface, positive electricity will accumulate. Thus the outer shell surface will be able to bring considerable amounts of positive electricity into action toward the externally located space, as the metal atom had done previously. If the shell was not completely neutral, but rather, for example, weakly positive, then as a result of this polarization the quantity of positive electricity thus coming into action toward the second sphere will appear to be strengthened in the same amount.[36]

In order to differentiate between the number of groups that could bind to an atom from the valence number proper, Werner introduced the concept of the 'coordination number' as the number of groups bound to the metal to form a complex. The valence number and coordination number were independent of one another, and both varied from element to element. For example, cobalt had a valence number of 3 and a coordination number of 6. Carbon had a valence of 4, and nitrogen and boron a valence of 3, but all three elements shared the same coordination number of 4, and 'it appears probable that this accidental coinciding of the two numerical values for carbon has prevented the differentiation of the two concepts'.[37] Compounds with saturated valences could also complete their coordination number: for example as Werner noted, trimethyl boron ($B(CH_3)_3$) combined with ammonia (NH_3) to form a stable complex, in which the nitrogen and boron atoms were bound to four atoms, $(CH_3)_3BNH_3$. Ammonia molecules were found in the coordination sphere of metal atoms precisely because this completed their coordination number. Werner suggested that 'perhaps this concept is destined to serve as a basis for the theory of the constitution of inorganic compounds, just as valence theory formed the basis for the constitutional theory of carbon compounds'.

The octahedron and coordination theory were natural extensions of the ideas Werner had already expressed in his *Habilitationsschrift*. Although devoted to problems in organic chemistry, Werner's theory of affinity was easily adapted to inorganic compounds, and the metal-ammonia compounds in particular. Assuming an octahedral arrangement was a way of maximizing the chemical affinity of the metal atom among six different groups, and Werner simply modified this concept to include two spheres of affinity, later termed the 'primary' and 'secondary' valence

---

[36] Ibid., p. 82.
[37] Ibid., p. 85.

(*Haupt-* and *Nebenvalenz*). Werner's initial aim in writing his *Habilitationsschrift* was to construct a model of affinity in which the valence number was the consequence of a non-directive attractive force, and Werner made clear that his new model for the metal-ammonia complexes could not be explained by traditional concepts of valence. Because the *Habilitationsschrift* was originally published in the somewhat obscure *Vierteljahrschrift der Zürcher Naturforschenden Gesellschaft*, it seems unlikely that many chemists were familiar with it, therefore the concluding remarks in the 'Beitrag' contained the first widely distributed version of Werner's theory of affinity.[38]

Having looked at the overall structure of Werner's theory as it first appeared, we can step back and illustrate some general characteristics. First, in method and aims, there is a broad resemblance between the 'Beitrag' and Van 't Hoff's works which we considered in Chapter 3. Like Van 't Hoff, Werner did not introduce any new experimental evidence for his theory, but reinterpreted the known constitutional formulas and properties of the metal-ammonia compounds and organized them under the new theoretical framework offered by the octahedral or square planar configuration of atoms. Werner also relied on the method of isomer counting used by Van 't Hoff – using chemical formulas to predict the existence of various compounds and comparing that number to the known compounds. As we have seen, Werner's prediction of isomers in the square planar complexes was identical to Van 't Hoff's prediction for disubstituted carbon atoms.

Yet Werner's paper was also quite different in scope. Unlike the *Voorstel*, Werner's paper should not be considered a modest request for chemists to consider a new explanation for puzzling isomers, but as a radical restructuring of the understanding of metal-ammines that entailed a new conception of valence. Werner certainly considered his new formulas as iconic, but he did not add these iconic meanings to an existing set of chemical formulas. Rather, he invented a new, consciously iconic visual language for the metal-ammines. It is also important to note Werner's extreme self-confidence in his reinterpretation. He explicitly invoked, for example, Jørgensen's known praseo and violeo salts of bis-(ethylenediamine) complexes as the first 'confirmation' (*Bestätigung*) of the existence of configurational *cis* and *trans* isomers. In other words, he did not offer his explanation as a possible *alternative* to Jørgensen's, but as the *correct* one.

---

[38] Kauffman notes that Werner had originally submitted the *Habilitation* to an unknown German journal only to have it rejected, and presumes that this journal was the *Berichte*. Rejection by the *Berichte* would have been likely, given its tendency not to publish either lengthy or overly theoretical papers. It is also possible, however, that Werner could have submitted it to the *Annalen*, which had a record of publishing published longer theoretical works.

Werner's stereoformulas had a similar meaning to those used earlier for carbon atoms, and like their carbon and nitrogen counterparts, they possessed their own set of curiosities. Because Werner ignored the individual valences and central atom of the octahedron, his drawings of the octahedron closely resemble Wislicenus' drawings reproduced in Chapter 5. Werner's formulas therefore did not depict individual bonds to the metal atom, but merely their spatial arrangement around the metal, reflecting his conviction that the valence could not be a pre-existing directed unit of chemical affinity. One of the curious aspects of Werner's theory was the concept of the octahedron itself. According to Werner, the metal-ammonia complexes possessed an 'octahedral' geometry that implied the importance of the faces of the octahedron, but it is actually the *vertices* that are important, as there are only six groups surrounding the central metal atom. The faces of a cube would have provided an equally valid visualization of the arrangement of six groups around the metal atom, but Werner never seems to have considered this possibility.

The second important point to notice is the origin of the octahedron itself, for which there seem to be two overall reasons. The first was Werner's view that affinity was shared equally when the groups arranged themselves around a central atom to receive the maximum amount of affinity. This would make the octahedron a natural arrangement for the six groups surrounding the metal atom. The second reason is related to his overall method of 'creating' several series of compounds by removing successive molecules of ammonia, in which the number 6 appears as a result of deliberate manipulation of formulas to bring that number out. One can see the deliberate bias towards the octahedron when Werner noted that the removal of the fourth ammonia molecule, when represented according to the Blomstrand-Jørgensen theory, resulted in a formula for a compound 'that did not exist'. There is therefore a broad similarity in Werner's thought experiment to the manipulation of Berzelian formulas by chemists in the 1830s to determine which radicals existed in a series of organic compounds.[39] The number 6 was not therefore discovered or self-evident: rather, it was brought to Werner's attention by manipulation of formulas on paper.

As significant and innovative as Werner's theory was, it also contained continuities to earlier theories. The acid residues could be closer and further from the metal atom, and this position determined their chemical behavior. Although more sophisticated and consistent, the separation of affinity into valence and coordination, and the conception of different spheres of influence resembles in its broad form, if not in detail, the very concept of a molecular compound. The forces acting between the metal complex and the ions in the second sphere were better defined as the electrostatic attractions to complete the valence of the metal atom, but still Werner

---

[39] This was discussed in Chapter 2. See also Ursula Klein, *Experimente, Modelle, Paper-Tools: Kulturen der organischen Chemie im 19. Jahrhundert*, Habilitationsschrift, University of Konstanz, 1999.

constructed his numerous formulas largely to maintain the constancy of the coordination number for a given compound, just as Kekulé had constructed molecular compounds to maintain the constancy of valence.

Confirmation of the Theory, 1893–1907

When the 'Beitrag' appeared, Werner was still a relatively unknown *Privatdozent* affiliated to Hantzsch's Zürich laboratory, but its publication was responsible in large part for Werner's appointment to the University of Zürich in 1893 as the successor to Victor Merz. Among the three finalists for the position – Johannes Thiele, Hans von Pechmann and Werner – Werner was the youngest and least known, and Pechmann and Thiele were both experienced students of Adolf von Baeyer. Werner's chemistry and abilities certainly played a crucial role in the decision, but it would also seem that his close association with Hantzsch (who expressed confidence that the relatively inexperienced Werner could direct a laboratory) and his presence in Zürich, including his impending Swiss citizenship, was responsible at least for Werner's inclusion in the list of finalists, if not the final appointment.[40] In August 1893, Werner was appointed as successor to Merz, but at the level of an extraordinary professor of organic and theoretical chemistry (most likely because of his relative inexperience). In 1895, he was promoted to ordinary professor at the University, and he would remain in Zürich until the end of his life.

The additional application of the coordination theory and inorganic chemistry and would eventually overtake Werner's previous training and interests as an organic chemist. The 'Beitrag' was only the first of a series of 20 lengthy papers published in the *Zeitschrift für anorganische Chemie* between 1893 and 1899, after which he increasingly turned to the *Berichte* for publication. The first five of these papers in the *Zeitschrift* contained some new experimental results, but they were primarily theoretical and addressed Jørgensen's criticisms, restated the central justification for the octahedron, refined the concept of coordination, and introduced a new nomenclature system for the octahedron.

Although Werner surely regarded the isolation of predicted compounds as important for the success of the theory, it was not a major component of his initial effort to support his theory. Between 1893 and 1896, he published three papers in the *Zeitschrift für physikalische Chemie* on the conductivity (*Leitungsfähigkeit*) of metal-ammine salts co-authored with Arturo Miolati, Werner's close friend and fellow

---

[40] Kauffman, *Alfred Werner*, p. 28. In letters to George Lunge and his mentor Rudolf Schmitt, Hantzsch mentioned that his former assistant at Zürich, Heinrich Goldschmidt, was also under consideration, and he was torn between Goldschmidt's experience and his confidence that Werner 'would develop brilliantly'. Hantzsch to G. Lunge, 9 June 1892; Hantzsch to Rudolf Schmitt, 21 October 1892, Sammlung Darmstadter G1 1883, Staatsbibliothek zu Berlin, Preußischer Kulturbesitz.

student in Hantzsch's laboratory.[41] Werner and Miolati had begun the conductivity studies almost immediately following the completion of the 'Beitrag', and the first paper in the series was the last Werner would publish from Hantzsch's laboratory. The precise origins of these experiments are unknown because Werner did not mention the implications of his coordination theory for conductivity in his 1893 article, but it seems likely that the increasing importance that Hantzsch had given to conductivity measurements at this time and the presence of the new conductivity apparatus in Hantzsch's laboratory in 1892 (discussed in Chapter 7) made Werner aware of the utility of such measurements and gave him the capability to make them. Werner certainly recognized immediately that his and Jørgensen's theories made significantly different predictions about the conductivity of the metal-ammines, and he regarded the results of his conductivity measurements – the measurement of a physical property – as a primary piece of evidence in favor of his theory.

Werner and Miolati opened their first paper with a general summary of the coordination theory, and related it to the predicted number of ions in each member of the series. As we saw above, Werner had argued that in the sequential loss of ammonia molecules between the luteo and purpureo compounds, the number of ions would decrease to zero in the $M(NH_3)_3X_3$ complex, and then increase as the replacement of ammonia continued. As the number of ions decreased from four to zero and increased to four again, the conductivity of the various metal-ammonia salts should show a corresponding decrease to zero, followed again by an increase as the replacement continued. The predicted non-conductance of the middle complex $M(NH_3)_3X_3$ was crucial to the success of the theory. According to Werner, this complex possessed no ions and should be non-conductive, but according to Jørgensen's theory, there should be two ions, therefore it should be conductive.

Werner and Miolati reported that the conductivity for the luteo ($[Co(NH_3)_6]Br_3$), roseo ($[Co(NH_3)_5H_2O]Br_3$) and tetraammineroseo ($[Co(NH_3)_4(H_2O)_2]Br_3$) salts, as expected, increased with increasing dilution (molecular conductivity $[\mu] = c.\ 350$). The presence of water in the metal complex only slightly decreased the overall conductivity, but did not fundamentally change its magnitude. The conductivity of bromopurpureocobalt chloride ($[Co(NH_3)_5Br]Br_2$) and xanthocobalt chloride ($[Co(NH_3)_5NO_2]Br_2$) displayed a conductivity one-third lower ($\mu = c.\ 230$) than the luteo and roseo salts, showing 'above all, in what a sweeping manner the molecular conductivity of the compounds has been changed by the escape of the ammonia or water molecule, respectively'.[42] This result confirmed 'the conclusion, derived from

---

[41] A fourth paper in the series appeared in 1901. Alfred Werner, 'Beiträge zur Konstitution anorganischer Verbindungen. IV', *Zeit. phys. Chem.*, 1901, *38*, 331–52.

[42] Alfred Werner and Arturo Miolati, 'Beiträge zur Konstitution anorganischer Verbindungen. I: Abhandlung', *Zeit. phys. Chem.*, 1893, *12*, 35–55, translation in Kauffman, *Classics*, p. 101.

chemical facts, that negative residues which are bound directly to the metal atom are not ions'.[43]

As predicted, three of the 'non-conducting salts', that is those with three removed ammonia molecules – ($Co(NH_3)_3(NO_2)_3$, $Pt(NH_3)_2Cl_2$ and $Co(NH_3)_3Cl_3$) – showed negligible conductivities ($\mu = c.$ 7.43) when compared to the luteo, roseo and the purpureo salts. As a control, Werner and Miolati measured the conductivities of other complexes with nitrites, that is luteocobalt nitrite, $Co(NH_3)_6(NO_2)_3$, which showed a conductivity equivalent to other luteo salts:

> The unusually small conductivity of hexaminecobalt nitrite can therefore be caused only by the direct bonding of the three nitrite groups, and thus a decisive proof is furnished for the constitution that has been proposed for it and for the view that acid residues bonded directly to the metal atom can then only act as ions if the metal salt is able to form a hydrate.[44]

Finally, Werner and Miolati reported that additional removal of ammonia groups gave compounds that displayed conductivity. Erdmann's salt, $Co(NH_3)_2(NO_2)_4K$, and Cossa's salt, $Pt(NH_3)Cl_3K$, both showed conductivities around 100, indicating the presence of two ions.

Werner and Miolati's second article appeared in 1894, and reported many more examples of conductivity as a function of metal complexes. Compounds with one ion outside of the coordination complex (Werner and Miolati listed 8 salts, 4 cobalt and 4 platinum) showed molecular conductivities between 96.7 and 108.5; compounds with two ions outside the coordination complex (4 cobalt, 4 platinum, and 4 chromium compounds) showed values between 234.4 and 267.6, and compounds with three ions outside the coordination complex (6 cobalt compounds and 1 chromium compound) showed values of 383.8 to 426.9.[45] This steady increase in conductivity corresponded well with the increasing number of ions. Werner and Miolati tabulated these results in a now relatively famous graph that showed dramatically the effect of the number of ions on conductivity (Figure 9.5). Curiously, although the non-ionic coordination compounds were included on the graph at zero conductivity, the article lacked any discussion of their conductivity. There were also several missing values on two of the graphs for compounds that did not yet exist, and Werner and Miolati projected values for their conductivity based on the other half of the graph. Despite the missing values, they concluded that:

> Such a complete agreement ... is shown between the theoretically expected behavior and the actual behavior which appears in the magnitude, pattern, and variation in the molecular

---

[43] Ibid., p. 102.
[44] Ibid., pp. 107–8.
[45] Alfred Werner and Arturo Miolati, 'Beiträge zur Konstitution anorganischer Verbindungen. II. Abhandlung', *Zeit. phys. Chem.*, 1894, *14*, 506–21, translation in Kauffman, *Classics*, pp. 120–3.

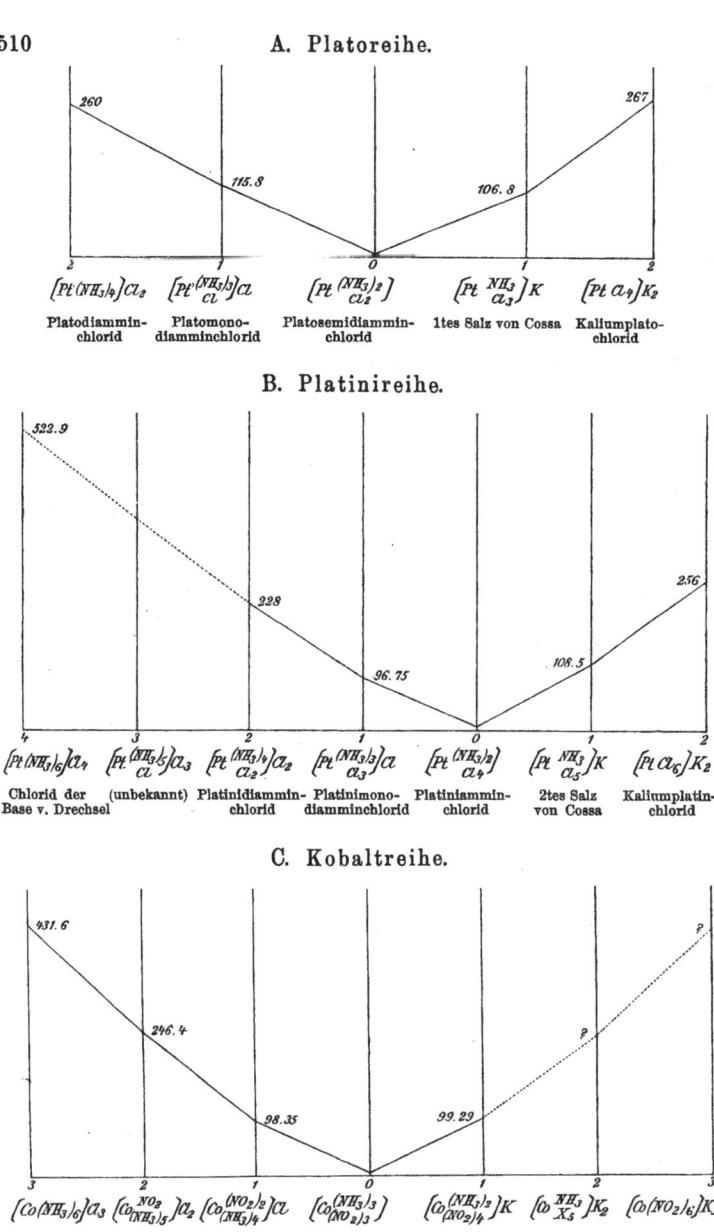

9.5 Werner and Miolati's plots of conductivity versus ammonia content in the ammine complexes of platinum (a) and (b) and cobalt (c). Note the missing values for the last two complexes in the cobalt series.

conductivity of the compounds that there can no longer be any doubt as to the correctness of the principle on which those theories are based.[46]

Because of the strong correlation that did exist between the number of ions and the conductivity values, they suggested that this method would also allow the simple determination of the number of ions in an unknown compound from its conductivity.

The other major effort to consolidate the evidence in favor of the coordination theory began in earnest in in 1897, when Werner and his students started to report the isolation of compounds predicted by theory. For our purposes, the most crucial compound predicted by Werner was the stereoisomeric *cis* dichloro violeo salt, known for the bis-(ethylene diamine) salts, but not for the simple ammine compound. Werner finally announced its isolation in 1907, noting the crucial place this compound occupied in his theory:

> I have repeatedly tried again and again to discover the theoretically predicted stereo-isomeric series among even the simplest diacido-tetrammine compounds – the dichlorotetrammine-cobalt salts $[Cl_2Co(NH_3)_4]X$. Of course, since there are no different possibilities in formulating the structural bonding of the chlorine atoms and the ammonia molecules in these salts, then the discovery of stereoisomeric dichloro-tetramminecobalti salts had to provide conclusive proof for the steric formulation of isomeric compounds with complex radicals $[CoA_4B_2]$. I have now finally reached this goal and in a manner indeed which simultaneously no longer leaves any doubt as to the configuration of the isomeric series.[47]

The new salt was prepared by cleaving an octammine-diol-dichlorodicobalti salt, a dinuclear metal complex first reported by Werner in an earlier paper, with concentrated hydrochloric acid:[48]

$[(NH_3)_4Co(OH)_2Co(NH_3)_4]Cl_4 + 2HCl \rightarrow [(NH_3)_4Co(OH_2)_2]Cl_3 + [Cl_2Co(NH_3)_4]Cl$

The immediate product mixture was a bluish-red mixture of the diaquo salts (the product on the left) and dichloro salts (the product on the right). The diaquo compound could be removed with water to leave the dichloro salt, which in turn consisted of the new violeo compound and very small amounts of its praseo isomer. Werner also reported the preparation of the violeo chloride, bromide, iodide, nitrate,

---

[46] Ibid., p. 119.

[47] Alfred Werner, 'Über 1.2-Dichloro-tetrammin-kobaltisalze (Ammoniakvioleosalze)', *Berichte*, 1907, *40*, 4817–25, translation in Kauffman, *Classics*, p. 145.

[48] The dinuclear compound was reported in the fifth paper in a series on polynuclear complexes. Alfred Werner, 'Über mehrkernige Metallammoniake. V: Über Ocatammin-dioldikobaltisalze', *Berichte*, 1907, *40*, 4434–41.

dithionate and sulfate salts, and pointed out that the new violeo compound and its known praseo isomer were analogs to the known bis-(ethylenediamine) cobalt isomers. The praseo isomer in both cases was green, and the violeo violet. The intense violet-blue color of the new salt varied with the counter-ion, but showed an overall shift toward the blue region of the spectrum from the ethylenediamine compounds. The new salt was also found to be less stable than its praseo isomer – the reason it had not yet been isolated. At 0° C, a solution of the violeo compound turned into a chloro-aquo-tetrammine salt, and concentrated hydrochloric acid converted the violeo compound into the more stable praseo isomer.

Configurational Assignments

In addition to suggesting the existence of *cis* and *trans* isomers in certain metal-ammines, Werner also attempted to assign specific configurations to pairs of isomers. Three specific examples will be sufficient to show his method. In the original 1893 paper, Werner assigned configurations to the isomeric square planar platinum complexes $Pt(NH_3)_2Cl_2$, using the six part argument summarized in Figure 9.6:

1 Werner assumed that platosemidammine chloride was the *cis* isomer and the platosammine chloride was the *trans* isomer (9.6(a)).
2 Reaction of $Pt(NH_3)_2Cl_2$ with pyridine and reaction of the pyridine complex $PtPy_2Cl_2$ with ammonia should give the same product (α), whereas reaction of the *trans* isomers should also give the same compound (β), which was different from α. The α and β compounds were *cis* and *trans* isomers (9.6(b) and 9.6(c)).
3 Heating α or β transformed them into the platosammines Pta_2x_2 with the *trans* configuration (9.6(d)).
4 Compound α will perform this transformation when it loses ammonia and pyridine together (from the *trans* position) to give the mixed compound $PtCl_2PyNH_3$ (with the *trans* configuration, 9.6(e)).
5 Compound β can perform this transformation in two different ways, resulting in two different products (9.6(f)): it can lose two molecules of ammonia, or lose two molecules of pyridine, resulting in either $PtCl_2(NH_3)_2$ or $PtCl_2Py_2$, both of which have the *trans* configuration.
6 The initial assumption given in (1) is correct, because, if we assume that the platosammines have the *cis* configuration, they should yield $[Pta_2b_2]Cl_2$ (*cis*), which could then transform into platosammine salts in three ways, resulting in three products. This was never observed in this series of compounds.[49]

---

[49] Werner, 'Beitrag zur Konstitution anorganischer Verbindungen', in Kauffman, *Classics*, pp. 69–72.

a

Cl–Pt(NH₃)(NH₃)–Cl   Cl–Pt(NH₃)(Cl)–NH₃
Platosemidiamminchlorid,   Platosamminchlorid.

b

$$\text{Cl–Pt(NH}_3\text{)}_2\text{–Cl} + \text{Py}_2 = (\text{Py–Pt(NH}_3\text{)}_2\text{–Py})\text{Cl}_2 \text{ (a)}$$

$$\text{Cl–Pt(Py)}_2\text{–Cl} + (\text{NH}_3)_2 = (\text{NH}_3\text{–Pt(Py)}_2\text{–NH}_3)\text{Cl}_2 \text{ (b)}$$

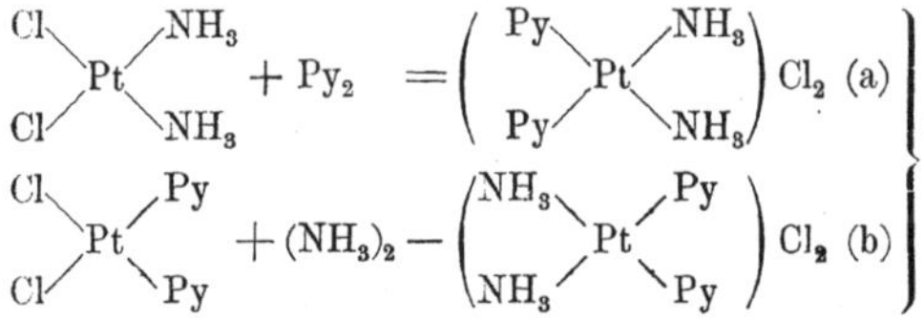

indentische Verbindung α.

c

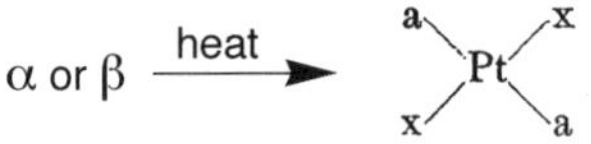

$$\text{Cl–Pt(NH}_3\text{)(NH}_3\text{)–Cl (trans)} + \text{Py}_2 = (\text{Py–Pt(NH}_3\text{)(NH}_3\text{)–Py})\text{Cl}_2 \text{ (c)}$$

$$\text{Cl–Pt(Py)(Py)–Cl (trans)} + (\text{NH}_3)_2 = (\text{NH}_3\text{–Pt(Py)(Py)–NH}_3)\text{Cl}_2 \text{ (d)}$$

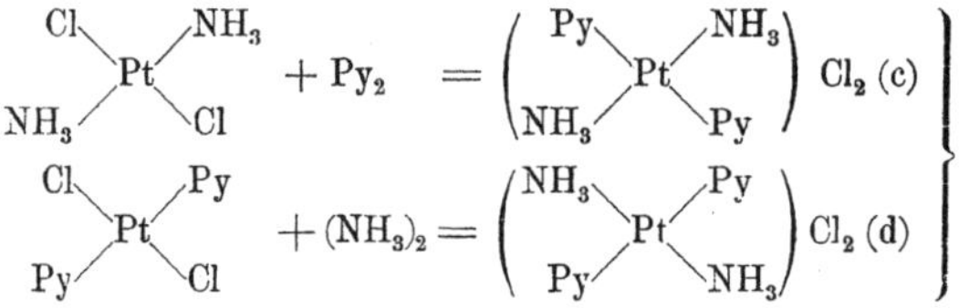

identische Verbindung β.

d

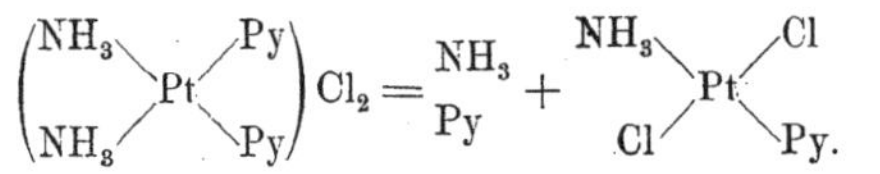

$$\alpha \text{ or } \beta \xrightarrow{\text{heat}} \text{a–Pt(x)(x)–a (trans)}$$

e

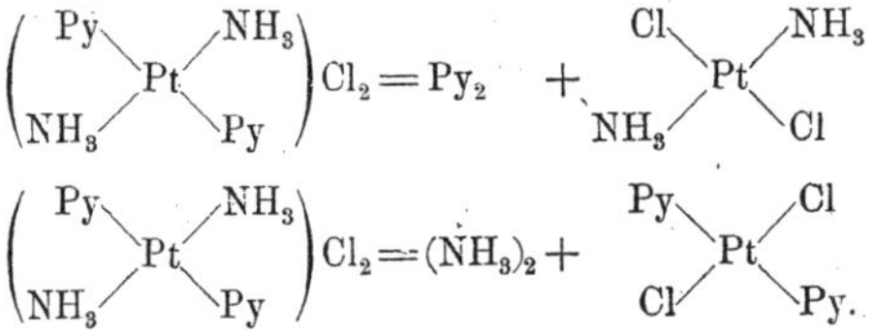

$$(\text{NH}_3\text{–Pt(Py)}_2\text{–NH}_3)\text{Cl}_2 = \frac{\text{NH}_3}{\text{Py}} + \text{NH}_3\text{–Pt(Cl)(Cl)–Py}$$

f

$$(\text{Py–Pt(NH}_3\text{)}_2\text{–Py})\text{Cl}_2 = \text{Py}_2 + \text{Cl–Pt(NH}_3\text{)}_2\text{–Cl}$$

$$(\text{Py–Pt(NH}_3\text{)}_2\text{–Py})\text{Cl}_2 = (\text{NH}_3)_2 + \text{Py–Pt(Cl)(Cl)–Py}$$

9.6 Werner's configurational assignment to the platosammine and platosemidiammine salts.

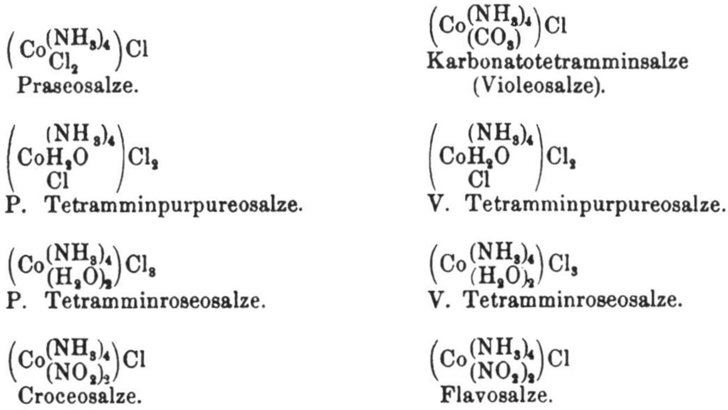

$\left(\text{Co}^{(\text{NH}_3)_4}_{\text{Cl}_2}\right)\text{Cl}$
Praseosalze.

$\left(\text{Co}^{(\text{NH}_3)_4}_{\text{(CO}_3)}\right)\text{Cl}$
Karbonatotetramminsalze
(Violeosalze).

$\left(\text{Co}^{(\text{NH}_3)_4}_{\substack{\text{H}_2\text{O}\\\text{Cl}}}\right)\text{Cl}_2$
P. Tetramminpurpureosalze.

$\left(\text{Co}^{(\text{NH}_3)_4}_{\substack{\text{H}_2\text{O}\\\text{Cl}}}\right)\text{Cl}_2$
V. Tetramminpurpureosalze.

$\left(\text{Co}^{(\text{NH}_3)_4}_{(\text{H}_2\text{O})_2}\right)\text{Cl}_3$
P. Tetramminroseosalze.

$\left(\text{Co}^{(\text{NH}_3)_4}_{(\text{H}_2\text{O})_2}\right)\text{Cl}_3$
V. Tetramminroseosalze.

$\left(\text{Co}^{(\text{NH}_3)_4}_{(\text{NO}_2)_2}\right)\text{Cl}$
Croceosalze.

$\left(\text{Co}^{(\text{NH}_3)_4}_{(\text{NO}_2)_2}\right)\text{Cl}$
Flavosalze.

9.7 Werner's chart showing the similarity in the genetic relationships during the synthesis of the flavo and croceo salts. The croceo salt can be derived from the praseo salt by the sequence of preparations shown in the first column. The flavo salt can be derived by a similar sequence of preparations from the carbonatotetrammine (violeo) salt at the top of the right column. Each column shows the same sequence of substitutions, and so the flavo and croceo salts must be structurally identical.

In a 1895 paper, the second in the 20-part series on the constitution of coordination compounds, Werner offered a six-page argument for assigning configurations to the isomeric dinitrotetrammines, or the flavo and croceo salts [Co(NH$_3$)$_4$(NO$_2$)$_2$]Cl.[50] Before he assigned configurations, however, Werner first argued that the two compounds were structurally identical by comparing the similar genetic relationships among the steps in their preparation (Figure 9.7). Because the flavo salt could be derived ultimately from the carbonatotetraammine salt and the croceo salt from the praseo salt by a similar series of reactions, they belonged to two isomerically related series. Like the other pairs in the series, the two compounds should therefore be stereoisomers, 'so the issue of configuration poses itself as a major problem'.[51] Werner offered three different ways of assigning configurations that all led to the same result, 'and in the one case where a true proof for the configurations is present, there can be no doubt that the problem is finally solved'.[52]

---

[50] Alfred Werner, 'Beitrag zur Konstitution anorganischer Verbindungen. II', *Zeit. anorg. Chem.*, 1895, *8*, 153–88, pp. 182–8.
[51] Ibid., p. 184.
[52] Ibid.

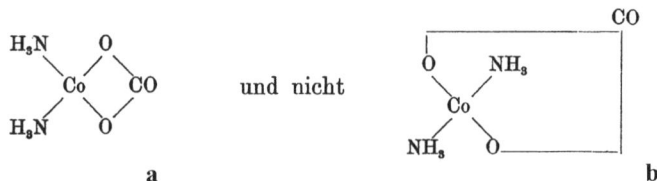

9.8 The two possible spatial isomers of the planar cobalt carbonato complexes. Werner ruled out the *trans* compound (b) because of the physical difficulty in the ability of the carbonato group to bridge the cobalt atom.

The first two methods were explicitly analogous to those used by Wislicenus and Van 't Hoff to assign configurations to the unsaturated acids. First, among geometric isomers, the *cis* form was ususally the 'unstable' and 'labile' form, while the *trans* form was more 'durable' and 'stable'. Because the violeo salts were the less stable of the two isomers, they must possess the *cis* configuration, and the more stable praseo isomer must possess the *trans* form. Second, Werner used the chemical property of ring formation to establish the configuration of carbonato complexes, echoing the arguments using anhydride formation for maleic and fumaric acid described in Chapter 3. In the carbonato complexes, the two oxygens of the carbonato group must combine with the central metal atom in the *cis* position, because placing the atoms in the *trans* position was physically impossible (Figure 9.8). This corroborated the first conclusion based on the stability of isomers: the violeo carbonato salts must therefore be *cis*, and the praseo salts *trans*.

Werner's third method for assigning configurations to the flavo and croceo salts was somewhat more complicated and used the sequence of reactions in which these salts were formed from cobalt triamminenitrite, $Co(NO_2)_3(NH_3)_3$, by the initial replacement of a nitro group with chlorine, followed by the insertion (*Einschieben*) of ammonia to give the complex $Co(NO_2)_2(NH_3)_4$. From the two starting materials A and B in Figure 9.9, it was possible to make the two isomeric dinitrotetrammine salts C and D. Compound C could be made only from compound A, however, whereas compound D could be made from *either* A or B. Considering the space formulas for this transformation (Figure 9.9), the dinitrotetrammine salts (either C or D) that can be formed from either triamminocobalt nitrite are 'those that have the two nitro groups in 'neighboring positions' (*in Nachbarstellung*) – that is, those that have the nitro groups *cis* to each other. Because the flavo salts could be obtained from A or B, they must have the *cis* configuration, and because Werner included the flavo salts as a violeo salt, the violeo salts must also be *cis*. This argument therefore led to the same

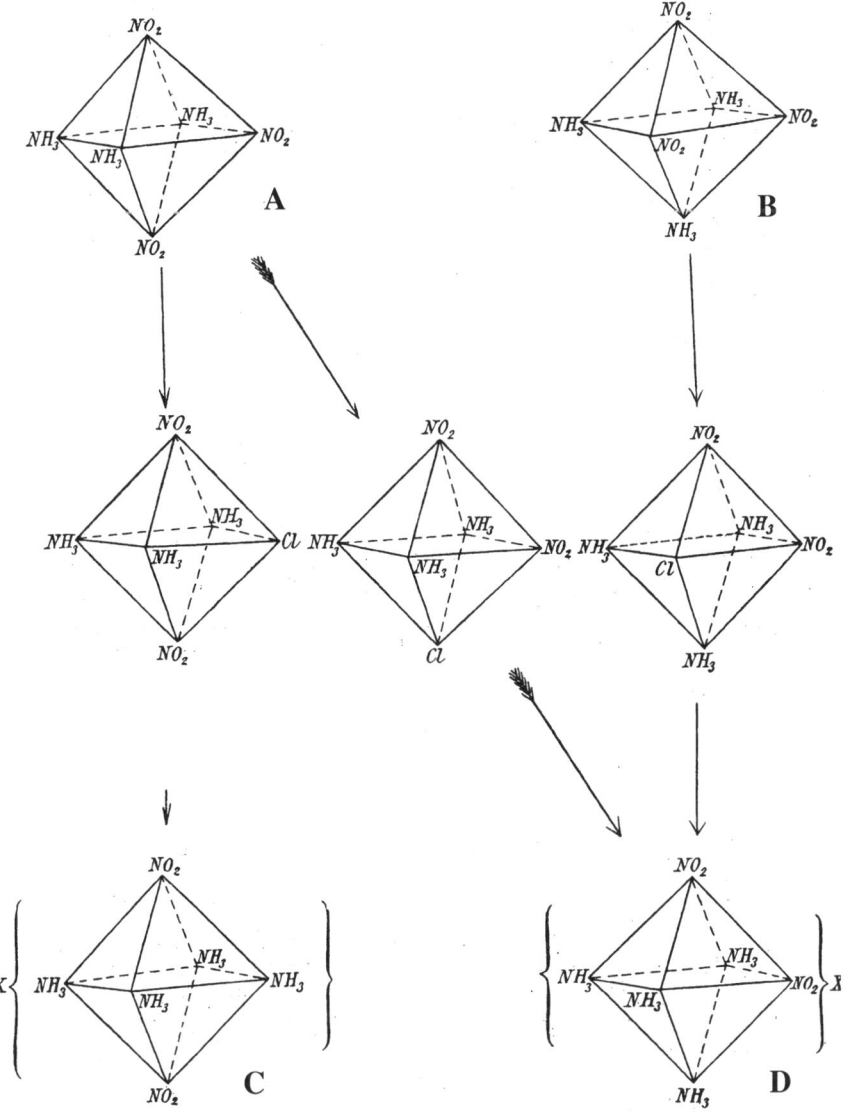

9.9 Werner's assignment of configurations to the flavo and croceo salts using octahedral formulas. C could be made only from A, while D could be made from either A or B.

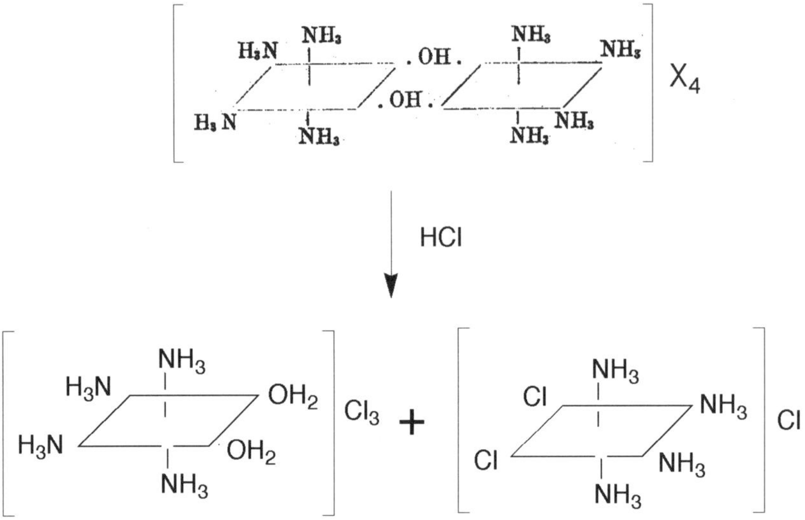

9.10 Werner's 1907 synthesis of the new violeo salt portrayed in stereoformulas. The cleavage of the dicobalt compound involves simple replacement of the bridging hydroxyl groups with chloride ions resulting in the *cis* arrangement of chlorine atoms in the violeo compound.

conclusion as the first two methods: the praseo salts were *trans*, the violeo salts were *cis*.

Werner's configurational assignments would also be vindicated in his 1907 synthesis of the violeo salts. Because cleavage of the dicobalt complex occurred on the common edge between the metal atoms containing the hydroxyl groups (Figure 9.10), simple substitution of the hydroxyl groups with chlorine atoms would yield the *cis* dichloro isomer. According to Werner: 'The method of formation of the violeo dichlorotetrammine cobalti salts explains their [*cis*] configuration.'[53] 'This result,' Werner noted, 'agrees completely with the configurational formulas derived earlier for the ethylenediamine compounds on the basis of quite different reactions so that the earlier configurational determinations are thereby confirmed in a welcome manner.'[54] It must be pointed out, however, that the 'agreement' between configurational assignments and the results of the cleavage of the dicobalt compound rested on Werner's assumption that the process of substitution took place simply by

---

[53] Alfred Werner, '1.2-Dichloro-tetrammin-kobaltisalze', in Kauffman, *Classics*, p. 147.
[54] Ibid., p. 148.

the replacement of atoms in the exact place where they were located, without any rearrangement of groups around the central atom. Werner's assumption was not unusual at the time, although the earlier discovery of the Walden inversion in 1895 had begun to erode the plausibility of this 'naïve' conception of substitution.

The Controversy with Jørgensen, 1893–1899

In 1893, shortly after reading Werner's initial paper on the coordination theory, Jørgensen candidly assessed Werner's theory in a letter to his friend Blomstrand:

> Much can be said about Werner's fantasies on the constitution of platinum bases. For example, with regard to the isomerism of the substituted platodiamminechlorides he merely presents my views in a new clothing, for he has to assume a successive substitution, just as I do. Yet I do not think I will elaborate the point. The matter cannot be made clear to a larger audience without going into details more lengthy than Werner's theory deserves. On the other hand, I believe I have refuted with sufficient clarity his main points and the central point in his own theory, so that everyone can see how frivolous he is throughout in his line of argument.[55]

It was natural that Jørgensen would find much to complain about in Werner's theory. Jørgensen was a naturally cautious chemist, inclined toward making hypotheses only as generalizations about the 'facts' that he had uncovered. On the other hand, Werner was not at all embarrassed about using working hypotheses to generate experiments, and like Wislicenus' enthusiasm for the tetrahedral carbon atom, Werner was extremely self-confident about the eventual success of his coordination theory. The publication of Werner's theory therefore prompted one of the more famous controversies in the history of chemistry between Jørgensen, the defender of Blomstrand's chain theory and conventional conceptions of valence, and Werner, the bold systematizer who was convinced that conventional valence theory was an 'unclear' concept in need of an overhaul.

The chemical details of this controversy were first summarized by Kauffman in 1960. More recently, Kragh has complemented Kauffman's analysis with a non-technical discussion of the the underlying methodological differences between Werner and Jørgensen hinted at by Kauffman, and suggested, *contra* Kauffman, that Jørgensen never accepted Werner's theory. Therefore, we do not need to recount the entire episode here, but we can draw attention to a few of its important characteristics in order to place the controversy in the context of the general aspects of Werner's theory outlined above

---

[55] Sophus Mads Jørgensen to Christian Blomstrand, 24 May 1893, translation in Helge Kragh, 'S.M. Jørgensen and His Controversy with A. Werner: A Reconsideration', *Brit. J. Hist. Sci.*, 1997, *30*, 203–19, p. 209.

and the earlier critiques of stereochemistry presented in previous chapters. To this end, I will focus on Jørgensen's view of Werner and not on Werner's defense.

Different Planes of Meaning

Jørgensen first replied to Werner's 'Beitrag' in August of 1893 in the fifth part of an ongoing series of papers on the constitution of cobalt, chromium and rhodium ammines. He noted early in the paper that Werner showed signs of 'unmistakable talent', and that the material had been presented in such a way as to arouse further interest among experts on metal-ammine chemistry. But Jørgensen immediately recognized that his and Werner's theories would be difficult to reconcile because they 'actually do not collide with the views developed by Blomstrand and myself. Rather, they lie on different planes'.[56] The metaphor of 'different planes' is useful, and can take on at least two different meanings. First, Jørgensen perceived correctly that Werner's theory 'dispenses with' valence theory, while his own was based solidly on it. Accepting Werner's theory meant 'giving up every theoretical explanation of atomistic composition'. Jørgensen also turned the tables on Werner. If valence theory was a 'confused' or unclear concept, as Werner claimed it was, 'Are the coordination numbers and coordination locations clear concepts?'.[57] In 1899, resigned to the fact that he would not convince Werner that his theory was incorrect, he ended his participation in the controversy by suggesting that perhaps the two theories would eventually 'merge into a higher unity'.[58]

As mentioned above, Werner and Jørgensen also occupied different planes regarding their fundamental attitudes toward the role of hypotheses. Werner used hypotheses freely as a tool for suggesting additional experiments and as the organizing principle for chemical compounds. We have seen this in the way Werner manipulated formulas to create series of compounds, and how he arranged compounds, known and unknown, into genetic sequences to argue for their structural identity. According to Jørgensen, however, this was 'ingenious speculation' (*geistreiche Spekulation*) for 'constructing [theories] a priori', whereas 'only relentless experiment will gradually achieve clarity over the tangled relationships in this extensive group of compounds'.[59] In 1896, he noted that the theory of metal-

---

[56] Sophus Mads Jørgensen, 'Zur Konstitution der Kobalt-, Chrom- und Rhodiumbasen. V', *Zeit. anorg. Chem.*, 1894, *5*, 147–96, p. 147.

[57] Ibid., p. 158.

[58] Sophus Mads Jørgensen, 'Zur Konstitution der Kobalt-, Chrom- und Rhodiumbasen. XI', *Zeit. anorg. Chem.*, 1899, *19*, 109–57, translation in Kauffman, *Classics in Coordination Chemistry, Part 2: Selected Papers 1798–1899*, New York, Dover, 1976, p. 164. All subsequent translations of this article are from Kauffman.

[59] Sophus Mads Jørgensen, 'Zur Constitution der Kobalt-, Chrom- und Rhodiumbasen. VIII', *Zeit. anorg. Chem.*, 1896, *13*, 172–90, pp. 172–3. Jørgensen, 'Konstitution V', p. 159.

ammines 'can of course not be decided at a desk, but only by experiment'. Like the choice between traditional valence and the coordination theory, Jørgensen also found that the conflict between the guiding roles of hypotheses and experiment was difficult to reconcile.

Isomers and Conductivity

Beyond the existence of the theories on different planes, Jørgensen found many specific problems in Werner's theory. Werner's use of the coordination theory as a predictive tool first of all resulted in 'a large number of isomers that have never been observed'. Whereas Werner found this predictive capacity a powerful incentive for further research, Jørgensen saw only needless multiplication of unknown compounds. Indeed, the numbers were somewhat daunting. Jørgensen listed 66 new metal-ammine compounds in 13 series predicted by Werner's system, only 14 of which were known at the time.[60] The role that Werner's theory played in shaping his predictions was also not lost on Jørgensen. In a footnote which appeared in his second critique of Werner, Jørgensen noted how Werner had declared on the one hand that 'a compound with the formula $M(NH_3)_2X_3$ does not exist at all', while on the other hand actively predicting the existence of unknown compounds with his own theory:

> If a compound is not known that can be deduced correctly or incorrectly from my theory, it does not exist, and its ability to exist is directed as a weighty argument against my theory. But when entire series of compounds are missing in Werner's own system, they are 'not yet observed', which in this underdeveloped area is not surprising.[61]

Quite rightly, Jørgensen also pointed out the that spatial isomers predicted by Werner in the praseo-violeo compounds did not exist, and that the 'octahedral formulas require types of isomerism which have never been observed'.[62] In other words, Van 't Hoff's spatial explanation for the unsaturated acids had been for known pairs of isomeric acids, but Werner was now proposing an explanation for isomers that did not exist.

Both Werner and Jørgensen recognized immediately that the conductivity results for the $M(NH_3)_3X_3$ series would be crucial to Werner's argument. According to Jørgensen, Werner's theory 'stands or falls with this prediction'.[63] Jørgensen was no stranger to using physical methods such as conductivity methods, and had used them

---

[60] Sophus Mads Jørgensen, 'Zur Konstitution der Kobalt-, Chrom- und Rhodiumbasen. IX', *Zeit. anorg. Chem.*, 1897, *14*, 404–22, pp. 409–10.
[61] Sophus Mads Jørgensen, 'Zur Konstitution der Kobalt-, Chrom- und Rhodiumbasen. VI', *Zeit. anorg. Chem.*, 1894, *7*, 289–330, p. 319.
[62] Jørgensen, 'Zur Konstitution XI', in Kauffman, *Classics, Part 2*, pp. 151–2.
[63] Jørgensen, 'Zur Konstitution V', p. 150. Jørgensen, 'Zur Konstitution VI', p. 318.

to establish consistent molecular formulas for the cobalt-ammines. He therefore generally accepted Werner's and Miolati's conductivity measurements for the metal-ammines where they agreed with the predictions made by his own formulas, but not surprisingly, he attempted to descredit the data for Werner's supposed non-conducting salts of the type $M(NH_3)_3X_3$. Jørgensen pointed out that Werner had made the crucial prediction of non-conductivity in his 1893 paper by using only two examples, both incompletely understood. He doubted the existence of Erdmann's triamminecobalt nitrite $(Co(NO_2)_3(NH_3)_3)$, the compound Werner and Miolati had found to have zero conductivity, because he had been unable to prepare it himself. It was possible, therefore, that Werner and Miolati had made an incorrect cobalt determination and conducted measurements on a related salt that was capable of small conductivities.

Jørgensen would later also point out an inconsistency in how Werner used conductivities. In their third paper, Werner and Miolati had reported that the conductivity of dinitrotriamminecobalt chloride at room temperature had a value consistent with that of a single ion, but at 0° C the value was consistent with a non-conducting complex, therefore 'proving' (*bewiesen*) the 'complete untenability' of Jørgensen's theory. Yet 'the chemical behavior of the salt', Jørgensen complained, indicated the presence of ions, but this chemical property had 'nothing to say: conductivity is the only property of importance'. Further, Werner's conductivity measurements of the luteo cobalt chloride at 1° C showed a conductivity equivalent to a compound with two ions at 25° C, meaning that the luteo salts should have two, not three ions. Jørgensen concluded:

> This cuts both ways. This is an either/or situation. Either Werner must assume that the luteo salts have two negative ions, and that agrees neither with the theory nor with the chemical properties of these salts; or he must assume that dinitrotriammine chloride has one negative ion. That agrees with the chemical properties of the salt but not with his theory. I leave him the choice.[64]

Logical Inconsistencies

Jørgensen also attempted to undermine Werner by showing other logical inconsistencies in his theory, and pointed out that Werner was required to make unnecessary secondary hypotheses in order to maintain the illusion of consistency in the octahedral formula. Jørgensen recognized the crucial importance Werner had assigned to the number 6 in his formulas for the metal-ammines. This was particularly acute for Werner's formulas for the hydrates, in which Werner had explained the existence of more than six molecules of water to the residual affinities of components

---

[64] Jørgensen, 'Zur Konstitution IX', p. 412.

within the complex. Jørgensen simply saw this as explaining away anomalies to maintain the importance of the number 6:

> Actually I am most repelled by Werner's explanations of hydrated compounds by the fact that they permit too much arbitrariness. For if there can occur not only more than 6 H_2O to 1 metal atom by residual affinities (or whatever I should call it) of acids, not only polybasic ones such as sulfuric acid but also monobasic ones such as hydrochloric or hydrobromic acids, and less than 6 H_2O by formation without recognizable cause of compounds poorer in water on an unlimited scale, then I cannot see what special value the number six could have. Here it may correctly be said: 'Quit nimium probat, nihil probat' [He who proves too much, proves nothing].[65]

Werner had relied on the consistency of the appearance of six groups because that coincided with the octahedral formula – 'the weak point of the theory', according to Jørgensen, that had a 'rather accidental or arbitrary character'.[66] Jørgensen went on to describe a fundamental contradiction in Werner's concept of coordination:

> If one atom of chlorine or another electronegative radical, for example, is thought of as being coordinated to the trivalent cobalt atom and does not saturate a valence of this atom, then it is incomprehensible why the radical Coa_5Cl formed in this manner is only divalent, combines with only two electronegative ions, and does not function trivalently just as well as Coa_6 does. On the other hand, if the chlorine atom saturates a valence of the cobalt atom, then it is not coordinated because coordinated atom groups do not, of course, change the valence of the metal atom. In this case, one can certainly understand that the radical $CoCla_5$ is only divalent, but now no octahedron can be formed because in this case only 5 coordinated atom groups are present.[67]

The principal result of Werner's thought experiment about the sequential replacement of ammonia molecules, characterized by Werner as a 'true substitution of ammonia molecules for acid residues' was, according to Jørgensen, a 'pure subreption' (*bloß Subreption*) – a deliberate mispresentation, or an inference drawn from it – because Werner had assumed, not proven, that all the ammonia molecules in the complex could be bound directly to the metal atom. Werner had argued that because the three ions in the complex $M(NH_3)_3X_3$ were bound directly to the metal, the ammonia molecules 'must' also be bound directly to the metal. Jørgensen found this argument specious (only a consequence of the octahedron), and compared it to an analogously constructed argument for an organic compound:

> Because in butyl chloride, $CH_3Cl(CH_2)_3$, both the three hydrogen atoms and the chlorine atom are directly bound to the indicated carbon atom, so the three methylene groups must be bound to it also. But putting in Werner's proof the correct premises, a completely different

---

[65] Jørgensen, 'Zur Konstitution VI', p. 36, translation in Kauffman, *Classics*, p. 36.
[66] Jørgensen, 'Zur Konstitution XI', in Kauffman, *Classics, Part 2*, pp. 161–2.
[67] Ibid.

conclusion is reached: Because in the single previously known compound $M(NH_3)_3X_3$, the dinitrotriammine cobalt salts, only two of the three negative radicals are bound directly to the metal atom, the third must, by means of the ammonia molecule, be bound to it as well.[68]

Another series of compounds, the croceo salts, put Werner in a 'very peculiar dilemma'. As we saw above in the analysis of Werner's configurational assignments, Werner claimed that the croceo salts were *trans*-dinitro complexes and belonged to the praseo series. The flavo salts were *cis*-dinitro complexes and belonged to the violeo series, a series whose parent compound was still completely unknown. This assignment was supported by the known transformation of praseo chloride salts ($[Co(NH_3)_4Cl_2]X$) into croceo salts with nitrous acid. Yet to Jørgensen, three things remained 'unintelligible' in accepting Werner's claim:

1. The two NO_2 groups in the croceo salts behaved differently towards acids, even though in Werner's formulas they were identical and bonded in exactly the same manner to the cobalt atom.
2. No praseo salt was formed even by boiling the violeo salt with hydrochloric acid.
3. When warmed with hydrochloric acid, the flavo salts form praseo salts in considerable amounts.[69]

In order to save his theory, Werner could have assumed that in the croceo compound, one NO_2 group was attached by the nitrogen atom (a nitro group) while the other was attached by the oxygen atom (a nitrito group). The isomeric flavo compound would contain two nitro groups, making the flavo and croceo salts structural isomers, but then Werner's presumed configurational isomerism between the croceo and flavo salts would be 'superfluous'. 'The [non-existent] ammonia-violeo salts,' Jørgensen wrote, 'have an extraordinary importance for Werner's theory, and therefore he tries to preserve their existence even if he has to sacrifice simplicity.'[70]

By 'sacrificing simplicity', Jørgensen meant that Werner's theory required him to propose complicated additional hypotheses to maintain the consistency of the octahedral arrangement of groups. As an example, Jørgensen used the formation of chloroaquotetrammine chloride from praseo chloride in which, according to Jørgensen's formulas, a water molecule was simply inserted between the metal and the chlorine atom (Figure 9.11), in a process that was completely analogous to the formation of aquopentammine chloride from chloropentammine chloride. According to Werner, however, the water initially replaced a chlorine atom to form an intermediate compound that rearranged to the chloroaquotetraammine chloride. Both

---

[68] Jørgensen, 'Zur Konstitution VI', pp. 319–20.
[69] These are paraphrases from Jørgensen, 'Zur Konstitution XI', in Kauffman, *Classics, Part 2*.
[70] Ibid., pp. 148–9.

Chloropentamminchlorid: $Co.a_4.\overset{.Cl}{\underset{.a.Cl}{Cl}} + H_2O = Co.a_4.\overset{.OH_2.Cl}{\underset{.a.Cl}{Cl}}$: Aquopentamminchlorid,

Praseochlorid: $Co.a_4.\overset{.Cl}{\underset{.Cl}{Cl}} + H_2O = Co.a_4.\overset{.Cl}{\underset{.OH_2.Cl}{Cl}}$: Chloroaquotetramminchlorid,

a

so soll das Salz nun¹ nach folgendem Schema gebildet werden:

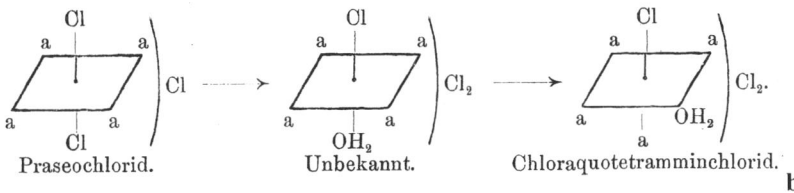

Praseochlorid.　　　　Unbekannt.　　　Chloraquotetramminchlorid.

b

9.11 According to Jørgensen's formulas (a) the formation of the chloroaquotetrammine chloride involved a simple insertion of the water molecule between the chlorine and cobalt atoms. Werner's spatial formulas (b), on the other hand, implied that a chlorine atom must be initially substituted to give an unknown intermediate compound ((b), middle) that then rearranged to the known chloroaquotetrammine chloride praseo chloride, with a simple insertion of an ammonia molecule.

the intermediate and the rearrangement were unknown, but Werner was forced to introduce them because the octahedron and the resultant configurational isomerism required them. At the core of this part of Jørgensen's argument was Werner's conflation of assumptions and proofs:

> And despite all these numerous rearrangements Werner is unable to prove the existence of ammonia-violeo salts. He *finds* that croceo salts are formed from praseo salts and therefore belong to the same series. But he *assumes* that the flavo salts, which are isomeric with the croceo salts, can be formed from an *unknown* series of violeo salts which is isomeric with the praseo salts, and from this *assumption* he further concludes that, since the flavo salts are formed from chloroaquotetraammine salts, the latter must belong to the ammonia-violeo series. This is a conclusion which rests on an assumption, but not a proof. In fact, no violeo chloride isomeric with praseo chloride and no praseo compounds isomeric with chloroaquotetrammine chloride or with the other salts considered as belonging to the violeo series are known.[71]

---

[71] Ibid., pp. 149–51.

Jørgensen and Other Critics of Stereochemistry

How do Jørgensen's analyses of Werner's theory compare to the other criticisms of stereochemistry that we have seen? The most obvious characteristic to note is Jørgensen's cautious empiricism that was similar to Michael, Fittig and others who disliked the use of hypotheses to direct research programs. Parts of Jørgensen's critique resemble Michael's critique examined in Chapter 5. Both Werner and Wislicenus were forced to invent hypothetical mechanisms that maintained the consistency of their claims about the three-dimensional properties of molecules. For Wislicenus, this meant constructing mechanisms to save the assumption of planesymmetric addition. For Werner, this meant constructing mechanisms to save the assumed existence of *cis-trans* isomerism in the praseo and violeo salts, a form of isomerism implied by the octahedron. Both Wislicenus and Werner were therefore required to embrace complexity, rather than simplicity in their explanations. In an 1897 letter to Blomstrand, Jørgensen complained that Werner '*has to construct new hypotheses all the time, such as he does in his latest works*; but as soon as one examines them [the hypotheses] closely, he abandons them and that with the same frivolity'.[72]

The controversy publicly came to an end without resolution in 1899, when Jørgensen wrote his last summary argument against Werner's theory. According to Kauffman, it was Werner's isolation of the long-predicted *cis* 1,2-dichloro violeo salts in 1907, and the isolation of the first optically active metal-ammine in 1911 that convinced Jørgensen that Werner was correct and marked Jørgensen's 'gracious withdrawal from the controversy'. As Kragh has argued, however, Kauffman's conclusion echoes those made in nearly all contemporary recollections of the controversy and have a common source in a eulogy of Jørgensen by his student S.P.L. Sørensen given in April 1914.[73] In the eulogy, Sørensen recalled a 1907 conversation with Jørgensen about Werner's synthesis of the stereoisomeric violeo salt, in which Jørgensen announced that 'the strife is now decided'. Although the precise interpretation of this statement is open, it does not seem likely, as Kragh has suggested, that Jørgensen meant that *he* then accepted Werner's theory, but rather that the large majority of chemists would now accept it, in effect ending the controversy. Kragh argues that it is likely that Jørgensen never adopted Werner's theory, and continued privately to reject it until his death in 1914. Kragh's argument is supported by Jørgensen's known cautious attitude toward theories, reluctance to speculate, his strong sense of empiricism, his devotion to traditional valence theory, and his realization that his and

---

[72] Jørgensen to Blomstrand, 1 September 1897, translation in Kragh, 'S.M. Jørgensen and His Controversy with A. Werner', p. 211. Emphasis added.

[73] Kragh, 'S.M. Jørgensen and His Controversy with A. Werner', pp. 203–19 and 213–15.

Werner's theories 'lay on different planes' that would 'perhaps merge in a higher unity'.

Moreover, the other objections to stereochemistry that we have seen in previous chapters would seem to provide additional support for the claim that Jørgensen did not accept Werner's theory. The critics of stereochemistry did not change their minds about the problems that they saw inherent in considering the spatial properties of molecules, and their distrust in them was primarily based on their prefence for generating 'facts' that are represented by theory, rather than allowing theory to guide the generation of facts. It might be natural to conclude at this point that the acceptance of Werner's new theory was split along generational lines, and that it was only with the passing of an older generation – exemplified by Jørgensen – that it became fully accepted within the chemical community. As we shall see in the next chapter, we can observe a general shift in the attitudes towards stereochemistry among generations, but also a transitional generation that was itself split over its value for chemistry.

Optically Active Coordination Compounds, 1897–1914

Exactly when Werner first realized that coordination compounds could be asymmetric and optically active is not clear. The original 1893 paper made no mention of optical activity or of possible asymmetric complexes, even for the *cis*-ethylenediamine compounds that he explicitly discussed. Werner's earliest known mention of optically active coordination compounds was in an 1897 letter to his friend Arturo Miolati: 'At present we are searching for asymmetrically constructed cobalt molecules. Will it be successful?'[74] In print, Werner first mentioned asymmetric complexes in 1899, in the seventeenth article on the constitution of inorganic compounds.[75]

In this article Werner reported the results of research by his student A. Vilmos on the oxalato bis-ethylenediaminocobalt salt $[Co(C_2O_4)en_2]X$, which 'offers some special interest [for the general group CoX_2A_4] since on the basis of the octahedral formula, the spatial consideration of the radical $Co(C_2O_4)en_2$ leads to interesting consequences relating to the occurrence of a possible new case of isomerism'. In binding to the cobalt atom, the two oxygen atoms in the bidentate carbonato group most likely had a *cis* and not a *trans* relationship to one another. The nitrogen atoms in each of the two bidentate ethylenediamine molecules therefore also likely had a *cis* relationship to one another, and the resulting complex, Werner realized, was

---

[74] Werner to Miolati, 20 February 1897, translation in Kauffman, *Classics*, p. 156.
[75] Alfred Werner, 'Beitrag zur Konstitution anorganischer Verbindungen. XVII: Über Oxalato-diäthylendiaminkobaltsalze', *Zeit. anorg. Chem.*, 1899, *21*, 145–58.

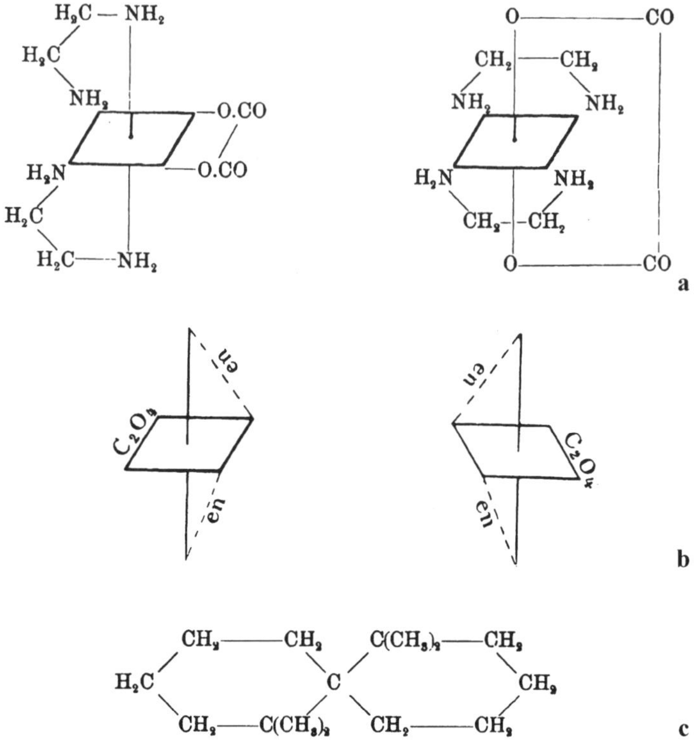

9.12 Werner's first depiction of the configuration of optically active coordination compounds, giving full structural details (a), and in abbreviated form (b). (c) The structure of an analogous optically active organic compound that does not contain an asymmetric carbon atom.

asymmetric (Figure 9.12). '[T]hat means,' Werner continued, using almost identical language to Van 't Hoff, 'it can be constructed in two spatial arrangements that are related as image and mirror image, and cannot be made congruent.'[76]

Werner also recognized that because the two ethylenediamine compounds were identical, the asymmetry was molecular, and not caused by the inorganic equivalent of an 'asymmetric' carbon atom. The closest analogous organic compound would be a spiro ring system with a molecular asymmetry, but without an asymmetric carbon atom (Figure 9.12). Werner was uncertain about how the properties of the two isomers would differ, but said it 'does not seem out of the question' that they could be

---

[76] Ibid., p. 147.

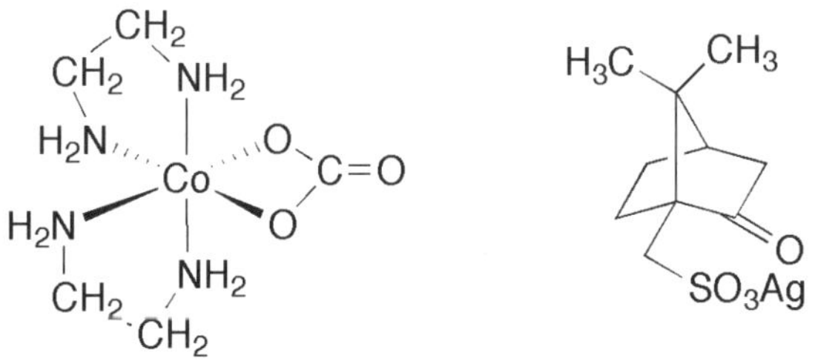

9.13 (a) The carbonato complex that Victor King initially attempted to resolve into enantiomers. (b) The two enantiomers of the cobalt complex that King succeeded in resolving using (+)-silver bromocamphorsulfonate (top right, the position of the bromine atom was unspecified by Werner). These formulas were not included in Werner's publication.

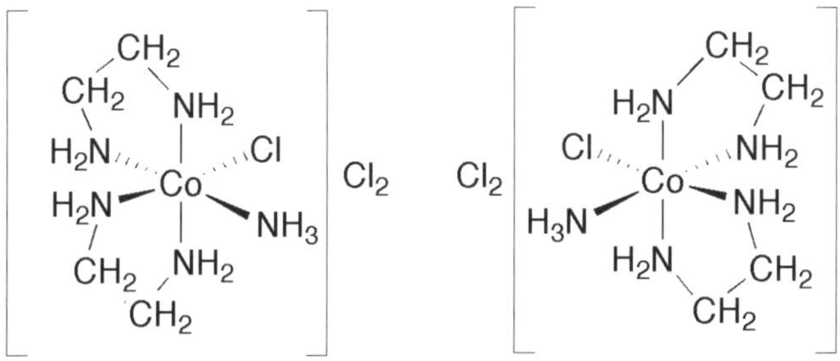

differentiated by an enantiomorphism in their crystal form. Although Vilmos reported ten new compounds of the carbonatoethylenediamine complex with different counter-ions, he was unsuccessful in producing the predicted isomers. Werner also reported trials on compounds where the carbonato group had been replaced with thiocyanate (SCN) and nitro (NO_2) groups.

Between 1897 and 1911, Werner continued to assign the project of isolating optically active coordination compounds to several students.[77] In 1910, he assigned the resolution of carbonatobis(ethylenediamine)cobalt (III) bromide to the American Victor King, whose laboratory notebooks reveal that he attempted this resolution for a full year before abandoning this compound for *cis*-chloroamminebis(ethylenediamine)cobalt (III) (Figure 9.13).[78] King later recalled

---

[77] V.L. King, 'A Rough but Brilliant Diamond', *J. Chem. Ed.*, 1942, *19*, 345.
[78] Victor L. King, *Über Spaltungsmethoden und ihre Anwendung auf komplexe Metal-Ammoniakverbindungen*, Dissertation, Universität Zürich, 1912. George B. Kauffman, 'The First

that he carried out 2000 fractional crystallizations while studying carefully Marie Curie's method of fractional crystallization used in her isolation of radium.[79] All attempts to resolve the compound were ineffective until King chose a camphor sulfonic acid (silver (+)-bromocamphor-π-sulfonate) of 'high optical activity', at which point the optically active stereoisomers 'literally fell apart'.[80] At almost the same time, Ernst Scholze, an assistant to Werner, resolved the bromine analog of King's compound, *cis*-bromo-amminebis(ethylenediamine)cobalt (III).

The results of both resolutions were submitted in June 1911 to the *Berichte* as a short article with the title 'Zur Kenntnis der asymmetrischen Kobaltatoms' (curious, in light of his emphasis in 1897 on the *molecular* asymmetry of the metal-ammines). In the introduction, Werner reviewed the three kinds of asymmetric complexes that the octahedral arrangement of groups made possible, and announced the isolation of the two optically active coordination compounds. The two compounds showed strong optical activity and unexpected stability even on heating and conversion to other members of the series. Werner listed three reasons why the existence of this compound was important:

> 1) the proof that *metal atoms can act as central atoms of stable, asymmetrically constructed molecules* and thereby lead to phenomena which agree with those produced by the asymmetric carbon atom; 2) the proof that *pure molecular compounds can also occur as stable mirror image isomers, whereby the difference between valence compounds and molecular compounds, which is still frequently maintained, disappears entirely*, and 3) the *confirmation of one of the most far-reaching conclusions from the octahedral formula*, by which the latter has found a new, important confirmation.[81]

Dozens of other optically active coordination compounds were to follow in the next three years. Werner's paper with King was the first of 12 papers of the same title published in the *Berichte* between 1911 and 1914. There were also three papers on active chromium compounds, two papers on active rhodium compounds, and individual papers on active compounds of iron, platinum and iridium. According to Kauffman, many more examples were also reported in the dissertations of Werner's students.[82] The speed with which these results were obtained and published was likely

---

Resolution of a Coordination Compound', pp. 126–42 in O.B. Ramsey, *Van 't Hoff-Le Bel Centennial*, Washington, DC, American Chemical Society, 1974. George B. Kauffman, 'The Discovery of Optically Active Coordination Compounds: A Milestone in Stereochemistry', *Isis*, 1975, 66, 38–62.

[79] King had apparently been working on the problem for so long that he was greeted on the streets of Zürich with the phrase 'Well, does it rotate yet?' (*Nun, dreht es schon?*) King, 'A Rough but Brilliant Diamond'.

[80] Ibid., p. 345.

[81] Alfred Werner, 'Zur Kenntnis des asymmetrischen Kobaltatoms. I', *Berichte*, 1911, 44, 1887–98, translation in Kauffman, *Classics*, p. 163.

[82] George B. Kauffman, 'Alfred Werner's Research on Optically Active Coordination Compounds',

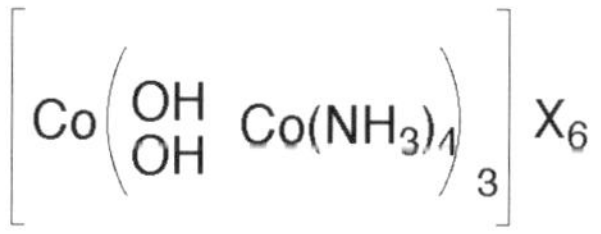

9.14 The formula (a) and configuration (b) of the first optically active inorganic complex that contained no carbon atoms. The stereoformula was not included in Werner's paper.

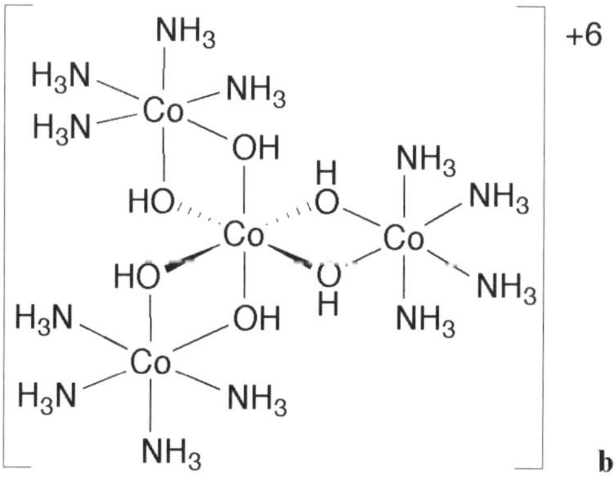

because many of these compounds had been prepared earlier in racemic form, and therefore required only resolution, and once the general method had been found, it could be relatively easily applied to other compounds. In paper 12 of the series, published in 1914, on active cobalt compounds, Werner reported the existence of an optically active complex that contained no carbon atoms, a dodecamminehexoltetracobalti salt (Figure 9.14), a compound first reported by Jørgensen in 1898. As Werner noted in the introduction to this paper, these compounds 'correspond constitutionally' to the triethylene diamine salts, and clearly were asymmetric. Although the optically pure compound was found to racemize relatively quickly in solution, the isolation of this compound and the measurement of its fleeting optical activity proved important, as critics of the theory could still point to the continued presence of carbon atoms in the compounds reported in the previous 11 articles. The resolution of an optically active coordination compound free of carbon would be Werner's last significant contribution to chemistry, as his health would decline quickly after 1914 until his forced retirement and death in 1919.

Coord. Chem. Rev., 1974, *12*, 105–49, p. 106. Kauffman lists 61 dissertations from Werner's students completed between 1898 and 1921. Unfortunately, he does not state which of these dissertations contained successful or unsuccessful attempts to isolate optically active compounds.

Conclusion

The adoption of Werner's system among chemists in Germany and elswhere is still relatively unexplored, but it does seem that it was not immediately regarded as a solution to the problem of representing the metal-ammine compounds. His initial 1893 article received an unremarkable eight-line abstract in the *Journal of the Chemical Society*, but in Germany the reception was somewhat warmer. The *Berichte* published a six-page review by F. Foerster, and on receiving advance word of Werner's essay, Meyer wrote to Emil Fischer that 'I am truly delighted with the theoretical-inorganic paper of A. Werner! These are truly new thoughts'.[83] Werner was still disappointed that chemists were slow to adopt his theory, however. In 1899, he wrote to a colleague that the slow acceptance of his theory derived from the:

> fear of leaving the firm ground of today's valence theory ... as soon as this fear is overcome, and according to all indications it will be, the molecular compounds with the constitutional formulas will also come into their own ... you, too, after studying them, will be convinced that the barrier between valence [compounds] and molecular compounds is a purely arbitrary one.[84]

On recognizing the reluctance of chemists to adopt the coordination theory, Werner adopted two strategies. First, in several articles following the initial 1893 article, Werner offered alternative arguments for justifying the octahedron, based on the isomer counting method used by Van 't Hoff in the *Voorstel* and later works. In these articles, Werner did not simply introduce the octahedron but suggested several alternative spatial arrangements of six groups around a central atom, including a pyramidal and hexagonal arrangements. Werner then predicted the number of possible isomers with each of these spatial arrangements and compared this number with the number of known compounds. Given that Werner changed his form of argument as early as 1895 and continued to employ it, it seems likely that Werner recognized that his initial *assumption* of the octahedron was a weak point of the theory, and he therefore devised a more rigorous argument to justify that assumption.

Werner's second tactic was to make chemists aware of the inherent problems with the predominant conception of valence as a directed force, and the inadequacies of the concept of molecular compounds. He wrote entries on inorganic chemistry for the 1902, 1903 and 1904 volumes of the *Jahrbuch der organischen Chemie*, and in 1905 he completed his influential textbook, *Neuere Anschauungen auf dem Gebiete der anorganischen Chemie*. While the *Jahrbuch* entries were broadly conceived and included discussion of analytical and radiochemistry, *Neuere Anschauungen*, despite its title, was largely a summary and compilation of Werner's own conception of

---

[83] Victor Meyer to Emil Fischer, 7 March 1893, translation in Kauffman, *Classics*, p. 7.
[84] Werner to unknown colleague, 9 January 1899, translation in Kauffman, *Classics*, p. 7.

inorganic compounds, to which he made additional revisions in the second (1909) and third (1913) editions. Werner also gave three public lectures on valence theory, in 1906 to the *Verein Deutscher Chemiker*, in 1907 to the British Association, and in 1911 to the *Deutsche Bunsen Gesellschaft*. In each of these works, Werner outlined the weaknesses that he saw inherent in the concept of directed affinity, and how his own coordination theory would solve this problem.

By 1913, Werner's stereochemistry of the metal-ammine compounds had been effectively established. It was not the isolation of the *cis* violeo compound in 1907, but the existence of optically active inorganic compounds that finally tipped the balance in Werner's favor, as evidenced by the awarding of the Nobel Prize in 1913. Werner had first been nominated for the Nobel Prize in 1907, along with Jørgensen, Mendeleev, Berthelot, Nernst, Ernest Rutherford, and Otto Wallach (Eduard Buchner was eventually awarded the prize for that year). According to the Nobel Committee, Werner was not selected in 1907 because his theory was still regarded as a 'working hypothesis', and:

> It appears that the time for awarding the orginator with a Nobel Prize will only be ready when his conception – which at the time is applicable only to certain special classes of compounds – has been developed into a theory of more general scope and when means have been obtained for testing experimentally the validity of this theory to a larger extent.[85]

Werner would receive 18 additional nominations for the prize in chemistry during the next six years, and he was finally selected in 1913, two years after King's first isolation of the optically active coordination compound.

Yet, as Jørgensen would agree, the system that Werner set up did not come without a price. Werner was forced to assume new hypotheses to maintain the consistency of the octahedron, and he was forced to manipulate formulas in order to maintain the constancy of the coordination number. The number of compounds predicted by Werner's theory was also extraordinarily large, and would have been daunting to anyone not absolutely convinced, as Werner was, that the coordination theory was correct about the possible number of metal-ammine compounds. Werner's strong methodology in which theory guided almost his every move, was in sharp contrast to Fischer's empiricism described in the last chapter. Returning to Rheinberger's metaphor of a labyrinth, if Fischer's establishment of a methodology for characterizing the carbohydrates was the *construction* of a labyrinth whose solution was unknown, Werner's 1893 paper was the presentation of a *constructed* labyrinth with its dimensions and paths already laid out. Werner required only the isolation of the predicted compounds to fill in the details.

---

[85] Memorandum of the Nobel Committee, quoted in Kragh, 'S.M. Jørgensen and His Controversy with A. Werner', pp. 216–17.

Chapter 10

Conclusion

> In modern organic chemistry there is presently no other theory so widely interesting, so rich with problems and so successful as stereochemistry.
>
> Paul Walden, 1900

By the beginning of the twentieth century, appeals to the spatial properties of molecules had become widespread, legitimate modes of explanation in organic chemistry, and Werner was well on his way to establishing a similar explanatory system for inorganic compounds. Wislicenus, Meyer, Hantzsch, Fischer and Werner, among others, had all become convinced that chemical theory would benefit by considering the spatial properties of organic molecules. Stereochemical principles have subsequently worked their way into modern chemical theory not as a separate area of study, but simply as one of the major factors that chemists must consider when planning a synthesis or explaining results. Modern organic chemists live and die by the axioms of stereochemistry, designing stereoselective, stereospecific and enantiospecific reactions. The term 'structure', first used for a concept so carefully defined by Butlerov and others during the 1860s, today *includes* the concept of 'configuration', and the two words have become almost synonymous. Part of the purpose of this book has been to show how the concepts of 'structure' and 'configuration' did *not* originally have the same meaning.

Stepping back from what has so far been primarily a historical narrative and analysis of the establishment of 'configuration' as a valid theoretical concept in chemistry, we can now conclude by turning to some of the general characteristics of the development of stereochemical theory that have appeared in the preceding chapters, moving outward in scope from the nature of stereochemistry itself, to the development of nineteenth-century chemistry, and finally, to broader issues concerning the general character of scientific change. What follows is as suggestive as it is explanatory, and room certainly remains for additional analysis of the philosophical, historical and sociological aspects of this episode.

The Nature of Stereochemistry

Structural Identity

We can return first to a principal epistemological question I posed in the Introduction. How did chemists become convinced that a spatial arrangement of atoms was a necessary component in certain chemical explanations? The main answer is that chemists employed spatial arrangement to explain certain cases of isomerism that were difficult or impossible to explain under conventional structure theory. Chemists regarded the use of spatial arrangement as an extension of the traditional use of 'arrangement' as an explanatory device for differentiating between isomeric compounds. Establishing the 'structural identity' of isomers, first accomplished by Wislicenus in 1873 for the lactic acids, proved to be the specific motivation for invoking spatial principles. Meyer began a serious inquiry into the stereochemistry of the benzildioximes only when he suspected that the two isomers were, in fact, structurally identical, and he and Auwers devoted the majority of their paper on the benzildioximes to a detailed justification for their claim of structural identity. Fischer became convinced that the known hexoses were structurally identical, and that Van 't Hoff's theory would allow him to differentiate between them.

An iron-clad 'proof' for the structural identity of isomers was not, however, an absolutely essential criterion for invoking spatial principles. Wislicenus, convinced that Van 't Hoff's theory of the tetrahedron was correct, applied it to the unsaturated acids with the assumption that the pairs of acids were structurally identical, and did not establish unequivocally that spatial causes were a necessary form of explanation. He justified this assumption on the basis of the theory's success in tying together a wide range of empirical data. Similarly, Hantzsch began his study of the stereoisomerism of the oximes in 1890 and the diazocompounds in 1894 by simply assuming their structural equivalence. According to Hantzsch, the carbon–nitrogen double bond in the oximes and the nitrogen–nitrogen double bond in the diazo compounds constituted true physical analogs to the carbon–carbon double bond of the unsaturated acids, and the analogy was convincing enough to Hantzsch that he did not make an explicit effort to establish their structural identity.

Although I have not yet explicitly pointed it out, there were, in fact, two different kinds of structural identity. The first kind was a true physical isomerism, in which two or more isomers differed only in their physical properties. For example, the lactic acids and the enantiomeric monosaccharides differed only in their optical activity. All of the chemical properties remained identical – hence the resulting identical chemical structures. In the second, more prevalent kind of structural identity, two or more isomers possessed different physical *and* chemical properties. Most of this book has dealt with the second kind of identity: the isomeric unsaturated acids, the benziloximes, the monosaccharides and certain metal-ammine complexes all differed in both physical and chemical properties, and any claims about the structural identity of these compounds precluded considering them as 'structurally identical' in the literal sense of the term. As Claus pointed out, stereochemists had made the fundamental assumption that differences in *degree* of reactivity were less important than differences in *kind* of reactivity. It did not matter, for example, whether γ or α benzilmonoxime yielded acetates that melted at different melting points, or whether they reacted with acetic anhydride or hydroxylamine at different rates. What counted was that both *yielded acetates*, and both *reacted with acetic anhydride or hydroxylamine in the same manner*. Spatial differences were then employed to explain these *particular*, *subtle* differences between the isomers' reactivity.

The concept of structural identity also arose in the case of metal-ammine chemistry, but in a way different from the tetrahedral carbon atom, and both Van 't Hoff and Le Bel correlated the presence of optical activity with a specific structural feature of a compound. Van 't Hoff expanded his theory to include cases of the second kind of structural identity (unsaturation), where no optical activity was involved. Werner, on the other hand, began with the manipulation of formulas, and was led to the octahedral and square planar arrangement of groups. Stereoisomerism and structural identity were *assumed* by Werner as a consequence of these spatial arrangements, and importantly, Werner did not initially predict the existence of optically active inorganic compounds. In other words, Werner was not attempting to explain the existence of structurally identical compounds. Van 't Hoff and Werner therefore approached molecular asymmetry from different directions. While Van 't Hoff used the existing property of optical activity to justify his theory, Werner predicted the existence of optically active coordination compounds as a consequence of, and justification for, the octahedron.

Although the broad epistemological basis for using spatial arrangement was the assumption of structural identity, chemists also adopted the principles of stereochemistry for personal reasons related to their own fundamental interests. For Van 't Hoff, the tetrahedron gave organic chemistry a mathematical or geometrical foundation that could allow a more precise prediction of the number of possible isomers and provide an explanation for existing cases of isomerism. For Wislicenus, the importance of stereochemistry lay in its ability to uncover the underlying structure of the atom. Meyer was intrigued because the tetrahedral arrangement of valences forced questions about the nature of valence itself. Hantzsch readily adopted the physical and chemical analogy between bonding in carbon and nitrogen atoms, and in 1896, had become convinced that spatial explanations of isomers would be predominant in inorganic chemistry. Fischer found the tetrahedron to be a convenient means of classifying the mono- and disaccharides, and Werner's stereochemistry was the consequence of his own coordination theory and consistent with his conviction that valence could not be a directed attractive force. Thus, although the general epistemic reason for invoking spatial properties of atoms remained similar among the stereochemists, we cannot tie the adoption of stereochemical principles to this single empirical factor – it was appealing to chemists for a variety of reasons beyond the simple ability to explain isomers.

The Visual Language of Chemistry and the Meaning of Chemical Formulas

Historians of science recently have begun to study the roles that visual images play in the construction of scientific knowledge, and shown how visual images are not simple representations of the world 'as it is', but as active tools, whose production and reproduction themselves shape the content of the theories they represent.[1] In one of the earliest and still most influential articles on the role of visual languages in science, Martin Rudwick drew attention to the extensive use of visual materials among geologists. In particular, he noted the centrality of visual materials to the scientific culture of geology:

> Whether in talks at conferences or in published scientific paper and books, modern geologists make extensive use of visual materials – maps sections, colour slides, diagrams of all kinds. They do this so much as a matter of course that to call these material 'visual aids' seems both inadequate and pretentious.[2]

Rudwick's observation for geology could equally be applied to chemistry. Like their geological counterparts, chemical formulas are an essential and ubiquitous tool for chemists in both discussion and publication, and referring to them merely as 'visual aids' would be equally pointless. The formative periods for these visual languages in chemistry and geology largely coincide, and geological images and chemical formulas are similar in the sense of showing graphically relationships that cannot easily (if at all) be expressed in words, although the relationship between image and phenomena are inverted. Chemical formulas represent microscopic entities in macrocosmic from, while geological diagrams represent macrocosmic entities in microcosmic form. As Rudwick pointed out, such visual languages depend on a largely conventional language that is not self-evident to novices, and reading maps and interpreting chemical formulas requires extensive practise to become familiar with their conventions and rules.

Van 't Hoff's suggestion that chemical formulas could represent the microscopic world of the molecule initiated the last major addition to the development of chemistry's visual language. As I described in Chapter 2, the crux of this transition in meaning in organic chemistry can be described as a shift from symbolic to iconic

---

[1] Alex Soojung-Kim Pang, 'Visual Representation and Post-constructivist History of Science', *Hist. Stud. Phys. Biol. Sci.*, 1997, *28*, 139–71.

[2] Martin Rudwick, 'The Emergence of a Visual Language for Geological Science', *Hist. Sci.*, 1976, *14*, 149–65, p. 149.

formulas. This transition is an excellent example of how the use of visual images, in this case structural formulas, can suggest a greater ontological meaning. In inorganic chemistry, there was no such transformation in meaning, as Werner, no doubt influenced by the success of Van 't Hoff's theory, consciously created iconic formulas for the metal ammines. The transformation in organic chemistry was quite significant in terms of the underlying ontological assumptions: the *same* chemical formulas, first developed as a convenient symbolic shorthand or mnemonic device to represent a compound's reactions, had by the end of the century become representations of the molecule as an object. Although the specific style of representing the three-dimensional properties of molecules on paper and in physical models continued to evolve over time, the general idea that the chemical structure is related to the physical molecule has remained part of chemists' thinking since the 1880s. Like their structural counterparts, stereoformulas also retained a degree of conventionality, as there were as many different stereoformulas as chemists. But stereoformulas became in another sense more 'realist', in the sense that they were now meant to portray the external properties of an object of the microworld.

There are many examples from the history of science in which visual tools, orginally developed as purely heuristic or pedagogic aids, became representations of real things. As Rudwick points out, geological images increasingly began to take on increased theoretical meaning beyond a representation of the static distribution of stratigraphic layers. In 1900, Paul Ehrlich first used visual diagrams to illustrate the side chain theory of immune action, a theory originally inspired by Fischer's lock and key hypothesis. Originally meant as a heuristic aid, these diagrams were soon interpreted to mean a literal picture of the entities in Ehrlich's immunological theory. During the 1960s, Geoffrey Chew and others invested new meaning and significance in Feynmann diagrams, which had been developed by Richard Feynmann in the 1940s as a mnemonic device for carrying out complex mathematical calculations.[3]

For our particular case study, there are three significant characteristics to the shift in the meaning given to chemistry's visual language. First, the ontological change was permanent – the meaning given to structural formulas by chemists since 1874 has never been replaced, although it was initially challenged. Second, the original meaning of chemical structure was retained for stereoformulas: that is, the concept of structure was not replaced wholesale, but chemists attributed another layer of theoretical meaning to the same chemical formulas. Another notable aspect, discussed in more detail below, is the absence of criticism regarding the addition of this layer of meaning to chemical formulas. In contrast, both Ehrlich's 'beautiful

---

[3] David Kaiser, 'Stick-figure Realism: Conventions, Reification, and the Persistence of Feynmann Diagrams, 1948–1964', *Representations*, 2000, *70*, 49–86. Alberto Cambrosio, Daniel Jacobi and Peter Keating, 'Ehrlich's "Beautiful Pictures" and the Controversial Beginnings of Immunological Imagery', *Isis*, 1993, *84*, 662–99.

pictures' and the reinterpretation of Feymann diagrams were hotly criticized at the time because immunologists and physicists had given a literal, realist interpretation to diagrams that others read as purely heuristic devices.

When developing their iconic visual language, chemists seemed to become more aware of the theoretical problem of how an atom could produce a divided and directed attractive force, and they addressed this problem with varying degrees of sophistication.[4] The easiest solution, adopted by Van 't Hoff, Hantzsch and Fischer, was to remain pragmatic and ignore it. Van 't Hoff implicitly raised the question of valence when he introduced the tetrahedron, but he was nearly silent about the exact nature of the carbon atom. His only suggestion about the physical location of the affinity unit was the public letter to Buys-Ballot, in which he located the sites of bonding with the faces of the tetrahedron. He also later confided to Wilhelm Ostwald that the carbon atom 'must consist of tetrahedral symmetry'.[5] It seems clear, however, that what he meant by the 'tetrahedral carbon atom' was not the atom itself, but the spatial distribution of valences *around* the atom. That he used different models to emphasize different aspects of his theory illustrates a pragmatic use of stereoformulas without an explicit interest in the actual appearance of the carbon atom. Hantzsch also practiced stereochemistry without any model of the atom beyond the simple tetrahedral arrangement of atoms, and he explicitly denied that any kind of theory of valence was necessary for the success of stereochemistry. Fischer was even more pragmatic, as he never addressed the issue at any level in any of his published papers on carbohydrates.

While Van 't Hoff, Hantzsch and Fischer essentially ignored the physical problem of valence, Wislicenus, Meyer, Wunderlich and Werner attempted with varying degrees of sophistication to reconcile the tetrahedron with known physical laws. At the theoretically most sophisticated end stood Meyer and Riecke, who proposed a unique, detailed physical concept of the carbon atom to account for its ability to bond in different ways. Wislicenus proposed the least sophisticated of the atomic models, in which the carbon atom was 'probably' a tetrahedron whose corners possessed a concentration of chemical affinity. Between Wislicenus and Meyer were the atomic models formulated by Wunderlich and Werner. In 1892, Hermann Sachse offered yet another sophisticated model to account for the appearance of directed valence

---

[4] The analysis in this section is a summary of Peter J. Ramberg, 'Pragmatism, Belief, and Reduction: Stereoformulas and Atomic Models in Early Stereochemistry', *HYLE*, 2000, *6*, 5–61. Similar observations were also made in Jost Weyer, 'A Hundred Years of Stereochemistry: The Principal Development Phases in Retrospect', *Ang. Chem. Int. Ed. Eng.*, 1974, *13*, 591–8.

[5] Van 't Hoff to Wilhelm Ostwald, 20 January 1888, in Hans-Günther Körber, ed., *Aus dem wissenschaftlicher Briefwechsel Wilhelm Ostwalds*, 2 vols, Berlin, Akademie-Verlag, 1969, vol. 2, p. 213.

in the tetrahedron. Expanding on earlier statements about the form of the carbon atom that resembled Wislicenus' 1887 model, Sachse proposed a mathematically and geometrically rigorous argument for the chemical bond that relied on circulating streams of solenoids that created attractive and repulsive forces depending on the distance between atoms.[6]

Yet it must be pointed out that not all chemists were compelled to construct atomic models, and those chemists who did suggest models often denied that they served any useful purpose. If this is true, then what role did these models serve? In simple terms, they explained the phenomena of valence, bonding and the tetrahedron at a higher level. But while they were explanatory (with varying degrees of success), they were not at all *predictive* for chemical theory. The models they presented were 'stories' meant simply to account for the physical characteristics of the carbon atom demanded by stereochemical theories. They were independent of chemical theory and irrelevant to the 'progress' of stereochemistry: that is, its capability of predicting isomers or postulating reaction mechanisms. Even the most influential of the models, those by Wunderlich and Werner, remained purely explanatory and had no predictive character. The concept of affinity in both was still rather vague physically, as it remained an *undefined* attractive force. Sachse developed the physically most sophisticated conception of valence and affinity, but his model was also an attempt to explain in more fundamental terms what stereochemists had already accepted: that the carbon atom was tetrahedral, that in carbon–carbon single bonds there was free rotation, and that in carbon–carbon double bonds there was no free rotation.

This tension between the active use of stereoformulas and the physical problems raised by them illustrates a general tendency of chemists to be pragmatic, in the simple sense of being practical, in adopting the tools and concepts necessary to reach their goals. That is, chemists have formed a pragmatic culture in which the adoption of useful concepts and tools occur readily even if they raise significant physical or philosophical questions. For example, chemists adopted the principle of valence almost without question, even though it raised crucial physical questions about its nature, because it helped to explain chemical behavior of substances and the appearance of isomers. Chemists adopted a similar pragmatic attitude toward the use of the atomic theory by excluding questions about the actual reality of atoms from their discussion, and simply proceeding to use them *as if* they existed.[7] Chemists constructed and used stereoformulas in a similar pragmatic way to give the explanation of isomers a graphic clarity by using the tetrahedral carbon atom or

---

[6] Hermann Sachse, 'Eine Deutung der Affinität', *Zeit. phys. Chem.*, 1893, *11*, 185–219, p. 186. Sachse's model is recounted in more detail in Ramberg, 'Pragmatism'.

[7] For the example of atomism, see Britta Görs, *Chemischer Atomismus: Anwendung, Veränderung, Alternativen im deutschprachigen Raum in der zweiten Hälfte des 19. Jahrhunderts*, Berlin, ERS Verlag, 1999.

octahedral metal atom – tools for portraying visually what could not be described verbally. Stereochemists used stereoformulas *as if* the tetrahedron were real, despite the fact that it forced questions about the nature of affinity and valence – questions that could be answered, but did not require an answer. We must also keep in mind that because these representations were meant to differentiate isomers by showing spatial differences in molecules, stereoformulas did not represent the three-dimensional characteristics of the *atom*. Therefore, a specific model for the carbon atom was unnecessary, as Fischer and Hantzsch (and, to a large extent, Wislicenus) carried out highly successful research programs without addressing the nature of valence.

The pragmatic use of stereoformulas and the independence of atomic models from chemical theory is also reinforced by doubts that chemists expressed about ever understanding the ultimate nature of matter. In a letter to Svante Arrhenius, Van 't Hoff noted the provisional nature of the tetrahedron:

> the representations themselves, atom, molecule, their dimensions, and perhaps their shapes, are after all something doubtful, as is the tetrahedron itself. But as long as something good comes *from* it, one can console oneself and believe that there is also something good *in* it.[8]

As we saw in Chapter 6, two years after the appearance of his paper with Riecke, Meyer remarked that their theory of valence had no relevance to his actual work on the benzildioximes, and none of the papers on the chemistry of the benzildioximes mentioned the Meyer/Riecke theory. Even Wislicenus, the strongest proponent of physical atomism among the stereochemists, did not offer any concrete suggestions about the ultimate nature of matter.

In describing the use of stereoformulas and atoms as 'pragmatic', however, we must be careful not to equate this pragmatism with pure instrumentalism. 'Pragmatism' in the sense described here does not mean, for example, that chemists considered stereoformulas as mere instruments that did not depict reality independent of human experience. Stereochemists believed that the groups around the carbon atom were arranged in a tetrahedron, and that stereoformulas represented in some fashion the molecule as a physical object. In short, they believed they had access to the physical appearance of the molecule, and had not simply invented instruments for prediction. Furthermore, as Wislicenus made evident, considering molecules as physical objects – and thereby implying a three-dimensional distribution of valences of the atoms in the molecule – necessitated considering the *atom's* properties as a physical object: specifically, how the valences could be directed in space.

---

[8] Van 't Hoff to Svante Arrhenius, undated, quoted in Andrew G. van Melsen, *From Atomos to Atom: A History of the Concept Atom*, New York, Harper, 1960.

The Reception of the Tetrahedron

We can turn now to the general question of the nature of the overall reception of Van 't Hoff's theory among chemists in Germany. As we have seen, there was very little application or extension of Van 't Hoff's ideas before 1885, even by Van 't Hoff himself. Baeyer's strain theory in 1885 and especially 'Spatial Arrangement', made chemists aware of the potential in Van 't Hoff's ideas, and after 1887 the number of papers discussing spatial properties of molecules continued to increase. In 1890, the pages of the *Berichte* were filled with papers on stereochemical topics, including Sachse's stereochemical theory of cyclohexane, Bischoff's dynamic hypothesis, and several of the spatial models of the nitrogen atom, and in 1891 the configuration of the fatty acids was established using Wislicenus' methods. Also during this period, Baeyer was occupied with research on the constitution of benzene and cyclohexane, in which he incorporated a significant stereochemical component.[9] The late 1880s and early 1890s therefore saw the largest initial spurt of activity in the application of Van 't Hoff's theory, and throughout the 1890s chemists continued to apply stereochemical principles to additional compounds and elements.

One of the most prominent aspects of this general increase in the application of stereochemical principles after 1885 is the relative lack of controversy. From the beginning, chemists had a cautious optimism towards the tetrahedral carbon atom and accepted it positively, albeit privately at first. As chemists began to apply the tetrahedron to various cases, no significant rivalries developed among chemists as to the best way to apply the theory, or about alternative interpretations of Van 't Hoff's theory. The only significant dispute between stereochemists was that between Hantzsch and Meyer, which on examination does not display the characteristics of a scientific controversy in the true sense of the word.[10] The reception of the tetrahedral carbon atom could best be described as an application or expansion of Van 't Hoff's basic ideas to additional compounds and elements, but without any discussion of alternative theories.

This is not to say that stereochemistry proceeded without criticism. We have seen several criticisms of stereochemistry, from Kolbe, Lossen, Claus, Fittig, Jørgensen and Michael. Kolbe thought the tetrahedral carbon atom was pure speculation with no grounding in the true principles of chemical research. Fittig resented Wislicenus'

---

[9] R. Benedikt, 'Technologie der Fette', in Richard Meyer, ed. *Jahrbuch der Chemie*, 1891, *1*, p. 383. Tonja A. Koeppel, 'Significance and Limitation of Stereochemical Benzene Models', pp. 97–113 in O. Bertrand Ramsay, ed., *Van't Hoff-Le Bel Centennial*, Washington, DC, American Chemical Society, 1975. Tonja A. Koeppel, *Benzene Structure Controversies 1865–1920*, PhD, University of Pennsylvania, 1973.

[10] Kragh has offered a useful discussion of what constitutes a true controversy in science. Helge Kragh, 'S.M. Jørgensen and his Controversy with A. Werner: a Reconsideration', *Brit. J. Hist. Sci.*, 1997, *30*, 203–19.

intrusion into his earlier work on unsaturated acids, destroying the facts that he and his students had so carefully gathered. Michael thought Wislicenus had made empirically ungrounded *a priori* assumptions that were a 'corruption' of the true natural laws of chemistry. Claus thought the use of the spatial properties of the oximes unnecessary for a chemical explanation, and Lossen found Van 't Hoff's model of the double bond physically unsatisfactory. While all valid in some respects, these criticisms were isolated and came from individual chemists who addressed specific aspects of other chemists' work. There was no attempt, systematic or otherwise, to offer an alternative explanation for structurally identical compounds that did not rely on the older principles of structural chemistry. Michael presented by far the strongest critique of stereochemistry, but offered no alternative explanation for structural identity of certain isomers. We would therefore be hard pressed to lump any of these chemists together into an organized movement or allied group of chemists in opposition to the use of stereochemical principles, or, with the exception of Kolbe and perhaps Michael, to attribute extreme anti-stereochemical views to any of them. Lossen, for example, distanced himself quite strongly from Kolbe's critique, and Fittig did not specifically criticize Wislicenus' theory.

If there was any hesitancy to accept the tetrahedral carbon atom, it tended to be towards Van 't Hoff's second and third hypotheses, and not towards the first. Denial of Van 't Hoff's hypothesis by chemists could therefore often mean only the denial of the second and third hypotheses. In fact, the first hypothesis – the tetrahedron as the explanation for optical activity – seems to have been accepted without any controversy at all, as there was little or no opposition to it. Claus and Lossen, for example, only commented on stereochemical theories of the double bond. Furthermore, to my knowledge, chemists offered no alternatives to the tetrahedral carbon atom for the explanation of the optical activity of organic compounds after 1874. This was either because most chemists accepted the asymmetric carbon atom as the explanation for optical activity, or because Van 't Hoff's general correlation between structure and optical activity could be accepted without necessarily adopting the tetrahedron. In any case, most objections to Van 't Hoff's theory were directed at the second and third hypotheses, and to what could be considered as excessive speculation in their application.

It appears that once Van 't Hoff had suggested the iconic nature of formulas, chemists not only had little or no opposition to it, they saw no alternatives for ascertaining the spatial properties of molecules. There were essentially no broad debates between chemists about the utility or the reality of the tetrahedron, and there were no suggested alternatives to it.

The overall acceptance of Van 't Hoff's theory with little controversy fits well within a general pattern of how dramatic changes in chemical theory tend to occur relatively quietly and quickly, without much discussion or fuss.[11] In his discussion of such 'quiet revolutions' in chemistry, Rocke has pointed to several key events of nineteenth- and twentieth-century chemistry. The crucial change in the 1850s from a dualist to unitary view of molecules took place extraordinarily quickly, in some cases virtually overnight, and with very little comment. During the 1830s, the transformation of Berzelian notation from simple abbreviations to active tools for exploring the composition and reactions of organic compounds also took place extraordinarily quickly almost without comment, at the time or since. Similarly, the adoption of Kekulé's benzene theory and the periodic table, the conversion from valence-bond to molecular orbital theory, and the dramatic increase in the use of physical instrumentation in chemistry in the second half of the twentieth century all occurred quickly and with little controversy or attention.

Van 't Hoff's theory could easily be added to this list of quiet revolutions. It has remained famous to this day, yet the fundamental nature of the ontological transformation in chemical formulas has been forgotten or never noticed, even at the time. The fact that there were no alternatives to the tetrahedral carbon atom has likewise escaped any attention or comment. The quietness and completeness of the transformation is further substantiated by the absence of any comment or criticisms about the shift to an iconic meaning of chemical formulas. Despite numerous caveats and warnings by nearly all chemists about the ontological status of formulas and models during the 1860s, most chemists seemed to forget them almost overnight, and embraced the shift almost intuitively. It was as if Van 't Hoff had pointed out the obvious – the epistemological distinction between chemical and physical arrangement of atoms had been unnecessary.

The reception of Van 't Hoff's theory also supports the results of Stephen Brush's recent efforts to understand the role that novel prediction plays in the adoption of theories by physicists and chemists. Brush has found that physicists tend to give a higher status to the role of theories as retrodictive, primarily explanatory devices than to their ability to make novel predictions. But the examples from chemistry that Brush has explored so far – Mendeleev's periodic table, Kekulé's benzene theory, and molecular orbital theory – suggest that the ability of theories to make predictions played a large role in their acceptance by chemists.[12] Because the tetrahedral carbon

---

[11] Alan J. Rocke, 'Quiet Revolutions in Chemistry', Dexter Award Symposium, Division of History of Chemistry, American Chemical Society Meeting, San Francisco, March 2000. I thank Alan Rocke for making the text of this talk available to me.

[12] Another factor Brush mentions in the acceptance of chemical theories is their ability to organize existing knowledge, a factor that also played a role in the acceptance of stereochemistry. Steven Brush, 'The Reception of Mendeleev's Periodic Law in America and Britain', *Isis*, 1996, *87*, 595–628. Brush 'Dynamics of Theory Change in Chemistry. Part 1: The Benzene Problem, 1865–1845', *Stud. Hist. Phil. Sci.*, 1999, *30*, 21–79. Brush, 'Dynamics of Theory Change in Chemistry. Part 2: Benzene and Molecular Orbitals, 1945–1980', *Stud. Hist. Phil. Sci.*, 1999, *30*, 263–93.

atom allowed chemists to predict the existence of novel compounds, its relatively fast and non-controversial adoption would substantiate Brush's existing case studies. Indeed, one of the roles Van 't Hoff desired for his theory was the more accurate prediction of the number of possible isomers. Although it is not yet entirely clear *why* chemists would prefer theories that make novel predictions, it may be related to the craft-like nature of chemistry in which the active creation of new things plays a central role.[13]

Although Baeyer's strain theory was the first novel application of the tetrahedral carbon atom, it was Wislicenus who gave perhaps the most important stimulus to the development of Van 't Hoff's hypotheses, first by sponsoring the translation of *La chimie dans l'espace* in 1875, and by the publication of 'Spatial Arrangement' in 1887. While Baeyer's theory was ingenious and important, it did not have much relevance to acyclic compounds. In 'Spatial Arrangement', however, Wislicenus made clear how Van 't Hoff's hypothesis could be used to organize information about the unsaturated acids and stimulate research into the genetic relationships between them in order to assign configurations. Although chemists did not accept all of Wislicenus' argument, after its publication there seemed little doubt among chemists that the unsaturated acids were spatial isomers.

Another way of emphasizing the importance of both Wislicenus in general and 'Spatial Arrangement' in particular is to consider a counterfactual scenario that removes him from the scene either in 1875 or 1887. What if Wislicenus had suddenly died in 1875, before he read *La chimie dans l'espace*, or in 1885, before composing 'Spatial Arrangement?' Or what if, for whatever reason, he had decided not to suggest the translation of Van 't Hoff's book and/or study the unsaturated acids? Such questions are impossible to answer, but they do drive home how the fortunes of a particular theory can be shaped by contingent factors, and suggest clearly the key roles that particular individuals can play in the reception and spread of new theories. What impact would *La chimie dans l'espace* have had without a German translation? Would Meyer have reconsidered the isomerism of the benzildioximes without the stimulus of 'Spatial Arrangement'? Would Hantzsch and Werner have proposed a spatial model of the nitrogen atom without Meyer and Auwers' establishment of the structural identity of the benzildioximes? It seems highly unlikely that Van 't Hoff's theory would *never* have taken root had Wislicenus not taken the role of its primary defender and promoter – indeed, there does not seem to have been any alternative to

---

[13] Peter J. Ramberg, 'Paper Tools and Fictional Worlds: Prediction, Synthesis and Auxiliary Hypotheses in Chemistry', pp. 61–78, in Ursula Klein, ed., *Tools and Modes of Representation in the Laboratory Sciences*, Dordrecht, Kluwer (2001).

it – but without him, the rate at which chemists adopted the hypothesis, at least in Germany, would likely have been much slower.[14]

Although Wislicenus' specific idea that chemical attractions determine a molecule's 'favored' configuration was from the beginning viewed with some skepticism, and was eventually abandoned altogether, the general concept of a favored configuration and the idea that the proportions of different configurations in a molecular aggregate are determined by the amount of heat have proven useful to this day for judging the reactivity of various compounds. More important than the specific hypotheses Wislicenus made about molecular dynamics, however, was the new way in which he approached and thought about chemical problems. In the history of chemistry, 'Spatial Arrangement' occupies a unique position. It was the first work of chemical theory that advocated a useful conception of molecules as objects governed by a set of inner dynamics. The idea that chemical phenomena could be explained by mechanical means was not unique to Wislicenus. Other chemists before him, as far back as Robert Boyle, had been sympathetic to chemistry founded on the principles of mechanics, but attempts to make chemistry purely mechanical had never been particularly effective. Wislicenus, however, could trace the fate of individual atoms within the molecule by following a line from starting materials to products. Such reasoning was possible by assuming that the *combination* of chemical and physical forces in a molecule caused a molecule's chemical behavior. He justified his specific ideas of a favored configuration and planesymmetric addition, and believed he had gained access to the intricate 'mechanisms' of chemical reactions because they enabled him to direct a highly successful experimental research program.

In any case, the overall acceptance of Wislicenus' chemistry derived not from the specific hypotheses he offered, but from its central organizing principles, in which a sufficient explanation involved an appeal to an underlying cause, and diverse but related phenomena (such as the behavior of the unsaturated acids) should be tied together by a 'unified viewpoint', preferably a few axioms that would give rise to testable predictions. This Whewell-like concilience of inductions is the primary reason, I believe, that Meyer, Auwers, Hantzsch and others, including Wislicenus'

---

[14] The available evidence does not suggest any direct influence of Wislicenus on Werner's coordination theory or Fischer's carbohydrate chemistry, but Fischer's case allows some speculation. As I noted in Chapter 8, Fischer's first paper on the reaction of sugars with phenylhydrazine appeared in 1884, but his project did not really begin until the appearance of the second paper in March 1887, and the joint paper with Julius Tafel the following August. It was only at this point that Fischer became increasingly occupied with sugar chemistry and sorting out the isomers by considering the chemical properties of the sugars and the genetic relationships between them. Part of this delay was certainly due to his move from Erlangen to Würzburg in 1885, but the renewed research on sugars also appeared within six months of the publication of 'Spatial Arrangement', which argued for the assignment of configurations based on the chemical behavior and genetic relationships of the unsaturated acids. After seeing the overall methodology employed by Wislicenus, Fischer may have been spurred to investigate further the genetic relationships between the carbohydrates.

colleague at Leipzig, Wilhelm Ostwald, praised 'Spatial Arrangement' so highly. Although most chemists, including those who drew their inspiration from Wislicenus' revitalization of Van 't Hoff's theory, found fault in *specific* explanations he offered, they became convinced of its validity because it offered a unified set of principles that tied together previously unrelated and unexplained chemical relationships. Although Michael argued successfully that in some cases these principles were unfounded, because he did not present a feasible theoretical alternative the tetrahedral model remained for most chemists the most plausible means of ascertaining the physical form of molecules.

Wislicenus' assumption of a dynamic molecule, and the adoption of this principle by other chemists, changes our current understanding of the 'origins' of conformational analysis and the emergence of a chemical dynamics in organic chemistry. Conformational analysis has been assumed to begin with Hermann Sachse's 1890 study of cyclohexane – a work, like many others, that followed Wislicenus' lead in using rotation about the carbon–carbon single bond to predict isomers.[15] Organic chemists during this period have also been assumed to use molecular structures that lacked dynamic characteristics. In her study of the disciplinary dynamics of nineteenth- and twentieth-century chemistry, Mary Jo Nye has assumed that chemical dynamics appeared only with the emergence of the new physical chemistry of the 1880s, and that only in the 1920s did chemists decide 'to join ranks with physicists in search of the dynamics of a chemical molecule indistinguishable from the physical molecule'.[16] Yet beginning with Van 't Hoff in 1874, chemists had already begun to assume that the chemical and physical molecules were identical. It also seems clear that the origins of conformational analysis lay within Wislicenus' chemistry (and in Van 't Hoff's own work), and were well founded and elaborated by the 1920s, although chemists remained skeptical about the precise nature of the factors governing the formation of favored configurations.

---

[15] Weyer, 'Hundred Years of Stereochemistry'. Colin A. Russell, 'The Origins of Conformational Analysis', pp. 159–78 in Ramsay, *Van 't Hoff-Le Bel Centennial*. O. Bertrand Ramsay, 'The Early History and Development of Conformational Analysis', pp. 54–77 in J.G. Traynham, ed., *Essays on the History of Organic Chemistry*, Baton Rouge, LA, Louisiana State University Press, 1987.

[16] Mary Jo Nye, *From Chemical Philosophy to Theoretical Chemistry: Dynamics of Matter and Dynamics of Disciplines, 1800–1950*, Berkeley, University of California Press, 1993, pp. 6 and 112. Nye, 'Chemical Explanation and Physical Dynamics: Two Research Schools at the First Solvay Conference, 1922–1928', *Ann. Sci.*, 1989, *46*, 461–80. The quote is from the article.

The Nature of Chemistry

The History of the Atomic Theory

Having summarized some of the major characteristics of early stereochemistry, we can turn now to the broader issue of what stereochemistry tells us about the development of chemistry in the nineteenth century. The first major point to make about stereochemistry is its place in the long and complex story of atomism in nineteenth-century chemistry. Chemists throughout the first three-quarters of the nineteenth century overwhelmingly supported and used the concept of chemical atoms, defined as the smallest unit of chemical combination, and kept this separate from any beliefs they might have held about physical atoms, defined as indivisible, massy, microscopic particles. It is difficult to imagine, however, accepting the premises of stereochemistry without accepting some form of explicit physical atomism. The basic premise of stereochemistry was that stereoformulas represented in some fashion a three-dimensional object, and stereochemists declared openly that it was no longer possible to consider molecules or atoms as dimensionless points with no physical significance. Wislicenus, for example, thought it impossible not to conceive of atoms as three-dimensional objects.[17] The emergence of stereochemistry was therefore the first explicit appearance of a full-blown physical atomism in organic and inorganic chemistry.

As Wislicenus pointed out, the 'reality' of the tetrahedron implied that the atom had parts, and that the carbon atom was composed of simpler primitive elements (*Urelemente*). Meyer's model of the carbon atom suggested a subatomic structure, and his pyrochemical experiments were explicit attempts to decompose the known elements into simpler substances. For the other stereochemists, the belief in primitive elements is not as evident, but the general conviction that the known elements were composed of still simpler substances permeated much of nineteenth-century chemistry, beginning with William Prout's 1815 hypothesis that all known elements were multiples of hydrogen.[18] The appearance of the periodic law in 1869 also suggested that the existing elements were composed of a smaller number of more fundamental substances, and Wislicenus, Victor Meyer, Lothar Meyer and Wilhelm Ostwald, among others, assumed that the relationship between chemical properties and the atomic weight indicated the presence of a more fundamental substance. The

---

[17] Peter J. Ramberg, 'Johannes Wislicenus, Atomism, and the Philosophy of Chemistry: A Translation and Commentary', *Bull. Hist. Chem.*, 1994, *15/16*, 45–53.

[18] William H. Brock, *From Protyle to Proton: William Prout and the Nature of Matter, 1785–1985*, Adam Hilger, Bristol, 1985. H. Kragh, 'Julius Thomsen and Nineteenth Century Speculations on the Complexity of the Elements', *Ann. Sci.*, 1982, *39*, 37–60. W.V. Farrar, 'Nineteenth Century Speculations on the Composition of the Elements', *Brit. J. Hist. Sci.*, 1965, *2*, 297–323; W.V. Farrar, ' "Chemistry in Space" and the Complex Atom', *Brit. J. Hist. Sci.*, 1968, *4*, 65–7.

appeal of Prout's hypothesis and the periodic law stimulated research into precise atomic weight measurements in the 1860s by Jean Servais Stas, modifications of Prout's hypothesis by Lothar Meyer in 1876 to account for Stas' fractional atomic weights, and experiments to test whether chemical reactions could change the atomic weight of the elements by Hans Landolt in 1883.[19] Beginning in the 1850s and the valence theory, organic chemists were already presenting models of the carbon atom that implied some sort subatomic structure that had become numerous enough for Lossen to write an extended article in 1880 on the shortcomings of making assumptions about the subatomic structure of the atom.[20] Among the stereochemists, Wislicenus and Meyer, at least, saw clearly how the tetrahedral carbon atom fitted within the current context of research about primitive matter and the earlier context of speculations on subatomic structure based on the existence of valence.

Mechanism and Physicalism in Chemical Theory

One of the significant changes between the concepts of structure and spatial arrangement was the incorporation of a distinctly mechanical conception of atoms and molecules that included both static and dynamic elements. The differentiation of isomers by the spatial arrangement of atoms was static, because it assumed that the simple position of the atoms determined the physical and chemical properties of the molecule. Although novel, this means of explanation was little different from that employed in the structure theory itself – it simply offered a more accurate prediction or explanation for the number of possible isomers. The application of Van 't Hoff's second hypothesis and planesymmetric addition and elimination, on the other hand, explicitly invoked a dynamic conception of the molecule in which chemical properties were determined by specific intramolecular motions of the atoms.

While all stereochemists relied on stereochemical statics, their attitude toward stereochemical dynamics varied widely. At one extreme stood Wislicenus, who did not hesitate to use the concepts of favored configurations and planesymmetric addition and elimination to postulate various 'mechanisms' for reactions. In his dynamic hypothesis, Carl Bischoff supposed that the amount of free rotation about a carbon–carbon single bond depended not on attractive forces, as Wislicenus had argued, but on 'the filling of space' of the radicals attached to the carbon atoms.[21] At the other extreme were Fischer and Meyer, who did not incorporate any degree of

---

[19] Britta Görs, *Chemischer Atomismus: Anwendung, Veränderung, Alternativen im deutschprachigen Raum in der zweiten Hälfte des 19. Jahrhunderts*, Berlin, ERS Verlag, 1999.

[20] Alan J. Rocke, 'Subatomic Speculations and the Origin of Structure Theory', *Ambix*, 1983, *30*, 1–18. Wilhelm Lossen, 'Über die Vertheilung der Atome im Raum', *Annalen*, 1880, *204*, 265–364.

[21] G.V. Bykov, 'The Conceptual Premises of Conformational Analysis in the Work of C.A. Bischoff', pp. 114–22 in Ramsay, *Van 't Hoff-Le Bel Centennial*, p. 117. Translation in Bykov.

stereochemical dynamics into their theories. Between these two extremes were Hantzsch, who used planesymmetric elimination to predict the outcome of the Beckmann rearrangement but did not often graphically portray reaction processes, and Werner, who assumed a specific mechanism in the reaction of metal ammines but did not use molecular dynamics to assign configurations. It is perhaps interesting to note that the stereochemists' level of commitment to mechanism was largely independent of their statements about physical atomic models.

This dichotomy between the static and dynamic components of Van 't Hoff's theory also helps us understand the principal successes and failures of early stereochemistry. The absolute success of stereochemical theory was in its use of chemical statics to count isomers and make the theoretical number of compounds agree with the number isolated. The accomplishment of this agreement demonstrated the usefulness of considering the molecule as a spatial object, from the explanation of optical activity, to the unsaturated acids, oximes, diazo compounds, monosaccharides and metal ammines.

The success of the new chemical dynamics, on the other hand, is not so easy to measure. Wislicenus and Hantzsch used the concept of planesymmetric addition, an admittedly naïve but reasonable assumption, to great success. Michael showed this assumption to be unfounded for the unsaturated acids, however, and in 1921 Meisenheimer would show that Hantzsch's similar assumption for the Beckmann rearrangement was also incorrect.[22] Axialsymmetric addition and elimination, as improbable as it seemed, and as mechanically inconceivable as it was, would become the only alternative. The concept of favored configurations also proved useful for Wislicenus and others, but as Michael pointed out, chemists were not in all in agreement about what factors governed the formation of a favored configuration. The concept of substitution, for which chemists had an equally naïve literal conception, would suffer the same fate as planesymmetric addition after the discovery of the Walden inversion in 1896.[23] In 1911, Emil Fischer concluded on the basis of his own study of the Walden inversion that 'van't Hoff's model of carbon is certainly incorrect for dynamic processes, but maintains it value for statics'.[24]

---

[22] Georg Wittig, *Stereochemie*, Leipzig, Akademische Verlagsgesellschaft, 1930, p. 190.

[23] Paul Walden, a student of Carl Bischoff, began the work that led to the discovery in 1893, and made the discovery itself in 1896. Paul Walden, 'Über die gegenseitige Umwandlungen optischer Antipoden', *Berichte*, 1896, 133–8; 1897, *30*, 2795–8; 1897, *30*, 3146–51; 1899, *32*, 1833–64.

[24] Emil Fischer to William Ramsay, 31 October 1911, Emil Fischer Papers, BANC MSS 71/95 z, The Bancroft Library, University of California, Berkeley, CA. Four other letters to colleagues describe this conclusion. In a letter to Theodore Richards, he described the tetrahedron as a 'total failure' for chemical dynamics.

Methods and Methodology in Organic Chemistry

We can also place the practise of stereochemistry within the overall development of chemical methods, defined as specific manual techniques in the laboratory, and overall chemical methodology, defined as modes of reasoning or categories of mental activity.[25] Broadly considered, the methods used by chemists to elucidate the spatial properties of molecules were no different from those used for ascertaining chemical structures. Chemists continued to use standard isolation techniques of crystallization and distillation, and used the physical properties of compounds merely as 'markers' for the occurrence of chemical (or stereochemical) transformations. The magnitude of any change in physical properties remained unimportant, and only the existence of a change was necessary to indicate that a chemical transformation had taken place. In other words, stereochemists used precisely those properties that Kekulé and Butlerov had considered epistemologically suspect for making claims about molecules as physical objects. In this transdictive process, chemists made sophisticated claims about the microscopic world of the individual molecule, specifically its *appearance*, on the basis of purely macroscopic manipulations of various substances.

Despite this relatively unchanging nature of chemical practice, we can, however, note the appearance of three new methods that took on increasing importance. The first is the measurement of optical activity, a physical property that Van 't Hoff used to justify the tetrahedron and that Fischer relied on for identification of individual monosaccharides. The introduction of optical activity was not crucial for the development of much of stereochemistry, however, as many of the compounds we have considered here – unsaturated acids, benzildioximes, many metal ammines – were not optically active. Furthermore, chemists used optical activity in the same sense as other physical properties – as markers for identification. The second new method was the measurement of conductivities by Hantzsch and Werner, used primarily to establish configurations of labile isomers that could not otherwise by isolated and characterized by traditional methods. Again, this method proved useful for certain compounds, but was not required for the practise of stereochemistry.

A third new method was the use of hand-held physical models to predict possible spatial isomers or suggest favored configurations. These models served a similar role to their paper counterparts, as a medium for three-dimensional thought experiments. That these models were invaluable is clear: numerous letters to Van 't Hoff in 1875 request sets of models, Fischer noted in his autobiography how Baeyer's improvised bread ball and toothpick models were insufficient and how only a good set of models allowed him to solve a problem in sugar chemistry. Yet the actual practice of using models remained almost invisible from the beginning, and references to physical

---

[25] This distinction was made by John H. Brooke, 'Methods and Methodology in the Development of Organic Chemistry', *Ambix*, 1987, *34*, 146–55.

models in the sterochemical literature are rare and extremely brief, limited to the publication of templates in *Die Lagerung*, Baeyer's description of Kekulé models in the strain theory, a single mention of physical models by Wislicenus in 'Spatial Arrangement', the brief mention of the Friedlander models by Meyer and Fischer, and Sachse's templates for constructing models of cyclohexane. There were no discussions, to my knowledge, of how chemists were to use them, nor of any dangers in using them. It also seems tacitly assumed in these brief mentions of models that chemists knew how to use them, or how to build and acquire model sets. In a study largely devoted to the use of molecular models in the twentieth century, Eric Franceour has called such models 'the forgotten tool', an aspect of chemical experimentation that is unmentioned and implicit within the chemical literature.[26] The early history of stereochemistry suggests that these models were 'forgotten' as soon as they were invented.

Within stereochemistry, we can identify three methodological strategies employed by chemists. There was, for example, the continued use of analogy between the stereochemistry of different compounds. This is especially notable in Werner, who first suggested the physical correspondence between carbon–carbon and carbon–nitrogen double bonds. His model of affinity, initially developed for carbon, was easily transferred to metal atoms for the coordination theory.

The growth of interest in stereochemistry during the last quarter of the nineteenth century also exemplifies the bifurcation of methodology that took place in the 1850s (outlined in Chapter 2), and reveals the persistence of the older tradition of an inductive, classificatory chemistry. The stereochemists as a whole emphasized strongly the predictive character of spatial arrangements, and the theory itself was explicitly hypothetico-deductive in character, making predictions about the existence of new compounds and, via reaction mechanisms and favored configurations, allowing the assignment of configurations. Many of the criticisms of sterochemistry we have encountered, especially Michael and Jørgensen, but also Fittig and Kolbe, were based primarily on their objection to the excessive use of speculation by stereochemists.

Importantly, however, this split between primarily inductive and primarily deductive methodologies is also found among the stereochemists themselves. Wislicenus, Hantzsch and Werner placed by far the most emphasis on the importance of predictive power in a theoretical system, and Van't Hoff's metaphor of military conquest described in Chapter 4 appropriately describes the methodology of these

---

[26] Eric Francoeur, 'The Forgotten Tool: the Design and Use of Molecular Models', *Social Studies of Science*, 1997, *27*, 7–40.

three chemists. A few fundamental principles drove their research programs, by which they 'shot down the fortress from different sides'. Later, they brought forth the evidence necessary for these principles. Chemical synthesis served the specific purpose of testing their hypotheses – in other words, it was a means to an end, not an end in itself.

Van 't Hoff's metaphor applies less well to Meyer or Fischer, as both were less bold in their use of predictive hypotheses, and worked in a style that remained more cautious, making generalizations (the advancement of Van 't Hoff's 'single battery') about the spatial properties of the benzildioximes and monosaccharides. They assigned configurations to stereoisomers, but did not rely on the same level of mechanical assumptions as Wislicenus or even Hantzsch. They tended to use only those aspects of Van 't Hoff's hypothesis that closely resembled the structure theory: the prediction of new compounds. Meyer predicted a third benzildioxime, and Fischer matched the growing number of synthetic sugars with the isomers predicted by Van 't Hoff's theory, but neither Meyer nor Fischer employed assumptions about the mechanism of chemical reactions or favored configurations, and neither was forced to invent additional hypotheses to maintain their own configurational assignments. Baeyer also exemplified this epistemic caution when he proposed his strain theory, and Meyer and Fischer, as the two most distinguished students of Baeyer, did not stray far from their master's own style.

The contrast between Wislicenus and Hantzsch on one hand, and the Baeyer school on the other, therefore illustrates in microcosm the overall century-long shift from regarding theories as 'conclusions' drawn from data, to theories as 'speculations' that predicted new data. The methodological split also demonstrates that, in this case at least, there was no simple dichotomy between deductive and inductive methodologies. Whether chemists used one or the other method depended on their level of their theoretical commitment. When employing the principles of stereochemistry, Meyer, Fischer and Baeyer used its ability to predict isomers, but nothing more. Wislicenus used the implications about mechanism and configurations to make predictions, and publicly maintained that chemical theory progressed only through the testing of predictions.

Although he used Van 't Hoff's theory as a guide for predicting the number of possible sugars, Fischer was not nearly so enthusiastic, at least in public, about the role of theory in guiding experiment. In his autobiography, he made a veiled and mildly disingenuous criticism of Wislicenus, who tended to use experiment 'to test preconceived theoretical views', rather than to 'follow the phenomena empirically'.[27] Wislicenus was perhaps the easiest to accuse of 'fixing' the data to fit his hypothesis, but he did not see any justification for those accusations. Fischer and Baeyer were

---

[27] Emil Fischer, *Aus Meinem Leben*, Berlin, Springer, 1922, p. 112.

also skeptical of Hantzsch's unorthodox methods, in which 'any conductivity is measured and a conclusion drawn from it without caring what is happening in the solution'.[28] Baeyer and Fischer thought it inadequate to say that substances existed if they had not actually been isolated and characterized.

To these two relatively well-known methodological characteristics – the use of analogy and the tension between the natural historical and experimental, predictive roles of chemical theory – I would like to add a third: the use of chemical compounds as exemplary models, in the same sense that individual organisms are used in biology. Historians and philosophers of biology have long examined the role of exemplary models in creating biological theories. As experimental systems, biological models are created by 'reconstructing' or 'standardizing' an organism to maximize the benefit for a given scientific project, and the process of producing a 'standard' organism can take years or decades, in a process that has been explicitly compared to the purification of chemicals. Milton Greenman, Director of the Wistar Institute which developed the standard laboratory rat, explicitly compared the standardization of the rat to the purification of chemicals by chemists. In his book on *Drosophila* genetics, *Lords of the Fly*, Robert Kohler also makes several comparisons between the standardization of *Drosophila* and chemical practice. For example, he says, '"Standard" drosophilas were, like chemical reagents or physical instruments, constructed artifacts of laboratory life.'[29]

For chemists, the 'standardization', or purification of compounds is a relatively simpler process, but serves the similar purpose of preparing a model substance that will provide insight into the general nature of chemical structure and configuration. As biologists have studied only a chosen few out of the thousands of possible organisms at their disposal, so organic chemists have relied on the chemistry of a select few compounds: alcohol, ether, benzene, and in the case of stereochemistry, lactic acid, the unsaturated acids, and the benzildioximes. In one of the only discussions of exemplary models in chemistry, Klein has noted the central importance of alcohol and its derivatives during the 1820s and 1830s in the transformation of organic chemistry from a natural-history oriented science to an activity based on artificial experimentation.[30]

---

[28] Adolf von Baeyer to Emil Fischer, 19 October 1904, Fischer to Baeyer, 2 November 1904, Emil Fischer Papers, BANC MSS 71/95 z, The Bancroft Library, University of California, Berkeley, CA.

[29] Bonnie Tocher Clause, 'The Wistar Rat as a Right Choice: Establishing Mammalian Standards and the Ideal of a Standardized Mammal', *J. Hist. Biol.*, 1993, *26*, 329–49, p. 337. Robert Kohler, *Lords of the Fly: Drosophilia Genetics and the Experimental Life*, Chicago, IL, University of Chicago Press, 1994, p. 88.

[30] Ursula Klein, *Experimente, Modelle, Paper-Tools: Kulturen der organischen Chemie im 19. Jahrhundert*, Habilitationsschrift, University of Konstanz, 1999, pp. 276–81.

We can note the similar use of compounds as models for constructing a general theory of the spatial characteristics of molecules. In Van 't Hoff's theory, optically active molecules gave a clue as to the arrangement of valences not only for the asymmetric carbon atoms in those molecules, but for all carbon atoms in all molecules. For Wislicenus, the unsaturated compounds explained by Van 't Hoff's theory served as exemplars for establishing the nature of carbon–carbon bonding, for understanding the character of addition/elimination processes, and subatomic structure. The highly specific chemistry of the benzildioximes led Meyer to modify Van 't Hoff's second hypothesis, a theory that turned out to be highly specialized, but also led Werner to the general stereochemistry of nitrogen. For Paul Walden, the halogenated optically active malic acids served as a model for understanding the nature of substitution.

Expanding on the specific role of organisms as exemplary models in biology, Geison and Laubichler have recently pointed out how the success of biological theories is based on the highly contingent, interacting elements of variation among those organisms with local laboratory techniques and the regional or national scientific culture or 'style'.[31] The success of chemical theories is also dependent on such variation. The success of chemists in constructing general theories was tied to the idiosyncratic chemical behavior of specific known compounds, the existence of which was highly contingent and often restricted to a single location. It might be helpful here to consider another counter-factual scenario. Suppose that Meyer had not worked extensively with hydroxylamine, or that Fischer had not discovered phenylhydrazine. Would either have made, or thought of making, the necessary derivatives of benzil or the monosaccharides? Suppose Kiliani had not published his chain extension technique that allowed much of Fischer's own work on the carbohydrates. While it is impossible to predict the course of events in these hypothetical cases, it illustrates the importance of chemists discovering and working with the right substance at the right time.

Comparisons of biology and chemistry have tended, quite rightly, to emphasize the common natural historical aspects of both sciences, or to focus on the issue of reductionism, but with the exception of Klein's discussion of alcohol chemistry, have not yet recognized parallels in experimental methodology. As a methodological strategy in chemistry, the existence of exemplary models has largely escaped the attention of historians of chemistry, perhaps because this is not a deliberate strategy on the part of chemists, and is largely a contingent factor in chemical practice. But treating compounds as exemplary models for the development of chemical theory addresses a fundamental problem in the emergence of scientific theories about how we obtain knowledge of the universal from the specific. How can we go from the

---

[31] Gerald L. Geison and Manfred D. Laubichler, 'The Varied Lives of Organisms and Variation in the Historiography of the Biological Sciences', *Stud. Hist. Phil. Biol. Sci.*, 2001, *32C*, 1–30.

properties of individual compounds to concepts applicable to all compounds? As Richard Burian notes for the parallel problem in biology, this is 'an especially acute version of the traditional philosophical problem of induction'.[32] In the case of chemistry, the use of exemplary models and a sophisticated visual language can also help explain how chemical theory can change dramatically with a relatively static, unchanging set of chemical methods.

Broader Issues and Further Questions

Generational Dynamics

In a 1974 overview of the general developments in stereochemistry since 1874, Jost Weyer noted an overall generational split among chemists regarding their acceptance of stereochemistry, and subsequently argued for the validity of Max Planck's hypothesis that scientific theories are adopted largely by members of a younger generation.[33] According to this view, new theories are not always adopted by convincing opponents of their correctness, but largely because the older opposing generation dies out, and because the members of a younger generation are more open and flexible to new ideas and methods. If true, 'Planck's principle', also endorsed by Thomas Kuhn in *The Structure of Scientific Revolutions*, implies that the empirical evidence in favor of theories does not play a significant role in their acceptance. While in its bare form the Planck-Kuhn thesis is somewhat simplistic, the perspective gained from studying the generational dynamics of a discipline has rich potential for understanding the nature of scientific change, and a brief generational analysis of stereochemistry is revealing.

Although Weyer's analysis was confined to one part of an overall short history of stereochemistry, and he did not differentiate between the acceptance of Van 't Hoff's three hypotheses, his general claim is correct: the acceptance of Van 't Hoff's theory was relatively swift, but it was not universal with respect to age. Not surprisingly, the older generation of chemists, roughly those born before 1830, were mostly indifferent, with the exception of Kolbe, who was openly hostile to it. There were effectively no strong supporters of the principles of stereochemistry from this generation, although Modderman, Wurtz and Buys-Ballot were warmly supportive of Van 't Hoff as a young scientist pursuing a novel idea. Those chemists born after 1840, and especially those born after 1850, on the other hand, were the strongest supporters and promoters of 'chemistry in space'. In this group we find Meyer and

---

[32] Richard M. Burian, 'How the Choice of Experimental Organism Matters: Epistemological Reflections on an Aspect of Biological Practice', *J. Hist. Biol.*, 1993, 26, 351–67, p. 367.
[33] Weyer, 'A Hundred Years of Stereochemistry'.

Le Bel (both born in 1847), Van 't Hoff and Fischer (both born in 1852), Carl Bischoff (born in 1855), Hantzsch (born in 1857), Paul Walden (born in 1863) and Werner (born in 1867). Other chemists who wrote supportive letters to Van 't Hoff in the 1870s were Theodore Zincke (born in 1844) and Otto Wallach (born in 1847). The most curious group of chemists includes those born between 1830 and 1840. Claus (born in 1838), Fittig (born in 1835), Lossen (born in 1838) and Jørgensen (born in 1837) were opponents of stereochemistry, but other major chemists born in the 1830s, including Landolt (born in 1831), Wislicenus (born in 1835) and Baeyer (born in 1835), were strong supporters. Those chemists born in the 1830s were therefore a transitional generation in which the opposition between the traditional 'structural' and the novel stereochemical meaning of chemical formulas came to a head.

As in all generational analyses of this sort, there are exceptions. Some of the opponents of stereochemistry were born after 1840, for example Anschütz (born in 1844), Michael (born in 1853) and Eugen Bamberger (born in 1857), and were in roughly the same generation as Hantzsch, Meyer and Fischer. There were no exceptions, to my knowledge, among those chemists born before 1830. This preliminary generational analysis, admittedly sketchy and confined to those chemists included in this study, indicates neither a sharp generational break nor a universal adoption of Van 't Hoff's theory among chemists of all age groups over a certain period. There does appear to be a 'generational inertia', or resistance to scientific innovation implied by Planck's principle, but it also does not appear to be a major factor in the overall dissemination of stereochemistry. Indeed, we can specify a significant particular influence of the younger generation on those chemists born during the 1830s. Van 't Hoff was 21 when he conceived of the tetrahedral carbon atom, and Wislicenus, 19 years older, very quickly became his champion. Baeyer, the same age as Wislicenus, also quickly and enthusiastically adopted Van 't Hoff's theory, and Landolt, 23 years Van 't Hoff's senior, was prompted to incorporate the implications of Van 't Hoff's theory in his ongoing study of optical activity. Werner, 10 years younger than his mentor Hantzsch, quickly convinced Hantzsch of the tetrahedral nitrogen atom.

Whether this episode itself is significant for understanding generational influences on the spread of scientific theories is unclear, as the current literature discussing generational issues is sparse and contradictory. A statistical analysis of the acceptance of Darwinism in 1978, for example, showed that there was a large generational difference between proponents and opponents of Darwinism, but that statistically, it was not true that Darwinism was only accepted by a younger generation, and significantly, members of the older generation who accepted Darwinism did so as readily as the younger generation.[34] Also in 1978, Lenoir came to the opposite

---

[34] David L. Hull et al., 'Planck's Principle', *Science*, 1978, *202*, 717–23.

conclusion for the acceptance of *romantische Naturphilosophie* among generations of German biologists, and argued that specific generational factors contributed to the acceptance or rejection of the concept of the prototype (*Urtyp*) among biologists.[35] Two more sophisticated recent studies have moved away from simply linking generational cohorts with acceptance or denial of specific theories. Jonathan Harwood has linked complex generational differences within genetics to the overall changes in the German professoriate in the Weimar period, and Lynn Nyhart has charted the changing fortunes of nineteenth-century German morphology to the emergence of generational groups within the German professoriate, the dispersion of Darwinism and the emergence of experimental biology.[36] From these few existing studies of the generational dynamics of disciplines, the most likely conclusion to draw is either that the results are inconclusive, or just as probable, highly dependent on the historical subject.

Nevertheless, in the case of stereochemistry, we can identify a distinct generational shift that results in two principal questions: what caused this apparent generational shift, and why is the generation born in the 1830s split? I would suggest three preliminary inter-related explanations. First, the adoption of stereochemistry among chemists of the later generation can be traced to the increasing reliance on structural formulas and physical modeling – tools that chemists increasingly considered as realistic depictions of the microworld of the molecule. Another more speculative factor could be the increasing role of the kindergarten in childhood education. Based on the idea that young children learn about the world by active manipulation of objects, in particular carefully designed geometric toys, Friedrich Fröbel opened the first kindergarten in 1837, and by the 1850s kindergartens were prominent throughout Germany, at the same time as chemists were introducing physical modeling techniques.[37] Those chemists born during the 1850s may therefore have begun to attend kindergarten in greater numbers, thus acclimatizing them to the role of three-dimensional modeling and construction. This is clearly speculative, as data on kindergarten attendance for chemists is admittedly sparse.[38]

The split between those chemists born during the 1830s can also be explained, I

---

[35] Timothy Lenoir, 'Generational Factors in the Origin of *Romantische Naturphilosophie*', *J. Hist. Biol.*, 1978, *11*, 57–100.

[36] Jonathan Harwood, *Styles of Scientific Thought: The German Genetics Community, 1900–1933*, Chicago, IL, University of Chicago Press, 1993.

[37] Meinel has noted the striking similarity between the toys designed by Fröbel and Hofmann's early demonstration models, Christoph Meinel, 'Modelling a Visual Language for Chemistry, 1860–1875', unpublished manuscript. The relationship between kindergartens and modeling is also noted in Christopher Ritter, 'The Impulse to Visualize and Meaning-in-Practice: Chemical Models, 1857–1874', 1999 History of Science Society Annual Meeting, Pittsburgh, PA.

[38] Meinel, in 'Modelling a Visual Language for Chemistry', notes that the kindergartens had spread throughout Protestant Germany by the 1850s. In this respect, a demographic survey of chemists from Protestant and Catholic Germany could be useful in characterizing their attitudes toward modeling.

would argue, by the continuing tension between the predictive and classificatory role of theories suggested above. By the 1860s and the 1870s, the hypothetico-deductive character of the structure theory had become more fully established, and chemists publicly embraced speculation as a useful role for theories. The generation born during the 1830s were trained in the 1850s, at a time when the major theories of organic chemistry were stabilizing around the central concept of structure, and it was not yet completely fashionable to embrace speculation publicly. The generations trained in the 1860s and after may therefore have attributed a greater significance to the speculative role of theories, while those trained in the 1850s were more divided. We might also consider the influence of local traditions on this methodological shift. It seems no accident that Claus, Fittig and Lossen were all students at Göttingen during the late 1850s, and all carried with them a strong sense of empiricism and a distaste for speculation. Explaining the split generation of the 1830s by invoking a shift to a new methodology is not entirely satisfactory, however, as the reason for the methodological shift remains unclear. Clearly, if we are to obtain any concrete generalizations about the influence of generational dynamics on the development of chemistry, studies of other episodes in chemistry and of chemists in different countries are necessary.

Another avenue for chemistry in Germany would be integrating generational studies into the broader changes in the German professoriate in late nineteenth and early twentieth centuries, along the lines suggested by Harwood in his fine-grained analysis of German geneticists in Weimar Germany. In *Styles of Scientific Thought*, Harwood identified two distinct groups of geneticists in Germany, labeled 'pragmatic' and 'comprehensive'. Those geneticists identified as pragmatic, Harwood noted, tended to be specialists rather than generalists, had little or no interest in philosophical issues or art, music and literature, and adopted a particular political orientation. Those geneticists who were 'comprehensive' tended to identify with the classical ideal of the German professoriate, valued education as a cultivation or *Bildung*, valued philosophy and high culture, and tended to be generalists in their science. Building on Fritz Ringer's well-known thesis on the decline of the German Mandarins (the 'comprehensives', according to Harwood's analysis), Harwood relates the changing German professoriate to the process of modernization involving the emerging middle class occupied with vocational education or *Ausbildung*, resulting in a shift among geneticists from comprehensive to pragmatic.

Harwood's rich thesis, which can only be alluded to in brief here, shows some potential for a comparative analysis of the changes in the German chemical professoriate that would complement and enhance the studies of academic chemistry in imperial Germany by Jeffrey Johnson.[39] Does Harwood's model of generational

---

[39] Jeffrey Johnson, 'Academic Chemistry in Imperial Germany', *Isis*, 1985, 76, 500–24. Johnson,

change between pragmatics and comprehensives apply to chemistry? I have already characterized chemists in general as pragmatic in terms of the theories and modes of representation they adopt. But can we identify chemists as pragmatic in Harwood's sense? In the sense that chemists are less inclined to speculate on the deeper underlying theoretical foundations of their discipline, all are pragmatic under this definition, but is there a dichotomy between pragmatic and comprehensive styles of thought among chemists? Harwood himself identified Emil Fischer as a pragmatic, and Werner could equally be identified as one. Wislicenus, a generation older than Fischer, could be identified as a comprehensive, based on his interest in the fundamental theories of chemistry, as could Hantzsch and Meyer, based on the generality of their research interests.

Research Groups and Research Schools

Another method for characterizing the nature of scientific change has been the 'research school', broadly defined as a group of scientists practicing similar methods, methodologies or instrumentation who are attempting to solve a common problem.[40] The science of chemistry seems naturally suited to exploiting the concept of research schools, and historians of chemistry have looked at the composition and administration of many of the large chemical research groups directed by Liebig, Fischer, Baeyer, Kolbe, Jean-Baptiste Dumas and August Wilhelm Hofmann.[41] These individual case studies have helped enormously in characterizing the formation of and roles played by various research schools in nineteenth-century chemistry, but by their very nature they emphasize the pedagogical aspects of research schools in the

---

'Hierarchy and Creativity in Chemistry, 1871–1914', *Osiris*, 1989, 5, 214–40. Johnson, *The Kaiser's Chemists: Science and Modernization in Imperial Germany*, Chapel Hill, NC, University of North Carolina Press, 1990.

[40] Gerald L. Geison, 'Scientific Change, Emerging Specialties, and Research Schools', *Hist. Sci.*, 1981, 19, 20–40. Other recent studies of research schools include: Gerald Geison and Frederic L. Holmes, eds, 'Research Schools: Historical Reappraisals', *Osiris*, 1993, 8; J.B. Morrell, 'The Chemist Breeders: The Research Schools of Liebig and Thomson', *Ambix*, 1972, 19, 1–46, and John W. Servos, 'Research Schools and their Histories', *Osiris*, 1993, 8, 3–15.

[41] Morrell, 'The Chemist Breeders'. Joseph Fruton, 'Contrasts in Scientific Style: Emil Fischer and Franz Hofmeister, Their Research Groups and Their Theory of Protein Structure', *Proc. Am. Phil. Soc.*, 1985, 129, 313–70. Fruton, 'The Liebig Research Group: A Reappraisal', *Proc. Am. Phil. Soc.*, 1988, 132, 1–66. Fruton, 'Contrasts in Scientific Style: Research Groups in Chemical and Biological Sciences', American Philosophical Society, Philadelphia, PA, 1990. L.J. Klostermann, 'A Research School of Chemistry in the Nineteenth Century: Jean-Baptiste Dumas and his Research Students', *Ann. Sci.*, 1985, 42, 1–40. Frederic L. Holmes, 'The Complementarity of Teaching and Research in Liebig's Laboratory', *Osiris*, 1989, 5, 121–64. Alan J. Rocke, 'Group Research in German Chemistry: Kolbe's Marburg and Leipzig Institutes', *Osiris*, 1993, 8, 53–79. Michael Keas, *The Structure and Philosophy of Group Research: August Wilhelm Hofmann's Research Program in London (1845–1865)*, PhD, University of Oklahoma, 1992.

sense of individual, geographically concentrated research *groups*, rather than schools in the broader sense.

Considering the individual research groups in stereochemistry, indeed we find thriving centers of activity in Leipzig, Würzburg and Zürich, directed by Wislicenus, Fischer, Hantzsch and Werner, all enthusiastic promoters of stereochemistry, who no doubt created dozens of adherents among their students.[42] But the concept of research *schools* in the broader sense also applies well to early stereochemistry. From the core text of *Die Lagerung* we can identify two distinct research schools. The first was the least controversial, and involved the 'isomer counting' school, in which the concept of spatial arrangement was suited to matching possible structures to known compounds, and nothing more. Because all stereochemists were committed to the spatial reality of molecules as an explanation for cases of isomerism, they all belonged in this school. The second research school can be characterized as the 'dynamic' or 'mechanistic' school, represented by Wislicenus, Bischoff and Hantzsch. This school derived from the enthusiastic adoption of Van 't Hoff's second hypothesis as a predictive tool and/or the assumption of planesymmetric addition. While *Die Lagerung* provided the initial impetus for this dynamic school, 'Spatial Arrangement' would prove to be just as influential.

The emergence of these broad 'schools' in stereochemistry fits well within our current understanding of how research schools operate to create new scientific knowledge. First, we have seen how a cluster of related research schools can form around the consequences of one central idea – the tetrahedral carbon atom – and how that idea could then be adapted to different chemical systems. Stereochemists as a group explored a similar set of chemical problems and provided similar answers, and created an identifiable, geographically disperse research school in the classical sense of the term.

Research schools have also been considered to be the site of disciplinary change, and most case studies of research schools have tended to focus on those schools that eventually, either deliberately or accidentally, created a new scientific discipline, and in the process often created sources of conflict within the larger community. This is not the case for stereochemistry. Although the stereochemists were clearly occupied with something new and shared a broadly common set of goals, stereochemistry did not emerge as a new discipline. Promoters of stereochemistry never worked for disciplinary autonomy by creating separate societies, journals or university chairs, and remained fundamentally organic or inorganic chemists. Nor was 'chemistry in space' confined to one area of chemistry – organic, inorganic and biochemistry all benefitted equally from its principles. While the chemists occupied with

---

[42] Although Meyer can be placed within the general stereochemical research school, he did not create a large research group in that topic. Rather, his project involved a few students, primarily Auwers and Demuth as assistants, and was highly localized to the benzildioximes.

stereochemical problems did not always have harmonious collaborations, the case of stereochemistry does show that research schools can emerge from within an established tradition, and enhance that tradition, without either the formation of a new discipline or the creation of conflict.[43]

Because stereochemists considered their research to be a logical development of traditional aims of chemical theory – that is, as an inevitable outgrowth of the structural theory of organic chemistry established during the 1860s – the early development of stereochemistry is an excellent example of how different research schools can take root and thrive within an established tradition. In his study of Kolbe's research group, Rocke has argued that research schools can be used for understanding the creation of various 'inflection points' within established disciplines.[44] While Rocke's study was applied to the character of one specific research group, the case of stereochemistry provides a broader example of an inflection point within the discipline as a whole, in which new ideas emerge and develop without concurrent changes in instrumentation and methods. In her study of nineteenth-century morphology, Nyhart has used the term 'orientation' (from the German *Richtung*) to describe morphology as an area within biology whose practitioners were spread over many different institutes and locations.[45] The practitioners of stereochemistry were similarly scattered among different universities, approaching the 'morphology' of molecules in different ways.

Finally, we can use the various stereochemical research groups for characterizing a particular national 'style' of German chemistry. Organic chemistry during the last quarter of the nineteenth century has traditionally been characterized as a 'golden age' of classical organic chemistry, dominated by the isolation and structural elucidation of organic compounds, and the synthesis of a seemingly endless number of new compounds. As organic chemistry became more routine, historians have noted, many aspiring organic chemists became bored with it, and actively chose to pursue other research projects, usually in the emerging subdiscipline of physical chemistry. In large part, the schools of physical chemistry established in the United States were products of students who had rejected organic chemistry for just that reason.[46] To a certain extent, these accusations about the routine nature of research in

---

[43] In a review of the concept of research schools, Servos noted the propensity for historians to focus on the emergence of controversial schools, and asked whether historians could find 'harmony and cooperation' among research schools if they were to look for it. Servos, 'Research Schools and their Histories'.

[44] Alan J. Rocke, 'Group Research'.

[45] Lynn K. Nyhart, *Biology Takes Form: Animal Morphology and the German Universities*, Chicago, IL, University of Chicago Press, 1995.

[46] Jeffrey Johnson, 'Hierarchy and Creativity in Chemistry, 1871–1914' and 'Academic Chemistry in Imperial Germany'. Robert Kohler, 'The Origin of G.N. Lewis's Theory of the Shared Pair Bond', *Hist. Stud. Phys. Sci.*, 1971, *3*, 343–76. Nye, *From Chemical Philosophy to Theoretical Chemistry*, note 16. Nye,

organic chemistry are true. Chemistry in many research groups became routine and unexciting, and in many projects, it was only a matter of time before a satisfactory structure was found, or a compound successfully purified. The movement toward other subdisciplines of chemistry must also have sharpened the need for the updated chemical institutes that appeared during the 1890s.

Yet chemists were very excited about the prospects for Van 't Hoff's theory because of its fundamental *theoretical* importance. In the foreword to the 1894 edition of *Die Lagerung*, for example, Wislicenus remarked that Van 't Hoff's theory had provided 'the stimulus of a meaningful movement, in certain ways indeed of a new epoch of our science'.[47] In 1900 Paul Walden remarked that 'in modern organic chemistry, there is presently no other theory so widely interesting, so rich with many problems, and so successful as stereochemistry'.[48] What provoked this excitement? Simply put, those chemists who developed 'chemistry in space' continued where structural chemistry, long considered by chemists and historians alike as the end point of nineteenth-century organic chemistry, had left off. Originally, the tetrahedral carbon atom offered an innovative way of accounting for isomerism, but as Wislicenus noted, 'chemistry in space' could offer deeper insights into the nature of atoms themselves.

Wislicenus was the most explicit and enthusiastic about the implications of stereochemical principles, but we cannot forget the other stereochemists. Hantzsch confided that current conceptions of valence and affinity must be modified, and he assumed that stereoisomerism would be the more common form of isomerism. Meyer's study of the benzildioximes resulted in a short-lived but detailed model of the carbon–carbon bond, and Werner's study of inorganic stereochemistry stemmed directly from his theoretical ideas on affinity and valence. The stereochemistry of the sugars led Fischer directly into one of the most intriguing subjects of all – the relationship between molecular geometry and the processes of life. Overall, we must conclude that stereochemistry was not a routine exercise in isolation and synthesis, characteristic of the traditional conception of 'classical' organic chemistry, but fundamental to research on the nature of atoms and affinity, and the chemical basis of life.

'National Styles? French and English Chemistry in the Nineteenth and Early Twentieth Centuries', *Osiris*, 1993, *8*, 30–49. John W. Servos, *Physical Chemistry from Ostwald to Pauling: The Making of a Science in America*, Princeton, NJ, Princeton University Press, 1990.

[47] Johannes Wislicenus, foreword in Jacobus Henricus Van 't Hoff, *Die Lagerung der Atome im Raume*, 2nd edn, Braunschwieg, Vieweg, 1894.

[48] Paul Walden, 'Fünfundzwanzig Jahre stereochemischer Forschung', *Naturwissenschaftliche Rundschau*, 1900, *15*, 145–8, 157–60, 169–73, 185–8 and 197–9.

Appendices

Appendix 1

Tjaden Modderman to Neighbors of Van 't Hoff's Parents, 1874

In the fall of 1874, Tjaden Modderman, Professor of Chemistry at the University of Groningen, received a copy of the Voorstel *from Van 't Hoff, and soon sent a letter to friends in Rotterdam who were neighbors of Van 't Hoff's parents. The original letter in Dutch is located in the Van 't Hoff papers, Johns Hopkins University. Ernst Cohen published a near complete German translation of this letter in his 1912 biography of Van 't Hoff, but did not comment on why Modderman wrote the letter, and it is not clear whether Modderman was at all personally acquainted with Van 't Hoff as a child or young man. This translation was prepared from the original Dutch.*

Groningen, 17 October 1874

Dear Friend,

Mr Van 't Hoff's brochure with the long title has been sent to me by its author. Should you speak with him, please express my gratitude. I read his 'proposal' with interest. Great sagacity lies within it, and it furnishes abundant evidence of the author's comprehensive knowledge and synthetic aptitude. If the author develops as well practically as theoretically, he shows more than a little promise for the future. All the same, it is very difficult to make a definite judgement about the value of his 'proposal'. One could argue for or against it, but *summa summarum*, experiment must decide. As for myself, I believe that structural formulas have had their day, at least that they will not stay in the foreground as boldly as they do today. Undoubtedly they have done much good, but also bad. It is so easy to believe you are finished with a compound when it is put together on paper. Mr Van 't Hoff can do no better than to demonstrate by experiment the value of his speculations. In this sense, his hypothesis will always be useful, whatever the outcome may be. But to have the necessary pluck [*opgewektheid*] and desire to do so, he must keep believing in his theory with unwavering conviction. So I am of the opinion that it would be much better for him if no critiques by me or others appeared. For this would probably raise doubts in him, even if he did not agree with the objections. As for myself, I would have advised him against publishing on the matter until such time as he could substantiate his assertions with experiments. I am curious how it will progress.

One thing is certain, there is something about your young friend.

Jeanne is afflicted with fever now and then, but is well otherwise, as is Adriane. Greet your wife etc. cordially on our behalf, and have faith in me,

Tjaden Modderman

Appendix 2

Van 't Hoff's Preface to *La chimie dans l'espace*, 1875

Van 't Hoff's preface to La chimie dans l'espace *is an expanded version of the first few sentences in the 1874 Dutch pamphlet about the limitations of current structural theory. He also displays his frustration at the indifference with which his theory has so far been greeted.* La chimie dans l'espace *has not yet been translated into English.*

It is organic chemistry – the chemistry of carbon – that gave birth to the beautiful theory of atomicity, allowing the representation of the molecule as a grouping of atoms connected among themselves according to certain laws to form a complete and stable system. It uses a very simple symbol, which indicates simultaneously the qualitative and quantitative composition, and the chemical character, a symbol that almost exactly allows the prediction of the role that this or that combination plays under the influence of some reagent. It is organic chemistry, I say, that gave birth to this theory, and it is that which will provide for its development by making it more accurate.

The hypothesis of atomic constitution gives not only concise and simple form to the observed facts; it is not only an ingenious notation, but a theory; it generalizes, it predicts; this is the mark of its accuracy.

We can debate the value of proofs expressed in terms of the qualities of the bodies, but those proofs that express themselves by means of numbers are indisputable. What, then, is the criterion for the present theory? The answer to this question is simple – it is found in the number of possible isomeric compounds: it is here that the atomic theory pronounces itself in a rigorous manner, it is here that it sees its views confirmed or invalidated.

Let us examine the facts: *we will see that in many cases the number of existing isomers surpasses those predicted by theory*. One does not dare admit it; one hides it under terms such as physical or geometrical isomers; one has recourse to the biatomicity of carbon; one closes the chain of one's atoms; but the truth remains: *current theory is powerless to predict certain isomers*.

Those who allow themselves to be guided by a hypothesis, basing themselves upon some exclusive numerical principles and pronouncing upon all things with the same intolerance, will encounter some contradictory facts. On the other hand, he who has the prudence to remain somewhat undecided will be able to employ those facts that its predecessor came up against, to develop a theory with the aid of those same facts which had apparently been fatal to its success.

In a pamphlet that has appeared in Holland, I have attempted to give a theory of structural formulas in space; I called for a discussion of my views, and I wanted to profit from its results.

I have been done the honor of having a French translation inserted into the *Archives Neérlandes*, but for that which I have so much desired – a judgment, a discussion – I have waited in vain. But each new hypothesis, if I may say so here, ought to pass through two very distinct phases: at first, it is a question of seeing if it offers an advantage over existing hypotheses in its interpretation of the known facts. Then, if it has received this support, it is still necessary that experiment demonstrate the truth of its predictions. It is for this first phase that the judgment of a scientist is so desirable. I have not been able to find any of this in Holland.

At the same time, M. Le Bel pronounced himself in favor of one part of my views in the 5 November 1874 meeting of the Chemical Society of Paris, and in the 19 March meeting of this year M. Berthelot was kind enough to present certain observations on our ideas: I am seizing this occasion to implore the eminent chemist to accept the expression of my very fervent gratitude.

It appears, then, that the moment has arrived to present my theory with the developments it has since received, and to ask for a judgment from foreign scientists.

Rotterdam, May 1875

Appendix 3

Felix Hermann and Johannes Wislicenus to Van 't Hoff, 1875

From the content of these letters, it appears that they were sent together to Van 't Hoff in November 1875. Cohen published a full transcription of Wislicenus' letter in his biography of Van 't Hoff. The original letter is located in the Van 't Hoff papers, Johns Hopkins University. Hermann's letter has remained unknown until now, and is also located in the Van 't Hoff papers.

Würzburg, 9 November 1875

Honored Sir!

In the month of June of this year, you had the kindness to send my supervisor, Herr Professor Wislicenus, a sample copy of your brochure entitled 'La chimie dans l'espace' along with the necessary models. Herr Professor Wislicenus, impressed by the importance of the contents of your manuscript, honored me with the task of presenting a report on your theory to the local 'chemical society' whose members include several lecturers and specialists as well as students. I began the reporting in last night's session, in which I initially simply conveyed the formal development of your hypothesis. As I had expected beforehand, your theory raised in no small way the interest of all those concerned with chemistry in the scholarly sense; and without doubt this interest became still greater on further acquaintance with the beautiful consequences of your theory, if I can be permitted to judge by my personal experience. For the more I have deepened my study of your manuscript, the more I have become convinced of the truly fundamental importance of your views. The same meaning will be clear in the opinion of Herr Professor Wislicenus, who promised to include a few lines with the present letter. I join heartily in the desire that your recent views be spread to the widest circles of our science. Rendering your manuscript into the German language could initiate the achievement of this goal in Germany, where presently the scientific preservation and advancement of chemistry in other countries is certainly de-emphasized. Because I must take for granted your comprehensive knowledge of the German professional literature and complete mastery of the German language, I don't know whether my inquiry to render your manuscript into German suits you. Should my intention be desired by you, I ask in this case to allow some freedom in the treatment of your work. Because of the variation in the manner of presentation in both languages, a literal translation of a scientific treatise is an almost impossible task. But then it occurred to me on study of your work, that it could be made somewhat more palatable for the chemical public, for whom your theory is primarily intended. Developments of a mathematical nature, which for an audience trained in this area are easily comprehensible with only fleeting clues, must be recast explicitly to achieve the full understanding of a readership deficient in mathematical preparation who only rarely consider mathematical thoughts. Forgive this remark, for it rests not only on my own view, but also on the opinions of competent persons to whom I have introduced the contents of your work.

Changes in the intended spirit would be confined to the deviations which I would allow myself in the revision of your work, if you should grant your kind permission. In this case I would send you within a short time the manuscript of the German treatment for your [illegible] approval.

Finally, permit me to inform you that if you grant permission for this, a trainee occupied with scientific work in our laboratory, Herr Dr Fabingi from Hungary, who shows the most lively interest in your views, would see to a translation of your work into the Hungarian language. In anticipation of a delightful response, I sign with highest esteem

Dr F. Hermann
Assistant in the chemical laboratory of the University of Würzburg.

Würzburg, 10 November 1875

Honored Colleague!

You should not blame deficient interest or insufficient appreciation for the failure of my thanks to arrive until today for the kind forwarding of your small work 'La chimie dans l'espace'. The reason lies in various fates which have befallen me since its arrival, along with the desire to study your work thoroughly.

Permit me to say that your theoretical development has given me great joy and great pleasure, and that I see in it not only an extraordinary intellectual attempt to explain previously incomprehensible facts, but also believe that it provides a great host of entirely new stimuli for our science, and subsequently will be of epoch-making importance. It therefore deeply satisfied me, and I welcomed it with the greatest joy. In a short time, you will see in my own projects, so I hope, the highest approving interest I take in your work.

I recommend your cordial agreement to the content of the enclosed letter from my assistant Dr Hermann, if you have not already accepted any other offers concerning a German edition. Dr Hermann is a clever, mathematically educated young chemist who will undertake the translation of your work into the German language with enthusiasm, and whose explanatory expansions I believe are desirable for the easier understanding of the greater part of the scientific chemical community, and by the look of it, may gain your approval (which will certainly be obtained first). If you then permit me, I would perhaps give a short introductory preface for the German edition.

I look forward to your decision on this matter, and I hope that this initial written contact between us will develop further in the future.

Once more with expression of my heartiest thanks for sending your beautiful work and the accompanying models, which I am now having prepared on a larger scale for my older pupils to see in the laboratory, I sign in great esteem

your most devoted colleague,
J. Wislicenus

Appendix 4

Wislicenus' Foreword to the First Edition of *Die Lagerung der Atome im Raume*, 1877

Wislicenus' foreword to the second edition of Die Lagerung *(1894) appeared in English in the 1898 English translation by the American chemist Arnold Eiloart, a student of Wislicenus in Leipzig, and has been relatively accessible. Wislicenus' foreword to the first edition (1877), on the other hand, has never been translated into English. Wislicenus' remarks were also accompanied by a two-page foreword by Hermann on the nature of the translation. The foreword is printed in the book as if by both Wislicenus and Hermann, rather than just Wislicenus.*

Translator's Foreword

The present manuscript is a free interpretation of the brochure 'La chimie dans l'espace', written by J.H. van 't Hoff that appeared from P.M. Bazendijk in Rotterdam in 1875. The original intended translation of this brochure gradually transformed the original by various deviations to the present version, which does attempt to present the essential content of the original, but contains only little of the original text. The mentioned deviations are of two kinds: on the one hand omissions, on the other hand additions, so that the entirety, despite having enhanced contents in some respects, will correspond little in scope to the original. The treatment of all speculations concerning bodies with closed carbon chains (aromatic compounds) have been omitted, as well as a series of very detailed and ingenious considerations about the carbon atom's divalence. The former did not appear to be well founded, and the latter I believe deviates a little too far from the consequences of the intended purpose. The additions consist of an extensive and somewhat more precise presentation of the formal development of the hypothesis, and a more explicit derivation of the analogy that exists between circular polarizable crystals and those carbon compounds endowed with optical activity.

Moreover, the choice was made to introduce the facts by citing the relevant original literature, while refraining from a discussion of the cited works, in light of the tight limits placed on the present small work to preserve thoroughly its original character as a brochure. An appendix added at the end contains a description of the models necessary for visual clarity, which no doubt can easily be constructed according to the outlines included in the text. At this point, I would like to express my heartiest thanks to Dr Van 't Hoff, not only for the graciousness with which he approved all of the changes he encountered from me, but also for the great amount of material he made available to me, which consisted, in part of corrections to the original text, and in part of extremely valuable additions, so that, in fact, the content of this version could undergo an enrichment with respect to the original.

F. Hermann

The time is not too far past in which loud protests were raised from the quarter of the most progressive representatives of theoretical-chemical views against the thought that chemistry would ever be able to advance toward an explanation of the properties of a compound by referring to the spatial arrangement of atoms. These protests arose from diverse stubborn misunderstandings of the intellectual content [*Ideeninhalt*] of so-called structural chemistry – they derived their legitimacy from the current state of empirical material knowledge [*Erkenntnismateriale*] and from the questions under consideration in the interest of research.

That the elementary atoms composing the molecule – so far as they are assumed at all – must be spatially arranged in some manner, that the same elementary atoms in the same sequence are still able to be differently grouped in their mutual bonding in complicated molecules, and therefore possibly give rise to small deviations in the properties of structurally identical molecules, had already at the time suggested itself in speculative thought, indeed there were isolated facts that already stimulated attempts at this kind of explanation. Nevertheless, such thoughts were either never mentioned, or only very shyly and vaguely.

Meanwhile, the current standpoint of investigations of isomeric organic compounds in chemical science went further along its natural course and led to unavoidable facts for whose understanding the structure theory no longer sufficed. I myself saw it necessary in my study of paralactic acid to express the statement that the facts force us to explain the difference between isomeric molecules of equivalent structural formulas by different arrangements in space, and therefore openly supported the justification for chemistry to bring geometrical ideas into the doctrine of the constitution of compound molecules.

Van 't Hoff deserves the credit for having taken this step in a quite definite and extremely fortunate manner. The fundamental idea of his theory lay in the evidence that the compounds of a carbon atom with four different simple or compound radicals must give rise in each case to spatial isomers. So astounding [*frappant*] was the effect of this idea on reading Van 't Hoff's little work, 'La chimie dans l'espace', so gripping [*fesselnd*] for me was his further mathematical development and the application to the ever more numerous growing cases designated by me as 'geometrical' isomerism, and to the optically active organic substances.

It might be that Van 't Hoff's discussion partly exceeds current needs, or that some of his particular applications will not later be fully confirmed: the theory of carbon compounds has still made a true and important step forward and this step is an organic and intrinsically necessary one. He develops the currently best-grounded views in a logically coherent manner, and supports them by extending them to actual observed cases that appeared beyond the capabilities of these older views.

Even though Dr Van 't Hoff has sent his work to numerous scholars who are closer to the issues concerned, its content has become less well known than the ideas within it deserve. Promoting the wider distribution of these concerns by a German translation would therefore appear a timely venture. Herr Dr Felix Hermann, who has thoroughly studied Van 't Hoff's brochure with understanding, keen interest and independent judgment, gladly took on the translation, subjecting himself to the task, and produced, with the agreeement of the esteemed author, and in part directly supported by him, the version herewith presented to German chemists, which in comparison to the original edition has several not inessential advantages.

I am quite convinced that Van 't Hoff's theory and this small work do not require my encouraging words. But if despite this I offer them gladly, I subsequently intend them chiefly to fulfill publicly my contribution to Van 't Hoff's desire for a discussion and judgment of his theory.

Würzburg, October 1876

Johannes Wislicenus

Appendix 5

Victor Meyer's Letters to Baeyer Concerning the Strain Theory, 1885

In October of 1885, Victor Meyer sent two postcards and a letter to his mentor Baeyer, in which he displayed an almost breathless enthusiasm for the strain theory. The two postcards contained a single message, but the second postcard is known only by its mention in the letter, and apparently has not survived. The first postcard and letter are located in the Victor Meyer papers in the Archives of the Deutsches Museum in Munich. In the biography of his brother, Richard Meyer published partial transcriptions of the letter and both postcards. The remainder of the message on the second card is missing, however, and the message is fragmentary. These are translations of the complete text of the letter, and the text of the existing postcard and the text published by Meyer from the second. The figures are reproduced directly from the documents.

Frankfurt, 5 October 1885

Dear Friend!

On reading your article, I was entirely stupefied for a few days, only now am I ready, as I sense the need, to write you. First: the experimental part is enchanting, like experiencing a fairy tale. I envy you the unceasing freshness that belongs to such unprecedented work, and wish you luck from all my heart. As far as the theory is concerned, I may not yet have a final judgment, also can't just yet, but am endlessly pleased that we are experiencing something truly new yet again. The ideas about bending and strain are in any case brilliant, and for now I see no appreciable objections. But I would like to ask: (I) How is it that benzene is so stable with its three double bonds (strains)? (II) I agree entirely to C_3; but isn't it audacious to think of C_4, C_5 in a plane, rather than in space? I would tend to consider C_5 as a double tetrahedron [Figure A5.1(a)] rather than a pentagon. (III) What distinction do you make between 'looseness' [*Lockersein*] and 'smaller stability' [*geringer Festigkeit*]? At the beginning, you emphasize correctly that dimethylene (ethylene) is the loosest ring. Later (bottom of p. 2280), you say: 'By conversion of the single bond into the double bond, a small increase in the stability occurs.' This I can't quite grasp, because in the single bond (ethane) there is no strain at all, meaning the stability is at its greatest, so how can an increase in stability take place in the conversion of ethane into (the so greatly strained) ethylene?

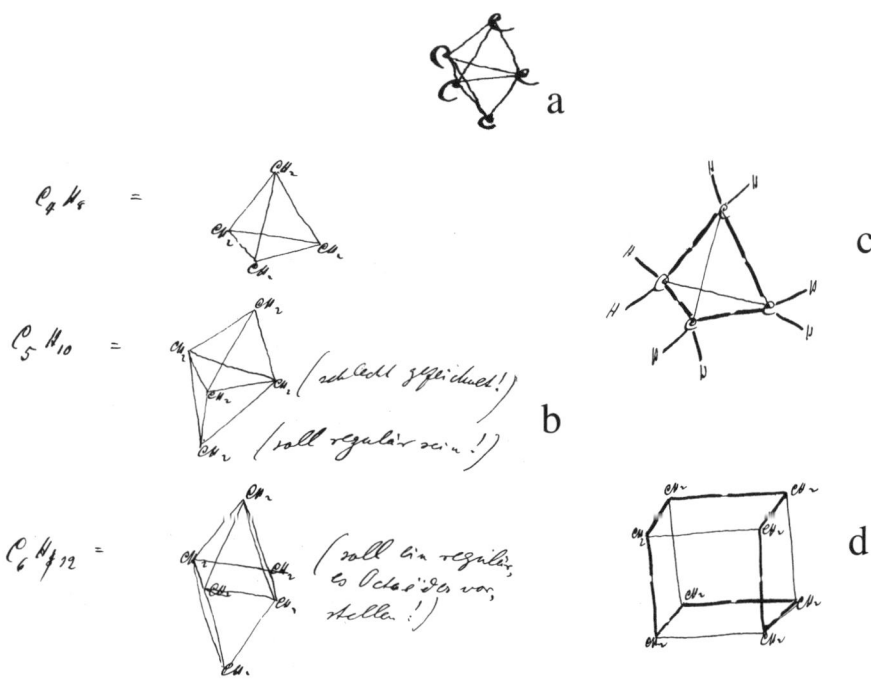

A5.1 (a) Meyer's double tetrahedron from the postcard of 5 October 1885. (b) Meyer's depiction of cyclobutane (top), cyclopentane (middle) and cyclohexane (bottom) as regular polyhedra. (c) Meyer's drawing of the tetrahedral geometry of cyclobutane. (d) Meyer's drawing of cyclooctane as a cube.

Appendix 5

Göttingen, 18 October 1885

Dear Friend!

Your physical (*körperliche*) theory of rings and multiple bonds prompted me recently, while traveling, to two fleeting postcards (Frankfurt) that you have probably already received? Permit me a few more words today on the matter that greatly interests me:

Would it not be worth the trouble to argue that here would be a *corporeal, not planar*, form of multiple-membered rings? You are thinking, if I understand you correctly, of trimethylene as a triangle, tetramethylene as a square, etc.? But I am compelled to conceive of tetramethylene, penta-, hexamethylene as tetrahedra, double tetrahedra (regular) and regular octahedra [Figure A5.1(b)].

One could try out your idea on such models. The simplest case, tetramethylene, gives [the results shown in Figure A5.1(c)] (the thick lines are supposed to be bonds, the thin only heuristic lines that we should ignore). There, the C- affinities will be brought out of the normal inclination (109° 28') into the angle 60°. And the same for C_3, C_4, C_5, to C_6, but *no further*. The ring C_8H_{16} for example, would be a cube [Figure A5.1(d)] there the deviation would be smaller, etc.

So we would come to somewhat different, but also interesting, result accessible by experiment. C_2H_4 has a strong strain, C_3H_6, C_4H_8, C_5H_{10}, and C_6H_{12} would be of smaller but *among each another equivalent* strain, and the following rings would again be of different strain, but smaller. You will still object that the facile opening of C_3H_6 and C_4H_8 is no longer explained – but the facts are already sufficient to establish quite certainly one such different stability between C_3H_6 and C_4H_8? As you note, above all it appears to me, then, your so fruitful thoughts also appear to be lasting, if the atoms in the rings are larger than C_3, are thought to be lying not in a plane, but in space, in any case at least the closer lying one. Please don't be angry with my meddling! It comes only from my deep interest in the matter. If my thoughts are correct, the heat of combustion for C_3H_6 must *of course* be exactly half that for C_6H_{12}, while you arrive at another result. I would be very pleased to hear how you respond to my remarks. I am extremely curious as to what the theoretical investigation begun by you will yield further; also what other chemists say about it; I haven't heard anything, but you have no doubt received all kinds of opinions.

I close for today with affectionate greetings for you and your dear ones.

Your old
Victor Meyer

Victor Meyer to Adolf van Baeyer, 5 October 1885, Deutsches Museum Archives, Standnummer 7154. Reproduced with kind permission.

Appendix 6

'Translation' of Van 't Hoff's Sign Notation for Compounds with Multiple Asymmetric Carbon Atoms into Modern Fischer Projections

As Fischer recognized, the four asymmetric carbon atoms in glucose allowed a maximum of 16 isomers. Fischer initially used Van 't Hoff's +/− notation to differentiate the 16 configurations, but quickly abandoned them in favor of his famous projections still familiar to chemists today. Because Van 't Hoff's original notation can be confusing for modern chemists familiar with Fischer's formulas, a 'translation' of Van 't Hoff's notation into Fischer projections is given below (Figure A6.1). The formulas are arranged in the same order as in Van 't Hoff's initial chart (see Figure 8.10). Van 't Hoff's configuration for each carbon (+ or −) is placed next to each asymmetric carbon atom in the Fischer projections. The aldehyde group is at the top in each case.

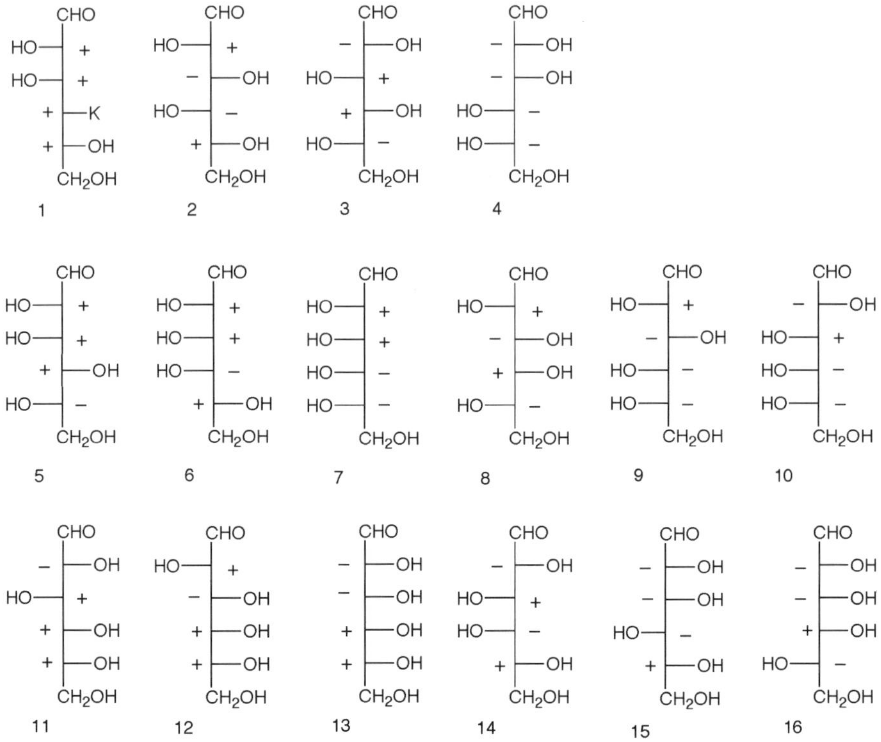

A6.1 Eight of the possible sixteen isomers of glucose, portrayed by Fischer projections. Van 't Hoff's notation is indicated at each asymmetric carbon atom.

Bibliography

Archival Sources

Archiv, Deutsches Museum, München
Emil Fischer Papers, Bancroft Library, University of California, Berkeley, California
Handschriftenabteilung, Zentralbibliothek Zürich
Staatsbibliothek zu Berlin, Preußischer Kulturbesitz
Van 't Hoff Papers, Special Collections, Milton S. Eisenhower Library, Johns Hopkins University, Baltimore, Maryland
Wissenschaftshistorische Sammlungen, ETH Bibliothek, ETH, Zürich

Abbreviations of Journals Cited in Text

Abh. math.-phys. Cl. kön. sächs. Ges. Wiss.	*Abhandlungen der mathematische-physische Classe der königliche sächsische Gesellschaft der Wissenschaften*
Adv. Carb. Chem.	*Advances in Carbohydrate Chemistry*
Ang. Chem. Int. Ed. Eng.	*Angewandte Chemie International Edition in English*
Ann. Sci.	*Annals of Science*
Annalen	*Annalen der Chemie und Pharmacie* (before 1873), and Justus Liebig's *Annalen der Chemie* (after 1873)
Ber. Verh. kön. sächs. Ges. Wiss. Leip.	*Berichte der Verhandlungen der königlichen sächsischen Gesellschaft der Wissenschaften zu Leipzig*
Ber. Verh. sächs. Akad. Wiss. Leip.	*Berichte über die Verhandlung der sächsische Akademie der Wissenschaften zu Leipzig*

Berichte	Berichte der Deutschen Chemischen Gesellschaft
Biog. Mem. Nat. Acad. Sci.	Biographical Memoirs of the National Academy of Sciences
Brit. J. Hist. Sci.	British Journal for the History of Science
Brit. J. Phil. Sci.	British Journal for the Philosophy of Science
Bull. Acad. Roy. Belg.	Bulletin de l'Académie Royale des Sciences, des Lettres, et des Beaux-Arts de Belgique
Bull. Hist. Chem.	Bulletin for the History of Chemistry
Bull. soc. chim.	Bulletin de la societé chimique de France (Paris)
Chem. Soc. Rev.	Chemical Society Reviews
Chem. Soc. Trans.	Chemical Society Transactions
Coord. Chem. Rev.	Coordination Chemistry Reviews
Forsch. Fort.	Forschung und Fortschritt
Hist. Sci.	History of Science
Hist. Stud. Phys. (Biol.) Sci.	Historical Studies in the Physical (and Biological) Sciences
HYLE	HYLE: An International Journal for the Philosophy of Chemistry
Ind. Eng. Chem.	Industrial and Engineering Chemistry
J. Am. Chem. Soc.	Journal of the American Chemical Society
J. Chem. Ed.	Journal of Chemical Education
J. Chem. Soc.	Journal of the Chemical Society (London)
J. Hist. Biol.	Journal for the History of Biology
J. prak. Chem.	Journal für praktische Chemie
Monatshefte	Monatshefte für Chemie
Nat. Prod. Rep.	Natural Product Reports
Naturwiss. Rund.	Naturwissenschaftliche Rundschau
Phil. Trans. Roy. Soc.	Philosophical Transactions of the Royal Society
Phys. Zeit.	Physikalische Zeitschrift
Proc. Am. Phil. Soc.	Proceedings of the American Philosophical Society
Proc. Chem. Soc.	Proceedings of the Chemical Society
Proc. Roy. Soc.	Proceedings of the Royal Society

Sitz. der phys.-med. Ges. Würz.	*Sitzungsberichte der physikalische-medicinische Gesellschaft zu Würzburg*
Stud. Hist. Biol.	*Studies in the History of Biology*
Stud. Hist. Phil. Biol. Sci.	*Studies in the History and Philosophy of the Biological Sciences*
Stud. Hist. Phil. Sci.	*Studies in the History and Philosophy of Science*
Trans. Am. Phil. Soc.	*Transactions of the American Philosophical Society*
Trans. Roy. Soc. Edin.	*Transactions of the Royal Society of Edinburgh*
Verh. Ges. Deut. Nat.	*Verhandlung der Gesellschaft Deutscher Naturforscher und Ärtzte*
Verh. sächs. Akad. Wiss.	*Berichte über die Verhandlungen der sächsische Akademie der Wissenschaften zu Leipzig*
Verh. Vers. Ges. Deut. Naturf. Artze.	*Verhandlungen der Versammlungen der Gesellschaft Deutscher Naturforscher und Ärtzte*
Vier. Zür. naturf. Ges.	*Vierteljahrschrift der Züricher naturforschenden Gesellschaft*
Zeit. angew. Chem.	*Zeitschrift für angewandte Chemie*
Zeit. anorg. Chem.	*Zeitschrift für anorganische Chemie*
Zeit. Chem.	*Zeitschrift für Chemie*
Zeit. Elek. angew. phys. Chem.	*Zeitschrift für Elektrochemie und angewandte physikalische Chemie*
Zeit. ges. Naturw.	*Zeitschrift für die gesammte Naturwissenschaften*
Zeit. phys. Chem.	*Zeitschrift für physikalische Chemie*
Zeit. phys. chem. Unter.	*Zeitschrift für den physikalischen und chemischen Unterricht*

Selected Primary Sources

Anschütz, Richard, 'Zur Geschichte der Isomerie der Maleinsäure und Fumarsäure', *Annalen*, 1887, *239*, 161–84.

——, 'Zur Geschichte der Isomerie der Fumarsäure und Maleinsäure. II. Abhandlung', *Annalen*, 1889, *254*, 168–82.

——, *August Kekulé*, 2 vols, Berlin, Verlag Chemie, 1929.

Auwers, Karl, *Die Entwickelung der Stereochemie*, Heidelberg, Carl Winter Verlag, 1890.

——, 'Über das Oxim des p-Tolylphenylketon', *Berichte*, 1890, *23*, 399–403.

Beckmann, Ernst, 'Johannes Wislicenus', *Berichte*, 1904, *37*, 4861–946.

Behrend, Robert, 'Zur Stereochemie stickstoffhältiger Körper', *Berichte*, 1890, *23*, 454–8.

Biehringer, J, 'Johannes Wislicenus', *Naturwiss. Rund.*, 1903, *18*, 192–4, 204–7.

Bischoff, Carl, 'Beiträge zur Stereochemie des Stickstoffs', *Berichte*, 1890, *23*, 1967–1972.

——, 'Organische Chemie', *Jahrbuch der Chemie*, Richard Meyer, ed., Braunschweig, Vieweg, 1892.

——, 'Organische Chemie', *Jahrbuch der Chemie*, Richard Meyer, ed., Frankfurt, Bechhold, 1893.

Bischoff, Carl and Paul Walden, *Handbuch der Stereochemie*, Frankfurt, H. Bechhold Verlag, 1894.

——, *Materialen Der Stereochemie*, 2 vols, Braunschweig, Vieweg, 1904.

Bloch, Ernst, *Alfred Werner's Theorie des Kohlenstoffatoms und die Stereochemie der Karbocyclischen Verbindungen*, Vienna, Carl Fromme, 1903.

Blomstrand, Christian W., *Die Chemie der Jetztzeit*, Heidelberg, Winter, 1869.

Bremer, G.J.W., 'Vorläufige Mittheilung über eine neue Aepfelsäure, welche die Polarisationsebene rechts dreht', *Berichte*, 1875, *8*, 861–3.

Brown, Alexander Crum, 'On the Theory of Isomeric Compounds', *Trans. Roy. Soc. Edin.*, 1864, *23*, 707–19.

Burch, G.J. and J.E. Marsh, 'The Dissociation of Amine Vapours', *J. Chem. Soc.*, 1899, *55*, 656–64.

Butlerov, Alexander, 'Einiges über die chemische Structur der Körper', *Zeit. Chem.*, 1861, *4*, 549–60.

Claus, Adolf, *Theoretische Betrachtungen und deren Anwendung zur Systematik der Organischen Chemie*, Freiburg, Poppen, 1866.

——, 'Zur Frage nach den Äffinitätsgrossen des Kohlenstoffs', *Berichte*, 1881, *14*, 432–5.

——, 'Zur Kenntniss der Oxime und der sogenannten Stereochemie', *J. prak. Chem.*, 1891, *44*, 312–35.

——, 'Über Oxime fettaromatischer Ketone, welche in Aromatischen Rest orthoständig zur Ketonbildung Halogen enthalten', *J. prak. Chem.*, 1892, *46*, 40–50.

——, 'Zur Kenntniss der Oxime und der sogenannten Stereochemie', *J. prak. Chem.*, 1892, *45*, 1–20.

——, 'Zur Kenntniss der gemischten fettaromatischen Ketone und ihrer Oxime', *J. prak. Chem.*, 1892, *45*, 377–97.

——, 'Zur Characteristik der sogenannten Stereochemie des Stickstoffs', *J. prak. Chem.*, 1892, *46*, 546–59.
——, 'Zur Theorie der Oxime', *J. prak. Chem.*, 1893, *47*, 139–51.
——, 'Die Isomerie asymmetrische Hydrazone', *J. prak. Chem.*, 1893, *47*, 267–73.
——, 'Über Stereoisomerie und die sogenannte Stereochemische Isomerie', *J. prak. Chem.*, 1894, *49*, 445–65.
——, 'Die sogenannte Stereochemie des Stickstoffs und J.H. Van't Hoff', *J. prak. Chem.*, 1894, *50*, 567–76.
Cohen, Ernst, *Jacobus Henricus van't Hoff: Sein Leben und Werken*, Leipzig, Akademische Verlagsgesellschaft, 1912.
Delépine, M. Marcel, ed., *Vie et Oeuvres de Joseph-Achille Le Bel*, Paris, Dupont, 1947.
Eiloart, Arnold, *A Guide to Stereochemistry*, New York, Wilson, 1894.
——, 'Progress in Stereochemistry', *Nature*, 1896, *54*, 321–4.
Fichter, F., 'Rudolf Fittig', *Chemiker-Zeitung*, 1910, *34*, 1277–8.
——, 'Rudolph Fittig', *Berichte*, 1911, *44*, 1339–83.
Fischer, Emil, 'Synthetical Chemistry in its Relation to Biology', *J. Chem. Soc.*, 1907, *91*, 1747–65.
——, *Aus Meinem Leben*, Berlin, Springer-Verlag, 1922.
——, *Gesammelte Werke*, Max Bergmann, ed., 9 vols, Berlin, Springer-Verlag, 1908–1924.
Fittig, Rudolph, 'Beiträge zur Kenntniss der sogenannten ungesättigten Verbindungen', *Berichte*, 1876, *9*, 119–23, 1189–99; 1876, *10*, 513–24.
——, 'Über die Beziehungen zwischen Fumar- und Maleinsäure und zwischen Citraconsäure und Maesaconsäure', *Berichte*, 1877, *10*, 516–18.
——, 'Untersuchungen über die ungesättigte Säuren. II. Abhandlung', *Annalen*, 1879, *195*, 56–179.
——, 'Über die Einwirkung von Brom auf die Angelicasäure und Maleinsäure', *Annalen*, 1890, *259*, 1–40.
——, 'Bemerkungen zu der Abhandlung von Johannes Wislicenus "Über die Bromadditionsproducte der Angelicasäure und Tiglinsäure"', *Annalen*, 1893, *273*, 127–32.
Frank, G., 'Gustav Wislicenus', *Allgemeine Deutsche Biographie*, vol. 43, 542–5, Leipzig, Dunecker and Humboldt, 1898.
Frankland, Percy Faraday, 'Johannes Wislicenus', *Proc. Roy. Soc.*, 1907, *A78*, iii–xii.
Goldschmidt, Heinrich, 'Über die Einwirkung von Hydroxlamin auf Diketone', *Berichte*, 1883, *16*, 2176–80.
Hantzsch, Arthur, 'Vorläufige Mittheilung über stereochemisch isomere Oxime des Phenylthienylketons und der Phenylglyoxyls', *Berichte*, 1890, *23*, 2332–3.
——, 'Versuche zur Stereochemie des Stickstoffs', *Berichte*, 1890, *23*, 2769–73.

——, 'Die stereochemische-isomere Oxime des p-Tolylphenylketons. II. Mittheilung', *Berichte*, 1890, *23*, 2776–80.
——, 'Die Bestimmung der räumlichen Configuration stereoisomerer Oxime', *Berichte*, 1891, *24*, 13–30.
——, 'Die Configuration asymmetrischer Oxime ohne stereoisomerie', *Berichte*, 1891, *24*, 31–6.
——, 'Über Oxime von Aldehyden und α-Ketosäuren', *Berichte*, 1891, *24*, 36–51.
——, 'Über stereoisomere Ketoxime', *Berichte*, 1891, *24*, 51–61.
——, 'Über die Einwirkung des Hydroxylamins auf β-Ketosäuren und β-Diketone', *Berichte*, 1891, *24*, 495–506.
——, 'Über die Isomerie der Oxime und ihr Auftreten in der Fettreihe', *Berichte*, 1891, *24*, 1192–8.
——, 'Zur Nomenclatur stereoisomerer Stickstoffverbindungen und stickstoffhältiger Ringe', *Berichte*, 1891, *24*, 3479–88.
——, 'Über die Configuration der fetten Ketoxime', *Berichte*, 1891, *24*, 4018–24.
——, 'Über die Einwirkung von Hydroxylamin auf Chloral', *Berichte*, 1892, *25*, 701–5.
——, 'Über Stereoisomerie bei Glyoximen der Fettreihe', *Berichte*, 1892, *25*, 705–12.
——, 'Über die Claus'sche Auffassung der isomeren Oxime und Hydrazone', *Berichte*, 1892, *25*, 1692–700.
——, 'Die Configuration der Aldoximessigsäure (β-Oximidopropionsäure)', *Berichte*, 1892, *25*, 1904–7.
——, 'Über Beziehungen zwischen Constitution, Configuration, und chemischen Verhalten der Oxime', *Berichte*, 1892, *25*, 2164–85.
——, *Grundriss der Stereochemie*, Breslau, Trewendt, 1893.
——, 'Stereochemie', pp. 169–245 in *Handwörterbuch der Chemie*, vol. XI, Albert Ladenburg, ed., Breslau, Trewendt, 1893.
——, 'Über Stereoisomerie bei asymmetrischer Hydrazon', *Berichte*, 1893, *26*, 9–17.
——, 'Über Stereoisomerie bei Diazoverbindungen und die Natur der "Isodiazokörper"', *Berichte*, 1894, *27*, 1702–26.
——, 'Über stereoisomerische Salze des Benzoldiazosulfonsäure', *Berichte*, 1894, *27*, 1726–9.
——, 'Zur Statik und Dynamik der Stickstoffverbindungen', pp. 186–202 in *Festschrift der Naturforschenden Gesellschaft in Zürich 1746–1896*, Zürich, Zürcher and Furrer, 1896.
Hantzsch, Arthur and F. Herrmann, 'Bemerkungen über Desmotropie', *Berichte*, 1888, *21*, 1754–8.
Hantzsch, Arthur and Friedrich Kraft, 'Zur Frage des asymmetrische Stickstoffatoms', *Berichte*, 1890, *23*, 2780–4.

——, 'Über das Auftreten von Stereoisomerie bei nicht oximartigen Stickstoffverbindungen', *Berichte*, 1891, *24*, 3511–28.
Hantzsch, Arthur and Arturo Miolati, 'Über die Beziehung zwischen der Configuration und den Affinitätsgrössen stereoisomerer Stickstoffverbindungen', *Zeit. phys. Chem.*, 1892, *10*, 1–33.
Hantzsch, Arthur and M. Schmiedel, 'Weiteres über Diazosulfonate und über freie Diazosulfonsäuren', *Berichte*, 1897, *30*, 71–88.
Hantzsch, Arthur and Alfred Werner, 'Über die räumliche Anordnung der Atome in stickstoffhaltigen Molekülen', *Berichte*, 1890, *23*, 11–30.
——, 'Bemerkungen über stereochemisch isomere Stickstoffverbindungen', *Berichte*, 1890, *23*, 1243–53.
——, 'Über stereochemische Isomerie asymmetrischer Monoxime', *Berichte*, 1890, *23*, 2322–5.
——, 'Die stereochemisch-isomere Oxime des p-Tolyl-phenylketon', *Berichte*, 1890, *23*, 2325–32.
——, 'Bermerkungen über stereochemisch isomere Stickstoffverbindungen', *Berichte*, 1890, *23*, 2764–9.
Helferich, B., 'Arthur Hantzsch', *Forschung und Fortschritt*, 1935, *11*, 152.
——, 'Nachruf auf Arthur Hantzsch', *Ber. Verh. sächs. Akad. Wiss. Leip.*, 1935, *87*, 213–22.
Hoesch, Kurt, *Emil Fischer: Sein Leben und sein Werk*, Berlin, Deutschen Chemischen Gesellschaft, 1921.
Hoff, Jacobus Henricus van 't, *Bijdrage tot de Kennis van Cyanzijnzuur en Malonzuur*, Utrecht, P.W. van de Weijer, 1874.
——, 'Sur les formules de structure dans l'espace', *Archives Néerlandaises des Sciences Exactes et Naturelles*, 1874, *9*, 445–54.
——, *Voorstel tot Uitbreiding der tegenwoordig in de scheikunde gebruikte Structuur-Formules in de ruimte; benevens een daarmeê samenhangende opmerkung omtrent het verband tusschen optisch actief Vermogen en Chemische Constitutie van Organische Verbindingen*, Utrecht, 1874.
——, *La chimie dans l'espace*, Rotterdam, Bazendijk, 1875.
——, 'Sur les formules de structure dans l'espace', *Bull. soc. chim.*, 1875, *23*, 295–301.
——, 'Styro-kamfer, een nieuw lichaam uit styrax', *Mandblaad voor Natuurwetenschappen*, 1876, *6*, 71.
——, 'Een rechtsdraaeind lichaam in styrax', *Mandblaad voor Natuurwetenschappen*, 1876, *7*, 4.
——, 'Die Identität von Styrol und Cinnamol, ein neuer Körper aus Styrax', *Berichte*, 1876, *9*, 5–6.
——, 'Beiträge zur Kenntniss des Styrax', *Berichte*, 1876, *9*, 1339.
——, 'Die Ladenburg'sche Benzolformel', *Berichte*, 1876, *9*, 1881–3.

——, 'Over die bindingsrichtingen van het stickstoofatoom', *Maandblad voor Natuurwetenschappen*, 1877, 7, 109.

——, 'Über den Zusammenhang zwischen optische Activität und Constitution', *Berichte*, 1877, *10*, 1620–3.

——, *Die Lagerung der Atome im Raume*, Felix Hermann, tr., Braunschweig, Vieweg, 1877.

——, *Ansichten über die organische Chemie*, 2 vols, Braunschweig, Vieweg, 1881.

——, *Dix années dans l'histoire d'une théorie*, Rotterdam, 1887.

——, *Chemistry in Space*, J.E. Marsh, tr., Oxford, Clarendon, 1891.

——, *Die Lagerung der Atome im Raume*, 2nd edn, Braunschwieg, Vieweg, 1894.

——, 'Les bases positives de la stéréochimie', *Revue général des sciences*, 1894, *5*, 265–74.

——, 'Aus der Stereochemie: Vorgetragen im naturwissenschaftlichen Ferienkursus zu Berlin am 5. und 8. Oktober 1896', *Zeit. phys. chem. Unter.*, 1898, *11*, 23–30.

——, *The Arrangement of Atoms in Space*, Arnold Eiloart, tr., New York, Longmans, 1898.

——, *Die Lagerung der Atome im Raume*, 3rd edn, Braunschweig, Vieweg, 1908.

——, 'Stereochemistry', *Encyclopaedia Britannica*, 1908 edn.

——, *Imagination in Science*, G.F. Springer, tr., Berlin, Springer-Verlag, 1967.

Jacobsen, Paul, 'Victor Meyer', *Naturwiss. Rund.*, 1897, *12*, 553–6, 564–7.

Jørgensen, Sophus Mads, 'Zur Konstitution der Kobalt-, Chrom- und Rhodiumbasen. V', *Zeit. anorg. Chem.*, 1894, *5*, 147–96.

——, 'Zur Konstitution der Kobalt-, Chrom- und Rhodiumbasen. VI', *Zeit. anorg. Chem.*, 1894, *7*, 289–330.

——, 'Zur Konstitution der Kobalt-, Chrom- und Rhodiumbasen. VIII', *Zeit. anorg. Chem.*, 1896, *13*, 172–90.

——, 'Zur Konstitution der Kobalt-, Chrom- und Rhodiumbasen. IX', *Zeit. anorg. Chem.*, 1897, *14*, 404–22.

——, 'Zur Konstitution der Kobalt-, Chrom- und Rhodiumbasen. XI', *Zeit. anorg. Chem.*, 1899, *19*, 109–57.

Kahlbaum, Georg W.A., *Claus, Adolf Karl Ludwig*, Berlin, Reimer, 1906.

Kekulé, August, *Lehrbuch der organische Chemie*, 2 vols, Erlangen, 1861–1866.

——, 'Untersuchungen über organische Säuren', *Annalen (Suppl.)*, 1862, *2*, 85–116.

Kiliani, Heinrich, 'Über die Einwirkung von Blausäure auf Dextrose', *Berichte*, 1886, *19*, 767–72.

——, 'Über die Zusammensetzung und Constitution der Arabinosecarbonsäure bezw. der Arabinose', *Berichte*, 1887, *20*, 339–46.

——, 'My Life and Work', *J. Chem. Ed.*, 1932, *9*, 1908–14.

King, V.L., *Über Spaltungsmethoden und ihre Anwendung auf komplexe Metal-Ammoniakverbindungen*, Dissertation, Universität Zürich, 1912.

——, 'A Rough but Brilliant Diamond', *J. Chem. Ed.*, 1942, *19*, 345.

Klinger, H., 'Zur Frage nach den Affinitätsgrössen des Kohlenstoffs', *Berichte*, 1881, *14*, 783–5.
Landolt, Hans, 'Untersuchungen über optische Drehungsvermögen. 1. Abhandlung', *Annalen*, 1877, *189*, 241–337.
——, *Das optische Drehungsvermögen organischen Substanzen und die praktischer Anwendungen desselben*, Braunschweig, Vieweg, 1879.
Lassar-Cohn, 'Wilhelm Lossen', *Berichte*, 1907, *40*, 5079–86.
Le Bel, Joseph Achille, 'Sur des relations qui existent entre les formules atomiques des corps organiques et le pouvoir rotatoire de leurs dissolutions', *Bull. soc. chim.*, 1874, *22*, 337–47.
Liebermann, Carl, 'Victor Meyer', *Berichte*, 1897, *30*, 2157–68.
Lossen, Wilhelm, 'Über die Vertheilung der Atome im Raume', *Annalen*, 1880, *204*, 265–364.
——, 'Über die sogenannte Verschiedenheit der Valenzen eines mehrwehrtiges Atoms', *Berichte*, 1881, *14*, 760–5.
——, 'Ueber die Lage der Atome im Raume', *Berichte*, 1887, *20*, 3306–10.
Lunge, G., 'Nachruf auf Victor Meyer', *Vier. Zür. naturf. Ges.*, 1897, *42*, 347–61.
Meyer, Lothar, *Die modernen Theorien der Chemie und ihre Bedeutung für die chemische Statik*, 2nd edn, Breslau, Maruschke and Berendt, 1872.
——, 'Über die Constitution des Benzols', *Annalen*, 1888, *247*, 251–4.
Meyer, Richard, *Victor Meyer: Leben und Wirken eines deutschen Chemikers und Naturforschers 1848–1897*, Leipzig, Akademische Verlagsgesellschaft, 1917.
Meyer, Victor, 'Zur Valenz und Verbindungsfähigkeit des Kohlenstoffs', *Annalen*, 1876, *180*, 192–206.
——, *Die Thiophengruppe*, Braunschweig, Vieweg, 1888.
——, 'The Chemical Problems of To-day', *J. Am. Chem. Soc.*, 1889, *11*, 101–20, tr. by L.H. Friedburg.
——, *Chemische Probleme der Gegenwart*, Heidelberg, Carl Winter, 1890.
——, 'Ergebnisse und Ziele der stereochemischen Forschung', *Berichte*, 1890, *23*, 567–619.
——, *Aus Natur und Wissenschaft: Wanderblätter und Skizze*, Heidelberg, 1892.
——, 'Grundzüge der Stereochemie', *Chemiker-Zeitung*, 1893, *17*, 1869–76.
——, 'Probleme der Atomistik', *Verh. Ges. Deut. Nat.*, 1895, 95–110.
Meyer, Victor and Karl Auwers, 'Untersuchungen über die zweite van't Hoff'sche Hypothese', *Berichte*, 1888, *21*, 790–817.
——, 'Über die Rauolt'schen Methode der Molekulargewichtbestimmung und das Acetoxime', *Berichte*, 1888, *21*, 1068–70.
——, 'Über die Einwirkung der Wärme auf Benzil-Dihydrazone', *Berichte*, 1888, *21*, 2806–7.
——, 'Weitere Untersuchungen über die Isomerie der Benzildioxime', *Berichte*, 1888, *21*, 3510–29.

——, 'Über zwei isomere Benzilmonoxime', *Berichte*, 1889, *22*, 537–51.
——, 'Bemerkung zu der Anhandlung von E. Beckmanns: "Zur Isomerie der Oximidoverbindungen – Isomerie monosubstiuirte Hydroxylamine" ', *Berichte*, 1889, *22*, 564–6.
——, 'Über das dritte Benzildioxime', *Berichte*, 1889, *22*, 705–20.
——, 'Über die Oxime des Phenanthrenchinons', *Berichte*, 1889, *22*, 1985–95.
——, 'Über Tetramethylbernsteinsäure', *Berichte*, 1889, *22*, 2011–15.
——, 'Über die Anhydridbildung bei den Säuren der Bernsteinsäurereihe', *Berichte*, 1890, *23*, 101–3.
——, 'Über Tetramethylbernsteinsäure und Trimethylglutarsäure', *Berichte*, 1890, *23*, 293–316.
——, 'Über Oxime halogenirter Benzophenone', *Berichte*, 1890, *23*, 2063–4.
——, 'Zur Stereochemie der Aethenderivate', *Berichte*, 1890, *23*, 2079–83.
——, 'Über die Isomeren Oxime unsymmetrische Ketone und die Configuration des Hydroxylamins', *Berichte*, 1890, *23*, 2403–9.
——, 'Über die Claus'sche Theorie der Benziloxime', *Berichte*, 1891, *24*, 3267–71.
——, 'Bemerkungen zu der Abhandlung von A. Hantzsch und Friedrich Kraft: Über das Auftreten von Stereoisomerie bei nicht oximartigen Stickstoffverbindungen', *Berichte*, 1891, *24*, 4225–30.
Meyer, Victor and R. Demuth, 'Zur Kenntniss der Isodibrombernsteinsäuren', *Berichte*, 1888, *21*, 264–70.
Meyer, Victor and Heinrich Goldschmidt, 'Über das Benzil', *Berichte*, 1883, *16*, 1616–17.
Meyer, Victor and Paul Jacobsen, *Lehrbuch der organischen Chemie*, Berlin, de Gruyter, 1893.
Meyer, Victor and Alois Janny, 'Über die Einwirkung von Hydroxylamin auf Acetone', *Berichte*, 1882, *15*, 1324–6.
——, 'Über stickstoffhaltiger Acetonderivate', *Berichte*, 1882, *15*, 1164–7.
Meyer, Victor and Eduard Riecke, 'Einige Bemerkungen über das Kohlenstoffatom und die Valenz', *Berichte*, 1888, *21*, 946–56.
——, 'Nachtrag', *Berichte*, 1888, *21*, 1620.
Michael, Arthur, 'Zur Kritik der Abhandlung von J. Wislicenus: "Über die räumliche Anordnung der Atome in organische Molekülen" ', *J. prak. Chem.*, 1888, *36*, 6–39.
——, 'Bemerkung zu der Abhandlung von J. Wislicenus: "Zur geometrischen Constitution der Krotonsäuren und ihrer Halogensubstitutionsprodukte" ', *J. prak. Chem.*, 1889, *40*, 29–44.
Moore, T.S., 'The Hantzsch Memorial Lecture', *J. Chem. Soc.*, 1936, 1051–66.
Ostwald, Wilhelm, 'Johannes Wislicenus', in *Abhandlungen und Vorträge allgemeine Inhalt*, Leipzig, Akademische Verlagsgesellschaft, 1904, 444–5.
——, 'Jacobus Henricus van't Hoff', *Berichte*, 1911, *44*, 2219.

Perkin, William Henry Jr., 'Wislicenus Memorial Lecture', *J. Chem. Soc.*, 1905, *87*, 501–34.

——, 'Adolf von Baeyer', *Nature*, 1917, *100*, 188–90.

Pickering, S.U., 'Atomic Valency', *Proc. Chem. Soc.*, 1885, *1*, 122–5.

——, 'On Water of Crystallization', *J. Chem. Soc.*, 1886, *49*, 411–32.

——, 'Note on the Stereoisomerism of Nitrogen Compounds', *J. Chem. Soc.*, 1893, *63*, 1069–75.

Rassow, B. 'Johannes Wislicenus', *Zeit. angew. Chem.*, 1903, *16*, 1–4.

Richardson, G.M., ed. *The Foundations of Stereochemistry: Memoirs by van't Hoff, Le Bel, and Wislicenus*, New York, American Book, 1901.

Sachse, Hermann, 'Über die Configuration des Benzolmoleküls', *Berichte*, 1888, *21*, 2530–8.

——, 'Über die geometrischen Isomerien der Hexamethylenederivate', *Berichte*, 1890, *23*, 1365–6.

——, 'Über die Konfigurationen der Polymethylenringe', *Zeit. phys. Chem.*, 1892, *10*, 203–41.

——, 'Eine Deutung der Affinität', *Zeit. phys. Chem.*, 1893, *11*, 185–219.

Skraup, Zdenko H., 'Über die Umwandlung der Maleinsäure in Fumarsäure', *Monatshefte*, 1891, *12*, 107–45.

——, 'Zur Theorie der Doppelbindung', *Monatshefte*, 1891, *12*, 146–50.

Sonne, Wilhelm, *Erinnerungen an Johannes Wislicenus aus den Jahren 1876–1881*, Leipzig, Wilhelm Engelmann Verlag, 1907.

Stewart, Alfred W., *Stereochemistry*, 2nd edn, London, Longmans, 1919.

Thorpe, T.E., 'Victor Meyer Memorial Lecture', *J. Chem. Soc.*, 1900, 194.

Tollens, Bernhard, *Kurzes Handbuch der Kohlenhydrate*, Leipzig, J.A. Barth, 1888.

Vaubel, Wilhelm, *Das Stickstoffatom*, Giessen, Ottmann, 1891.

Vis, G.N., 'Adolf Claus', *J. prak. Chem.*, 1900, *62*, 127–33.

Voigt, W., 'Eduard Riecke als Physiker', *Phys. Zeit.*, 1915, *16*, 219–21.

Von Baeyer, Adolf, *Gesammelte Werke*, 2 vols, Brauschweig, Vieweg, 1903.

Walden, Paul, 'Fünfundzwanzig Jahre stereochemischer Forschung', *Naturwiss. Rund.*, 1900, *15*, 145–8, 157–60, 169–73, 185–8, 197–9.

——, 'Carl Bischoff', *Chemiker-Zeitung*, 1908, *32*, 1053.

——, *Optische Umkehrerscheinungen*, Braunschweig, Vieweg, 1919.

Walker, James, 'Van't Hoff Memorial Lecture', pp. 255–71 in *Chemical Society Memorial Lectures*, vol. II, London, Gurney and Jackson, 1913.

Wallach, Otto, 'Bernhard Tollens', *Berichte*, 1918, 51, 1539–55.

Wedekind, Edgar, 'Die Grundlagen und Aussichten der Stereochemie', *Phys. Zeit.*, 1899–1900, *1*, 213–15, 30–2, 38–43.

——, *Zur Stereochemie des fünfwertigen Stickstoffes: Mit besonderer Berücksichtigung des asymmetrischen Stickstoffes in der anatomischen Reihe*, Leipzig, Viet, 1899.

Weissberger, Arnold, 'Arthur Hantzsch', pp. 1067–83 in Eduard Farber, ed., *Great Chemists*, New York, Interscience, 1961.

Werner, Alfred, 'Beiträge zur Theorie der Affinität und Valenz', *Vier. Zür. naturf. Ges.*, 1891, *36*, 129–69.

——, 'Beitrag zur Konstitution anorganischer Verbindungen', *Zeit. anorg. Chem.*, 1893, *3*, 267–330.

——, 'Beitrag zur Konstitution anorganischer Verbindungen. III. Über Beziehungen zwischen Koordinations- und Valenzverbindungen', *Zeit. anorg. Chem.*, 1895, *8*, 189–97.

——, 'Beitrag zur Konstitution anorganischer Verbindungen. XVII. Über Oxalatodiäthylendiaminkobaltisalze', *Zeit. anorg. Chem.*, 1899, *21*, 145–58.

——, 'Der Stand der Chemie am Beginne des XX. Jahrhunderts: Die theoretischen Bestrebungen auf organischem Gebiete', *Chem. Zeit.*, 1901, *1*, 1–5, 25–30.

——, 'Über Isomerien bei anorganischen Verbindungen. I. Über stereoisomere Kobaltverbindungen', *Berichte*, 1901, *34*, 1705–19.

——, 'Über Isomerien bei anorganischen Verbindungen. II. Über stereoisomere Dinitrodiaethylendiaminkobalsalze', *Berichte*, 1901, *34*, 1719–32.

——, 'Über Haupt- und Nebenvalenzen und die Constitution der Ammoniumverbindungen', *Annalen*, 1902, *322*, 261–96.

——, 'Anorganische Chemie', in *Jahrbuch der Chemie*, vol. 12, Richard Meyer, ed., Braunschweig, Vieweg, 1903.

——, *Lehrbuch der Stereochemie*, Jena, Gustav Fischer Verlag, 1904.

——, 'Anorganische Chemie', in *Jahrbuch der Chemie*, vol. 14, Richard Meyer, ed., Braunschweig, Vieweg, 1905.

——, *Neuere Anschauungen auf dem Gebiete der anorganischen Chemie*, Braunschweig, Vieweg, 1905.

——, 'Untersuchungen über anorganische Konstitutions- und Konfigurations-Fragen', *Berichte*, 1906, *40*, 15–69.

——, 'Zur Valenzfrage', *Zeit. anorg. Chem.*, 1906, *19*, 1345, lecture at the Verein Deutscher Chemiker, 8 June 1906.

——, 'Über 1.2-Dichloro-tetrammin-kobaltisalze (Ammoniakvioleosalze)', *Berichte*, 1907, *40*, 4817–25.

——, 'Valency', *Chemical News*, 1907, *96*, 128–31.

——, 'Theorie der Valenz', *Zeit. Elek. angew. phys. Chem.*, 1911, *17*, 601–9.

——, 'Zur Kenntnis des asymmetrischen Kobaltatoms. I', *Berichte*, 1911, *44*, 1887–98.

——, 'Über die raumisomeren Kobaltverbindungen', *Annalen*, 1912, *386*, 1–272.

——, 'Valenzlehre', in *Handwörterbuch der Naturwissenschaften*, vol. 10, Jena, Gustav Fischer, 1913.

——, 'S.M. Jørgensen', *Chemiker-Zeitung*, 1914, *38*, 557–9.

——, 'Zur Kenntnis des asymmetrischen Kobaltatoms. XII. Über optische Aktivität bei Kohlenstofffreien Verbindungen', *Berichte*, 1914, *47*, 3087–94.

Werner, Alfred and Arturo Miolati, 'Beiträge zur Konstitution anorganischer Verbindungen. I. Abhandlung', *Zeit. phys. Chem.*, 1893, *12*, 35–55.
——, 'Beiträge zur Konstitution anorganischer Verbindungen. II. Abhandlung', *Zeit. phys. Chem.*, 1894, *14*, 506–21.
Willgerodt, C., 'Vorläufige Mittheilungen zur Kenntniss der Hydrazine', *J. prak. Chem.*, 1888, *37*, 449–54.
——, 'Zur Kenntniss der räumlichen Anordnung der Atome in stickstoffhältige Verbindungen', *J. prak. Chem.*, 1890, *41*, 291–300.
——, 'Beitrag zur Kenntniss der Stereochemie von Verbindungen der Elemente der Stickstoffgruppe', *J. prak. Chem.*, 1890, *41*, 526–8.
——, 'Zur Kenntniss der Stereochemie isomerer Stickstoffverbindungen', *J. prak. Chem.*, 1890, *42*, 63–4.
Willgerodt, C. and M. Ferko, 'Beiträge zur Kenntniss des Phenylhydrazins', *J. prak. Chem.*, 1888, *37*, 345–58.
Wislicenus, Johannes, 'Theorie der Gemischten Typen', *Zeit. Ges. Naturw.*, 1859, *14*, 96–175.
——, 'Vorläufige Notiz über eine neue Synthese der Milchsäure', *Zeit. Ges. Naturw.*, 1862, *19*, 76–7.
——, 'Synthese der Paramilchsäure', *Zeit. Ges. Naturw.*, 1862, *19*, 448–50.
——, 'Studien zur Geschichte der Milchsäure und ihrer Homologen', *Annalen*, 1863, *128*, 1–67.
——, 'Über die isomeren Milchsäuren', *Annalen*, 1872, *166*, 3–64.
——, 'Über die isomeren Milchsäuren. II. Abhandlung, Über die optisch-activ Milchsäure der Fleischflüssigkeit, die Paramilchsäure', *Annalen*, 1873, *167*, 302–46.
——, 'Über die isomeren Milchsäuren. Über die Aethylenmilchsäure, III. Abhandlung', *Annalen*, 1873, *167*, 346–56.
——, *Adolph Strecker's Lehrbuch der organische Chemie*, Braunschweig, Vieweg, 1876.
——, 'Wilhelm Heintz', *Berichte*, 1882, *16*, 3121–40.
——, 'Abhängigkeit des optische Drehung von Constitution', *Sitz. der phys.-med. Ges. Würz.*, 1883, 37–40.
——, 'Über Chlorderivate der Krotonsäure', *Berichte*, 1887, *20*, 1008–10.
——, 'Über die Entwickelung der Lehre von der Isomerie chemischer Verbindungen', *Tageblatt der Versammlung deutscher Naturforscher und Ärzte*, 1887, *60*, 47–56.
——, 'Über die räumliche Anordnung der Atome in organischen Molekülen und ihre Bestimmung in geometrisch-isomeren ungesättigten Verbindungen', *Abh. math.-phys. Cl. kön. sächs. Ges. Wiss.*, 1887, *14*, 1–77.
——, 'Über die Lage der Atome in Raume: Antwort auf Lossen's Frage', *Berichte*, 1888, *21*, 581–5.

——, 'Untersuchungen zur Bestimmung der räumlichen Atomlagerung. Erste Abhandlung: Fumarsäure und Maleinsäure', *Annalen*, 1888, *246*, 53–96.

——, 'Untersuchungen zur Bestimmung der räumlichen Atomlagerung. Dritte Abhandlung: Über einige Glieder der Stilbengruppe', *Annalen*, 1888, *248*, 1–34.

——, 'Untersuchungen zur Bestimmung der räumlichen Atomlagerung. Dritte Abhandlung: Geometrische Constitution der Crotonsäure und ihrer Halogensubstitutionsproducte', *Annalen*, 1888, *248*, 281–355.

——, 'Chemische Wirkung der in einer Kohlenstoffkette an das erste Atom gebundenen Elemente auf das fünfte', *Berichte über die Verhandlungen der königliche sächsichen Gesellschaft der Wissenschaften zu Leipzig*, 1889, *41*, 232–7.

——, 'Untersuchungen zur Bestimmung der räumlichen Atomlagerung. Vierte Abhandlung: Pseudobutylens, der Angelicasäure und Tiglinsäure', *Annalen*, 1889, *250*, 224–54.

——, *Die Umsetzung stereoisomerer, ungesättiger, organischer Verbindungen bei höherer Temperatur*, Leipzig, 1890.

——, 'Die wichtigsten Errungschaften der Chemie im letzen Vierteljahrhundert', *Berichte*, 1892, *25*, 3398–410.

——, 'Ueber die Bromadditionsproducte der Angelica- und Tiglin-Säure', *Annalen*, 1892, *272*, 1–99.

——, *Die Chemie und das Problem von der Materie*, Leipzig, Alexander Edelmann Verlag, 1893.

——, 'Natur der Gemische von Angelicasäuredibromür und Tiglinsäuredibromür: Antwort auf Rudolph Fittig', *Annalen*, 1893, *274*, 99–119.

——, 'Umlagerung stereoisomerer, ungesättigter Verbindungen durch Halogen in Sonnenlicht', *Berichte über die Verhandlungen der königliche sächsische Gesellschaft der Wissenschaften zu Leipzig*, 1895, *47*, 489–93.

Wislicenus, Johannes Adolf, *Zur Kenntnis der geometrisch-isomeren Crotonsäure und einiger Derivate*, Stuttgart, 1892.

Wislicenus, Wilhelm, 'Johannes Wislicenus', pp. 500–12 in *Lebensläufe aus Franken*, vol. 2, Würzburg, Kommisions Verlag, 1922.

Wittig, Georg, *Stereochemie*, Leipzig, Akademische Verlagsgesellschaft, 1930.

Wunderlich, Aemilius, *Configuration organischer Moleküle*, Würzburg, Leitholdt, 1886.

Secondary sources

Barkan, Diana, *Walther Nernst and the Transition to Modern Physical Science*, Cambridge, Cambridge University Press, 1999.

Benfey, O.T., 'The Role of Imagination in Science: Van't Hoff's Inaugural Address', *J. Chem. Ed.*, 1960, *37*, 467–70.

——, ed., *Classics in the Theory of Chemical Combination*, New York, Dover Publications, 1963.
——, *From Vital Force to Structural Formulas*, Washington, DC, American Chemical Society, 1975.
Brock, William H., *From Protyl to Proton: William Prout and the Nature of Matter, 1785–1985*, Bristol, Adam Hilger, 1985.
——, *The Norton History of Chemistry*, New York, Norton, 1993.
Brooke, J.H., 'Wöhler's Urea, and its Vital Force: A Verdict from the Chemists', *Ambix*, 1968, *15*, 84.
——, 'Organic Synthesis and the Unification of Chemistry: A Reappraisal', *Brit. J. Hist Sci.*, 1971, *5*, 363.
——, 'Chlorine Substitution and the Future of Organic Chemistry: Methodological Issues in the Laurent-Berzelius Correspondence (1843–1844)', *Stud. Hist. Phil. Sci.*, 1973, *4*, 47.
——, 'Laurent, Gerhardt, and the Philosophy of Chemistry', *Hist. Stud. Phys. Sci.*, 1975, *6*, 405–29.
——, 'Avogadro's Hypothesis and its Fate: A Case Study in the Failure of Case Studies', *Hist. Sci.*, 1981, *19*, 235–73.
——, 'Methods and Methodology in the Development of Organic Chemistry', *Ambix*, 1987, *34*, 146–55.
Browne, C.A., 'Bernhard Tollens (1841–1918) and Some American Students of His School of Agricultural Chemistry', *J. Chem. Ed.*, 1942, *19*, 253–9.
Brush, Steven, 'The Reception of Mendeleev's Periodic Law in America and Britain', *Isis*, 1996, *87*, 595–628.
——, 'Dynamics of Theory Change in Chemistry. Part 1: The Benzene Problem, 1865–1845', *Stud. Hist. Phil. Sci.*, 1999, *30*, 21–79.
——, 'Dynamics of Theory Change in Chemistry. Part 2: Benzene and Molecular Orbitals, 1945–1980', *Stud. Hist. Phil. Sci.*, 1999, *30*, 263–93.
Burian, Richard M., 'How the Choice of Experimental Organism Matters: Epistemological Reflections on an Aspect of Biological Practice', *J. Hist. Biol.*, 1993, *26*, 351–67.
Bykov, Georgii Vladimirovich, *Istoriia stereokhimii organicheskikh soedinenii*, Moscow, Nauka, 1966.
——, 'The Conceptual Premises of Conformational Analysis in the Work of C.A. Bischoff', pp. 114–22 in Ramsay, *Van't Hoff-Le Bel Centennial*.
Cambrosio, Alberto, Daniel Jacobi and Peter Keating, 'Ehrlich's "Beautiful Pictures" and the Controversial Beginnings of Immunological Imagery', *Isis*, 1993, *84*, 662–99.
Cameron, Margaret Davis, 'Victor Meyer and the Thiophene Compounds', *J. Chem. Ed.*, 1949, *26*, 521–4.
Carneiro, Ana, 'Adolphe Wurtz and the Atomism Controversy', *Ambix*, 1993, *40*, 75–95.

Clause, Bonnie Tocher, 'The Wistar Rat as a Right Choice: Establishing Mammalian Standards and the Ideal of a Standardized Mammal', *J. Hist. Biol.*, 1993, *26*, 329–49.

Cohen, P.S., 'Effect of the Fixity of Ideas on the Werner-Jørgensen Controversy', *Advances in Chemistry*, 1967, *62*, 8–40.

Costa, Albert, 'Arthur Michael (1853–1942) – the Meeting of Thermodynamics and Organic Chemistry', *J. Chem. Ed.*, 1971, *28*, 243–6.

——, 'Arthur Michael', *Dictionary of American Biography*, Edward T. James, ed., Suppl. 3, New York, Scribner's, 1973.

Crawford, Elisabeth, *Arrhenius: From the Ionic Theory to the Greenhouse Effect*, Canton, OH, Science History Publications, 1995.

Eliel, E.L., 'Perspectives in Stereochemistry', *Chemical Technologist*, 1974, 758.

Engel, Michael, 'A Projection on Fischer', *Chemistry in Britain*, 1992, *28*, 1106–9.

Farrar, W.V., 'Nineteenth Century Speculations on the Composition of the Elements', *Brit. J. Hist. Sci.*, 1965, *2*, 297–323.

——, '"Chemistry in Space" and the Complex Atom', *Brit. J. Hist. Sci.*, 1968, *4*, 65–7.

Fieser, Louis F., 'Arthur Michael', *Biog. Mem. Nat. Acad. Sci.*, 1975, *XLVI*, 331–66.

Fischmann, E., 'A Reconstruction of the First Experiments in Stereochemistry: Letters From Van't Hoff to Bremer in a New Chronological Sequence', *Janus*, 1985, *72*, 131.

Fisher, N.W., 'Organic Classification Before Kekulé, parts I and II', *Ambix*, 1973, *20*, 106–31, 209–33.

——, 'Kekulé and Organic Classification', *Ambix*, 1974, *21*, 29–52.

——, 'Wislicenus and Lactic Acid: The Chemical Background to van't Hoff's Hypothesis', pp. 33–54 in Ramsay, *Van't Hoff-Le Bel Centennial*.

Forster, Martin O. 'Emil Fischer Memorial Lecture', *J. Chem. Soc.*, 1920, *117*, 1157–201.

Francoeur, Eric, 'The Forgotten Tool: The Design and Use of Molecular Models', *Social Studies of Science*, 1997, *27*, 7–40.

——, 'Beyond Dematerialization and Inscription: Does the Materiality of Molecular Models Matter?', *HYLE*, 2000, *6*, 63–84.

Freudenberger, Karl, 'Emil Fischer and his Contribution to Carbohydrate Chemistry', *Adv. Carb. Chem.*, 1966, *21*, 1–38.

Fruton, Joseph, *Molecules and Life: Historical Essays on the Interplay of Chemistry and Biology*, New York, Wiley, 1972.

——, 'Contrasts in Scientific Style: Emil Fischer and Franz Hofmeister, their Research Groups and their theory of Protein Structure', *Proc. Am. Phil. Soc.*, 1985, *129*, 313–70.

——, 'The Liebig Research Group: A Reappraisal', *Proc. Am. Phil. Soc.*, 1988, *132*, 1–66.

——, 'Contrasts in Scientific Style: Research Groups in Chemical and Biological Sciences', Philadelphia, PA, American Philosophical Society, 1990.
——, *Proteins, Enzymes, Genes: the Interplay of Chemistry and Biology*, New Haven, CT, Yale University Press, 1999.
Galison, Peter, *How Experiments End*, Chicago, IL, University of Chicago Press, 1987.
Gay, Hannah, 'The Asymmetric Carbon Atom: (a) A Case Study in Independent Discovery; (b) An Inductivist Model for Scientific Method', *Stud. Hist. Phil. Sci.*, 1978, *9*, 207.
Geison, Gerald, 'Pasteur on Vital Versus Chemical Ferments: A Previously Unpublished Paper on the Inversion of Sugar', *Isis*, 1981, *72*, 425–45.
——, 'Scientific Change, Emerging Specialties, and Research Schools', *Hist. Sci.*, 1981, *19*, 20–40.
——, *The Private Science of Louis Pasteur*, Princeton, Princeton University Press, 1995.
Geison, Gerald L. and Manfred D. Laubichler, 'The Varied Lives of Organisms and Variation in the Historiography of the Biological Sciences', *Stud. Hist. Phil. Biol. Sci.*, 2001, *32C*, 1–30.
Geison, Gerald, and James A. Secord, 'Pasteur and the Process of Discovery: The Case of Optical Isomerism', *Isis*, 1988, *79*, 7–36.
Gillespie, Charles, ed., *Dictionary of Scientific Biography*, New York, Scribner's, 1973.
Görs, Britta, *Chemischer Atomismus: Anwendung, Veränderung, Alternativen im deutschprachigen Raum in der zweiten Hälfte des 19. Jahrhunderts*, Berlin, ERS Verlag, 1999.
Grossman, R.B., 'Van't Hoff, Le Bel, and the Development of Stereochemistry: A Reassessment', *J. Chem. Ed.*, 1989, *66*, 30–3.
Harwood, Jonathan, *Styles of Scientific Thought: The German Genetics Community, 1900–1933*, Chicago, IL, University of Chicago Press, 1993.
Hilgetag, Günter, and Paul Heintz, 'Zur Wissenschaftliche Leistung Fischers', *Zeit. Chem.*, 1970, *10*, 281–9.
Hoffmann, Roald, and Pierre Laszlo, 'Representation in Chemistry', *Ang. Chem. Int. Ed. Eng.*, 1991, *30*, 1–16
Holmes, Frederic L., *Claude Bernard and Animal Chemistry*, Cambridge, MA, Harvard University Press, 1974.
——, 'The Fine Structure of Scientific Creativity', *Hist. Sci.*, 1981, *19*, 60–70.
——, 'Lavoisier and Krebs: The Individual Scientist in the Near and Deeper Past', *Isis*, 1984, *75*, 131–42.
——, *Lavoisier and the Chemistry of Life: An Exploration of Scientific Creativity*, Madison, WI, University of Wisconsin Press, 1985.
——, 'The Complementarity of Teaching and Research in Liebig's Laboratory', *Osiris*, 1989, *5*, 121–64.

——, *Hans Krebs: The Formation of a Scientific Life, 1900–1933*, New York, Oxford University Press, 1991.

——, *Hans Krebs: Architect of Intermediary Metabolism, 1933–1937*, New York, Oxford University Press, 1993.

——, 'Justus Liebig and the Construction of Organic Chemistry', pp. 119–34 in Mauskopf, *Chemical Sciences*.

Hudson, C.S., 'Emil Fischer's Discovery of the Configuration of Glucose', *J. Chem. Ed.*, 1941, *18*, 353–7.

——, 'Historical Aspects of Emil Fischer's Fundamental Conventions for Writing Stereo-formulas in a Plane', *Adv. Carb. Chem.*, 1948, *3*, 1–22.

——, 'The Basic Work of Fischer and Van't Hoff in Carbohydrate Chemistry', *J. Chem. Ed.*, 1953, *30*, 120–1.

Huisgen, Rolf., 'Adolf von Baeyer's Scientific Achievements – a Legacy', *Ang. Chem. Int. Ed. Eng.*, 1986, *25*, 297–311.

Hull, David L. et al., 'Planck's Principle', *Science*, 1978, *202*, 717–23.

Ihde, Aaron J., 'The Unraveling of Geometric Isomerism and Tautomerism', *J. Chem. Ed.*, 1959, *30*, 330.

——, *The Development of Modern Chemistry*, New York, Harper and Row, 1964.

Johnson, Jeffrey A., 'Academic Chemistry in Imperial Germany', *Isis*, 1985, *76*, 500–24.

——, 'Hierarchy and Creativity in Chemistry, 1871–1914', *Osiris*, 1989, *5*, 214–40.

——, *The Kaiser's Chemists: Science and Modernization in Imperial Germany*, Chapel Hill, NC, University of North Carolina Press, 1990.

Jones, Paul R., 'The First Half Century of Chemistry at Clark University', *Bull. Hist. Chem.*, 1991, *9*, 15–19.

——, 'The Young Johannes Wislicenus in America', *Bull. Hist. Chem.*, 1997, *20*, 28–32.

Jorissen, W.P., 'Eenige brieven van van't Hoff (1874–1875)', *Chemische Weekblad*, 1924, *21*, 495–501.

Jungnickel, Christa, and Russel McCormmach, *Intellectual Mastery of Nature: Theoretical physics from Ohm to Einstein*, 2 vols, Chicago, IL, University of Chicago Press, 1986.

Kaiser, David, 'Stick-figure Realism: Conventions, Reification, and the Persistence of Feynmann Diagrams, 1948–1964', *Representations*, 2000, *70*, 49–86.

Kauffman, George B., 'Sophus Mads Jørgensen (1837–1914): A Chapter in Coordination Chemistry History', *J. Chem. Ed.*, 1959, *36*, 521–7.

——, 'Sophus Mads Jørgensen and the Werner-Jørgensen Controversy', *Chymia*, 1960, *6*, 180–204.

——, *Alfred Werner: Founder of Coordination Chemistry*, Berlin, Springer-Verlag, 1966.

——, 'Alfred Werner's Inaugural Dissertation', *J. Chem. Ed.*, 1966, *43*, 155–65.

——, 'Alfred Werner's Habilitationsschrift', *Chymia*, 1967, *12*, 183–216, p. 191.
——, *Classics in Coordination Chemistry, Part 1: The Selected Papers of Alfred Werner*, New York, Dover, 1968.
——, 'Stereochemistry of Trivalent Nitrogen Compounds: Alfred Werner and the Controversy over the Structure of Oximes', *Ambix*, 1972, *19*, 129–44.
——, 'Quinquevalent Nitrogen and the Structure of Ammonium Salts: Contributions of Werner and Others', *Isis*, 1972, *62*, 78–95.
——, 'The Discovery of Optically Active Coordination Compounds: A Milestone in Stereochemistry', *Isis*, 1973, *65*, 38–62.
——, 'Alfred Werner's Research on Optically Active Coordination Compounds', *Coord. Chem. Rev.*, 1974, *12*, 105–49.
——, 'Christian Wilhelm Blomstrand (1826–1897): Swedish Chemist and Mineralogist', *Ann. Sci.*, 1975, *32*, 13–37.
——, 'The First Resolution of a Coordination Compound', pp. 126–42 in Ramsay, *Van't Hoff-Le Bel Centennial*.
——, 'An Ingenious Impudence: Alfred Werner's Coordination Theory', *J. Chem. Ed.*, 1976, *53*, 445–6.
——, *Classics in Coordination Chemistry, Part 2: Selected Papers 1798–1899*, New York, Dover, 1976.
——, 'Christian Wilhelm Blomstrand (1826–1897) and Sophus Mads Jørgensen (1837–1914): their Correspondence from 1870 to 1897', *Centaurus*, 1977, *21*, 44–63.
——, 'Early Experimental Studies of Cobalt-ammines', *Isis*, 1977, *68*, 393–403.
Kauffman, George B. and Richard P. Ciula, 'Emil Fischer's Discovery of Phenylhydrazine', *J. Chem. Ed.*, 1977, *54*, 295.
Kauffman, George B. and Robin D. Myers, 'The Resolution of Racemic Acid: A Classic Stereochemical Experiment for the Undergraduate Laboratory', *J. Chem. Ed.*, 1975, *52*, 777–81.
Keas, Michael, *The Structure and Philosophy of Group Research: August Wilhelm Hofmann's Research Program in London (1845–1865)*, PhD, University of Oklahoma, 1992.
Kipnis, A.Y., B.E. Yavelov and J.S. Rowlinson, *Van der waals and Molecular Science*, New York, Oxford University Press, 1996.
Klein, Ursula, *Experimente, Modelle, Paper-Tools: Kulturen der organischen Chemie im 19. Jahrhundert*, Habilitationsschrift, University of Konstanz, 1999.
Klooster, H.S. van, 'Van't Hoff (1852–1911) in Retrospect', in *Proceedings of the International Symposium on the Reactivity of Solids, 1952*, Goetberg, 1954, 1095–100.
Klostermann, L.J., 'A Research School of Chemistry in the Nineteenth Century: Jean-Baptiste Dumas and his Research Students', *Ann. Sci.*, 1985, *42*, 1–40.

Knight, David, *Ideas in Chemistry*, New Brunswick, NJ, Rutgers University Press, 1992.
Koeppel, Tonja A., *Benzene Structure Controversies 1865–1920*, PhD, University of Pennsylvania, 1973.
——, 'Significance and Limitation of Stereochemical Benzene Models', pp. 97–113 in Ramsay, *Van't Hoff-Le Bel Centennial*.
Kohler, Robert, 'The Background to Eduard Buchner's Discovery of Cell-free Fermentation', *J. Hist. Biol.*, 1971, *4*, 35–61.
——, 'The Origin of G.N. Lewis's Theory of the Shared Pair Bond', *Hist. Stud. Phys. Sci.*, 1971, *3*, 343–76.
——, 'The Enzyme Theory and the Origin of Biochemistry', *Isis*, 1973, *64*, 181–96.
——, 'The Lewis Langmuir Theory of Valence and the Chemical Community', *Hist. Stud. Phys. Sci.*, 1975, *6*, 431–68.
——, *Lords of the Fly: Drosophilia Genetics and the Experimental Life*, Chicago, IL, University of Chicago Press, 1994.
Körber, Hans-Günther, ed., *Aus dem wissenschaftlicher Briefwechsel Wilhelm Ostwalds*, 2 vols, Berlin, Akademie-Verlag, 1969.
Kottler, Dorian B., 'Louis Pasteur and Molecular Disymmetry, 1844–1857', *Stud. Hist. Biol.*, 1978, *2*, 57–98.
Kragh, Helge, 'Julius Thomsen and Nineteenth Century Speculations on the Complexity of the Elements', *Ann. Sci.*, 1982, *39*, 37–60.
——, 'S.M. Jørgensen and his Controversy with A. Werner: A Reconsideration', *Brit. J. Hist. Sci.*, 1997, *30*, 203–19.
Larder, David F., 'Historical Aspects of the Tetrahedron in Chemistry', *J. Chem. Ed.*, 1967, *44*, 661–6.
——, 'A Dialectical Consideration of Butlerov's Theory of Chemical Structure', *Ambix*, 1971, *18*, 26–48.
Lenoir, Timothy, 'Generational Factors in the Origin of *Romantische Naturphilosophie*', *J. Hist. Biol.*, 1978, *11*, 57–100.
Levere, Trevor H., 'Affinity or Structure: An Early Problem in Organic Chemistry', *Ambix*, 1970 *17*, 111.
——, *Affinity and Matter: Elements of Chemical Philosophy, 1800–1865*, Oxford, Clarendon, 1971.
Lichtenthaler, Frieder W., 'Emil Fischer's Proof of the Configuration of Sugars: A Centennial Tribute', *Ang. Chem. Int. Ed. Eng.*, 1992, *31*, 1541–56.
——, '100 Years "Schlüssel-Schloss-Prinzip": What Made Emil Fischer Use This Analogy?', *Ang. Chem. Int. Ed . Eng.*, 1994, *33*, 2364–74.
Mauskopf, Seymour H., *Crystals and Compounds: Molecular Structure and Composition in Nineteenth Century French Science*, Philadelphia, PA, American Philosophical Society, 1976.

——, ed., *Chemical Sciences in the Modern World*, Philadelphia, PA, University of Pennsylvania Press, 1993.

Melhado, Evan, 'Mitscherlich's Discovery of Isomorphism', *Hist. Stud. Phys. Sci.*, 1980, *11*, 87–123.

Melsen, Andrew G. van, *From Atomos to Atom: A History of the Concept Atom*, New York, Harper, 1960.

Morrell, J.B., 'The Chemist Breeders: The Research Schools of Liebig and Thomson', *Ambix*, 1972, *19*, 1–46.

Nye, Mary Jo, 'Philosophies of Chemistry Since the Eighteenth Century', pp. 3–24 in Mauskopf, *Chemical Sciences*.

——, 'The Nineteenth Century Atomic Debates and the Dilemma of an Indifferent Hypothesis', *Stud. Hist. Phil. Sci.*, 1976, *7*, 245–68.

——, 'Chemical Explanation and Physical Dynamics: Two Research Schools at the First Solvay Conference, 1922–1928', *Ann. Sci.*, 1989, *46*, 461–80.

——, 'Explanation and Convention in Nineteenth-century Chemistry', pp. 171–86 in R.P.W. Visser, ed, *New Trends in the History of Science*, Amsterdam, Editions Rodepi, 1989.

——, 'Physics and Chemistry: Commensurate or Incommensurate Sciences?', pp. 205–24 in Mary Jo Nye, Joan L. Richards and Roger H. Stuewer, eds, *The Invention of Physical Science: Intersections of Mathematics, Theology and Natural Philosophy Since the Seventeenth Century*, Dordrecht, Kluwer, 1992.

——, *From Chemical Philosophy to Theoretical Chemistry: Dynamics of Matter and Dynamics of Disciplines, 1800–1950*, Berkeley, CA, University of California Press, 1993.

——, 'National Styles? French and English Chemistry in the Nineteenth and Early Twentieth Centuries', *Osiris*, 1993, *8*, 30–49.

Nyhart, Lynn K., *Biology Takes Form: Animal Morphology and the German Universities*, Chicago, IL, University of Chicago Press, 1995.

Palladino, Paolo, 'Stereochemistry and the Nature of Life: Mechanist, Vitalist, and Evolutionary Perspectives', *Isis*, 1990, *81*, 44–67.

Pang, Alex Soojung-Kim, 'Visual Representation and Post-constructivist History of Science', *Hist. Stud. Phys. Biol. Sci.*, 1997, *28*, 139–71.

Paoloni, Leonello, 'Stereochemical Models of Benzene, 1869–1875', *Bull. Hist. Chem.*, 1992, *12*, 10–24.

Partington, J.R., *A History of Chemistry*, 4 vols, London, Macmillan, 1961–1970.

Ramberg, Peter J., 'Johannes Wislicenus, Atomism, and the Philosophy of Chemistry: A Translation and Commentary', *Bull. Hist. Chem.*, 1994, *15/16*, 45–53.

——, 'Arthur Michael's Critique of Stereochemistry, 1887–1900', *Hist. Stud. Phys. Biol. Sci.*, 1995, *26*, 89–138.

——, 'Pragmatism, Belief, and Reduction: Stereoformulas and Atomic Models in Early Stereochemistry', *HYLE*, 2000, *6*, 5–61.

——, 'Paper Tools and Fictional Worlds: Prediction, Synthesis and Auxiliary Hypotheses in Chemistry', in Ursula Klein, ed., *Tools and Modes of Representation in the Laboratory Sciences*, Dordrecht, Kluwer, 2001, pp. 61–78.

Ramberg, Peter J. and Geert J. Somsen, 'The Young J. H. van 't Hoff: The Background to the Publication of his 1874 Pamphlet on the Tetrahedral Carbon Atom, Together with a New English Translation', *Ann. Sci.*, 2001, *58*, 51–74.

Ramsay, O.B., ed., *Van't Hoff-Le Bel Centennial*, Washington, DC, American Chemical Society, 1975.

Ramsay, O.B., 'Molecules in Three Dimensions', *Chemistry*, 1974, *47* (1), 6–9.

——, 'Molecules in Three Dimensions. Part II', *Chemistry*, 1974, *47* (2), 6–11.

——, 'Molecular Models in the Early Development of Stereochemistry. I: The van't Hoff Model; II: The Kekulé Models and the Baeyer Strain Theory', pp. 74–96 in Ramsay, *Van't Hoff-Le Bel Centennial*.

——, *Stereochemistry*, Heyden, London, 1981.

——, 'The Early History and Development of Conformational Analysis', pp. 54–77 in Traynam, *Essays*.

Reinhardt, Carsten, *Forschung in der chemischen Industrie: Die Entwickelung synthetischer Farbstoffe bei BASF und Hoechst, 1863 bis 1914*, Freiberg, TU Bergakademie, 1997.

Rheinberger, Hans-Jörg, *Towards a History of Epistemic Things: Synthesizing Proteins in the Test Tube*, Palo Alto, CA, Stanford University Press, 1997.

Robinson, M.J.T., 'Studies of the Structure of Tartaric Acid before 1874', *Tetrahedron*, 1974, *30*, 1499.

Robinson, M.J.T. and F.G. Riddell, 'J.H. Van't Hoff and J.A. Le Bel – Their Historical Context', *Tetrahedron*, 1974, *30*, 2001.

Rocke, Alan J., 'Kekulé, Butlerov, and the Historiography of the Theory of Chemical Structure', *Brit. J. Hist. Sci.*, 1981, *14*, 27–57.

——, 'Subatomic Speculations and the Origin of Structure Theory', *Ambix*, 1983, *30*, 1–18.

——, *Chemical Atomism in the Nineteenth Century: From Dalton to Canizarro*, Columbus, OH, Ohio State University Press, 1984.

——, 'Hypothesis and Experiment in the Early Development of Benzene Theory', *Ann. Sci.*, 1985, *42*, 355–81.

——, 'Convention Versus Ontology in Nineteenth-century Organic Chemistry', pp. 1–20 in Traynham, *Essays*.

——, 'Kolbe vs. the "Transcendental Chemists": The Emergence of Classical Organic Chemistry', *Ambix*, 1987, *34*, 156–68.

——, 'Methodology and Its Rhetoric in Nineteenth Century Chemistry: Induction Versus Hypothesis', pp. 137–55 in Elizabeth Garber, ed., *Beyond History of Science: Essays in Honor of Robert E. Schofield*, Bethlehem, PA, Lehigh University Press, 1990.

——, 'Group Research in German Chemistry: Kolbe's Marburg and Leipzig Institutes', *Osiris*, 1993, *8*, 53–79.
——, *The Quiet Revolution: Hermann Kolbe and the Science of Organic Chemistry*, Berkeley, CA, University of California Press, 1993.
——, 'The Quiet Revolution of the 1850s: Social and Empirical Sources of Scientific Theory', pp. 87–118 in Mauskopf, *Chemical Sciences*.
——, 'Organic Analysis in Comparative Perspective', pp. 273–310 in Frederic L. Holmes and Trevor Levere, eds, *Instruments and Experimentation in the History of Chemistry*, Cambridge, MA, MIT Press, 2000.
——, *Nationalizing Science: Adolphe Wurtz and the Battle for French Chemistry*, Cambridge, MA, MIT Press, 2000.
Root-Bernstein, Robert Scott, 'The Ionists: Founding Physical Chemistry, 1872–1890', PhD, Princeton University, 1980.
Rudwick, Martin, 'The Emergence of a Visual Language for Geological Science', *Hist. Sci.*, 1976, *14*, 149–65.
Russell, C.A., *The History of Valency*, New York, Humanities Press, 1971.
——, 'The Origins of Conformational Analysis', pp. 159–78 in Ramsay, *Van't Hoff-Le Bel Centennial*.
——, 'The Changing Role of Synthesis in Organic Chemistry', *Ambix*, 1987, *34*, 168–80.
——, *Edward Frankland*, Cambridge, Cambridge University Press, 1996.
Schmidt, Gustav, 'The Discovery of Nitroparaffins by Victor Meyer', *J. Chem. Ed.*, 1950, *27*, 557–9.
Sementsov, Anatol, 'The Eightieth Anniversary of the Asymmetrical Carbon Atom', *American Scientist*, 1955, *43*, 97–100.
Servos, John W., *Physical Chemistry from Ostwald to Pauling: The Making of a Science in America*, Princeton, NJ, Princeton University Press, 1990.
——, 'Research Schools and their Histories', *Osiris*, 1993, *8*, 3–15.
Shapin, Steven and Simon Schaffer, *Leviathan and The Air Pump: Hobbes, Boyle, and the Experimental Life*, Princeton, NJ, Princeton University Press, 1985.
Snelders, H.A.M., 'The Birth of Stereochemistry: An Analysis of the 1874 Papers of J.H. Van't Hoff and J.A. Le Bel', *Janus*, 1973, *60*, 261–78.
——, 'The Reception of J.H. Van't Hoff's Theory of the Asymmetric Carbon Atom', *J. Chem. Ed.*, 1974, *51*, 2–7.
——, 'Practical and Theoretical Objections to J.H. van't Hoff's Stereochemical Ideas', pp. 55–65 in Ramsay, *Van't Hoff-Le Bel Centennial*.
——, 'J.A. Le Bel's Stereochemical Ideas Compared with Those of J.H. van't Hoff', pp. 66–73 in Ramsay, *Van't Hoff-Le Bel Centennial*.
——, 'J. H. Van't Hoff's Research School in Amsterdam', *Janus*, 1984, *71*, 1–30.
Stewart, Alfred W., *Stereochemistry*, 2nd edn, London, Longmans, 1919.
Stocklöv, Joachim, *Arthur Rudolf Hantzsch im Briefwechsel mit Wilhelm Ostwald*, Berlin, ERS-Verlag, 1998.

Tansjö, L., 'While waiting for Werner: Chemistry in Chains', pp. 35–40 in George B. Kauffman, ed., *Coordination Chemistry: A Century of Progress*, Washington, DC, American Chemical Society, 1994.

Tarbell, Douglas and Ann Tracy Tarbell, *Essays on the History of Organic Chemistry in the United States, 1875–1955*, Nashville, TN, Folio Publishers, 1986.

Traynam, James G., ed., *Essays on the History of Organic Chemistry*, Baton Rouge, LA, Louisiana State University Press, 1987.

Ward, Robert, *The Development of the Polarimeter in Relation to Problems in Pure and Applied Chemistry: An Aspect of Nineteenth Century Scientific Instrumentation*, PhD, University of London, 1980.

Weininger, Stephen J., 'Contemplating the Finger: Visuality and the Semiotics of Chemistry', *HYLE*, 1998, *4*, 3–27.

Weyer, Jost, 'A Hundred Years of Stereochemistry: The Principal Development Phases in Retrospect', *Ang. Chem. Int. Ed. Eng.*, 1974, *13*, 591.

——, 'Die Aufnahme der van't Hoff'schen Hypothese vom asymmetrischen Kohlenstoffatomen (1874) in Deutschland', 311–20 in Gunter Mann and Rolf Winau, eds, *Medizin, Naturwissenschaft, Technik und das Zweite Kaiserreich*, Göttingen, Vandenhoeck and Ruprecht, 1977.

——, 'Van't Hoff, Kekulé, und die Stereochemie: Zwei unveröffentliche Briefe von J.H. van't Hoff an A. Kekulé', *Janus*, 1977, *64*, 217–30.

Index

Name Index

Anschütz, Richard 94, 134, 178
Arrhenius, Svante 328
Auwers, Karl von 6, 168–73, 322

Baeyer, Adolf 244, 248, 264, 294, 365, reaction to Van 't Hoff 88–9, strain theory 102–5, modeling 105
Behrend, Robert 226–7
Berthelot, Marcelin 77–8, 90–1, 356
Berzelius, Jakob 15, 17–18, 282
Biot, Jean-Baptiste 30–1
Bischoff, Carl 8, 227, 336
Blomstrand, Christian Wilhelm 281, 283
Bremer, Gustav 74–7
Brown, Alexander Crum 27, 41
Burch, G.J. 226
Butlerov, Aleksandr 24–5
Buys-Ballot, C.H.D. 83, 87–8

Claus, Adolf 92, 231–6, 323, 330
Curie, Marie 316

Demuth, Robert 162
Dumas, Jean-Baptiste 18–19

Ehrlich, Paul 272, 325

Fischer, Emil 7, 88, 318, 322, 323, 326, 336, 350, 369, early career 243–4, study of carbohydrates 249–60, use of tetrahedral carbon atom 254, 257, 259, synthesis of d-glucose 256, synthesis of l-glucose 257, configuration of glucose 260–2, Fischer projections 263–4 fig. 8.11, use of models 263, reception of 264–5, methodology of 265–7, sugars and enzymes 268–72, lock and key hypothesis 271, role of theory 272–3, use of chemical formulas 273, importance of biological compounds 273–4, role of synthesis 274, asymmetric synthesis 275, on Walden inversion 337
Fischer, Otto 244
Fittig, Rudolf 94, 248, 329, angelic acid dibromide 135–40, use of hypothesis 140–1
Frankland, Edward 20, 26
Fremy, Edmond 281
Fröbel, Friedrich 345

Hantzsch, Arthur 6, 322, 323, 326, 332, 350, early career 194–6, theory of benzildioximes 196–201, confirmation of theory 206, influence of Wislicenus 201, stereoformulas 201–2, and V. Meyer 202–13, analogy to Van 't Hoff's third hypothesis 213, establishing configurations of oximes 214–22, stable and labile isomers 218–22, use of physical properties 220–2, response to Claus 235–7, importance of stereochemistry 241
Heintz, Wilhelm 30n, 42
Hermann, Felix 79–80, 357, 361–2
Hofmann, A.W. 38, 244

Janny, Alois 163
Jørgensen, Sophus Mads 283, 305–13, 317

Kekulé, August 244, structure theory 21–4, work on unsaturated acids 40–1, reaction to Van 't Hoff 100
Kiliani, Heinrich 248, 249

King, Victor 315–16
Kolbe, Hermann 5, 14, 18, 38, 95–6, 237, 329

Landolt, Hans 89–90, 336
Le Bel, J.A. 1, 356, argument for tetrahedral carbon atom 60–3, on saturated compounds 60–1, on unsaturated compounds 61–3, comparison with Van 't Hoff's theory 63–5, influence on Van 't Hoff 72–4
Loschmidt, Josef 26, 40
Lossen, Wilhelm 92–3, 96, 135, 330

Marsh, J.E. 226
Merz, Victor 294
Meyer, Ernst von 237
Meyer, Lothar 26, 178–9, 335, 336
Meyer, Richard 365
Meyer, Victor 6, 84, 193, 264, 318, 322, 323, 332, 335, 336, 350, 365, early career 158–9, reaction to Van 't Hoff 89, 159–61, reaction to strain theory 161–2 fig. 6.3, early tests of Van 't Hoff's theory 164, influence of Wislicenus 166–8, structural identity of the benzildioximes 168–71, Van 't Hoff's second hypothesis 171–3, configurational analysis of the benzildioximes 173, confirmation of theory 175–7, defense of stereochemistry 179, introduction of 'stereochemistry' 179–80, 'Aims and Achievements' lecture 180–2, on valence 182–4, atomic model 184–8, on reduction and autonomy of chemistry 188–91, motives for adopting stereochemistry 191, and Hantzsch 202–13, analogy to Van 't Hoff's first hypothesis 213, response to Claus 232–3
Michael, Arthur 141–7, 178, 330, 337
Miolati, Arturo 215, 221, 277, 294–8, 313
Modderman, Tjaden 87, 100, 353

Ostwald, Wilhelm 240–1, 326, 335

Pachenstecher, Alexander 135
Pasteur, Louis 33–5, 59, 64, 76–7, 85, 268
Pickering, S.W.U. 117, 227–8
Pückert, Maximilian 135

Riecke, Eduard 184–8

Sachse, Hermann 189, 326, 334
Schiff, Hugo 248
Schmitt, Rudolf 194, 237
Skraup, Zdenko 134–5
Stas, Jean-Servais 89, 336

Tafel, Julius 251
Thierfelder, Hans 268
Tollens, Bernhard 247, 248, 264

Van der Waals, J.D. 96
Van 't Hoff, J.H. 1, 3, 41, 222, 323, 353, 355, 357, 362, early career 55–7, 67, proposal of tetrahedral carbon atom 57, argument in *Voorstel* 57–9, comparison to Le Bel's theory 63–5, and Pasteur 59, 64, 76–7, 85, dissertation 66–7, revisions to original theory 67–71, 80–1, influence of Le Bel on 72–4, and Berthelot 77–8, and molecular models 68, 81–4 fig. 3.10, reaction to Berthelot 91, reaction to Kolbe 97
Vaubel, Wilhelm 228–9, 230 fig. 7.23
Vilmos, A. 313
Volhard, Jakob 141, 244

Walden, Paul 350
Werner, Alfred 6, 323, 332, 350, early career 278–80, theory of benzildioximes 196–201, on valence 279, and coordination theory 284–94, origins of coordination theory 284, introduction of octahedron 288, similarity to Van 't Hoff 290, 292, 302, 325, stereoformulas 293, use of conductivity 294–8, use of synthesis 298–9, configurational assignments 299–305, Jørgensen's criticism of 305–13, optically active coordination compounds 313–17, publicizing theory 318–19
Wiedemann, Georg 196
Willgerodt, Conrad 223–5
Wislicenus, Gustav 42
Wislicenus, Johannes 1, 223, 322, 323, 350, 357, 361, 362–3, early life 42, research on lactic acid 42–50, geometrical isomerism 48, meaning of structural formulas 49–50, and Van 't Hoff, 78–9,

early development of tetrahedral carbon
   atom 112, 'Spatial Arrangement'
   113–21, and maleic and fumaric acids
   121–8, and crotonic acids 128–33, and
   Baeyer 133, and Rudolf Fittig 135–41,
   and Arthur Michael 141–7, on atomism
   148–51, methodology 151–2, response
   to Lossen 149–50, on subatomic
   structure 149–50, reaction to Kolbe
   153, 154, general influence of 332–4,
   333n
Wunderlich, Aemelius 105, 105n, 188, on
   carbon atom 106, on bonding 106–7,
   on strain theory 107, on benzene 107–8
Wurtz, Adolphe, 56, 89

Subject Index

Acrose, α and β, synthesis by oxidation of
   glycerin 252, synthesis from acrolein
   252, 252 fig. 8.6, relationship with
   glucose 253, identity with mannose
   255
Affinity, *see* valence
Alloisomerism 142
Angelic and tiglic acid 135–41
Arabinose 248, 259, 262
Archives Neérlandaises 67, 87, 356
Arrangement 17, 35, 59, 322
Asymmetric carbon atom, *see* tetrahedral
   carbon atom
Atomic models 326–7, Wislicenus 149,
   Meyer 185–8, 186 fig. 6.17
Atomism, 148–51, 328, 335–6

Beckmann rearrangement 215–18
Benzildioximes, isomerism of 164–6,
   Meyer's explanation 171–5, prediction
   of fourth isomer 180, 181 fig. 6.16,
   Hantzsch/Werner theory 199–201,
   determination of configuration 214–22
Benzilmonoximes 176, 231–2
Berzelian formulas 15–17

Carbohydrates, early history of 247–9,
   Fischer's study of 249–60,
   configuration of 260–4, first synthesis
   of 252
Chemical formulas, *see* structural formulas,
   stereoformulas

Chemical structure, *see* structural formulas
Conductivity 220–2, 294–8, 307–8, 338, 341
Configuration 9, 108, 321
Configuration organischen Molekülen 105–8
Configurational assignments 114–15,
   maleic and fumaric acid 122–4,
   crotonic acids 128–33, Michael's
   criticism 141–7, Meyer and
   benzildioximes 174, 177, Hantzsch and
   oximes 214–22, via planesymmetric
   elimination 214, via Beckmann
   rearrangement 215–18, of
   carbohydrates 260–2, of metal
   ammines 299–305
Contingency, historical 332, 342
Coordination number 290–1
Crotonic acids 128–33, 142–4

Dextrose, *see* glucose
Die Lagerung der Atome im Raume 5,
   78–81, 361–3, reaction to 87–97
Dix années dans l'histoire d'une théorie,
   109
Dualism 16, 18

Enzymes 271
Erlangen, University of 244–5
Exemplary models in chemistry 341–3
Explanation, modes of 36–7, 336

Favored configurations 116, 118 fig. 5.2, 333
Ferments, *see* enzymes
Feynmann diagrams 325
Fischer projections 263–4 fig. 8.11, 369, 370
   fig. A6.1
Fructose, *see* levulose

Generational dynamics 343–7
'Genetic' relationships 114–15, 265, 266
   fig. 8.12
Gluconic acid 247
d-Glucose 246, 250, synthesis of 256,
   configuration of 260–2
l-Glucose, synthesis of 257, configuration
   of 260–2
Göttingen, University of 166, 346
Grape sugar, *see* d-Glucose
d-Gulose 258, 262

Hydrazones 210–12
Hydroxylamine 162–4

Hypothesis, role of 35–6, 96–8, 139–40, 145, 151–3, 192, 283, 319, 339–40, 346

Iconic vs. symbolic meaning of chemical formulas 50–2, 105, 324–5, 330, 331
Isomerism 17, 355, absolute 40, as stimulus for stereochemical theory 59, 322

Journal für praktische Chemie 14, 237–8
Journals, chemical 13–14

Kiliani chain extension 249
Kindergartens 345

La chimie dans l'espace 5, 67–72, 355–6, reaction to 87–97
Lactic acid 43–8, Wislicenus' synthesis of 44 fig. 2.3
Levulose 250
Lock and key hypothesis 7, 271
Luteo compounds 281

Maleic and fumaric acid 121–8
Mannitol 247
Mannose, synthetic and natural 254, identity with α-acrose 255, configuration of 262
Metal ammines, early history 280–4, Blomstrand's theory of 282, Jørgensen's theory 283, Werner's treatment of 284–94
Molecular models, origins in 1860s 37–9, Kekulé models and Van 't Hoff 56–7, Van 't Hoff's models 68, 81–4 fig. 3.10, Baeyer's use of 105, Fischer's use of 263, as new method 338–9
Multiple bonding, under structural theory 40–1, under stereochemistry 57–9, critiques of 93–4, of carbon with nitrogen 199–201, 213
Munich, University of 244

Naturforscherversammlungen 13, 45, 112, 113, 189, 255, 277
Nitrogen, spatial models of, by Hantzsch and Werner 196–201, by others 222–30
Nobel Prize 264, 319

Octahedral geometry of metal ammines 288, 293
Optical activity 30–1, 252, 338, use by Pasteur 33–5, study by Landolt 89–90, in metal ammines 313–17
Organic chemistry, institutional structure 12–13, early history 15–20, 'classical' 349
Osazones 250

Periodic law 150, 335
Phenylglucazone, *see* osazone
Phenylhydrazine 244, 249–51
Physical properties, use of by chemists 30–2, 195, 341
'Planck's principle' 343
Planesymmetric addition 9, 80–1 fig. 3.9, 119–20 fig. 5.3, 336
Planesymmetric elimination 214, 215, 216
Polarimetry, *see* optical activity
Polyacetylenes, synthesis of 101–2, explosive character 102
Pragmatic vs. comprehensive geneticists 346–7
Pragmatism 327
Praseo compounds 281
Prout's hypothesis 336
Purpureo compounds 281
Pyroelectricity 184, 184n

'Quiet revolutions' 20, 331

Radical theory 17, 18
Research groups, *see* research schools
Research schools 347–9
Residual affinity 117

Saccharic acid 247, 261, 263
Sorbitol, synthesis of 258
'Spatial Arrangement' 113–21
Stereochemical formulas, meaning of 50–2, 85–6, 120, early reception 100–1, by Wislicenus 116, by V. Meyer 183–4, by Hantzsch and Werner 201–2, by Werner 293, and the visual language of chemistry 324–6
Strain theory 102–5, 365–7
Structural formulas, meaning according to Kekulé 23–4, meaning according to Butlerov 24–5, 28, 50–2, Crum Brown's notation for 27–8, formulation in the laboratory 28–30, iconic vs. symbolic meaning 50–2
Structural identity 171, 233–5, 322

Index 399

Structure theory 4, origins of 20–3, limitations of 39–41
Subatomic structure, *see* atomism
Sugars, *see* carbohydrates

Tartaric acid 33–5
Tetrahedral carbon atom 2, 3, proposal by Van 't Hoff 57, proposal by Le Bel 60–3, early experiments on 74–8, private reaction to 87–90, public reaction to 90–7, use by Fischer 254, 257, 259, general reception 329–32, acceptance of first vs. second and third hypotheses 330, success vs. limitations as theory 337
Type theory 18–19

Valence 20, 39, 278, 279, 326, 327, 328, Claus on 92, Lossen on 92–4, V. Meyer on 182–4, Hantzsch on 239–41, Werner on 279, and coordination theory 290–1
Violeo salts 298, 304
Voorstel 57–9, 87

Walden inversion 305, 337
Würzburg, University of 13, 78, 195, 333n

Xylose 247, 258, 259, 262

Yeasts 268–9

Zürich 13, University of 13, 294, Polytechnical Institute 13, 158, 195